RULES OF THUMB

FOR

MECHANICAL

ENGINEERS

Gulf Publishing Company
Houston, Texas

RULES OF THUMB
FOR MECHANICAL ENGINEERS

A manual of quick, accurate solutions

to everyday mechanical engineering problems

J. Edward Pope, Editor

RULES OF THUMB FOR
MECHANICAL ENGINEERS

10   9   8   7   6   5   4   3

Gulf Publishing Company
Book Division
P.O. Box 2608 □ Houston, Texas 77252-2608

Library of Congress Cataloging-in-Publication Data

Rules of thumb for mechanical engineers : a manual of
    quick, accurate solutions to everyday mechanical
    engineering problems / J. Edward Pope, editor ; in
    collaboration with Andrew Brewington . . . [et al.].
        p.   cm.
    Includes bibliographical references and index.
    ISBN 0-88415-790-3 (acid-free paper)
    1. Mechanical engineering—Handbooks, manuals,
etc.   I. Pope, J. Edward, 1956–   .   II. Brewington,
Andrew.
TJ151.R84   1996
621—dc20   96-35973
                                                      CIP

Printed on acid-free paper (∞).

Contents

11: Vibration, 238

12: Materials, 259

13: Stress and Strain, 294

1
Fluids

Bhabani P. Mohanty, Ph.D., Development Engineer, Allison Engine Company

FLUID PROPERTIES

A *fluid* is defined as a "substance that deforms continuously when subjected to a shear stress" and is divided into two categories: ideal and real. A fluid that has zero viscosity, is incompressible, and has uniform velocity distribution is called an *ideal fluid. Real fluids* are called either Newtonian or non-Newtonian. A Newtonian fluid has a linear relationship between the applied shear stress and the resulting rate of deformation; but in a non-Newtonian fluid, the relationship is nonlinear. Gases and thin liquids are Newtonian, whereas thick, long-chained hydrocarbons are non-Newtonian.

Density, Specific Volume, Specific Weight, Specific Gravity, and Pressure

The *density* ρ is defined as mass per unit volume. In inconsistent systems it is defined as lbm/cft, and in consistent systems it is defined as slugs/cft. The density of a gas can be found from the *ideal gas law:*

$$\rho = p/RT \qquad (1)$$

where p is the absolute pressure, R is the gas constant, and T is the absolute temperature.
The density of a liquid is usually given as follows:

- The *specific volume* v_s is the reciprocal of density:
  $v_s = 1/\rho$
- The *specific weight* γ is the weight per unit volume:
  $\gamma = \rho g$

- The *specific gravity* s of a liquid is the ratio of its weight to the weight of an equal volume of water at standard temperature and pressure. The s of petroleum products can be found from hydrometer readings using API (American Petroleum Institute) scale.

- The fluid *pressure* p at a point is the ratio of normal force to area as the area approaches a small value. Its unit is usually lbs/sq. in. (psi). It is also often measured as the equivalent height h of a fluid column, through the relation:

  $p = \gamma h$

Surface Tension

Near the free surface of a liquid, because the cohesive force between the liquid molecules is much greater than that between an air molecule and a liquid molecule, there is a resultant force acting towards the interior of the liquid. This force, called the *surface tension,* is proportional to the product of a surface tension coefficient and the length of the free surface. This is what forms a water droplet or a mercury globule. It decreases with increase in temperature, and depends on the contacting gas at the free surface.

Vapor Pressure

Molecules that escape a liquid surface cause the evaporation process. The pressure exerted at the surface by these free molecules is called the *vapor pressure.* Because this is caused by the molecular activity which is a function of the temperature, the vapor pressure of a liquid also is a function of the temperature and increases with it. Boiling occurs when the pressure above the liquid surface equals (or is less than) the vapor pressure of the liquid. This phenomenon, which may sometimes occur in a fluid system network, causing the fluid to locally vaporize, is called *cavitation.*

Gas and Liquid Viscosity

Viscosity is the property of a fluid that measures its resistance to flow. Cohesion is the main cause of this resistance. Because cohesion drops with temperature, so does viscosity. The coefficient of viscosity is the proportionality constant in Newton's law of viscosity that states that the shear stress τ in the fluid is directly proportional to the velocity gradient, as represented below:

$$\tau = \mu \frac{du}{dy} \tag{2}$$

The μ above is often called the absolute or dynamic viscosity. There is another form of the viscosity coefficient called the kinematic viscosity ν, that is, the ratio of viscosity to mass density:

$$\nu = \mu/\rho$$

Remember that in U.S. customary units, unit of mass density ρ is *slugs per cubic foot*.

Bulk Modulus

A liquid's compressibility is measured in terms of its *bulk modulus* of elasticity. Compressibility is the percentage change in unit volume per unit change in pressure:

$$C = \frac{\delta v/v}{\delta p}$$

The bulk modulus of elasticity K is its reciprocal:

$$K = 1/C$$

K is expressed in units of pressure.

Compressibility

Compressibility of liquids is defined above. However, for a gas, the application of pressure can have a much greater effect on the gas volume. The general relationship is governed by the *perfect gas law:*

$$pv_s = RT$$

where p is the absolute pressure, v_s is the specific volume, R is the gas constant, and T is the absolute temperature.

Units and Dimensions

One must always use a consistent set of units. Primary units are mass, length, time, and temperature. A unit system is called *consistent* when unit force causes a unit mass to achieve unit acceleration. In the U.S. system, this system is represented by the (pound) force, the (slug) mass, the (foot) length, and the (second) time. The slug mass is defined as the mass that accelerates to one ft/sec^2 when subjected to one pound force (lbf). Newton's second law, F = ma, establishes this consistency between force and mass units. If the mass is ever referred to as being in lbm (inconsistent system), one must first convert it to slugs by dividing it by 32.174 before using it in any consistent equation.

Because of the confusion between weight (lbf) and mass (lbm) units in the U.S. inconsistent system, there is also a similar confusion between density and specific weight units. It is, therefore, always better to resort to a consistent system for engineering calculations.

FLUID STATICS

Fluid statics is the branch of fluid mechanics that deals with cases in which there is no relative motion between fluid elements. In other words, the fluid may either be in rest or at constant velocity, but certainly not accelerating. Since there is no relative motion between fluid layers, there are no shear stresses in the fluid under static equilibrium. Hence, all free bodies in fluid statics have only normal forces on their surfaces.

Manometers and Pressure Measurements

Pressure is the same in all directions at a point in a static fluid. However, if the fluid is in motion, pressure is defined as the average of three mutually perpendicular normal compressive stresses at a point:

$$p = (p_x + p_y + p_z)/3$$

Pressure is measured either from the zero absolute pressure or from standard atmospheric pressure. If the reference point is absolute pressure, the pressure is called the *absolute* pressure, whereas if the reference point is standard atmospheric (14.7 psi), it is called the *gage* pressure. A barometer is used to get the absolute pressure. One can make a simple barometer by filling a tube with mercury and inverting it into an open container filled with mercury. The mercury column in the tube will now be supported only by the atmospheric pressure applied to the exposed mercury surface in the container. The equilibrium equation may be written as:

$$p_a = 0.491(144)h$$

where h is the height of mercury column in inches, and 0.491 is the density of mercury in pounds per cubic inch. In the above expression, we neglected the vapor pressure for mercury. But if we use any other fluid instead of mercury, the vapor pressure may be significant. The equilibrium equation may then be:

$$p_a = [(0.0361)(s)(h) + p_v](144)$$

where 0.0361 is the water density in pounds per cubic inch, and s is the specific gravity of the fluid. The consistent equation for variation of pressure is

$$p = \gamma h$$

where p is in lb/ft^2, γ is the specific weight of the fluid in lb/ft^3, and h is in *feet*. The above equation is the same as p $= \gamma_w$sh, where γ_w is the specific weight of water (62.4 lb/ft^3) and s is the specific gravity of the fluid.

Manometers are devices used to determine differential pressure. A simple U-tube manometer (with fluid of specific weight γ) connected to two pressure points will have a differential column of height h. The differential pressure will then be $\Delta p = (p_2 - p_1) = \gamma h$. Corrections must be made if high-density fluids are present above the manometer fluid.

Hydraulic Pressure on Surfaces

For a horizontal area subjected to static fluid pressure, the resultant force passes through the centroid of the area. If the plane is inclined at an angle θ, then the local pressure will vary linearly with the depth. The average pressure occurs at the average depth:

$$p_{avg} = \frac{1}{2}(h_1 + h_2)\sin\theta \tag{3}$$

However, the center of pressure will not be at average depth but at the centroid of the triangular or trapezoidal pressure distribution, which is also known as the *pressure prism.*

Buoyancy

The resultant force on a submerged body by the fluid around it is called the *buoyant force,* and it always acts upwards. If v is the volume of the fluid displaced by the submerged (wholly or partially) body, γ is the fluid specific weight, and $F_{buoyant}$ is the buoyant force, then the relation between them may be written as:

$$F_{buoyant} = v \times \gamma \qquad (4)$$

The principles of buoyancy make it possible to determine the volume, specific gravity, and specific weight of an unknown odd-shaped object by just weighing it in two different fluids of known specific weights γ_1 and γ_2. This is possible by writing the two equilibrium equations:

$$W = F_1 + v\gamma_1 = F_2 + v\gamma_2 \qquad (5)$$

BASIC EQUATIONS

In derivations of any of the basic equations in fluids, the concept of *control volume* is used. A control volume is an arbitrary space that is defined to facilitate analysis of a flow region. It should be remembered that all fluid flow situations obey the following rules:

1. Newton's Laws of Motion
2. The Law of Mass Conservation (Continuity Equation)

3. 1st and 2nd Laws of Thermodynamics
4. Proper boundary conditions

Apart from the above relations, other equations such as Newton's law of viscosity may enter into the derivation process, based on the particular situation. For detailed procedures, one should refer to a textbook on fluid mechanics.

Continuity Equation

For a continuous flow system, the mass within the fluid remains constant with time: $dm/dt = 0$. If the flow discharge Q is defined as $Q = A.V$, the continuity equation takes the following useful form:

$$\dot{m} = \rho_1 A_1 V_1 = \rho_2 A_2 V_2 \qquad (6)$$

Euler's Equation

Under the assumptions of: (a) frictionless, (b) flow along a streamline, and (c) steady flow; *Euler's equation* takes the form:

$$\frac{dp}{\rho} + g.dz + v.dv = 0 \qquad (7)$$

When ρ is either a function of pressure p or is constant, the Euler's equation can be integrated. The most useful relationship, called *Bernoulli's equation,* is obtained by integrating Euler's equation at constant density ρ.

Bernoulli's Equation

Bernoulli's equation can be thought of as a special form of energy balance equation, and it is obtained by integrating Euler's equation defined above.

$$z + \frac{v^2}{2g} + \frac{p}{\rho g} = \text{constant} \tag{8}$$

The constant of integration above remains the same along a streamline in steady, frictionless, incompressible flow. The term z is called the potential head, the term $v^2/2g$ is the dy-namic head, and the $p/\rho g$ term is called the static head. All these terms represent energy per unit weight. The equation characterizes the specific kinetic energy at a given point within the flow cross-section. While the above form is convenient for liquid problems, the following form is more convenient for gas flow problems:

$$\gamma z + \frac{\rho v^2}{2} + p = \text{constant} \tag{9}$$

Energy Equation

The energy equation for steady flow through a control volume is:

$$q_{heat} + \frac{p_1}{\rho_1} + gz + \frac{v_1^2}{2} u_1 = w_{shaft} + \frac{p_1}{\rho_2} + gz$$

$$+ \frac{v_2^2}{2} + u_2 \tag{10}$$

where q_{heat} is heat added per unit mass and w_{shaft} is the shaft work per unit mass of fluid.

Momentum Equation

The linear momentum equation states that the resultant force F acting on a fluid *control volume* is equal to the rate of change of linear momentum inside the control volume plus the net exchange of linear momentum from the control boundary. Newton's second law is used to derive its form:

$$F = \frac{d(mv)}{dt} = \int_{cv} \rho v dV + \int_{cs} \rho v v . dA \tag{11}$$

Moment-of-Momentum Equation

The moment-of-momentum equation is obtained by taking the vector cross-product of F detailed above and the position vector r of any point on the line of action, i.e., $r \times F$. Remember that the vector product of these two vectors is also a vector whose magnitude is $Fr \sin\theta$ and direction is normal to the plane containing these two basis vectors and obeying the cork-screw convention. This equation is of great value in certain fluid flow problems, such as in turbomachineries. The equations outlined in this section constitute the fundamental governing equations of flow.

ADVANCED FLUID FLOW CONCEPTS

Often in fluid mechanics, we come across certain terms, such as Reynolds number, Prandtl number, or Mach number, that we have come to accept as they are. But these are extremely useful in unifying the fundamental theories in this field, and they have been obtained through a mathematical analysis of various forces acting on the fluids. The mathematical analysis is done through Buckingham's Pi Theorem. This theorem states that, in a physical system described by n quantities in which there are m dimensions, these n quantities can be rearranged into (n-m) nondimensional parameters. Table 1 gives dimensions of some physical variables used in fluid mechanics in terms of basic mass (M), length (L), and time (T) dimensions.

Table 1
Dimensions of Selected Physical Variables

Physical Variable	Symbol	Dimension
Force	F	MLT^{-2}
Discharge	Q	L^3T^{-1}
Pressure	p	$ML^{-1}T^{-2}$
Acceleration	a	LT^{-2}
Density	ρ	ML^{-3}
Specific weight	γ	$ML^{-2}T^{-2}$
Dynamic viscosity	μ	$ML^{-1}T^{-1}$
Kinematic viscosity	ν	L^2T^{-1}
Surface tension	σ	MT^{-2}
Bulk modulus of elasticity	K	$ML^{-1}T^{-2}$
Gravity	g	LT^{-2}

Dimensional Analysis and Similitude

Most of these nondimensional parameters in fluid mechanics are basically ratios of a pair of fluid forces. These forces can be any combination of gravity, pressure, viscous, elastic, inertial, and surface tension forces. The flow system variables from which these parameters are obtained are: velocity V, the density ρ, pressure drop Δp, gravity g, viscosity μ, surface tension σ, bulk modulus of elasticity K, and a few linear dimensions of l.

These nondimensional parameters allow us to make studies on scaled models and yet draw conclusions on the prototypes. This is primarily because we are dealing with the ratio of forces rather than the forces themselves. The model and the prototype are dynamically similar if (a) they are geometrically similar and (b) the ratio of pertinent forces are also the same on both.

Nondimensional Parameters

The following five nondimensional parameters are of great value in fluid mechanics.

Reynolds Number

Reynolds number is the ratio of inertial forces to viscous forces:

$$\mathbf{R} = \frac{\rho \, Vl}{\mu} \tag{12}$$

This is particularly important in pipe flows and aircraft model studies. The Reynolds number also characterizes different flow regimes (laminar, turbulent, and the transition between the two) through a *critical* value. For example, for the case of flow of fluids in a pipe, a fluid is considered turbulent if **R** is greater than 2,000. Otherwise, it is taken to be laminar. A turbulent flow is characterized by random movement of fluid particles.

Froude Number

Froude number is the ratio of inertial force to weight:

$$\mathbf{F} = \frac{V}{\sqrt{gl}} \tag{13}$$

This number is useful in the design of spillways, weirs, channel flows, and ship design.

Weber Number

Weber number is the ratio of inertial forces to surface tension forces.

$$W = \frac{V^2 l \rho}{\sigma} \qquad (14)$$

This parameter is significant in gas-liquid interfaces where surface tension plays a major role.

Mach Number

Mach number is the ratio of inertial forces to elastic forces:

$$M = \frac{V}{c} = \frac{V}{\sqrt{kRT}} \qquad (15)$$

where c is the speed of sound in the fluid medium, k is the ratio of specific heats, and T is the absolute temperature. This parameter is very important in applications where velocities are near or above the local sonic velocity. Examples are fluid machineries, aircraft flight, and gas turbine engines.

Pressure Coefficient

Pressure coefficient is the ratio of pressure forces to inertial forces:

$$C_p = \frac{\Delta p}{\rho V^2 / 2} = \frac{\Delta h}{\rho V^2 / 2g} \qquad (16)$$

This coefficient is important in most fluid flow situations.

Equivalent Diameter and Hydraulic Radius

The *equivalent diameter* (D_{eq}) is defined as four times the *hydraulic radius* (r_h). These two quantities are widely used in open-channel flow situations. If A is the cross-sectional area of the channel and P is the *wetted perimeter* of the channel, then:

$$r_h = \frac{A}{P} \qquad (17)$$

Note that for a circular pipe flowing full of fluid,

$$D_{eq} = 4r_h = \frac{4(\pi D^2 / 4)}{\pi D} = D$$

and for a square duct of sides and flowing full,

$$D_{eq} = 4r_h = \frac{4a^2}{4a} = a$$

If a pipe is not flowing full, care should be taken to compute the wetted perimeter. This is discussed later in the section for open channels. The hydraulic radii for some common channel configurations are given in Table 2.

Table 2
Hydraulic Radii for Common Channel Configurations

Cross-Section	r_h
Circular pipe of diameter D	D/4
Annular section of inside dia d and outside dia D	(D – d)/4
Square duct with each side a	a/4
Rectangular duct with sides a and b	a/4
Elliptical duct with axes a and b	(ab)/K(a + b)
Semicircle of diameter D	D/4
Shallow flat layer of depth h	h

PIPE FLOW

In internal flow of fluids in a pipe or a duct, consideration must be given to the presence of frictional forces acting on the fluid. When the fluid flows inside the duct, the layer of fluid at the wall must have zero velocity, with progressively increasing values away from the wall, and reaching maximum at the centerline. The distribution is parabolic.

Friction Factor and Darcy Equation

The pipe flow equation most commonly used is the Darcy-Weisbach equation that prescribes the head loss h_f to be:

$$h_f = f \frac{L}{D} \frac{V^2}{2g} \qquad (18)$$

where L is the pipe length, D is the internal pipe diameter, V is the average fluid velocity, and f is the Moody friction factor (nondimensional) which is a function of several nondimensional quantities:

$$f = f\left(\frac{\rho V D}{\mu}, \frac{\varepsilon}{D}\right) \qquad (19)$$

where ($\rho V D/\mu$) is the Reynolds number **R,** and ε is the specific surface roughness of the pipe material. The Moody fric-

tion chart is probably the most convenient method of getting the value of f (see Figure 1). For laminar pipe flows (Reynolds number **R** less than 2,000), $f = \frac{64}{\mathbf{R}}$, because head loss in laminar flows is independent of wall roughness.

If the duct or pipe is not of circular cross-section, an equivalent hydraulic diameter D_{eq} as defined earlier is used in these calculations.

The Swamy and Jain empirical equation may be used to calculate a pipe design diameter directly. The relationship is:

$$D = 0.66\left[\varepsilon^{1.25}\left(\frac{LQ^2}{gh_f}\right)^{4.75} + vQ^{9.4}\left(\frac{L}{gh_f}\right)^{5.2}\right]^{0.04} \qquad (20)$$

where ε is in ft, Q is in cfs, L is in ft, v is in ft^2/s, g is in ft/s^2, and h_f is in ft.lb/lb units.

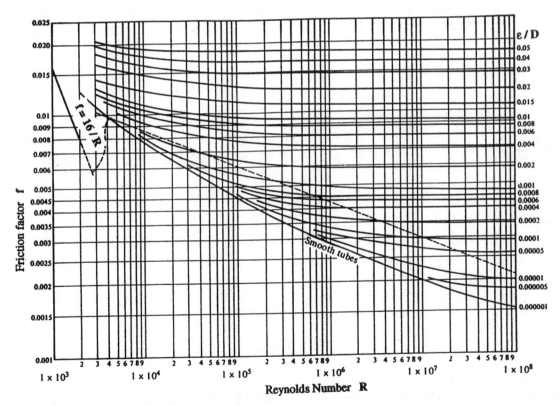

Figure 1. Friction factor vs. Reynolds number.

Losses in Pipe Fittings and Valves

In addition to losses due to friction in a piping system, there are also losses associated with flow through valves and fittings. These are called *minor losses,* but must be accounted for if the system has a lot of such fittings. These are treated as equivalent frictional losses. The *minor loss* may be treated either as a pressure drop $\Delta p = -K\rho V^2/2$ or as a head loss $\Delta h = -KV^2/(2g)$. The value of the loss coefficient K is obtained through experimental data. For valves and fittings, manufacturers provide this value. It may also be calculated from the equivalent length concept: $K = fL_e/D$, where L_e is the equivalent pipe length that has the same frictional loss. Table 3 gives these values for some common fittings.

For sudden enlargements in a pipe from diameter D_1 to a larger diameter D_2, the K value is obtained from:

$$K = [1 - (D_1/D_2)^2]^2$$

For sudden contractions in the pipeline from a larger diameter D_2 to a smaller diameter D_1, the value of the loss coefficient is:

Table 3
K Values for Common Fittings

Type of Fitting	K	L_e/D
45-degree elbow	0.35	17
90-degree bend	0.75	35
Diaphragm valve, open	2.30	115
Diaphragm valve, half open	4.30	215
Diaphragm valve, ¼ open	21.00	1050
Gate valve, open	0.17	9
Gate valve, half open	4.50	225
Globe valve, wide open	6.40	320
Globe valve, half open	9.50	475
Tee junction	1.00	50
Union and coupling	0.04	2
Water meter	7.00	350

$$K = [1 - (D_1/D_2)^2]/2$$

The above relations should serve as guidelines. Corrections should be made for enlargements and contractions that are gradual. Use values of K for fittings whenever furnished by the manufacturer.

Pipes in Series

Pipes connected in tandem can be solved by a method of equivalent lengths. This procedure lets us replace a series pipe system by a single pipeline having the same discharge and the same total head loss. As an example, if we have two pipes in series and if we select the first section as reference, then the *equivalent length* of the second pipe is obtained by:

$$L_2 = L_1 \frac{f_1}{f_2}\left(\frac{D_2}{D_1}\right)^5 \tag{21}$$

The values of f_1 and f_2 are approximated by selecting a discharge within the range intended for the two pipes.

Pipes in Parallel

A common way to increase capacity of an existing line is to install a second one parallel to the first. The flow is divided in a way such that the friction loss is the same in both (in series pipes, these losses are cumulative), but the discharge is cumulative. For an illustration of three pipes in parallel:

$$h_{f1} = h_{f2} = h_{f3} \frac{p_{entry}}{\gamma} + z_{entry} - \left(\frac{p_{exit}}{\gamma} + z_{exit}\right) \tag{22}$$

$$Q = Q_1 + Q_2 + Q_3 \tag{23}$$

where z_{entry} and z_{exit} are elevations at the two points.

If discharge Q is known, then the solution procedure uses this *equal loss* principle iteratively to find the solution (flow distribution and head loss).

The pipe network system behaves in an analogous fashion to a DC electrical circuit, and can be solved in an analogous manner by those familiar with the electrical circuit analysis.

OPEN-CHANNEL FLOW

Study of open channels is important in the study of river flow and irrigation canals. The mechanics of flow in open channels is more complicated than that in pipes and ducts because of the presence of a free surface. Unlike closed conduit flow, the *specific roughness factor* ε for open-channel flows is dependent on the hydraulic state of the channel. The flow is called *uniform* if the cross-section of the flow doesn't vary along the flow direction. Most open-channel flow situations are of turbulent nature. Therefore, a major part of the empirical and semi-empirical study has been done under the full turbulence assumption (Reynolds number **R** greater than 2,000 to 3,000).

Frictionless Open-Channel Flow

Flow in Venturi Flume

In the case of flow in a Venturi flume (Figure 2), where the width of the channel is deliberately changed to measure the flow rate, we can obtain all relations by applying Bernoulli's equation at the free surface, and the continuity equation, which are:

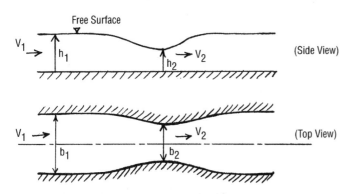

Figure 2. Flow in a Venturi flume.

$$V_1^2/2 + gh_1 = V_2^2/2 + gh_2$$

$$Q = V_1 b_1 h_1 = V_2 b_2 h_2$$

$$\frac{Q^2}{2g} = \frac{h_1 - h_2}{1/b_2^2 h_2^2 - 1/b_1^2 h_1^2}$$

Flow Over a Channel Rise

In the case of flow in a constant width (b) horizontal rectangular channel with a small rise on the floor (Figure 3), the relations are:

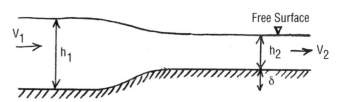

Figure 3. Open-channel flow over a rise.

$$V_1^2/2 + gh_1 = V_2^2/2 + g(h_2 + \delta)$$

$$Q = V_1 bh_1 = V_2 bh_2$$

$$\frac{Q^2}{2gb^2 h_1^2} = h_1 \frac{Q^2}{2gb^2 h_2^2} + (h_2 + \delta)$$

Note that the sum of the two terms $h + Q^2/(2gb^2 h^2)$ is called the *specific head*, H. The critical specific head H_c and the critical depth h_c can be found by taking the derivative of above term and equating it to zero.

$$h_c^3 = \left(\frac{Q^2}{gb^2} \right)$$

$$H_c = 3h_c/2$$

Note that for a given specific head and flow rate, two different depths of h are possible. The *Froude number* $V^2/(gh)$ specifies the flow characteristics of the channel flow. If it is less than unity, it is called *subcritical,* or tranquil, flow. If it is more than unity, it is called *supercritical,* or rapid, flow.

Flow Through a Sluice Gate

In the case of flow generated when a sluice gate that retains water in a reservoir is partially raised (Figure 4), by combining Bernoulli's equation and the continuity equation we get:

$$h_0 = \frac{Q^2}{2gb^2h^2} + h$$

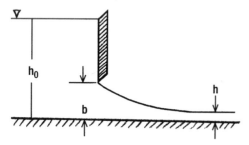

Figure 4. Flow through a sluice gate.

The maximum flow rate is given by:

$$\left(\frac{Q}{b}\right)^2_{max} = \frac{8}{27} gh_0^3$$

Here too, the Froude number is a measure of flow rate. Maximum flow rate is present when the Froude number is unity. By raising the gate from its closed position, the flow discharge is increased until a maximum discharge is obtained, and the depth downstream is two-thirds of the reservoir depth. If the gate is raised beyond this critical height, the flow rate actually drops.

The above analysis and observation is also true for flow over the crest of a dam, and the same equation for max flow rate is valid—where h_0 is the water level in the reservoir measured from the crest level, and h is the water level above the crest.

Laminar Open-Channel Flow

Considering the effects of viscosity, the steady laminar flow down an inclined plane (angle α), the velocity distribution is given by:

$$u = \frac{\rho g}{2u} (2h - y) y \sin \alpha$$

where y is the distance from the bottom surface of the channel (in a direction perpendicular to the flow direction). The volume flow per unit width (q) is given by:

$$q = \int_0^h u dy = \frac{g}{3v} h^3 \sin \alpha$$

Turbulent Open-Channel Flow

The wall shear stress τ_w due to friction in a steady, uniform, one-dimensional open-channel flow is given by:

$$\tau_w = \rho g (\sin \alpha) A/P$$

where A is the cross-sectional area of the channel, P is the wetted perimeter, and α is the downward sloping angle.

Hydraulic Jump

When a rapidly flowing fluid suddenly comes across a slowly flowing channel of a larger cross-sectional area, there is a sudden jump in elevation of the liquid surface. This happens because of conversion of kinetic energy to potential energy, the transition being quite turbulent. This phenomenon of steady nonuniform flow is called the *hydraulic jump*.

By applying the continuity and momentum equations, the increased depth y_2 can be expressed as:

$$y_2 = -\frac{y_1}{2} + \sqrt{\frac{y_1^2}{4} + \frac{2V_1^2 y_1}{g}}$$

The subscripts 1 and 2 represent flow conditions before and after the hydraulic jump. Through the energy equation, the losses due to this hydraulic jump as represented by h_{jump} can be found:

$$h_{jump} = \frac{(y_2 - y_1)^3}{4y_1y_2}$$

This phenomenon is often used at the bottom of a spillway to diffuse most of the fluid kinetic energy, and also as an effective way of mixing in a mixing chamber.

The Froude numbers $\mathbf{F_1}$ and $\mathbf{F_2}$ for a rectangular channel section before and after the jump are related by:

$$\mathbf{F_2} = \frac{2\sqrt{2}\,\mathbf{F_1}}{\left(\sqrt{1 + 8\mathbf{F_1^2}} - 1\right)^{1.5}}$$

where the dimensionless Froude number $\mathbf{F} = V/\sqrt{gy}$. The Froude number before the jump is greater than 1, and is less than 1 after the jump.

FLUID MEASUREMENTS

Total energy in a fluid flow consists of pressure head, velocity head, and potential head:

$$H = \frac{p}{\rho} + \frac{V^2}{2g} + z$$

The gravitational head is negligible; hence, if we know two of the three remaining variables (H, p, and V), we can find the other. In addition to the above, flow measurement also involves flow discharge, turbulence, and viscosity.

Pressure and Velocity Measurements

Static pressure is measured by either a piezometer opening or a static tube (Figure 5 [a]). The piezometer tap, a smooth opening on the wall normal to the surface, can measure the pressure head directly in feet of fluid: $h_p = p/\rho$. In the flow region away from the wall, the static tube probe may be introduced, directed upstream with the end closed. The static pressure tap must be located far enough downstream from the nose of the probe. The probe must also be aligned parallel to the flow direction.

Stagnation pressure (or "total pressure") is measured by a pitot tube (Figure 5 [b]), an open-ended tube facing directly into the flow, where the flow is brought to rest isentropically (no loss). At this point of zero velocity, $p_t = p + \rho V^2/2$. Often, both the static tube and the pitot tube are combined to make one "pitot-static" probe (Figure 5 [c]), which will in effect measure velocity of the flow. The two ends are connected to a manometer whose fluid has a specific gravity S_0. By applying Bernoulli's equation between the two points:

$$V = \sqrt{2(p_t - p)/\rho} = \sqrt{2g\Delta h\,(S_0 - S)/S}$$

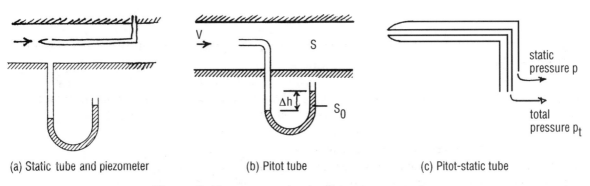

(a) Static tube and piezometer   (b) Pitot tube   (c) Pitot-static tube

Figure 5. Pressure and velocity measurements.

If the pitot probe is used in subsonic compressible flow, the compressible form of the stagnation pressure should be used:

$$p_t = p\left(1 + \frac{\gamma - 1}{2} M^2\right)^{\gamma/(\gamma-1)}$$

By knowing the stagnation and static pressures, and also the static temperature, the mach number at that point can also be found:

$$M = V/a = V/\sqrt{\gamma RT}$$

Rate Measurement

Flow rate in a Venturi meter (Figure 6 [a]) is given by:

$$Q = C_v A_2 \sqrt{2\left[\frac{(p_1 - p_2)/\rho + g(z_1 - z_2)}{(1 - A_2^2/A_1^2)}\right]}$$

Flow rate in a flow nozzle (Figure 6 [b]) is given by:

$$Q = C_n A_2 \sqrt{\frac{2}{\rho}\left[\frac{(p_1 - p_2)}{(1 - A_2^2/A_1^2)}\right]}$$

Flow rate in an orifice meter (Figure 6 [c]) is given by:

$$Q = C_o A_0 \sqrt{\frac{2}{\rho}\left[\frac{(p_1 - p_2)}{(1 - A_2^2/A_1^2)}\right]}$$

where C_v, C_n, and C_0 are the corresponding discharge coefficients for the three types of meters, and are functions of the Reynolds number. These are obtained through experimental tests.

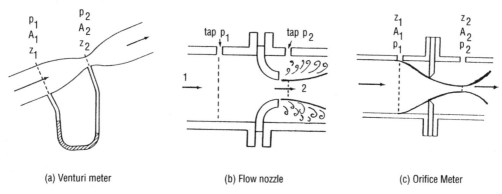

(a) Venturi meter   (b) Flow nozzle   (c) Orifice Meter

Figure 6. Flow measurement devices.

Hot-Wire and Thin-Film Anemometry

Air velocities may be measured by vane anemometers where the vanes drive generators that directly indicate the air velocity. They can be made very sensitive to extremely low air currents. Gas velocities may be measured with hot-wire anemometers. The principle of operation of these devices is the fact that the resistance to flow of electricity through a thin platinum wire is a function of cooling due to air around it.

$$R_{wire} = R_{ref}[1 + \alpha(T_{wire} - T_{ref})]$$

$$I^2 R_{wire} = hA(T_{wire} - T_{fluid})$$

where h is the convective heat transfer coefficient between wire and gas, A is wire surface area, I is the current in amperes, R is resistance of the wire, and T is temperature.

The same principle is applied in hot-film anemometers to measure liquid velocities. Here the probe is coated with a thin metallic film that provides the resistance. The film is usually coated with a very thin layer of insulating material to increase the durability and other problems associated with local boiling of the liquid.

Open-Channel Flow Measurements

A *weir* is an obstruction in the flow path, causing flow to back up behind it and then flow over or through it (Figure 7). Height of the upstream fluid is a function of the flow rate. Bernoulli's equation establishes the weir relationship:

$$Q = C\frac{2}{3}L\sqrt{2gH^3} = C_cLH^{1.5}$$

where C_c is the contraction coefficient (3.33 in U.S. units and 1.84 in metric units), L is the width of weir, and h is the head of liquid above the weir. Usually, a correction coefficient is multiplied to account for the velocity head. For a V-notch weir, the equation may be written as:

$$Q_{theoretical} = \frac{8}{15}\sqrt{2g}\tan(\phi/2)H^{2.5}$$

For a 90-degree V-notch weir, this equation may be approximated to $Q = C_vH^{2.5}$, where C_v is 2.5 in U.S. units and 1.38 in metric units.

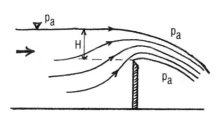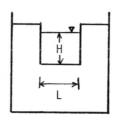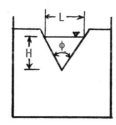

Figure 7. Rectangular and V-notch weirs.

Viscosity Measurements

Three types of devices are used in viscosity measurements: capillary tube viscometer, Saybolt viscometer, and rotating viscometer. In a capillary tube arrangement (Figure 8),

$$\mu = \frac{\Delta p\pi D^4}{128QL}$$

The reservoir level is maintained constant, and Q is determined by measuring the volume of flow over a specific time period. The Saybolt viscometer operates under the same principle.

In the rotating viscometer (Figure 9), two concentric cylinders of which one is stationary and the other is rotating (at constant rpm) are used. The torque transmitted from one to the other is measured through spring deflection.

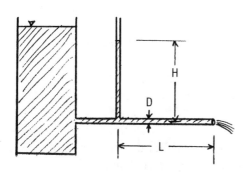

Figure 8. Capillary tube viscometer.

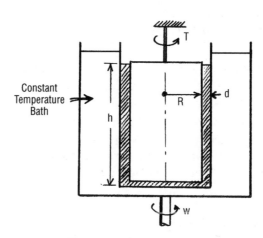

Figure 9. Rotating viscometer.

The shear stress τ is a function of this torque T. Knowing shear stress, the dynamic viscosity may be calculated from Newton's law of viscosity.

$$\mu = \frac{Td}{2\pi R^3 h\omega}$$

OTHER TOPICS

Unsteady Flow, Surge, and Water Hammer

Study of *unsteady flow* is essential in dealing with hydraulic transients that cause noise, fatigue, and wear. It deals with calculation of pressures and velocities. In closed circuits, it involves the unsteady linear momentum equation along with the unsteady continuity equation. If the nonlinear friction terms are introduced, the system of equations becomes too complicated, and is solved using iterative, computer-based algorithms.

Surge is the phenomenon caused by turbulent resistance in pipe systems that gives rise to oscillations. A sudden reduction in velocity due to flow constriction (usually due to valve closure) causes the pressure to rise. This is called *water hammer.* Assuming the pipe material to be inelastic, the time taken for the water hammer shock wave from a fitting to the pipe-end and back is determined by: $t = (2L)/c$; the corresponding pressure rise is given by: $\Delta p = (\rho c \Delta v)/g_c$.

In open-channel systems, the surge wave phenomenon usually results from a gate or obstruction in the flow path. The problem needs to be solved through iterative solution of continuity and momentum equations.

Boundary Layer Concepts

For most fluids we know (water or air) that have low viscosity, the Reynolds number $\rho U L/\mu$ is quite high. So inertia forces are predominant over viscous ones. However, near a wall, the viscosity will cause the fluid to slow down, and have zero velocity at the wall. Thus the study of most real fluids can be divided into two regimes: (1) near the wall, a thin viscous layer called the *boundary layer;* and (2) outside of it, a nonviscous fluid. This boundary layer may be laminar or turbulent. For the classic case of a flow over a flat plate, this transition takes place when the Reynolds number reaches a value of about a million. The boundary layer thickness δ is given as a function of the distance x from the leading edge of the plate by:

$$\delta = \frac{5.0x}{\sqrt{\rho U\, x/\mu}}$$

where U and μ are the fluid velocity and viscosity, respectively.

Lift and Drag

Lift and drag are forces experienced by a body moving through a fluid. Coefficients of lift and drag (C_L and C_D) are used to determine the effectiveness of the object in producing these two principal forces:

$$L = \frac{1}{2}\rho V^2 A C_L$$

$$D = \frac{1}{2}\rho V^2 A C_D$$

where A is the reference area (usually projection of the object's area either parallel or normal to the flow direction), ρ is the density, and V is the flow velocity.

Oceanographic Flows

The pressure change in the ocean depth is $dp = \rho gD$, the same as in any static fluid. Neglecting salinity, compressibility, and thermal variations, that is about 44.5 psi per 100 feet of depth. For accurate determination, these effects must be considered because the temperature reduces nonlinearly with depth, and density increases linearly with salinity.

The periods of an ocean wave vary from less than a second to about 10 seconds; and the wave propagation speeds vary from a ft/sec to about 50 ft/sec. If the wavelength is small compared to the water depth, the wave speed is independent of water depth and is a function only of the wavelength:

$$c = \sqrt{g/2\pi}\,\sqrt{L}$$

Tide is caused by the combined effects of solar and lunar gravity. The average interval between successive high waters is about 12 hours and 25 minutes, which is exactly one half of the lunar period of appearance on the earth. The lunar tidal forces are more than twice that of the solar ones. The *spring tides* are caused when both are in unison, and the *neap tides* are caused when they are 90 degrees out of phase.

2

Heat Transfer

Chandran B. Santanam, Ph.D., Senior Staff Development Engineer, GM Powertrain Group
J. Edward Pope, Ph.D., Senior Project Engineer, Allison Advanced Development Company
Nicholas P. Cheremisinoff, Ph.D., Consulting Engineer

INTRODUCTION

This chapter will cover the three basic types of heat transfer: conduction, convection, and radiation. Additional sections will cover finite element analysis, heat exchangers, and two-phase heat transfer.

Parameters commonly used in heat transfer analysis are listed in Table 1 along with their symbols and units. Table 2 lists relevant physical constants.

Table 1
Parameters Commonly Used in Heat Transfer Analysis

Parameters	Units	Symbols
Length	feet	L
Mass	pound mass	m
Time	hour or seconds	τ
Current	ampere	I
Temperature	Fahrenheit or Rankine	T
Acceleration	feet/secs^2	a
Velocity	ft/sec	V
Density	pound/cu. ft	ρ
Area	sq. feet	A
Volume	cubic feet	v
Viscosity	lbm/ft/sq. sec	μ
Force	pound	F
Kinematic viscosity	feet^2/sec^2	υ
Specific heat	Btu/hr/lb/°F	Cp
Thermal conductivity	Btu. in/ft^2/hr/°F	k
Heat energy	Btu	Q
Convection coefficient	Btu/sq. ft/hr/°F	h
Hydraulic diameter	feet	D_h
Gravitational constant	lbm.ft/lbf.sec^2	g

Table 2
Physical Constants Important in Heat Transfer

Constant Name	Value	Units
Avagadro's number	$6.022169*10^{26}$	Kmol^{-1}
Gas constant	53.3	ft-lb/lbm/°F
Plank's constant	$6.626196*10^{-34}$	J.s
Boltzmann constant	$1.380622*10^{-23}$	J/K
Speed of light in vacuum	91372300	ft/sec
Stefan-Boltzmann constant	$1.712*10^{-9}$	Btu/hr/sq.ft/°R^4
1 atm pressure	14.7	psi

CONDUCTION

Single Wall Conduction

If two sides of a flat wall are at different temperatures, conduction will occur (Figure 1). Heat will flow from the hotter location to the colder point according to the equation:

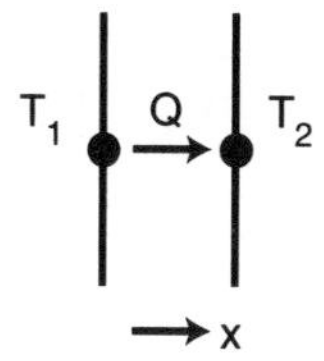

$$Q = \frac{k}{\text{Thickness}} (T_1 - T_2) \, \text{Area}$$

For a cylindrical system, such as in pipes (Figure 2), the equation becomes:

$$Q = 2\pi \, (k) \, (\text{length}) \frac{(T_o - T_i)}{\ln (r_o/r_i)}$$

Figure 1. Conduction through a single wall.

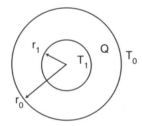

Figure 2. Conduction through a cylinder.

The equation for cylindrical coordinates is slightly different because the area changes as you move radially outward. As Figure 3 shows, the temperature profile will be a straight line for a flat wall. The profile for the pipe will flatten as it moves radially outward. Because area increases with radius, conduction will increase, which reduces the thermal gradient. If the thickness of the cylinder is small, relative to the radius, the cartesian coordinate equation will give an adequate answer. Thermal conductivity is a material property, with units of

$$\frac{Btu}{(hr)\,(foot)\,(^{\circ}F)}$$

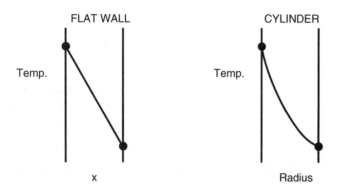

Figure 3. Temperature profile for flat wall and cylinder.

Tables 3 and 4 show conductivities for metals and common building materials. Note that the materials that are good electrical conductors (silver, copper, and aluminum), are also good conductors of heat. Increased conduction will tend to equalize temperatures within a component.

Example. Consider a flat wall with:
Thickness = 1 foot

Table 3
Thermal Conductivity of Various Materials at 0°C

Material	Thermal conductivity k	
	W/m·°C	Btu/h·ft·°F
Metals:		
Silver (pure)	410	237
Copper (pure)	385	223
Aluminum (pure)	202	117
Nickel (pure)	93	54
Iron (pure)	73	42
Carbon steel, 1% C	43	25
Lead (pure)	35	20.3
Chrome-nickel steel	16.3	9.4
(18% Cr, 8% Ni)		
Nonmetallic solids:		
Quartz, parallel to axis	41.6	24
Magnesite	4.15	2.4
Marble	2.08–2.94	1.2–1.7
Sandstone	1.83	1.06
Glass, window	0.78	0.45
Maple or oak	0.17	0.096
Sawdust	0.059	0.034
Glass wool	0.038	0.022
Liquids:		
Mercury	8.21	4.74
Water	0.556	0.327
Ammonia	0.540	0.312
Lubricating oil, SAE 50	0.147	0.085
Freon 12, CCl_2F_2	0.073	0.042
Gases:		
Hydrogen	0.175	0.101
Helium	0.141	0.081
Air	0.024	0.0139
Water vapor (saturated)	0.0206	0.0119
Carbon dioxide	0.0146	0.00844

Source: Holman [1]. Reprinted with permission of McGraw-Hill.

Area = 1 foot^2
Q = 1,000 Btu/hour

For aluminum, k = 132, $\Delta T = 7.58^{\circ}F$
For stainless steel, k = 9, $\Delta T = 111.1^{\circ}F$

Sources

1. Holman, J. P., *Heat Transfer.* New York: McGraw-Hill, 1976.
2. Cheremisinoff, N. P., *Heat Transfer Pocket Handbook.* Houston: Gulf Publishing Co., 1984.

Table 4
Thermal Conductivities of Typical Insulating
and Building Materials

Material	Temperature (°C)	Thermal Conductivity (kcal/m-hr-°C)
Asbestos	0	0.13
Glass wool	0	0.03
	300	0.09
Cork in slabs	0	0.03
	50	0.04
Magnesia	50	0.05
Slag wool	0	0.05
	200	0.07
Common brick	25	0.34
Porcelain	95	0.89
	1,100	1.70
Concrete	0	1.2
Fresh earth	0	2.0
Glass	15	0.60
Wood (pine):		
Perpendicular fibers	15	0.13
Parallel fibers	20	0.30
Burnt clay	15	0.80
Carborundum	600	16.0

Source: Cheremisinoff [2]

Composite Wall Conduction

For the multiple wall system in Figure 4, the heat transfer rates are:

$$Q_{1-2} = \frac{k_1}{\text{Thickness}_1}(T_1 - T_2)\,\text{Area}$$

$$Q_{2-3} = \frac{k_2}{\text{Thickness}_2}(T_2 - T_3)\,\text{Area}$$

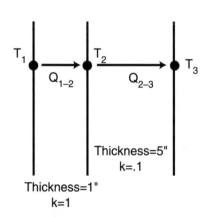

Figure 4. Conduction through a composite wall.

Obviously, Q and Area are the same for both walls. The term *thermal resistance* is often used:

$$R_{th} = \frac{\text{thickness}}{k}$$

High values of thermal resistance indicate a good insulation. For the entire system of walls in Figure 4, the overall heat transfer becomes:

$$Q = \frac{1}{\sum \dfrac{\text{thickness}_i}{k_i}} A\,(T_1 - T_{n+1})$$

The effective thermal resistance of the entire system is:

$$R_{eff} = \sum R_i = \sum \frac{\text{thickness}_i}{k_i}$$

For a cylindrical system, effective thermal resistance is:

$$R_{eff} = \sum R_i = \sum \frac{\ln (r_{i+1}/r_i)}{k_i}$$

Note that the temperature difference across each wall is proportional to the thermal effectiveness of each wall. Also note that the overall thermal effectiveness is dominated by the component with the largest thermal effectiveness.

Wall 1
    Thickness = 1. foot
    k = 1. Btu/(Hr*Foot*F)
    R = 1./1. = 1.

Wall 2
    Thickness = 5. foot
    k = .1 Btu/(Hr*Foot*F)
    R = 5./.1 = 50.

The overall thermal resistance is 51.
    Because only 2% of the total is contributed by wall 1, its effect could be ignored without a significant loss in accuracy.

The Combined Heat Transfer Coefficient

An overall heat transfer coefficient may be used to account for the combined effects of convection and conduction. Consider the problem shown in Figure 5. Convection occurs between the gas (T_1) and the left side of the wall (T_2). Heat is then conducted to the right side of the wall (T_3). Overall heat transfer may be calculated by:

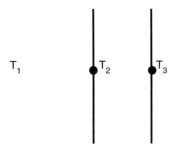

Figure 5. Combined convection and conduction through a wall.

$$Q = \frac{T_1 - T_3}{1/(hA) + thickness/(kA)}$$

The overall heat transfer coefficient is:

$$U = \frac{1}{(1/h) + (thickness/k)}$$

Heat transfer may be calculated by:

$$Q = UA (T_1 - T_3)$$

Although the overall heat transfer coefficient is simpler to use, it does not allow for calculation of T_2. This approach is particularly useful when matching test data, because all uncertainties may be rolled into one coefficient instead of adjusting two variables.

Critical Radius of Insulation

Consider the pipe in Figure 6. Here, conduction occurs through a layer of insulation, then convects to the environment. Maximum heat transfer occurs when:

$$r_{outer} = \frac{k}{h}$$

This is the critical radius of insulation. If the outer radius is less than this critical value, adding insulation will cause an increase in heat transfer. Although the increased insulation reduces conduction, it adds surface area, which increases convection. This is most likely to occur when convection is low (high h), and the insulation is poor (high k).

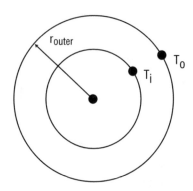

Figure 6. Pipe wrapped with insulation.

CONVECTION

While conduction calculations are straightforward, convection calculations are much more difficult. Numerous correlation types are available, and good judgment must be exercised in selection. Most correlations are valid only for a specific range of Reynolds numbers. Often, different relationships are used for various ranges. The user should note that these may yield discontinuities in the relationship between convection coefficient and Reynolds number.

Dimensionless Numbers

Many correlations are based on dimensionless numbers, which are used to establish similitude among cases which might seem very different. Four dimensionless numbers are particularly significant:

Reynolds Number

The Reynolds number is the ratio of flow momentum rate (i.e., inertia force) to viscous force.

$$Re = \frac{\rho u_m D_h}{\mu} = \frac{G D_h}{\mu}$$

The Reynolds number is used to determine whether flow is laminar or turbulent. Below a critical Reynolds number, flow will be laminar. Above a critical Reynolds number, flow will be turbulent. Generally, different correlations will be used to determine the convection coefficient in the laminar and turbulent regimes. The convection coefficients are usually significantly higher in the turbulent regime.

Nusselt Number

The Nusselt number characterizes the similarity of heat transfer at the interface between wall and fluid in different systems. It is basically a ratio of convection to conductance:

$$N = \frac{hl}{k}$$

Prandtl Number

The Prandtl number is the ratio of momentum diffusivity to thermal diffusivity of a fluid:

$$Pr = \frac{\mu C_p}{k}$$

It is solely dependent upon the fluid properties:

- For gases, $Pr = .7$ to 1.0
- For water, $Pr = 1$ to 10
- For liquid metals, $Pr = .001$ to $.03$
- For oils, $Pr = 50.$ to $2000.$

In most correlations, the Prandtl number is raised to the .333 power. Therefore, it is not a good investment to spend a lot of time determining Prandtl number for a gas. Just using .85 should be adequate for most analyses.

Grashof Number

The Grashof number is used to determine the heat transfer coefficient under free convection conditions. It is basically a ratio between the buoyancy forces and viscous forces.

$$Gr = \frac{g(T_w - T_{air})\rho^2}{T_{abs}\,\mu^2}$$

Heat transfer requires circulation, therefore, the Grashof number (and heat transfer coefficient) will rise as the buoyancy forces increase and the viscous forces decrease.

Correlations

Heat transfer correlations are empirical relationships. They are available for a wide range of configurations. This book will address only the most common types:

- Pipe flow
- Average flat plate
- Flat plate at a specific location
- Free convection
- Tube bank
- Cylinder in cross-flow

The last two correlations are particularly important for heat exchangers.

Pipe Flow

This correlation is used to calculate the convection co-efficient between a fluid flowing through a pipe and the pipe wall [1].

$$Re = \frac{W \times D_h}{A\mu}$$

For turbulent flow (Re > 10,000):

$$h = .023 K Re^{.8} \times Pr^n$$

n = .3 if surface is hotter than the fluid
= .4 if fluid is hotter than the surface

This correlation [1] is valid for $0.6 \leq P_r \leq 160$ and $L/D \geq 10$. For laminar flow [2]:

$$N = 4.36$$

$$h = \frac{N \times K}{D_h}$$

Average Flat Plate

This correlation is used to calculate an average convection coefficient for a fluid flowing across a flat plate [3].

$$Re = \frac{\rho V L}{\mu}$$

For turbulent flow (Re > 50,000):

$$h = .037 \, KRe^{.8} \times Pr^{.33}/L$$

For laminar flow:

$$h = .664 KRe^{.5} \times Pr^{.33}/L$$

Flat Plate at a Specific Location

This correlation is used to calculate a convection coefficient for a fluid flowing across a flat plate at a specified distance (X) from the start [3].

$$Re = \frac{\rho V X}{\mu}$$

For turbulent flow (Re > 50,000):

$$h = .0296 KRe^{.8} \times Pr^{.33}/X$$

For laminar flow:

$$h = .332 KRe^{.5} \times Pr^{.33}/X$$

Static Free Convection

Free convection calculations are based on the product of the Grashof and Prandtl numbers. Based on this product, the Nusselt number can be read from Figure 7 (vertical plates) or Figure 8 (horizontal cylinders) [6].

Tube Bank

The following correlation is useful for in-line banks of tubes, such as might occur in a heat exchanger [5]:

$$V_{max} = \frac{Flow}{\rho A_{min}}$$

It is valid for Reynolds numbers between 2,000 and 40,000 through tube banks more than 10 rows deep. For less than 10 rows, a correction factor must be applied (.64 for 1 row, .80 for 2 rows, .90 for 4 rows) to the convection coefficient.

Obtaining C and CEXP from the table (see also Figure 9, in-line tube rows):

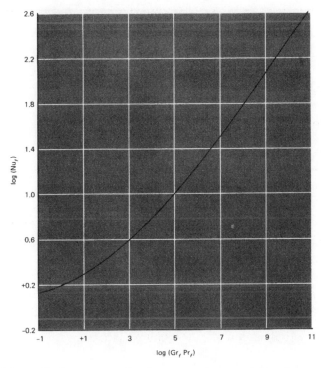

Figure 7. Free convection heat transfer correlation for vertical plates [6]. (*Reprinted with permission of McGraw-Hill.*)

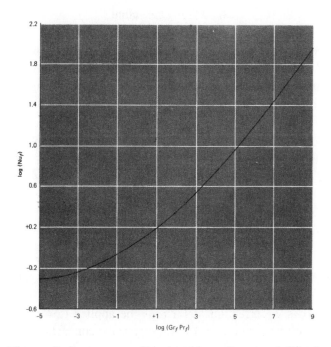

Figure 8. Free convection heat transfer correlation for horizontal cylinders [6]. (*Reprinted with permission of McGraw-Hill.*)

$$H = (CK/D) \, (Re)^{CEXP} \, (Pr/.7)^{.333}$$

	Sn/D							
	1.25		1.50		2.00		3.00	
Sp/D	C	CEXP	C	CEXP	C	CEXP	C	CEXP
1.25	.386	.592	.305	.608	.111	.704	.0703	.752
1.5	.407	.586	.278	.620	.112	.702	.0753	.744
2.0	.464	.570	.332	.602	.254	.632	.220	.648
3.0	.322	.601	.396	.584	.415	.581	.317	.608

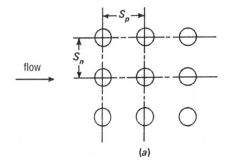

(a)

Figure 9. Nomenclature for in-line tube banks [6]. (*Reprinted with permission of McGraw-Hill.*)

Cylinder in Cross-flow

The following correlation is useful for any case in which a fluid is flowing around a cylinder [6]:

$$Re = \rho V 2r/\mu$$

Re < 4	C = .989	CEXP = .330
4 < Re < 40	C = .911	CEXP = .385
40 < Re < 4000	C = .683	CEXP = .466
4000 < Re < 40,000	C = .193	CEXP = .618
40,000 < Re < 400,000	C = .0266	CEXP = .805

$$N = C \times re^{CEXP} \, Pr^{.333}$$

$$h = \frac{K}{(2r)} N$$

Sources

1. Dittus, F. W. and Boelter, L. M. K., University of California Publications on Engineering, Vol. 2, Berkeley, 1930, p. 443.

2. Kays, W. M. and Crawford, M. E., *Convective Heat and Mass Transfer.* New York: McGraw-Hill, 1980.
3. Incropera, F. P. and Dewitt, D. P., *Fundamentals of Heat and Mass Transfer.* New York: John Wiley and Sons, 1990.
4. McAdams, W. H., *Heat Transmission.* New York: McGraw-Hill, 1954.

5. Grimson, E. D., "Correlation and Utilization of New Data on Flow Resistance and Heat Transfer for Cross Flow of Gases over Tube Banks," *Transactions ASME,* Vol. 59, 1937, pp. 583–594.
6. Holman, J. P., *Heat Transfer.* New York: McGraw-Hill, 1976.

Typical Convection Coefficient Values

After calculating convection coefficients, the analyst should always check the values and make sure they are reasonable. This table shows representative values:

Air, free convection	1–5
Water, free convection	5–20
Air or steam, forced convection	5–50
Oil or oil mist, forced convection	10–300
Water, forced convection	50–2,000
Boiling water	500–10,000
Condensing water vapor	900–100,000

RADIATION

The radiation heat transfer between two components is calculated by:

$$Q = A_1 F_{1-2} \sigma (E_1 T_1^4 - E_2 T_2^4)$$

σ is the Stefan-Boltzmann constant and has a value of 1.714×10^{-8} Btu /(hr $\times$ ft^2 $\times$ °R^4). A_1 is the area of component 1, and F_{1-2} is the view factor (also called a shape factor), which represents the fraction of energy leaving component 1 that strikes component 2. By the reciprocity theorem:

$$A_1 F_{1-2} = A_2 F_{2-1}$$

E_1 and E_2 are the emissivities of surfaces 1 and 2, respectively. These values will always be between 1 (perfect absorption) and 0 (perfect reflection). Some materials, such as glass, allow transmission of radiation. In this book, we will neglect this possibility, and assume that all radiation is either reflected or absorbed.

Before spending much time contemplating radiation heat transfer, the analyst should first decide whether it is significant. Since radiation is a function of absolute temperature to the fourth power, its significance increases rapidly as temperature increases. The following table shows this clearly. Assuming emissivities and view factors of 1, the equivalent h column shows the convection coefficient required to give the same heat transfer. In most cases, radiation can be safely ignored at temperatures below 500°F. Above 1,000°F, radiation must generally be accounted for.

Temperatures	Equivalent h
200–100	1.57
500–400	5.18
1,000–900	19.24
1,500–1,400	47.80
2,000–1,900	96.01

Emissivity

Table 5 shows emissivities of various materials. Estimation of emissivity is always difficult, but several generalizations can be made:

- Highly polished metallic surfaces usually have very low emissivities.
- Emissivity increases with temperature for all metallic surfaces.
- Emissivity for nonmetallic surfaces are much higher than for metallic surfaces, and decrease with temperature.
- Emissivity is very dependent upon surface conditions. The formation of oxide layers and increased surface roughness increases emissivity. Therefore, new components will generally have lower emissivities than ones that have been in service.

Source

Cheremisinoff, N. P., *Heat Transfer Pocket Handbook.* Houston: Gulf Publishing Co., 1984.

Table 5
Normal Total Emissivities of Different Surfaces

Surface	t (°F)	Emissivity
Metals		
Aluminum (highly polished, 98.3% pure)	440 ~ 1070	0.039 ~ 0.057
Brass (highly polished)		
73.2% Cu, 26.7% Zn	476 ~ 674	0.028 ~ 0.031
82.9% Cu, 17.0% Zn	530	0.030
Copper		
Polished	242	0.023
Plate heated @ 1110°F	390 ~ 1110	0.57
Molten-state	1970 ~ 2330	0.16 ~ 0.13
Gold	440 ~ 1160	0.018 ~ 0.035
Iron and steel:		
Polished, electrolytic iron	350 ~ 440	0.052 ~ 0.064
Polished iron	800 ~ 1800	0.144 ~ 0.377
Sheet iron	1650 ~ 1900	0.55 ~ 0.60
Cast iron	1620 ~ 1810	0.60 ~ 0.70
Lead (unoxidized)	260 ~ 440	0.057 ~ 0.075
Mercury	32 ~ 212	0.09 ~ 0.12
Nickel (technically pure, polished)	440 ~ 710	0.07 ~ 0.087
Platinum (pure)	440 ~ 1160	0.054 ~ 0.104
Silver (pure)	440 ~ 1160	0.0198 ~ 0.0324
Refractories and miscellaneous materials		
Asbestos	74 ~ 700	0.93 ~ 0.96
Brick, red	70	0.93
Carbon		
Filament	1900 ~ 2560	0.526
Candle soot	206 ~ 520	0.952
Lampblack	100 ~ 700	0.945
Glass	72	0.937
Gypsum	70	0.903
Plaster	50 ~ 190	0.91
Porcelain, glazed	72	0.924
Rubber	75	0.86 ~ 0.95
Water	32 ~ 212	0.95 ~ 0.963

View Factors

Exact calculation of view factors is often difficult, but they can often be estimated reasonably well.

Concentric Cylinders

Neglecting end effects, the view factor from the inner cylinder to the outer cylinder is always 1, regardless of radii (Figure 10). The view factor from the outer cylinder to the inner one is the ratio of the radii r_{inner}/r_{outer}. The radiation which does not strike the inner cylinder $1 - (r_{inner}/r_{outer})$ strikes the outer cylinder.

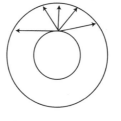

All radiation from inside cylinder strikes outside cylinder

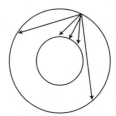

Radiation from outside cylinder strikes inside cylinder and outside cylinder.

Figure 10. Radiation view factors for concentric cir-

Parallel Rectangles

Figure 11 shows the view factors for parallel rectangles. Note that the view factor increases as the size of the rectangles increase, and the distance between them decreases.

Perpendicular Rectangles

Figure 12 shows view factors for perpendicular rectangles. Note that the view factor increases as A_1 becomes long and thin (Y/X = .1) and A_2 becomes large (Z/X = 10). In this arrangement, the view factor can never exceed .5, because at least half of the radiation leaving A_1 will go towards the other side, away from A_2.

Source

Holman, J. P., *Heat Transfer.* New York: McGraw-Hill, 1976.

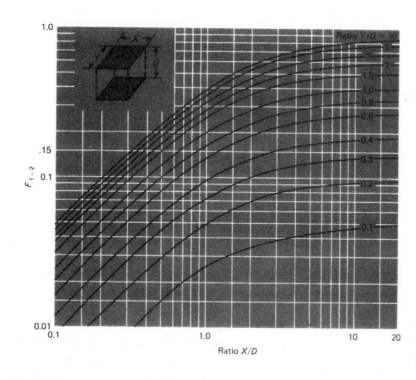

Figure 11. Radiation view factors for parallel rectangles. (*Reprinted with permission of McGraw-Hill.*)

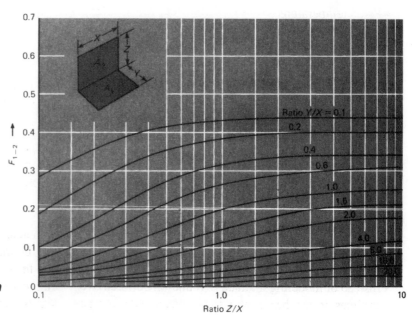

Figure 12. Radiation view factors for perpendicular rectangles. (*Reprinted with permission of McGraw-Hill.*)

Radiation Shields

In many designs, a radiation shield can be employed to reduce heat transfer. This is typically a thin piece of sheet metal which blocks the radiation path from the hot surface to the cool surface. Of course, the shield will heat up and begin to radiate to the cool surface. If we assume the two surfaces and the shield all have the same emissivity, and all view factors are 1, the overall heat transfer will be cut in half.

FINITE ELEMENT ANALYSIS

With today's computers and software, finite element analysis (FEA) can be used for most heat transfer analysis. Heat transfer generally does not require as fine a model as is required for stress analysis (to obtain stresses, derivatives of deflection must be calculated, which is an inherently inaccurate process). While FEA can accurately analyze complex geometries, it can also generate garbage if used improperly. Care should be exercised in creating the finite element model, and results should be checked thoroughly.

Boundary Conditions

Convection coefficients must be assigned to all element faces where convection will occur. Temperatures may be assigned in two ways:

• Fixed temperature
• Channels

Channels are flowing streams of fluid. As they exchange heat with the component, their temperature will increase or decrease. The channel temperatures will be applied to the element faces exposed to that channel. Conduction properties for all materials must be provided. Material density and specific heats must also be provided for a transient analysis. Precise calculation of radiation with FEA may be difficult, because view factors must be calculated between every set of radiating elements. This can add up quickly, even for a small model. Three options are available:

• Software is available to automatically calculate view factors for finite element models.
• Instead of modeling interactive radiation between two surfaces, it may be possible to have each radiate to an environment with a known temperature. Each environment temperature should be an average temperature of the opposite surface. This may require an iteration or two to get the environment temperature right. This probably is not a good option for transient analysis, because the environment temperatures will be constantly changing.
• For problems at low temperatures, or with high convection coefficients, radiation may be eliminated from the model with little loss in accuracy.

Some problems require modeling internal heat generation. The most common cases are bearing races, which generate heat due to friction, and internal heating due to electric currents.

Where two components contact, the conduction across this boundary is dependant upon the contact pressures, and the roughness of the two surfaces. For most finite element analyses, the two components may be joined so that full conduction occurs across the boundary.

2D Analysis

For many problems, 2D or axisymmetric analysis is used. This may require adjusting the heat transfer coefficients. Consider the bolt hole in Figure 13. The total surface area of the bolt hole is πDL, but in the finite element model, the surface area is only DL. In FEA, it is important the total hA product is correct. Therefore, the heat transfer coefficient should be multiplied by π. Similarly, for transient analysis, it is necessary to model the proper mass. If the wrong mass is modeled, the component will react too quickly (too little mass), or too slowly (too much mass) during a transient.

The user should keep in mind the limitations of 2D FEA. Consider the turbine wheel in Figure 14. The wheel is a solid of revolution, with 40 discontinuous blades attached to it. These blades absorb heat from the hot gases coming out of the combuster and conduct it down into the wheel. 2D FEA assumes that temperature does not vary in the tangential direction. In reality, the portions of the wheel directly under the blades will be hotter than those portions between the blades. Therefore, Location A will be hotter than Location B. Location A will also respond more quickly during a transient. If accurate temperatures in this region are desired, then 3D FEA is required. If the analyst is only interested in accurate bore temperatures, then 2D analysis should be adequate for this problem.

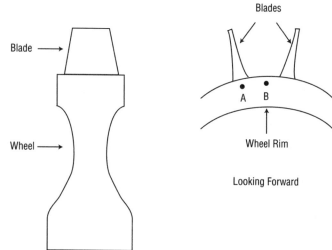

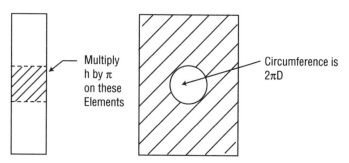

Figure 13. Convection coefficients must be adjusted for holes in 2D finite element models.

Figure 14. 2D finite element models cannot account for variation in the third dimension. Point A will actually be hotter than point B due to conduction from the blades.

Transient Analysis

Transient FEA has an added degree of difficulty, because boundary conditions vary with time. Often this can be accomplished by scaling boundary temperatures and convection coefficients.

Consider the problem in Figure 15. A plate is exposed to air in a cavity. This cavity is fed by 600°F air and 100°F air. Test data indicate that the environment temperatures range from 500°F at the top to 400°F at the bottom. The environment temperatures at each location (1–8) may be considered to be a function of the source (maximum) and sink (minimum) temperatures:

$$F_i = (T_i - T_{sink})/(T_{source} - T_{sink})$$

Here, the source temperature is 600°F and the sink temperature is 100°F. The environment temperatures at locations 1, 2, 3, and 4 are 90%, 80%, 70%, and 60%, respectively, of this difference. These percentages may be assumed to be constant, and the environment temperatures throughout the mission may be calculated by merely plugging in the source and sink temperatures.

$$T_i = T_{sink} + F_i (T_{source} - T_{sink})$$

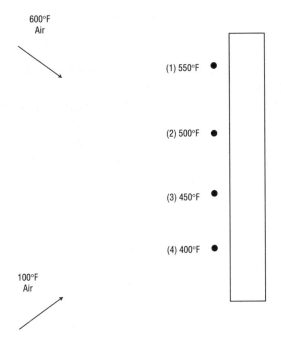

Figure 15. The environment temperatures (1–4) may be considered to be a function of the source (600°F) and the sink (100°F) temperatures.

For greater accuracy, F_i may be allowed to vary from one condition to another (i.e., idle to max), and linearly interpolate in between.

Two approaches are available to account for the varying convection coefficients:

- h may be scaled by changes in flow and density.
- The parameters on which h is based (typically flow, pressure, and temperature) are scaled, and the appropriate correlation is evaluated at each point in the mission.

Evaluating Results

While FEA allows the analyst to calculate temperatures for complex geometries, the resulting output may be difficult to interpret and check for errors. Some points to keep in mind are:

- Heat always flows perpendicular to the isotherms on a temperature plot. Figure 16 shows temperatures of a metal rod partially submerged in 200°F water. The rest of the rod is exposed to 70°F air. Heat is flowing upward through the rod. If heat were flowing from side to side, the isotherms would be vertical.
- Channels often show errors in a finite element model more clearly than the component temperatures. Temperatures within the component are evened out by conduction and are therefore more difficult to detect.
- Temperatures should be viewed as a function of source and sink temperatures ($F_i = [T_i - T_{sink}]/[T_{source} - T_{sink}]$). Figure 17 shows a plot of these values for the problem in Figure 18. These values should always be between 0 and 1. If different conditions are analyzed (i.e., max and idle), F_i should generally not vary greatly from one condition to the other. If it does, the analyst should examine why, and make sure there is no error in the

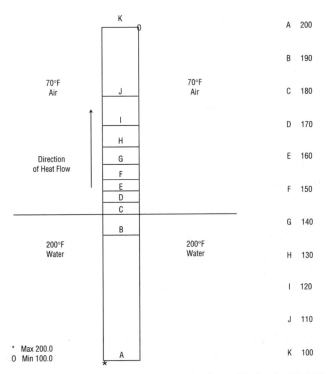

Figure 16. Finite element model of a cylinder in 200°F water and 70°F air. Isotherms are perpendicular to the direction of heat flow.

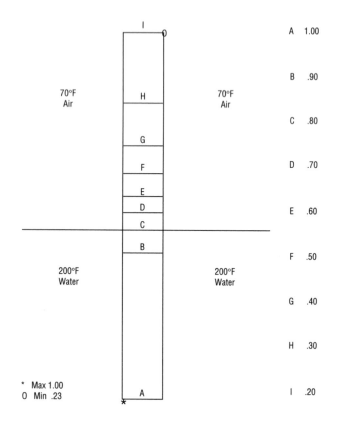

Figure 17. Component temperatures should always be between the sink and source temperatures.

model. When investigating these differences, the analyst should keep two points in mind:

1. Radiation effects increase dramatically as temperature increases.
2. As radiation and convection effects decrease, conduction becomes more significant, which tends to even out component temperatures.

• For transients, it is recommended that selected component and channel temperatures be plotted against time. The analyst should examine the response rates. Those regions with high surface area-to-volume ratios and high convection coefficients should respond quickly.

• To check a model for good connections between components, apply different temperatures to two ends of the model. Verify that the temperatures on both sides of the boundaries are reasonable. Figures 18a and 18b show two cases in which 1,000 degrees has been applied on the left, and 100 degrees on the right. Figure 18a shows a flange where contact has been modeled along the mating surfaces, and there is little discontinuity in the isotherms across the boundary. Figure 18b shows the same model where contact has been modeled along only the top section of the mating surfaces. Note that the temperatures at the lower mating surfaces differ by over 100 degrees.

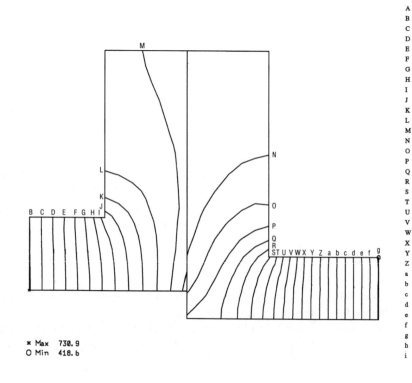

Figure 18a. 1,000°F temperatures were applied to the left flange and 100°F to the right flange. Shown here is good mating of the two flanges with little temperature difference across the boundary.

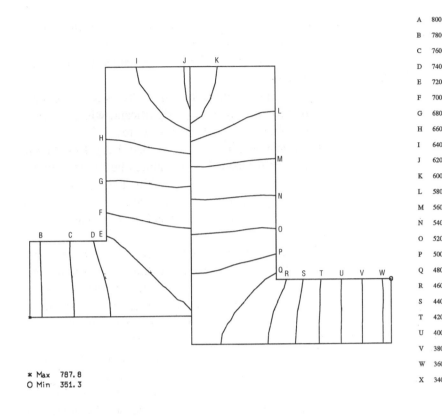

A	800
B	780
C	760
D	740
E	720
F	700
G	680
H	660
I	640
J	620
K	600
L	580
M	560
N	540
O	520
P	500
Q	480
R	460
S	440
T	420
U	400
V	380
W	360
X	340

✳ Max 787.8
O Min 351.3

Figure 18b. 1,000°F temperatures were applied to the left flange and 100°F to the right flange. Shown here is poor mating with a large temperature difference across the

HEAT EXCHANGER CLASSIFICATION

Types of Heat Exchangers

Heat transfer equipment can be specified by either service or type of construction. Only principle types are briefly described here. Table 6 lists major types of heat exchangers.

The most well-known design is the *shell-and-tube heat exchanger*. It has the advantages of being inexpensive and easy to clean and available in many sizes, and it can be designed for moderate to high pressure without excessive cost. Figure 19 illustrates its design features, which include a bundle of parallel tubes enclosed in a cylindrical casing called a shell.

The basic types of shell-and-tube exchangers are the fixed-tube sheet unit and the partially restrained tube sheet. In the former, both tube sheets are fastened to the shell. In this type of construction, differential expansion of the shell and tubes due to different operating metal temperatures or different materials of construction may require the use of an expansion joint or a packed joint. The second type has only one restrained tube sheet located at the channel end. Differential expansion problems are avoided by using a freely riding floating tube sheet or U-tubes at the other end. Also, the tube bundle of this type is removable for maintenance and mechanical cleaning on the shell side.

Shell-and-tube exchangers are generally designed and fabricated to the standards of the Tubular Exchanger Manufacturers Association (TEMA) [1]. The TEMA standards list three mechanical standards classes of exchanger construction: R, C, and B.

There are large numbers of applications that do not require this type of construction. These are characterized by low fouling and low corrosivity tendencies. Such units are considered low-maintenance items.

Services falling in this category are water-to-water exchangers, air coolers, and similar nonhydrocarbon appli-

Table 6
Summary of Types of Heat Exchangers

Type	Major Characteristics	Application
Shell and tube	Bundle of tubes encased in a cylindrical shell	Always the first type of exchanger to consider
Air cooled heat exchangers	Rectangular tube bundles mounted on frame, with air used as the cooling medium	Economic where cost of cooling water is high
Double pipe	Pipe within a pipe; inner pipe may be finned or plain	For small units
Extended surface	Externally finned tube	Services where the outside tube resistance is appreciably greater than the inside resistance. Also used in debottlenecking existing units
Brazed plate fin	Series of plates separated by corrugated fins	Cryogenic services: all fluids must be clean
Spiral wound	Spirally wound tube coils within a shell	Cryogenic services: fluids must be clean
Scraped surface	Pipe within a pipe, with rotating blades scraping the inside wall of the inner pipe	Crystallization cooling applications
Bayonet tube	Tube element consists of an outer and inner tube	Useful for high temperature difference between shell and tube fluids
Falling film coolers	Vertical units using a thin film of water in tubes	Special cooling applications
Worm coolers	Pipe coils submerged in a box of water	Emergency cooling
Barometric condenser	Direct contact of water and vapor	Where mutual solubilities of water and process fluid permit
Cascade coolers	Cooling water flows over series of tubes	Special cooling applications for very corrosive process fluids
Impervious graphite	Constructed of graphite for corrosion protection	Used in very highly corrosive heat exchange services

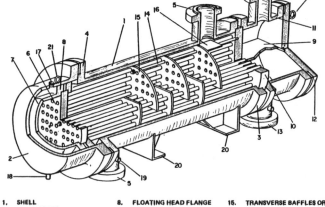

1. SHELL
2. SHELL COVER
3. SHELL CHANNEL
4. SHELL COVER END FLANGE
5. SHELL NOZZLE
6. FLOATING TUBESHEET
7. FLOATING HEAD

8. FLOATING HEAD FLANGE
9. CHANNEL PARTITION
10. STATIONARY TUBESHEET
11. CHANNEL
12. CHANNEL COVER
13. CHANNEL NOZZLE
14. TIE RODS AND SPACERS

15. TRANSVERSE BAFFLES OR SUPPORT PLATES
16. IMPINGEMENT BAFFLE
17. VENT CONNECTION
18. DRAIN CONNECTION
19. TEST CONNECTION
20. SUPPORT SADDLES
21. LIFTING RING

Figure 19. Design features of shell-and-tube exchangers [3].

cations, as well as some light-duty hydrocarbon services such as light ends exchangers, offsite lube oil heaters, and some tank suction heaters. For such services, Class C construction is usually considered. Although units fabricated to either Class R or Class C standards comply with all the requirements of the pertinent codes (ASME or other national codes), Class C units are designed for maximum economy and may result in a cost saving over Class R.

Air-cooled heat exchangers are another major type composed of one or more fans and one or more heat transfer bundles mounted on a frame [2]. Bundles normally consist of finned tubes. The hot fluid passes through the tubes, which are cooled by air supplied by the fan. The choice of air cool-

ers or condensers over conventional shell-and-tube equipment depends on economics.

Air-cooled heat exchangers should be considered for use in locations requiring cooling towers, where expansion of once-through cooling water systems would be required, or where the nature of cooling causes frequent fouling problems. They are frequently used to remove high-level heat, with water cooling used for final "trim" cooling.

These designs require relatively large plot areas. They are frequently mounted over pipe racks and process equipment such as drums and exchangers, and it is therefore important to check the heat losses from surrounding equipment to evaluate whether there is an effect on the air inlet temperature.

Double-pipe exchangers are another class that consists of one or more pipes or tubes inside a pipe shell. These exchangers almost always consist of two straight lengths connected at one end to form a U or "hair-pin." Although some double-pipe sections have bare tubes, the majority have longitudinal fins on the outside of the inner tube. These units are readily dismantled for cleaning by removing a cover at the return bend, disassembling both front end closures, and withdrawing the heat transfer element out the rear.

This design provides countercurrent or true concurrent flow, which may be of particular advantage when very close temperature approaches or very long temperature ranges are needed. They are well suited for high-pressure applications, because of their relatively small diameters. Designs incorporate small flanges and thin wall sections, which are advantageous over conventional shell-and-tube equipment. Double-pipe sections have been designed for up to 16,500 kPa gauge on the shell side and up to 103,400 kPa gauge on the tube side. Metal-to-metal ground joints, ring joints, or confined O-rings are used in the front end closures at lower pressures. Commercially available single tube double-pipe sections range from 50-mm through 100-mm nominal pipe size shells, with inner tubes varying from 20-mm to 65-mm pipe size.

Designs having multiple tube elements contain up to 64 tubes within the outer pipe shell. The inner tubes, which may be either bare or finned, are available with outside diameters of 15.875 mm to 22.225 mm. Normally only bare tubes are used in sections containing more than 19 tubes. Nominal shell sizes vary from 100 mm to 400 mm pipe.

Extended surface exchangers are composed of tubes with either longitudinal or transverse helical fins. An extended surface is best employed when the heat transfer properties of one fluid result in a high resistance to heat flow and those of the other fluid have a low resistance. The fluid with the high resistance to heat flow contacts the fin surface.

Spiral tube heat exchangers consist of a group of concentric spirally wound coils, which are connected to tube sheets. Designs include countercurrent flow, elimination of differential expansion problems, compactness, and provision for more than two fluids exchanging heat. These units are generally employed in cryogenic applications.

Scraped-surface exchangers consist of a rotating element with a spring-loaded scraper to wipe the heat transfer surface. They are generally used in plants where the process fluid crystallizes or in units where the fluid is extremely fouling or highly viscous.

These units are of double-pipe construction. The inner pipe houses the scrapers and is available in 150-, 200-, and 300-mm nominal pipe sizes. The exterior pipe forms an annular passage for the coolant or refrigerant and is sized as required. Up to ten 300 mm sections or twelve of the smaller individual horizontal sections, connected in series or series/parallel and stacked in two vertical banks on a suitable structure, is the most common arrangement. Such an arrangement is called a "stand."

A *bayonet-type exchanger* consists of an outer and inner tube. The inner tube serves to supply the fluid to the annulus between the outer and inner tubes, with the heat transfer occurring through the outer tube only. Frequently, the outer tube is an expensive alloy material and the inner tube is carbon steel. These designs are useful when there is an extremely high temperature difference between shell side and tube side fluids, because all parts subject to differential expansion are free to move independently of each other. They are used for change-of-phase service where two-phase flow against gravity is undesirable. These units are sometimes installed in process vessels for heating and cooling purposes. Costs per unit area for these units are relatively high.

Worm coolers consist of pipe coils submerged in a box filled with water. Although worm coolers are simple in construction, they are costly on a unit area basis. Thus they are restricted to special applications, such as a case where emergency cooling is required and there is but one water-supply source. The box contains enough water to cool liquid pump-out in the event of a unit upset and cooling water failure.

A *direct contact condenser* is a small contacting tower through which water and vapor pass together. The vapor is condensed by direct contact heat exchange with water droplets. A special type of direct contact condenser is a barometric condenser that operates under a vacuum. These units should be used only where coolant and process fluid mutual solubilities are such that no water pollution or product contamination problems are created. Evaluation of process fluid loss in the coolant is an important consideration.

A *cascade cooler* is composed of a series of tubes mounted horizontally, one above the other. Cooling water from a distributing trough drips over each tube and into a drain. Generally, the hot fluid flows countercurrent to the water. Cascade coolers are employed only where the process fluid is highly corrosive, such as in sulfuric acid cooling.

Impervious graphite heat exchangers are used only in highly corrosive heat exchange service. Typical applications are in isobutylene extraction and in dimer and acid concentration plants. The principal construction types are cubic graphite, block type, and shell-and-tube graphite exchangers. Cubic graphite exchangers consist of a center cubic block of impervious graphite that is cross drilled to provide passages for the process and service fluids. Headers are bolted to the sides of the cube to provide for fluid distribution. Also, the cubes can be interconnected to obtain additional surface area. Block-type graphite exchangers consist of an impervious graphite block enclosed in a cylindrical shell. The process fluid (tube side) flows through axial passages in the block, and the service fluid (shell side) flows through cross passages in the block. Shell-and-tube-type graphite exchangers are like other shell-and-tube exchangers except that the tubes, tube sheets, and heads are constructed of impervious graphite.

Sources

1. *Standards of Tubular Exchanger Manufacturer's Association,* 7th Ed., TEMA, Tarrytown, NY, 1988.

2. API Standard 661, "Air-Cooled Heat Exchangers for General Refinery Services."
3. Cheremisinoff, N. P., *Heat Transfer Pocket Handbook.* Houston: Gulf Publishing Co., 1984.

Shell-and-Tube Exchangers

This section provides general information on shell-and-tube heat exchanger layout and flow arrangements. Design details are concerned with several issues—principal ones being the number of required shells, the type and length of tubes, the arrangement of heads, and the tube bundle arrangement.

The total number of shells necessary is largely determined by how far the outlet temperature of the hot fluid is cooled below the outlet temperature of the other fluid (known as the "extent of the temperature cross"). The "cross" determines the value of F_n, the temperature correction factor; this factor must always be equal to or greater than 0.800. (The value of F_n drops slowly between 1.00 and 0.800, but then quickly approaches zero. A value of F_n less than 0.800 cannot be predicted accurately from the usual information used in process designs.) Increasing the number of shells permits increasing the extent of the cross and/or the value of F_n.

The total number of shells also depends on the total surface area since the size of the individual exchanger is usually limited because of handling considerations.

Exchanger tubes are commonly available with either smooth or finned outside surfaces. Selection of the type of surface is based on applicability, availability, and cost.

The conventional shell-and-tube exchanger tubing is the smooth surface type that is readily available in any material used in exchanger manufacture and in a wide range of wall thicknesses. With low-fin tubes, the fins increase the outside area to approximately 2½ times that of a smooth tube.

Tube length is affected by availability and economics. Tube lengths up to 7.3 m are readily obtainable. Longer tubes (up to 12.2 m for carbon steel and 21.3 m for copper alloys) are available in the United States.

The cost of exchanger surface depends upon the tube length, in that the longer the tube, the smaller the bundle diameter for the same area. The savings result from a decrease in the cost of shell flanges with only a nominal increase in the cost of the longer shell. In the practical range of tube lengths, there is no cost penalty for the longer tubes since length extras are added for steel only over 7.3 m and for copper alloys over 9.1 m.

A disadvantage of longer tubes in units (e.g., condensers) located in a structure is the increased cost of the longer platforms and additional structure required. Longer tube bundles also require greater tube pulling area, thereby possibly increasing the plot area requirements.

Exchanger tubing is supplied on the basis of a nominal diameter and either a minimum or average wall thickness. For exchanger tubing, the nominal tube diameter is the outside tube diameter. The inside diameter varies with the nominal tube wall thickness and wall thickness tolerance. Minimum wall tubing has only a plus tolerance on the wall thickness, resulting in the nominal wall thickness being the minimum thickness. Since average wall tubing has a plus-or-minus tolerance, the actual wall thickness can be greater or less than the nominal thickness. The allowable tolerances vary with the tube material, diameter, and fabrication method.

Tube inserts are short sleeves inserted into the inlet end of a tube. They are used to prevent erosion of the tube itself due to the inlet turbulence when erosive fluids are handled, such as streams containing solids. When it is suspected that the tubes will be subject to erosion by solids in the tube side fluid, tube inserts should be specified. Insert material, length, and wall thickness should be given. Also, inserts are occasionally used in cooling-water service to prevent oxygen attack at the tube ends. Inserts should be cemented in place.

The recommended TEMA head types are shown in Figure 20. The *stationary front head* of shell-and-tube exchangers is commonly referred to as the channel. Some common TEMA stationary head types and their applications are as follows:

Type A—Features a removable channel with a removable cover plate. It is used with fixed-tube sheet, U-tube, and removable-bundle exchanger designs. This is the most common stationary head type.

Type B—Features a removable channel with an integral cover. It is used with fixed-tube sheet, U-tube, and removable-bundle exchanger design.

Types C and N—The channel with a removable cover is integral with the tube sheet. Type C is attached to the shell by a flanged joint and is used for U-tube and re-

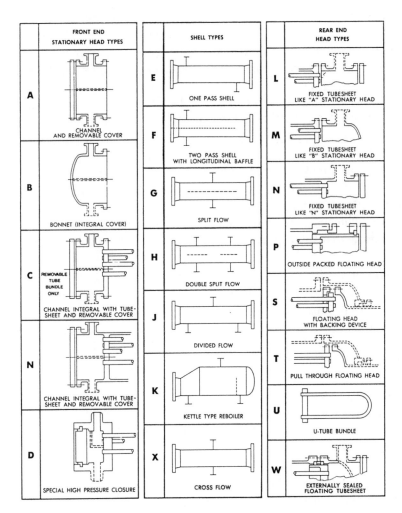

Figure 20. TEMA heat exchanger head types. (*Copyright © 1988 by Tubular Exchanger Manufacturers Association.*)

movable bundles. Type N is integral with the shell and is used with fixed-tube sheet designs. The use of Type N heads with U-tube and removable bundles is not recommended since the channel is integral with the tube bundle, which complicates bundle maintenance.

Type D—This is a special high pressure head used when the tube-side design pressure exceeds approximately 6,900 kPa gauge. The channel and tube sheet are integral forged construction. The channel cover is attached by special high pressure bolting.

The TEMA rear head nomenclature defines the exchanger tube bundle type and common arrangements as follows:

Type L—Similar in construction to the Type A stationary head. It is used with fixed-tube sheet exchangers when mechanical cleaning of the tubes is required.

Type M—Similar in construction to the Type B stationary head. It is used with fixed-tube sheet exchangers.

Type N—Similar in construction to the Type N stationary head. It is used with fixed-tube sheet exchangers.

Type P—Called an outside packed floating head. The design features an integral rear channel and tube sheet with a packed joint seal (stuffing box) against the shell. It is not normally used due to the tendency of packed joints to leak. It should not be used with hydrocarbons or toxic fluids on the shell side.

Type S—Constructed with a floating tube sheet contained between a split-ring and a tube-sheet cover. The tube sheet assembly is free to move within the shell cover. (The shell cover must be a removable design to allow access to the floating head assembly.)

Type T—Constructed with a floating tube sheet bolted directly to the tube sheet cover. It can be used with either an integral or removable (common) shell cover.

Type U—This head type designates that the tube bundle is constructed of U-tubes.

Type W—A floating head design that utilizes a packed joint to separate the tube-side and shell-side fluids. The packing is compressed against the tube sheet by the shell/rear cover bolted joint. It should *never* be used with hydrocarbons or toxic fluids on either side.

Tube bundles are designated by TEMA rear head nomenclature (see Figure 20). Principal types are briefly described below.

Fixed-tube sheet exchangers have both tube sheets attached directly to the shell and are the most economical exchangers for low design pressures. This type of construction should be considered when no shell-side cleaning or inspection is required, or when in-place shell-side chemical cleaning is available or applicable. Differential thermal expansion between tubes and shell limits applicability to moderate temperature differences.

Welded fixed-tube sheet construction cannot be used in some cases because of problems in welding the tube sheets to the shells. Some material combinations that rule out fixed-tube sheets for this reason are carbon steel with aluminum or any of the high copper alloys (TEMA—Rear Head Types L, M, or N).

U-tube exchangers represent the greatest simplicity of design, requiring only one tube sheet and no expansion joint or seals while permitting individual tube differential thermal expansion. U-tube exchangers are the least expensive units for high tube-side design pressures. The tube bundle can be removed from the shell, but replacement of individual tubes (except for ones on the outside of the bundle) is impossible.

Although the U-bend portion of the tube bundle provides heat transfer surface, it is ineffective compared to the straight tube length surface area. Therefore, when the effective surface area for U-tube bundles is calculated, only the surface area of the straight portions of the tubes is included (TEMA—Rear Head Type U).

A *pull-through floating head exchanger* has a fixed tube sheet at the channel end and a floating tube sheet with a separate cover at the rear end. The bundle can be easily removed from the shell by disassembling only the front cover. The floating head flange and bolt design require a relatively large clearance between the bundle and shell, particularly as the design pressures increase. Because of this clearance, the pull-through bundle has fewer tubes per given shell size than other types of construction do. The bundle-to-shell clearance, which decreases shell-side heat transfer capability, should be blocked by sealing strips or dummy tubes to reduce shell-size fluid bypassing. Mechanical cleaning of both the shell and tube sides is possible (TEMA—Rear Head Type T).

A *split-ring floating head exchanger* has a fixed-tube sheet at the channel end and a floating tube sheet that is sandwiched between a split-ring and a separate cover. The floating head assembly moves inside a shell cover of a larger diameter than that of the shell. Mechanical cleaning of both the shell and tube is possible (TEMA—Rear Head Type S).

There are two variations of *outside packed floating head designs:* the lantern ring type and the stuffing box type. In the lantern ring design, the floating head slides against a lantern ring packing, which is compressed between the shell flange and the shell cover. The stuffing box design is similar to the lantern ring type, except that the seal is against an extension of the floating tube sheet and the tube sheet cover is attached to the tube sheet extension by means of a split-ring. (TEMA—Rear Head Types P or W).

Sources

1. *Standards of Tubular Exchanger Manufacturer's Association,* 7th Ed., TEMA, Tarrytown, NY, 1988.
2. Cheremisinoff, N. P., *Heat Transfer Pocket Handbook.* Houston: Gulf Publishing Co., 1984.

Tube Arrangements and Baffles

The following are some general notes on tube layout and baffle arrangements for shell-and-tube exchangers. There are four types of tube layouts with respect to the shell-side crossflow direction between baffle tips: square (90°), rotated square (45°), triangular (30°), and rotated triangular (60°). The four types are shown in Figure 21.

Use of triangular layout (30°) is preferred (except in some reboilers). An exchanger with triangular layout costs less per square meter and transfers more heat per square meter than one with a square or rotated square layout. For this reason, triangular layout is preferred where applicable.

Rotated square layouts are preferable for laminar flow, because of a higher heat transfer coefficient caused by induced turbulence. In turbulent flow, especially for pressure-drop limited cases, square layout is preferred since the heat transfer coefficient is equivalent to that of rotated square layout while the pressure drop is somewhat less.

Tube layout for removable bundles may be either square (90°), rotated square (45°), or triangular (30°). Nonremovable bundles (fixed-tube sheet exchangers) are *always* triangular (30°) layout.

The *tube pitch* (PT) is defined as the center-to-center distance between adjacent tubes (see Figure 21). Common pitches used are given in Table 7.

The column "Heaviest Recommended Wall" is based on the maximum allowable tube sheet distortion resulting from rolling the indicated tube into a tube sheet having the minimum permissible ligament width at the listed pitch. The ligament is that portion of the tube sheet between two adjacent tube holes.

Tubes are supported by baffles that restrain tube vibration from fluid impingement and channel fluid flow on the shell side. Two types of baffles are generally used: segmental and double segmental. Types are illustrated in Figure 22.

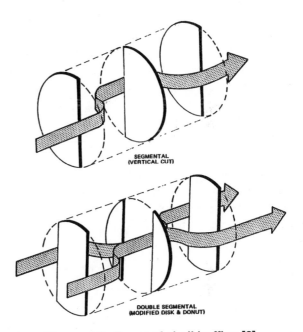

Figure 22. Types of shell baffles [2].

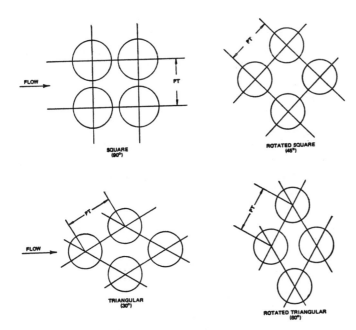

Figure 21. Tube layouts [2].

Table 7
Common Tube Pitch Values

Tube Size	Triangular (mm)	Square (mm)	Heaviest Recommended Wall (mm)
19.05 mm O.D.	23.81	—	2.41
19.05 mm O.D.	25.40	25.40	2.77
25.4 mm O.D.	31.75	31.75	3.40
38.1 mm O.D.	47.63	47.63	4.19
> 38.1 mm	Use 1.25 times the outside diameter		

The *baffle cut* is the portion of the baffle "cut" away to provide for fluid flow past the chord of the baffle. For segmental baffles, this is the ratio of the chord height to shell diameter in percent. Segmental baffle cuts are usually about 25%, although the maximum practical cut for tube support is approximately 48%.

Double segmental baffle cut is expressed as the ratio of window area to exchanger cross sectional area in percent. Normally the window areas for the single central baffle and the area of the central hole in the double baffle are equal and are 40% of the exchanger cross-sectional area. This allows a baffle overlap of approximately 10% of the exchanger cross-sectional area on each side of the exchanger. However, there must be enough overlap so that at least one row of tubes is supported by adjacent segments.

Baffle pitch is defined as the longitudinal spacing between baffles. The maximum baffle pitch is a function of tube size and, for no change of phase flow, of shell diameter. If there is no change of phase in the shell-side fluid, the baffle pitch should not exceed the shell inside diameter. Otherwise, the fluid would tend to flow parallel with the tubes, rather than perpendicular to them, resulting in a poorer heat transfer coefficient.

Impingement baffles are required on shell-side inlet nozzles to protect the bundle against impingement by the incoming fluid when the fluid:

1. is condensing
2. is a liquid vapor mixture
3. contains abrasive material
4. is entering at high velocity

In addition, TEMA requires bundle impingement protection when nozzle values of ρu^2 (fluid density, kg/m^3, times velocity squared m^2/s^2) exceed:

1. 2,230 kg/m-s^2 for noncorrosive, nonabrasive, single-phase fluids
2. 744 kg/m-s^2 for all other liquids

Also, the minimum bundle entrance area should equal or exceed the inlet nozzle area and should not produce a value of ρu^2 greater than 5,950 kg/m-s^2, per TEMA. Impingement baffles can be either flat or curved. In order to maintain a maximum tube count, the impingement plate is sometimes located in a conical nozzle opening or in a dome cap above the shell. The impingement plate material should be at least as good as that of the tubes.

Sources

1. *Standards of Tubular Manufacturers Association,* 7th Ed., TEMA, Tarrytown, NY, 1988.
2. Cheremisinoff, N. P., *Heat Transfer Pocket Handbook.* Houston: Gulf Publishing Co., 1984.

Shell Configurations

The following notes summarize design features of shells for shell-and-tube heat exchangers.

The *single-pass shell* is the most common shell construction used. The shell-side inlet and outlet nozzles are located at opposite ends of the shell. The nozzles can be placed on opposite or adjacent sides of the shell, depending on the number and type of baffles used. A typical one-shell pass exchanger with horizontal segmental baffles is illustrated in Figure 23 [A] (TEMA E).

A *two-pass shell* uses a longitudinal baffle to direct the shell-side flow. An exchanger with two shell passes is shown in Figure 23 [B]. Note that both the shell inlet and outlet nozzles are adjacent to the stationary tube sheets. A shell-side temperature range exceeding 195°C should be avoided, since greater temperature ranges result in excessive heat leakage through the baffle, as well as thermal stresses in the baffle, shell, and tube sheet.

The longitudinal baffle can be either welded or removable. Since there are severe design and cost penalties associated with the use of welded baffles in floating head exchangers, this type of design should be used only with fixed-tube sheet units that do not require expansion joints. Removable longitudinal baffles require the use of flexible light gauge sealing strips or a packing device between the baffle and the shell, to reduce fluid leakage from one side to the other (TEMA F).

A *divided flow shell* has a central inlet nozzle and two outlet nozzles, or vice-versa. A divided flow exchanger is illustrated in Figure 23 [C]. This type is generally used to reduce pressure drop in a condensing service. In minimizing pressure drop the shell fits in as follows:

• E shell with segmental baffles
• E shell with double segmental baffles

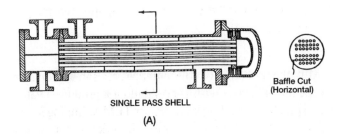

SINGLE PASS SHELL

(A)

Baffle Cut
(Horizontal)

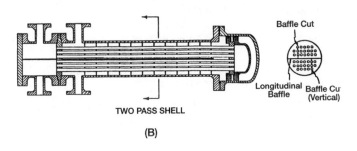

Baffle Cut

Longitudinal Baffle   Baffle Cut (Vertical)

TWO PASS SHELL

(B)

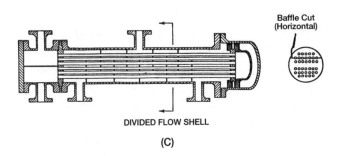

Baffle Cut
(Horizontal)

DIVIDED FLOW SHELL

(C)

Figure 23. (A) Single-pass shell; (B) two-pass shell; (C) divided flow shell [2].

- J shell with segmental baffles
- J shell with double segmental baffles
- E shells in parallel with segmental baffles
- E shells in parallel with double segmental baffles

- J shells in parallel with segmental baffles
- J shells in parallel with double segmental baffles

Generally, for most designs, double segmental baffles are used with J shells.

Double segmental baffles in a divided-flow exchanger normally have a vertical cut. This baffle arrangement also requires that there be an odd number of total baffles, but there must also be an odd number of baffles in each end of the shell. The center baffle for this arrangement is normally similar to the center baffle used with segmental cut. The baffles on each side of the central baffle and the last baffle toward the ends of the shell have solid centers with cutaway edges.

The choice of whether to stack shells depends on maintenance considerations, as well as on the amount of plot area available. Stacking shells requires less area and frequently less piping. Normally, shells are not stacked more than two high. However, stacked heat exchangers are more costly to maintain, because of accessibility.

If sufficient plot area is available, the following guidelines apply:

1. If the fluids are known to be clean and noncorrosive, the shells should usually be stacked.
2. If the fluids are moderately clean or slightly corrosive, the shells may be stacked.
3. If the fluids are very dirty or corrosive, the shells should not be stacked, to allow for ease of maintenance.

Sources

1. *Standards of Tubular Exchanger Manufacturer's Association,* 7th Ed., TEMA, Tarrytown, NY, 1988.
2. Cheremisinoff, N. P., *Heat Transfer Pocket Handbook.* Houston: Gulf Publishing Co., 1984.

Miscellaneous Data

Design-related data are given in Tables 8 through 10. Table 8 provides typical tube dimensions and tube surface areas per unit length. Table 9 gives thermal conductivities of materials commonly used for exchanger construction. Table 10 gives recommended maximum number of tube passes as a function of tube size.

Source

Cheremisinoff, N. P., *Heat Transfer Pocket Handbook.* Houston: Gulf Publishing Co., 1984.

Table 8
Tube Dimensions and Surface Areas Per Unit Length

$d_o =$ O.D. of Tubing (mm)	$\delta =$ Wall Thickness (mm)	$d_i =$ I.D. of Tubing (mm)	Internal Area (mm^2)	External Surface Per Ft Length (sq ft/ft)
19.05	2.77	13.51	143.8	0.0598
19.05	2.11	14.83	172.9	0.0598
19.05	1.65	15.75	194.8	0.0598
19.05	1.24	16.56	215.5	0.0598
25.40	3.40	18.59	271.6	0.0798
25.40	2.77	19.86	309.0	0.0798
25.40	2.11	21.18	352.3	0.0798
25.40	1.65	22.10	383.2	0.0798
38.10	3.40	31.29	769.0	0.1197
38.10	2.77	32.56	832.9	0.1197
38.10	2.11	33.88	901.3	0.1197

Table 9
Thermal Conductivities of Materials of Construction

Material	Composition	Thermal Conductivity, k, Wt/m-°C
Admiralty	71 Cu-28 Zn-1 Sn	111
Type 316 stainless steel	17 Cr-12 Ni-2 Mo	16
Type 304 stainless steel	18 Cr-8 Ni	16
Brass	70 Cu-30 Zn	99
Red brass	85 Cu-15 Zn	159
Aluminum brass	76 Cu-22 Zn-2 Al	100
Cupro-nickel	90 Cu-10 Ni	71
Cupro-nickel	70 Cu-30 Ni	29
Monel	67 Ni-30 Cu-1.4 Fe	26
Inconel		19
Aluminum		202
Carbon steel		45
Carbon-moly	0.5 Mo	43
Copper		386
Lead		35
Nickel		62
Titanium		19
Chrome-moly steel	1 Cr-0.5 Mo	42
	2¼ Cr-0.5 Mo	38
	5 Cr-0.5 Mo	35
	12 Cr-1 Mo	28

Table 10
Maximum Number of Tube Passes

Shell I.D. (mm)	Recommended Maximum Number of Tube Passes
<250	4
250 − <510	6
510 − <760	8
760 − <1,020	10
1,020 − <1,270	12
1,270 − <1,520	14

FLOW REGIMES AND PRESSURE DROP IN TWO-PHASE HEAT TRANSFER

Flow Regimes

Standard practice for heat exchanger analysis is to first identify the flow regimes and then employ the appropriate correlations.

Vertical Upward Concurrent Flow

Flows of this type are shown in Figure 24.

Bubbly flow. In this type, the gas or vapor phase is distributed as discrete bubbles in a continuous liquid phase. At one extreme, the bubbles may be small and spherical, and at the other extreme, the bubbles may be large with a spherical cap and a flat tail.

Slug flow. The gas or vapor bubbles are approximately the diameter of the pipe. The nose of the bubble has a charac-

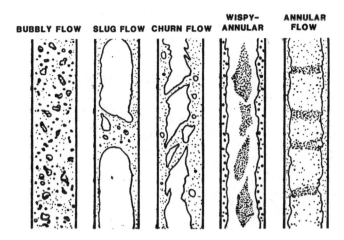

Figure 24. Flow patterns in vertical concurrent flow [1].

teristic spherical cap, and the gas in the bubble is separated from the pipe wall by a slowly descending liquid film. The liquid flow is contained in liquid slugs that separate successive gas bubbles. Slugs may or may not contain smaller entrained gas bubbles carried in the wake of the large bubble. The length of the main gas bubble varies.

Churn flow. Formed by the breakdown of the large vapor bubbles in slug flow. The gas or vapor flows chaotically through the liquid that is mainly displaced to the channel wall. The flow has a time-varying character and hence is called "churn flow." This region is also sometimes referred to as semi-annular or slug-annular flow.

Wispy-annular flow. The flow takes the form of a relatively thick liquid film on the walls of the pipe together with a considerable amount of liquid entrained in a central gas or vapor core. The liquid in the film is aerated by small gas bubbles and the entrained liquid phase appears as large droplets which have agglomerated into long irregular filaments or wisps. This generally occurs at high mass velocities.

Annular flow. A liquid film forms at the pipe wall with a continuous central gas or vapor core. Large amplitude coherent waves are usually present on the surface of the film, and the continuous break up of these waves forms a source for droplet entrainment, which occurs in varying amounts in the central gas core.

Vertical Heated Channel Upward Flow

Heat flux through the channel wall alters the flow pattern from that which would have occurred in a long unheated channel at the same local flow conditions. These changes occur due to:

1. The departure from thermodynamic equilibrium coupled with the presence of radial temperature profiles in the channel.
2. The departure from local hydrodynamic equilibrium throughout the channel.

Figure 25 shows a vertical tubular channel heated by a uniform low heat flux and fed with liquid just below the saturation temperature.

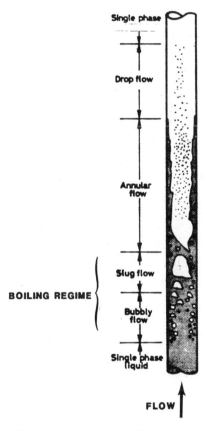

Figure 25. Flow patterns in a vertical evaporator tube [2].

In the initial single-phase region, the liquid is heated to the saturation temperature. A thermal boundary layer forms at the wall, and a radial temperature profile forms. At some distance from the inlet, the wall temperature and the conditions for the formation of vapor (nucleation) at the wall are satisfied. Vapor forms at preferred positions on the tube surface. Vapor bubbles grow from these sites finally detaching to form a bubbly flow. With the production of more vapor, the bubble population increases with length and coalescence occurs, forming slug flow, which in turn gives way to annular flow further along the channel. Close to this point the formation of vapor at sites on the wall may cease and further vapor formation will result from evaporation at the liquid-film vapor-core interface. Increasing velocities in the vapor core cause entrainment of liquid in the form of droplets. The depletion of the liquid from the film by this entrainment and by evaporation finally causes the film to dry out completely. Droplets continue to exist and are slowly evaporated until only single-phase vapor is present.

Figure 26 shows the flow patterns of liquid-vapor flow in a heated pipe as a function of wall heat flux. Liquid enters the pipe at a constant flow rate and at a temperature lower than the saturation temperature. As the heat flux increases, the vapor appears closer and closer to the pipe inlet. The local boiling length is the extent of pipe where bubbles form at the wall and condense in the liquid core where the liquid temperature is still lower than the saturation temperature. Vapor forms by:

1. Wall nucleation
2. Direct vaporization on the interfaces located in the flow itself

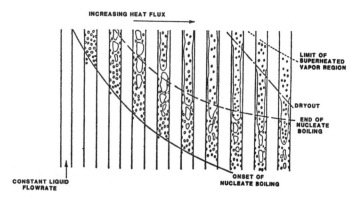

Figure 26. Convective boiling in a heated channel [3]. *(With permission of Elsevier Science Ltd.)*

There is progressively less liquid between the wall and the interfaces. Consequently, the thermal resistance decreases along with the wall temperature, resulting in an end to wall nucleation. In annular flow, the liquid film flow rate decreases through evaporation and entrainment of droplets, although some droplets are redeposited. In heat flux controlled systems, when the film is completely dried out, the wall temperature rises very quickly and can exceed the melting temperature of the wall (called dryout). Flow patterns are shown in Figure 27.

In upward bubbly flow, bubbles are spread over the entire pipe cross-section whereas in the downward flow bubbles gather near the pipe axis.

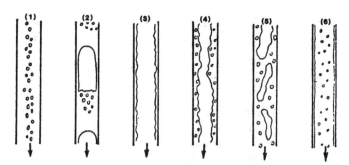

Figure 27. Air-water flow patterns in a downward concurrent flow in a vertical pipe: (1) bubbly, (2) slug, (3) falling film, (4) bubbly falling film, (5) churn, and (6) dispersed annular flow [4].

At higher gas flow rates (but a constant liquid flow rate) the bubbles agglomerate into large gas pockets. The tops of these gas plugs are dome-shaped whereas the lower extremity is flat with a bubbly zone underneath. This *slug flow* is generally more stable than in the upward case.

With annular flow, at small liquid and gas flow rates, a liquid film flows down the wall (*falling film flow*). If the liquid flow rate is higher, the bubbles are entrained within the film (*bubbly falling film*). At greater liquid and gas flow rates *churn flow* exists, which can evolve into dispersed annular flow for very high gas flow rates.

Horizontal Concurrent Flow

The flow patterns for this type of flow are shown in Figure 28.

Bubbly flow (froth flow). This resembles the case in vertical flow except that the vapor bubbles tend to travel in the upper half of the pipe. At moderate gas and liquid velocities, the entire pipe cross-section contains bubbles. At higher velocities, a flow pattern equivalent to the wispy-annular pattern exists.

Plug flow. This is similar to slug flow in the vertical direction. Again, the gas bubbles tend to travel in the upper half of the pipe

Stratified flow. This pattern only occurs at very low liquid and vapor velocities. The two phases flow separately with a relatively smooth interface.

Wavy flow. As the vapor velocity is increased, the interface becomes disturbed by waves traveling in the direction of flow.

Slug flow. At higher vapor velocities the waves at the interface break up to form a frothy slug which is propagat-

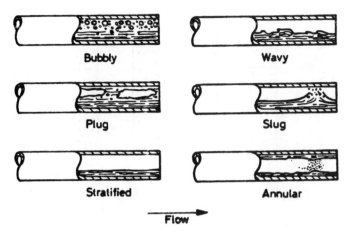

Figure 28. Flow patterns in horizontal flow [1].

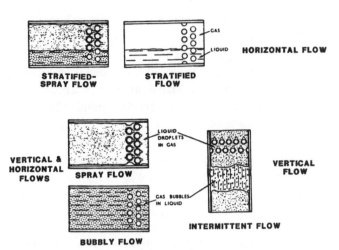

Figure 30. Shell-side two-phase flow patterns [5]. (*With permission of ASME.*)

ed along the channel at a high velocity. The upper surface of the tube behind the wave is wetted by a residual film, which drains into the bulk of the liquid.

Annular flow. At higher vapor velocities a gas core forms with a liquid film around the periphery of the pipe. The film may or may not be continuous around the entire circumference but it will be thicker at the base of the pipe.

Flow patterns formed during the generation of vapor in horizontal tubular channels are influenced by departures from thermodynamic and hydrodynamic equilibrium. Figure 29 shows a horizontal tubular channel heated by a uniform low heat flux and fed with liquid just below the saturation temperature. The sequence of flow patterns corresponds to a relatively low inlet velocity (<1 m/s). Note the intermittent drying and rewetting of the upper surfaces of the tube in wavy flow and progressive drying out over long tube lengths of the upper circumference of the tube wall in annular flow. At higher inlet liquid velocities, the influence of gravity is less obvious, the phase distribution becomes more symmetrical, and the flow patterns become closer to those in vertical flow.

Flow Normal to Tube Banks

The flow patterns in the crossflow zones are shown in Figure 30.

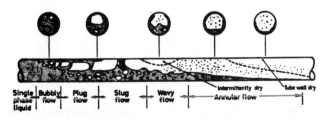

Figure 29. Flow patterns in a horizontal tube evaporator [2].

Spray flow. This occurs at high mass flow qualities with liquid carried along by the gas as a spray.

Bubbly flow. This occurs at low mass flow qualities with the gas distributed as discrete bubbles in the liquid.

Intermittent flow. Intermittent slugs of liquid are propelled cyclically by the gas.

Stratified-spray flow. The liquid and gas tend to separate with liquid flowing along the bottom. The gas-phase is entrained as bubbles in the liquid layer and liquid droplets are carried along by the gas as a spray.

Stratified flow. The liquid and gas are completely separated.

Spray and bubbly flows occur for either vertical up-and-down flow or horizontal side-to-side flow. Intermittent flow only occurs with vertical up-and-down flow and stratified-spray and stratified flow with horizontal side-to-side flow.

Sources

1. Cheremisinoff, N. P., *Heat Transfer Pocket Handbook,* Houston: Gulf Publishing Co., 1984.
2. Collier, J. G., *Convective Boiling and Condensation.* New York: McGraw-Hill, 1972.
3. Hewitt, G. F. and Hall-Taylor, N. S., *Annular Two-Phase Flow.* London: Pergamon Press, 1970.
4. Oshinowo, T. and Charles, M. E., in *Can. Journ. of Chem. Engrg., 52: 25–35, 1974.*
5. Grant, I. D. R. and Chisholm, D. in *Trans ASME, Journal of Heat Transfer,* 101 (Series C): 38–42, 1979.

Flow Maps

A flow pattern map is a two-dimensional representation of the flow pattern existence domains. The respective patterns may be represented as areas on a graph, the coordinates of which are the actual superficial-phase velocities (j_l or j_g). The coordinate systems are different according to various authors, and so far there is no agreement on the best coordinate system.

Vertical Upward Flow

Figure 31 shows a flow pattern map based on observations on low-pressure air-water and high-pressure steam-water flow in small diameter (1–3 cm) vertical tubes [4]. The axes are the superficial momentum fluxes of the liquid ($\rho_l j_l$) and vapor ($\rho_g j_g^2$) phases, respectively. These superficial momentum fluxes can also be expressed in terms of mass velocity G and the vapor quality x:

$$\rho_l j_l^2 = \frac{[G(1-x)]^2}{\rho_l} ; \rho_g j_g^2 = \frac{(Gx)^2}{\rho_g} \qquad (1)$$

Figure 31 should be considered as a rough guide only.

Vertical Downward Flow

Figure 32 shows one investigator's chart [2]. Data are based on two-component mixtures of air and different liquids flowing in a pipe 25.4 mm in diameter at a pressure of around 1.7 bar. The abscissa and ordinate are the quantities $Fr/\sqrt{\Lambda}$ and $\sqrt{\beta(1-\beta)}$ (where β is the liquid holdup fraction) which are calculated at the test section pressure and temperature. The Froude number, Fr, is defined by:

$$Fr = (j_g + j_l)^2 / g d_i \qquad (2)$$

where g = acceleration due to the gravity
d_i = pipe diameter
Λ = a coefficient that accounts for the liquid physical properties

$$\Lambda = (\mu/\mu_w) [(\rho_l/\rho_w)(\sigma/\sigma_w)^3]^{-1/4} \qquad (3)$$

where μ = liquid viscosity
ρ = liquid density
σ = liquid surface tension

Subscript w refers to water at 20°C and 1 bar.

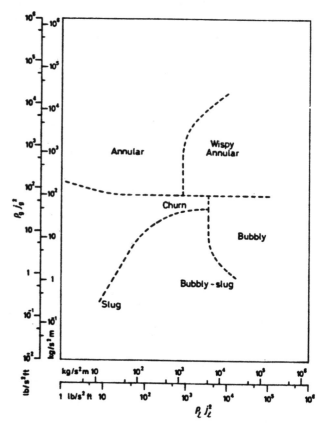

Figure 31. Flow pattern map for vertical upward flow [4]. (*With permission of AEA Technology plc.*)

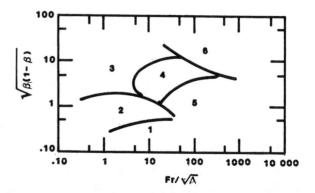

Figure 32. Flow pattern map for vertical downward flow: (1) bubbly, (2) slug falling film, (3) falling film, (4) bubbly falling film, (5) churn, and (6) dispersed annular flow [2].

Horizontal Flow

The well-known Baker plot consists of a plot of G_g/λ and $G_1\lambda\psi/G_g$, where G_g and G_1 are the superficial mass velocities of the vapor and liquid phases, respectively [5]. The factors λ and ψ are:

$$\lambda = \left[\left(\frac{\rho_g}{\rho_A}\right)\left(\frac{\rho_1}{\rho_w}\right)\right]^{1/2} \qquad (4)$$

and

$$\psi = \left(\frac{\sigma_w}{\sigma}\right)\left[\left(\frac{\mu_1}{\mu_w}\right)\left(\frac{\rho_w}{\rho_1}\right)^2\right]^{1/3} \qquad (5)$$

Baker's map has been modified by many investigators. Mandhane et al. [6] based a map upon 5,935 data points, 1,178 of which concern air-water flows. Its coordinates are the superficial velocities j_1 and j_g calculated at the test section pressure and temperature. The map is shown in Figure 33 and is valid for the parameter ranges given in Table 11.

Flow Normal to Tube Banks

Flow pattern maps for both vertical and horizontal flow normal to the tube banks are given in Figure 34. The parameters of these maps are those of Baker [5], modified according to Bell, et al. [7]. It is a plot of superficial gas ve-

Table 11
Parameter Ranges for the Flow Map
Proposed by Mandhane et al.

Conditions	Range of Values	
Pipe inner diameter	12.7 – 165.1	mm
Liquid density	705 – 1,009	kg-m^{-3}
Gas density	0.80 – 50.5	kg-m^{-3}
Liquid viscosity	$3 \times 10^{-4} - 9 \times 10^{-2}$	Pa
Gas viscosity	$10^{-5} - 2.2 \times 10^{-5}$	Pa
Surface tension	24 – 103	mN-m^{-1}
Liquid superficial velocity	0.09 – 731	cm-s^{-1}
Gas superficial velocity	0.04 – 171	m-s^{-1}

Source: Mandhane [6].

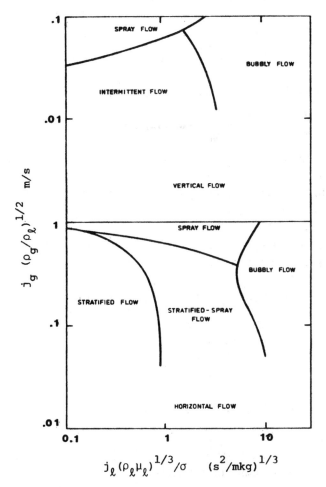

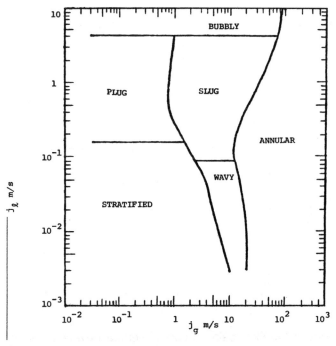

Figure 33. Flow map proposed by Mandhane et al. [6]. *(With permission of Elsevier Science Ltd.)*

Figure 34. Shell-side flow pattern maps [3]. *(With permission of ASME.)*

locity vs. superficial liquid velocity with physical property terms attached. Superficial is used in the sense that the total flow area and not the actual phase flow area is used to evaluate the phase velocity. The flow area referred to is the minimum cross-sectional area for flow through the tube bank.

Sources

1. Cheremisinoff, N. P., *Heat Transfer Pocket Handbook*. Houston: Gulf Publishing Co., 1984.
2. Oshinowo, T. and Charles, M. E., in *Can. Journ. of Chem. Engrg.*, 52: 25–35, 1974.
3. Grant, I. D. R. and Chisholm, D. in *Transactions ASME, Journal of Heat Transfer*, 101 (Series C): 38–42, 1979.
4. Hewitt, G. F. and D. N. Roberts, "Studies of Two-Phase Flow Patterns by Simultaneous X-Ray and Flash Photography," AERE-M2159, H.M.S.O., 1969. Copyright AEA Technology plc.
5. Baker, O. in *Oil and Gas Journ.*, 53 (12): 185–190, 1954.
6. Mandhane, J. M., Gregory, G. A., and Aziz, K. in *Intl. Journ. of Multi Flow*, 1: 533–537, 1974.
7. Bell, K. J., Taborek, J., and Fenoglio, F., *Chem. Engrg. Progress*, Symposium Series (Heat Transfer—Minneapolis), 66 (102): 150–165, 1970.

Estimating Pressure Drop

Two-phase drop in a shell-and-tube heat exchanger consists of friction, momentum change, and gravity:

$$\Delta P = \Delta P_f + \Delta P_m + \Delta P_g \qquad (6)$$

The entrance and exit pressure losses, usually considered in a compact heat exchanger application, are neglected because of

1. The lack of two-phase data for these pressure losses
2. Their small contribution to the total pressure drop for tubular exchangers

The evaluation of ΔP due to momentum and gravity effects is generally based on a homogeneous model.

Homogeneous Flow Model

This is the simplest two-phase flow model. The basic premise is that a real two-phase flow can be replaced by a single-phase flow with the density of the homogeneous mixture defined by:

$$v_{hm} = v_l (1 - x) + v_g x \qquad (7A)$$

$$\frac{1}{\rho_{hom}} = \frac{1 - x}{\rho_l} + \frac{x}{\rho_g} \qquad (7B)$$

where v is the specific volume. Subscripts l and g denote liquid and gas phases and x is the quality (the ratio of gas mass flow rate to total [gas + liquid] mass flow rate).

The pressure drop/rise due to an elevation change is:

$$\Delta P_g = \pm \rho_{hom} \left(\frac{g}{g_c} \right) L \sin \theta \qquad (8)$$

Angle θ is measured from the horizontal. The + sign stands for a downflow, and the − sign stands for an upflow. Gravity pressure drop predictions from this theory are good for high quality and high pressure applications. When ΔP_g is predominant (one half to two thirds of the ΔP), such as for low velocities and low pressure applications, the following equation, which takes into account the velocity slip between two phases via the void fraction α (the ratio of gas volume to total volume), should be used:

$$\Delta P_g = \pm (\rho_l (1 - \alpha) + \rho_g \alpha) (g/g_c) L \sin \theta;$$
$$\text{for } \Delta P_g > 0.5 \ \Delta P_{total} \qquad (9)$$

The momentum pressure drop/rise from the homogeneous model is:

$$\Delta P_m = \frac{G^2}{g_c} \left(\frac{1}{\rho_2} - \frac{1}{\rho_1} \right) \qquad (10)$$

ρ_2 and ρ_1 are the densities of homogeneous mixtures at the exchanger (tube) outlet and inlet, respectively. They are individually evaluated using Equation 7. G is the mass velocity.

Separated Flow Model

Here, the two phases are artificially segregated into two streams. Each stream (vapor and liquid) is under the same pressure gradient but not necessarily with the same velocity. The separated flow model reduces to the homogeneous flow model if the mean velocities of the two streams are the same. The best known separated flow model is the Lockhart and Martinelli correlation [2]. In the Lockhart-Martinelli method, the two fluid streams are considered segregated. The conventional pressure-drop friction-factor relationship is applicable to individual streams. The liquid- and gas-phase pressure drops are considered equal irrespective of the flow patterns. ϕ_l^2 denotes the ratio of a two-phase frictional pressure drop to a single-phase frictional pressure drop for the *liquid* flowing alone in the tube:

$$\phi_l^2 = \frac{\Delta P_f}{\Delta P_l} \qquad (11)$$

And for vapor:

$$\phi_g^2 = \frac{\Delta P_f}{\Delta P_g} \qquad (12)$$

where ΔP_g is the single-phase frictional pressure drop for the *gas* flowing alone in the tube.

χ^2 is the ratio of a single-phase pressure drop for the liquid phase flowing alone in the tube to that for the gas phase flowing alone in the tube.

$$\chi^2 = \frac{\Delta P_l}{\Delta P_g} \qquad (13)$$

The correlation is shown in Figure 35 and the curves can be represented in equation form as:

$$\phi_l^2 = 1 + \frac{C}{\chi} + \frac{1}{\chi^2} \qquad (14)$$

or

$$\phi_g^2 = 1 + C\chi + \chi^2 \qquad (15)$$

where the value of C is dependent upon the four possible single-phase flow regimes.

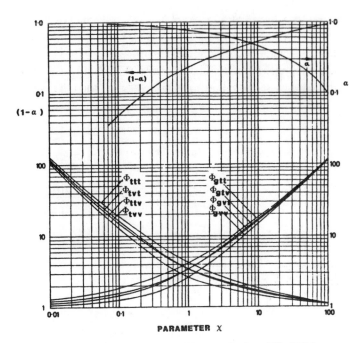

Figure 35. Lockhart-Martinelli correlation [2]. (*With permission of ASME.*)

Liquid		Gas		C	
Turbulent	–	Turbulent	(tt)	20	
Viscous	–	Turbulent	(vt)	12	
Turbulent	–	Viscous	(tv)	10	(16)
Viscous	–	Viscous	(vv)	5	

The two-phase frictional pressure drop by the Lockhart-Martinelli method is determined as follows. First, from the amount of liquid and gas flow rates, and using corresponding friction factors or appropriate correlations, ΔP_l and ΔP_g are calculated. The liquid flow is considered to occupy the entire cross-section for the ΔP_l evaluation, and the gas flow occupies the whole cross-section for the ΔP_g evaluation. The parameter χ is then calculated from Equation 13. The value of C is determined from Equation 16 and ϕ_l or ϕ_g are computed from Equations 14 and 15. The two-phase frictional pressure drop is then calculated from the definition of ϕ.

The Lockhart-Martinelli method was developed for two component adiabatic flows at a pressure close to atmospheric. Martinelli and Nelson [3] extended this method for forced convection boiling for all pressures up to the critical point. The mixture of steam and water was considered "turbulent-turbulent." They presented ϕ_{lo}^2 graphically as a function of the quality x and the system pressure as shown in Figure 36.

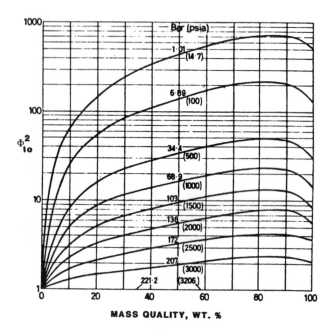

Figure 36. Martinelli-Nelson correlation [3].

$$\phi_l^2 = \frac{\Delta P_f}{\Delta P_{lo}} \qquad (17)$$

where ΔP_{lo} is the frictional pressure drop for the liquid flow alone, in the same tube, with a mass flow rate equal to the *total* mass flow rate of the two-phase flow.

The Martinelli-Nelson experimental curves of ϕ_{lo} vs. x show breaks in the slope due to changes in flow regimes. Surface tension is not included although it may have a significant influence at high pressure near the critical point. The Martinelli-Nelson method provides more correct results than the homogeneous model for low mass velocities (G < 1,360 kg/m²s). In contrast, the homogeneous model provides better results for high mass velocities.

Chisholm gives the following correlation for flow of evaporating two-phase mixtures that accounts for some of the effects neglected in other methods [4].

$$\phi_{lo}^2 = \frac{\Delta P_f}{\Delta P_{lo}}$$

$$= 1 + (\Gamma^2 - 1)\,[Bx^{(2-n)/2}\,(1-x)^{(2-n)/2} + x^{2-n}] \qquad (18)$$

where
$$B = (C\Gamma - 2^{2-n} + 2)/(\Gamma^2 - 1) \qquad (19)$$
$$\Gamma^2 = \Delta P_{go}/\Delta P_{lo} \qquad (20)$$
$$C = (\rho_l/\rho_g)^{1/2}/K + K\,(\rho_g/\rho_l)^{1/2} \qquad (21)$$
$$K = \text{velocity ratio} = j_g/j_l \qquad (22)$$

n is the exponent in the Blasius relation for friction factor $f = C_1/Re^n$, with n = 0.25 for the turbulent flow. These discussions are inclusive of tube flow only.

Two-phase pressure-drop correlations for the shell-side flow are available for a segmentally baffled shell-and-tube exchanger. The frictional pressure drop consists of two components, one associated with the crossflow zone and the other with the window zone. Grant and Chisholm determined the components of the pressure drop [4]. The two-phase crossflow zone and window zone frictional pressure drops are given by Equation 18 with values of B given in Table 12. Values of exponent n for the crossflow zone are: n = 0.46 for horizontal side-to-side flow, and n = 0.37 for vertical up-and-down flow.

Table 12
Values of B for Two-Phase Frictional Pressure-Drop Evaluation in Crossflow and Window-Flow Zones by Equation 18

Zone	Horizontal	Vertical Up and Down Flow
Crossflow		
Spray and bubble	0.75	1.0
Stratified and Stratified spray	0.25	—
Window (n = 0)	$2/(\Gamma + 1)$	$(\rho_l/\rho_{hom})^{0.25}$

Sources

1. Cheremisinoff, N. P., *Heat Transfer Pocket Handbook.* Houston: Gulf Publishing Co., 1984.
2. Lockhart, R. W. and Martinelli, R. C., in *Chem. Engrg. Prog.,* 45: 39–48, 1949.
3. Martinelli, R. C. and Nelson, D. B., in *Transactions ASME,* 70: 695, 1948.
4. Chisholm, D., *Intl. Journ. of Heat and Mass Transfer,* 16: 347–358, 1973.

3

Thermodynamics

Bhabani P. Mohanty, Ph.D., Development Engineer, Allison Engine Company

THERMODYNAMIC ESSENTIALS

Thermodynamics is the subject of engineering that predicts how much energy can be extracted from a working fluid and the various ways of achieving it. Examples of such areas of engineering interest are steam power plants that generate electricity, internal combustion engines that power automobiles, jet engines that power airplanes, and diesel locomotives that pull freight. The working fluid that is the medium of such energy transfer may be either steam or gases generated by fuel-air mixtures.

Phases of a Pure Substance

The process of energy transfer from one form to another is dependent on the properties of the fluid medium and phases of this substance. While we are aware of basically three phases of any substance, namely *solid, liquid,* and *gaseous,* for the purposes of thermodynamic analysis we must define several other intermediate phases. They are:

- *Solid:* The material in solid state does not take the shape of the container that holds it.
- *Subcooled liquid:* The liquid at a condition below its boiling point is called *subcooled* because addition of a little more heat will not cause evaporation.
- *Saturated liquid:* The state of liquid at which addition of any extra heat will cause it to vaporize.
- *Saturated vapor:* The state of vapor that is at the verge of condensing back to liquid state. An example is steam at 212°F and standard atmospheric pressure.
- *Liquid vapor mix:* The state at which both liquid and vapor may coexist at the same temperature and pressure. When a substance exists in this state at the saturation temperature, its *quality* is a mass ratio defined as follows:

$$\text{quality} = \frac{m_{vapor}}{m_{vapor} + m_{liquid}}$$

- *Superheated vapor:* The state of vapor at which extraction of any small amount of heat will not cause condensation.

- *Ideal gas:* At a highly superheated state of vapor, the gas obeys certain ideal gas laws to be explained later in this chapter.
- *Real gas:* At a highly superheated state of vapor, the gas is in a state that does not satisfy ideal gas laws.

Because the phase of a substance is a function of three properties, namely *pressure, temperature,* and *volume,* one can draw a three-dimensional phase diagram of the substance. But in practice, a two-dimensional phase diagram is more useful (by keeping one of the three properties constant). Figure 1 is one such example in the pressure-volume plane. The region of interest in this figure is the liquid-vapor regime.

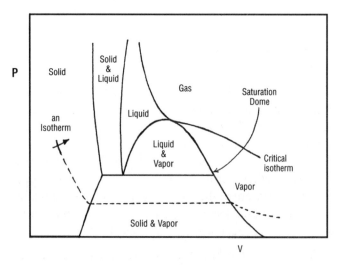

Figure 1. The p-v diagram.

Thermodynamic Properties

There are two types of thermodynamic properties: extensive and intensive. *Extensive properties,* such as mass and volume, depend on the total mass of the substance present. Energy and entropy also fall into this category. *Intensive properties* are only definable at a point in the substance. If the substance is uniform and homogeneous, the value of the intensive property will be the same at each point in the substance. Specific volume, pressure, and temperature are examples of these properties.

Intensive properties are independent of the amount of matter, and it is possible to convert an extensive parameter to an intensive one. Following are the properties that govern thermodynamics.

Mass (m) is a measure of the amount of matter and is expressed in pounds-mass (lbm) or in pound-moles.

Volume (V) is a measure of the space occupied by the matter. It may be measured directly by measuring its physical dimensions, or indirectly by measuring the amount of a fluid it displaces. Unit is ft^3.

Specific volume (v) is the volume per unit mass. The unit is given in ft^3/lbm.

$$v = \frac{V}{m} = \frac{1}{\rho}$$

Density (ρ) is the mass per unit volume. It is reciprocal of the specific volume described above.

Temperature (T) is the property that depends on the energy content in the matter. Addition of heat causes the temperature to rise. The *Zeroth Law of Thermodynamics* defines temperature. This law states that heat flows from one source to another only if there is a temperature difference between the two. In other words, two systems are in *thermal equilibrium* if they are at the same temperature. The temperature units are established by familiar freezing and boiling points of water (32°F and 212°F, respectively).

The relationship between the Fahrenheit and Celsius scales is:

$$T\,°F = 32 + \left(\frac{9}{5}\right) T\,°C$$

In all thermodynamic calculations, absolute temperatures must be used unless a temperature difference is involved. The absolute temperature scale is independent of properties of any particular substance, and is known as Rankine and Kelvin scales as defined below:

$$T\,°R = 460 + T\,°F$$

$$T\,°K = 273 + T\,°C$$

The pressure, volume, and temperature are related by the so-called *ideal gas law,* which is:

$$pV = RT$$

where R is the proportionality constant.

Pressure (p) is the normal force exerted per unit surface area:

$$p = \frac{F_n}{A}$$

Pressure measured from the surrounding atmosphere is called the *gage pressure,* and if measured from the absolute vacuum, it is called the *absolute pressure.*

$$p_{abs} = p_{atm} + p_{gage}$$

Its unit is either *psi* or *inches of water:*

1 atm = 14.7 psi = 407 inches of water = 1 bar

Internal energy (u, U) is the energy associated with the existence of matter and is unrelated to its position or velocity (as represented by potential and kinetic energies). It is a function of temperature alone, and does not depend on the process or path taken to attain that temperature. It is also known as *specific internal energy.* Its unit is *Btu/lbm.* Another form of internal energy is called the *molar internal* *energy,* and is represented as *U.* Its unit is *Btu/pmole.* These two are related by u = U/M; u is an intensive property like p, v, and T.

Enthalpy (h, H) is a property representing the total useful energy content in a substance. It consists of *internal energy* and *flow energy* pV. Thus,

H = U + pV/J (Btu/pmole)

h = u + pv/J (Btu/lbm)

Like internal energy, enthalpy also has the unit of energy, which is force times length. But they are expressed in the heat equivalent of energy, which is Btu in the U.S. customary system and Joule in the metric system.

The J term above is called the *Joule's constant.* Its value is 778 ft.lbf/Btu. It is used to cause the two energy components in enthalpy to have equivalent units. Enthalpy, like internal energy, is also an intensive property that is a function only of the state of the system.

Entropy (s, S) is a quantitative measure of the degradation that energy experiences as a result of changes in the universe. In other words, it measures *unavailable energy.* Like energy, it is a conceptual property that cannot be measured directly. Because entropy is used to measure the degree of irreversibility, it must remain constant if changes in the universe are reversible, and must always increase during irreversible changes.

For an isothermal process (at constant temperature T_0), the change in entropy is a function of energy transfer. If Q is the energy transfer per lbm, then the change in entropy is given by:

$$\Delta s = \frac{Q}{T_0}$$

Nonisothermal processes follow these relationships:

$$\Delta s = \int \frac{dQ}{T}$$

$$s_2 - s_1 = c_p \ln (T_2/T_1) - \frac{R}{J} \ln (p_2/p_1)$$

$$s_2 - s_1 = c_v \ln (T_2/T_1) + \frac{R}{J} \ln (v_2/v_1)$$

Specific heat (C): The slope of a constant pressure line on an h-T plot is called *specific heat at constant pressure,* and the slope at constant volume on a u-T plot is called *specific heat at constant volume.*

$$C_p = dh/dT, \ C_v = du/dT$$

$$R = C_p - C_v, \ k = C_p/C_v$$

because du = dh − RdT for an ideal gas. Values of C_p, C_v, k, and R for a few gases are given in Table 1. R is in ft − lbf/lbm − °R, and C_p, C_v are in Btu/lbm − °F.

Latent heats is defined as the amount of heat added per unit mass to change the phase of a substance at the same pressure. There is no change in temperature during this phase change process. The heat released or absorbed by a mass m is Q = m(LH), where LH is the latent heat. If the phase change is from solid to liquid, it is called the *latent heat of fusion.* When it is from liquid to vapor, it is called the *latent heat of vaporization.* Solid-to-vapor transition is known as the *latent heat of sublimation.* Fusion and vaporization values for water at 14.7 psi are 143.4 and 970.3 Btu/lbm, respectively.

Table 1
Gas Properties

Gas	Mol. Wt.	C_p	C_v	k	R
Acetylene	26.00	0.350	0.2737	1.30	59.4
Air	29.00	0.240	0.1714	1.40	53.3
Ammonia	17.00	0.523	0.4064	1.32	91.0
Carbon dioxide	44.00	0.205	0.1599	1.28	35.1
Carbon monoxide	28.00	0.243	0.1721	1.40	55.2
Chlorine	70.90	0.115	0.0865	1.33	21.8
Ethane	30.07	0.422	0.3570	1.18	51.3
Helium	4.00	1.250	0.7540	1.66	386.3
Hydrogen	2.00	3.420	2.4350	1.41	766.8
Methane	16.00	0.593	0.4692	1.32	96.4
Nitrogen	28.00	0.247	0.1761	1.40	55.2
Oxygen	32.00	0.217	0.1549	1.40	48.3
Propane	44.09	0.404	0.3600	1.12	35.0
Steam	18.00	0.460	0.3600	1.28	85.8
Sulphur dioxide	64.10	0.154	0.1230	1.26	24.0

Determining Properties

Ideal Gas

A gas is considered *ideal* when it obeys certain laws. Usually, the gas at very low pressure/high temperature will fall into this state. One of the laws is Boyle's law: pV = constant; the other is Charles' law: V/T = constant. Combining these two with Avogadro's hypothesis, which states that "equal volumes of different gases with the same temperature and pressure contain the same number of molecules," we arrive at the general law for the ideal gas (equation of state):

$$\frac{pV}{T} = R*$$

where R* is called the *universal gas constant.* Note that R* = MR, where M is the molecular weight and R is the specific gas constant. If there are n moles, the above equation may be reformatted:

$$pV = nR*T = mRT$$

where m is the mass: m = nM.

Table 2 provides the value of R*, the universal gas constant, in different units:

Table 2
Universal Gas Constant Values

Value of R*	Unit
1.314	(atm ft³)/(lb – mol °K)
1.9869	BTU/(lb – mol °R)
1545	(lbf – ft)/(lb – mol °R)
0.7302	(atm ft³)/(lb – mol °R)
8.3144	J/(gm – mol °K)
1.9872	cal/(gm – mol °K)

Van der Waals Equation

The ideal gas equation may be corrected for its two worst assumptions, i.e., infinitesimal molecular size and no intermolecular forces, by the following equation:

$$\left(p + \frac{a}{v^2}\right)(v - b) = RT$$

where a/v^2 accounts for the intermolecular attraction forces and b accounts for the finite size of the gas molecules.

In theory, equations of state may be developed that relate any property of a system to any two other properties. However, in practice, this can be quite cumbersome. This is why engineers resort to property tables and charts that are readily available. Following are some of the most widely used property charts:

- *p-v diagram:* Movement along an isotherm represents expansion or compression and gives density or specific volume as a function of pressure (see Figure 1). The region below the critical isotherm $T = T_c$ corresponds to temperatures below the critical temperature where it is possible to have more than one phase in equilibrium.
- *T-s diagram:* This is the most useful chart in representing the heat and power cycles (see Figure 2). A line of constant pressure *isobar* is shown along with the critical isobar $P = P_c$. This chart might also include lines of constant volume (*isochores*) or constant enthalpy (*isenthalps*).

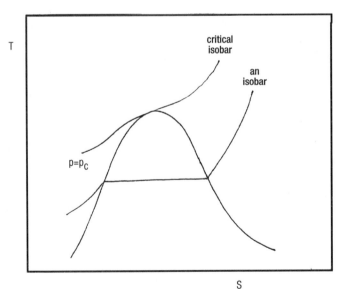

Figure 2. The T-s diagram.

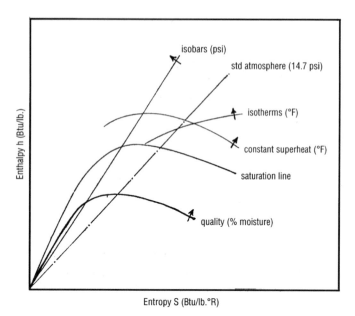

Figure 3. The Mollier chart (h-s diagram).

• *h-s diagram:* This is also called the *Mollier chart* (Figure 3). It is used to determine property changes between the superheated vapor and the liquid-vapor regions. Below the saturation line, lines of quality (constant moisture content) are shown. Above it are the lines of constant superheat and constant temperature. Isobars are also superimposed on top.

The properties may also be found through various tables with greater accuracy. These are:

• Steam tables, which give specific volume, enthalpy, entropy, and internal energy as functions of temperature.
• Superheat tables, which give specific volume, enthalpy, and entropy for combinations of pressure and temperature. These are in the superheated regime.
• Compressed liquid tables, which give properties at the saturation state and corrections to these values for various pressures.
• Gas tables, which are essentially superheat tables for various gases. Properties are given as functions of temperature alone.

Types of Systems

Matter enclosed by a well-defined boundary is called a *thermodynamic system.* Everything outside is called the *environment.* The volume of the enclosed region is called the *control volume,* and its surface is the *control surface.* If there is no mass exchange across the boundary, it is called a *closed* system as opposed to an *open* system. The most important system is a "steady flow open system," where the rate of mass exchange at the entry and exit are the same. Pumps, turbines, and boilers fall into this category.

Types of Processes

A process is defined in terms of specific changes to be accomplished. Two types of energy transfers may take place across a system boundary: thermal energy transfer (heat) and mechanical energy transfer (work). Any process must have a well-defined objective for energy transfer. Below are definitions of well-known processes and the relationships between variables in the processes. The equations are in a per lbm basis, but can be converted to a lb − mol basis by substituting V for v, H for h, and R* for R:

Isothermal: a constant temperature process ($T_2 = T_1$).
$$p_2 = p_1(v_1/v_2)$$
$$v_2 = v_1(p_1/p_2)$$
$$Q = W = T (s_2 - s_1) = RT \ln (v_2/v_1)$$
$$W = Q = T (s_2 - s_1) = RT \ln (v_2/v_1)$$
$$u_2 = u_1$$
$$s_2 = s_1 + (Q/T) = R \ln (v_2/v_1) = R \ln (p_1/p_2)$$
$$h_2 = h_1$$

Isobaric: a constant pressure process ($p_2 = p_1$).

$$T_2 = T_1 (v_2/v_1)$$
$$v_2 = v_1 (T_2/T_1)$$
$$Q = (h_2 - h_1) = c_p (T_2 - T_1) = c_v (T_2 - T_1) + p (v_2 - v_1)$$
$$W = p (v_2 - v_1) = R(T_2 - T_1)$$
$$u_2 = u_1 + c_v (T_2 - T_1) = c_v p (v_2 - v_1)/R = p (v_2 - v_1)/(k - 1)$$
$$s_2 = s_1 + c_p \ln(T_2/T_1) = c_p \ln(v_2/v_1)$$
$$h_2 - h_1 = Q = c_p (T_2 - T_1) = kp (v_2 - v_1)/(k - 1)$$

Isochoric: a constant volume process ($v_2 = v_1$).

$$p_2 = p_1 (T_2/T_1)$$
$$T_2 = T_1 (p_2/p_1)$$
$$Q = (u_2 - u_1) = c_v (T_2 - T_1)$$
$$W = 0$$
$$u_2 - u_1 = Q = c_v (T_2 - T_1) = c_v v (p_2 - p_1)/R = v (p_2 - p_1)/(k - 1)$$
$$s_2 = s_1 + c_v \ln (T_2/T_1) = c_v \ln (p_2/p_1)$$
$$h_2 = h_1 + c_p (T_2 - T_1) = kv (p_2 - p_1)/(k - 1)$$

Adiabatic: a process during which no heat is transferred between the system and its surroundings ($Q = 0$). Many real systems in which there is little time for heat transfer may be assumed to be adiabatic. Adiabatic processes can further be divided into two categories: isentropic and isenthalpic.

$$p_2 = p_1 (v_1/v_2)^k = p_1 (T_2/T_1)^k$$
$$v_2 = v_1 (p_1/p_2)^{1/k} = v_1 (T_1/T_2)^{\frac{1}{k-1}}$$
$$T_2 = T_1 (v_1/v_2)^{k-1} = T_1 (p_2/p_1)^{\frac{k-1}{k}}$$
$$Q = 0$$
$$W = u_1 - u_2 = c_v (T_1 - T_2) = \frac{p_1 v_1}{k - 1} [1 - (p_2/p_1)^{\frac{k-1}{k}}]$$
$$u_2 - u_1 = -W = c_v (T_2 - T_1) = (p_2 v_2 - p_1 v_1)/(k - 1)$$
$$s_2 - s_1 = 0$$
$$h_2 - h_1 = c_p (T_2 - T_1) = k (p_2 v_2 - p_1 v_1)/(k - 1)$$

Isentropic: a constant entropy process (steady flow) ($Q = 0$, $s_2 = s_1$).

$$p_2 = p_1 (v_1/v_2)^k = p_1 (T_2/T_1)^k$$
$$v_2 = v_1 (p_1/p_2)^{1/k} = v_1 (T_1/T_2)^{\frac{1}{k-1}}$$
$$T_2 = T_1 (v_1/v_2)^{k-1} = T_1 (p_2/p_1)^{\frac{k-1}{k}}$$
$$Q = 0$$
$$W = h_1 - h_2 = c_p T_1 [1 - (p_2/p_1)^{\frac{k-1}{k}}]$$
$$u_2 = u_1 + c_v (T_2 - T_1)$$
$$s_2 = s_1$$
$$h_2 = h_1 - W = h_1 + c_p (T_2 - T_1) = k (p_2 v_2 - p_1 v_1)/(k - 1)$$

Isenthalpic: a constant enthalpy process (steady flow). Also known as a throttling process ($Q = 0$, $W = 0$).

$$p_2 v_2 = p_1 v_1, \quad p_2 < p_1, \quad v_2 > v_1, \quad T_2 = T_1$$
$$u_2 = u_1$$
$$s_2 = s_1 + R \ln (p_1/p_2) = s_1 + R \ln (v_2/v_1)$$
$$h_2 = h_1$$

Polytropic: a process in which the working fluid properties obey the polytropic law: $p_1 v_1^n = p_2 v_2^n$.

$$p_2 = p_1 (v_1/v_2)^n = p_1 (T_2/T_1)^n$$
$$v_2 = v_1 (p_1/p_2)^{1/n} = v_1 (T_1/T_2)^{\frac{1}{n-1}}$$
$$T_2 = T_1 (v_1/v_2)^{n-1} = T_1 (p_2/p_1)^{\frac{n-1}{n}}$$
$$Q = c_v (n - k) (T_2 - T_1)/(n - 1)$$
$$W = R (T_1 - T_2)/(n - 1) = (p_1 v_1 - p_2 v_2)/(n - 1) = \frac{p_1 v_1}{n - 1} [1 - (p_2/p_1)^{\frac{n-1}{n}}]$$
$$u_2 = u_1 + c_v (T_2 - T_1) = (p_2 v_2 - p_1 v_1)/(n - 1)$$
$$s_2 = s_1 + c_v (n - k) (T_2 - T_1)/(n - 1) [\ln (T_2/T_1)]$$
$$h_2 = h_1 + c_p (T_2 - T_1) = n (p_2 v_2 - p_1 v_1)/(n - 1)$$

The Zeroth Law of Thermodynamics

The Zeroth Law of Thermodynamics defines temperature. This law states that heat flows from one source to another only if there is a temperature difference between the two. Therefore, two systems are in *thermal equilibrium* if they are at the same temperature.

FIRST LAW OF THERMODYNAMICS

The first law of thermodynamics establishes the principle of conservation of energy in thermodynamic systems. In thermodynamics, unlike in purely mechanical systems, trans- formation of energy takes place between different sources, such as chemical, mechanical, and electrical. The two basic forms of energy transfer are *work done* and *heat transfer.*

Work

Work may be done by (W_{out}) or on (W_{in}) a system. In thermodynamics, we are more interested in work done by a system W_{out}, considered *positive,* which causes the energy of the system to reduce. Work is a path function. Since it does not depend on the state of the system or of the sub- stance, it is not a property of the system. In a p-v diagram, work is the following integral:

$$W_{out} = \int_1^2 pdV$$

Heat

Heat is the thermal energy transferred because of tem- perature difference. It is considered positive if it is added to the system, that is, Q_{in}. A unit of heat is the same as en- ergy, that is, ft.lbf; but a more popular format is Btu:

$1 Btu = 778.17/ft.lbf = 252/calories = 1,055/Joules$

Like work, it is a path function, and not a property of the system. If there is no heat transferred between the system and the surroundings, the process is called *adiabatic.*

First Law of Thermodynamics for Closed Systems

Briefly, the first law states that "energy can not be cre- ated or destroyed." This means that all forms of energy (heat and work) entering or leaving a closed system must be ac- counted for. This also means that heat entering a closed sys- tem must either increase the temperature (in the form of U) and/or be used to perform useful work W:

$$Q = \Delta U + \frac{W}{J}$$

Note that the Joule's constant was used to convert work to its heat equivalent (ft.lbf to Btu).

First Law of Thermodynamics for Open Systems

The law for open systems is basically Bernoulli's equa- tion extended for nonadiabatic processes. For systems in which the mass flow rate is constant, it is known as the *steady flow energy equation.* On a per unit mass basis, this equation is:

$$Q = (h_2 - h_1) + \frac{v_2^2 - v_1^2}{2gJ} + \frac{g(z_2 - z_1)}{gJ} + \frac{W_{shaft}}{J}$$

Both sides may be multiplied by the mass flow rate ($m_d$ot) to get the units in Btu or be multiplied by ($m_d$otJ) to get the units in ft − lbf.

The above equation may be applied to any thermody- namic device that is continuous and has steady flow, such as turbines, pumps, compressors, boilers, condensers, noz- zles, or throttling devices.

SECOND LAW OF THERMODYNAMICS

All thermodynamic systems adhere to the principle of conservation of energy (the first law). The second law describes the restrictions to all such processes, and is often called the Kelvin-Planck-Clausius Law. The statement of this law: "It is impossible to create a cyclic process whose only effect is to transfer heat from a lower temperature to a higher temperature."

Reversible Processes and Cycles

A reversible process is one that can be reversed without any resultant change in either the system or the surroundings; hence, it is also an ideal process. A reversible process is always more efficient than an irreversible process. The four phenomena that may render a process irreversible are: (1) friction, (2) unrestrained expansion, (3) transfer of heat across a finite temperature difference, and (4) mixing of different substances.

A cycle is a series of processes in which the system always returns to the same thermodynamic state that it started from. Any energy conversion device must operate in a cycle. Cycles that produce work output are called *power cycles,* and ones that pump heat from lower to higher temperature are called *refrigeration cycles.* Thermal efficiency for a power cycle is given by:

$$\eta_{thermal} = \frac{W_{out}}{Q_{in}}, \; q_{in} = W_{out} + Q_{out}$$

whereas the *coefficients of performance* for refrigerators and heat pumps are defined as q_{in}/W_{in} and Q_{out}/W_{in}, respectively.

Thermodynamic Temperature Scale

If we run a Carnot cycle engine between the temperatures corresponding to boiling water and melting ice, it can be shown that the efficiency of such an engine will be 26.8%. Although water is used as an example, the efficiency of such an engine is actually independent of the working fluid used in the cycle. Because $\eta = 1 - (Q_l/Q_H)$, the value of Q_L/Q_H is 0.732. This sets up both our Kelvin and Rankine scales once we establish the differential. In Kelvin scale, it is 100 degrees; in Rankine scale, it is 180 degrees.

Useful Expressions

- Change in internal energy:

  $du = T \, ds - P \, dv$

- Change in enthalpy:

  $dh = T \, ds + v \, dP$

- Change in entropy:

  $ds = c_v dT/T + R dv/v$

- Volumetric efficiency is a measure of the ability of an engine to "breathe," and may be determined from the following equation:

  $$\eta_v = \frac{\text{volume of air brought into cylinder at ambient conditions}}{\text{piston displacement}}$$

- Mean effective pressure (mep) is net work output, in inch-lbf per cubic inch of piston displacement. It is applicable only to reciprocating engines, and effectively is the average gage pressure acting on the piston during a power stroke.
- Work done:

  $$(\text{mep}) \, (V_{dsp}) = (\text{mep}) \, \pi \, (\text{bore})^2 \, (\text{stroke})/4$$

- Brake-specific fuel consumption (bsfc):

  $$\text{bsfc} = \frac{\text{fuel rate in lbm/hr}}{\text{bhp}}$$

THERMODYNAMIC CYCLES

A thermodynamic cycle can be either open or closed. In an open cycle, the working fluid is constantly input to the system (as in an aircraft jet engine); but in a closed cycle, the working fluid recirculates within the system (as in a re- frigerator). A *vapor* cycle is one in which there is a phase change in the working fluid. A *gas* cycle is one in which a gas or a mixture of gases is used, as the fluid that does not undergo phase change.

Basic Systems and Systems Integration

While, in theory, a cycle diagram explains the thermo- dynamic cycle, it takes a real device to achieve that ener- gy extraction process. The change of property equations for these devices can be derived from the steady-flow energy equation. Following is a list of these devices:

- *Heat exchangers:* transfer energy from one fluid to another.
- *Pumps:* considered adiabatic devices that elevate the total energy content of the fluid.

- *Turbines:* adiabatic extraction of energy from the fluid with a drop in temperature and pressure.
- *Compressors:* similar to pumps in principle.
- *Condensers:* remove heat of evaporation from fluid and reject to the environment.
- *Nozzles:* convert the fluid energy to kinetic energy; adiabatic; no work is performed.

Carnot Cycle

The Carnot cycle (Figure 4) is an ideal power cycle, but it cannot be implemented in practice. Its importance lies in the fact that it sets the maximum attainable thermal effi- ciency for any heat engine. The four processes involved are:

1-2 Isothermal expansion of saturated liquid to saturated gas
2-3 Isentropic expansion
3-4 Isothermal compression
4-1 Isentropic compression

The heat flow in and out of the system and the turbine and compressor work terms are:

$$Q_{in} = T_{high} (s_2 - s_1) = h_2 - h_1$$

$$Q_{out} = T_{low} (s_3 - s_4) = h_3 - h_4$$

$$W_{turb} = h_2 - h_3$$

$$W_{comp} = h_1 - h_4$$

The thermal efficiency of the cycle is:

$$\eta_{thermal} = \frac{Q_{in} - Q_{out}}{Q_{in}} = \frac{T_{high} - T_{low}}{T_{high}}$$

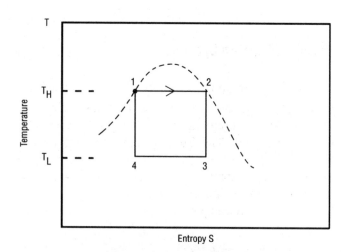

Figure 4. Carnot cycle.

Rankine Cycle: A Vapor Power Cycle

The Rankine cycle (Figure 5) is similar to the Carnot cycle. The difference is that compression takes place in the liquid region. This cycle is implemented in a steam power plant. The five processes involved are:

1-2 Adiabatic compression to boiler pressure
2-3 Heating to fluid saturation temperature
3-4 Vaporization in the boiler
4-5 Adiabatic expansion in the turbine
5-1 Condensation

The heat flow in and out of the system and the turbine and compressor work terms are:

$$q_{in} = h_4 - h_2 \qquad q_{out} = h_5 - h_1$$

$$W_{turb} = h_4 - h_5 \qquad W_{comp} = h_2 - h_1 = v_1 (p_2 - p_1)/J$$

The thermal efficiency of the cycle is:

$$\eta_{thermal} = \frac{q_{in} - q_{out}}{q_{in}} = \frac{(h_4 - h_5) - (h_2 - h_1)}{h_4 - h_2}$$

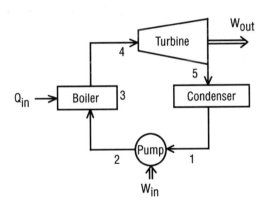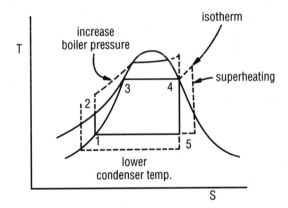

Figure 5. Rankine cycle.

Reversed Rankine Cycle: A Vapor Refrigeration Cycle

The reversed Rankine cycle (Figure 6) is also similar to the Carnot cycle. The difference is that compression takes place in the liquid region. This cycle is implemented in a steam power plant. The four processes involved are:

1-2 Isentropic compression; raise temperature and pressure
2-3 Reject heat to high temperature
3-4 Expander reduces pressure and temperature to initial value
4-1 Fluid changes dry vapor at constant pressure; heat added

The heat flow in and out of the system and the turbine and compressor work terms are:

$$q_{in} = h_1 - h_4 \qquad q_{out} = h_2 - h_3$$

$$W_{turb} = h_1 - h_2 \qquad W_{comp} = h_4 - h_3$$

The thermal efficiency of the cycle is:

$$\eta_{thermal} = \frac{q_{in} - q_{out}}{q_{in}} = \frac{(h_1 - h_4) - (h_2 - h_3)}{h_1 - h_4}$$

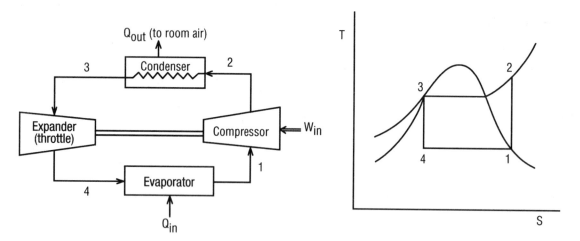

Figure 6. Reversed Rankine cycle (vapor refrigeration system).

Brayton Cycle: A Gas Turbine Cycle

The Brayton cycle (Figure 7) uses an air-fuel mixture to keep the combustion temperature as close to the metallurgical limits as possible. A major portion of the work output from the turbine is used to drive the compressor. The remainder may be either shaft output (perhaps to drive a propeller, as in a turboprop, or drive a fan, as in a turbofan) or nozzle expansion to generate thrust (as in a turbojet engine). The four processes involved are:

1-2 Adiabatic compression (in compressor)
2-3 Heat addition at constant pressure (in combustor)
3-4 Adiabatic expansion (in turbine)
4-1 Heat rejection at constant pressure

The heat flow into the system and the turbine and compressor work output terms are:

$$q_{in} = c_p (T_3 - T_2) = h_3 - h_2$$

$$W_{turb} = c_p (T_3 - T_4) = h_3 - h_4,$$

$$W_{comp} = c_p (T_2 - T_1) = h_2 - h_1$$

The thermal efficiency of the cycle is:

$$\eta_{thermal} = \frac{(T_3 - T_2)(T_4 - T_1)}{T_3 - T_2}$$

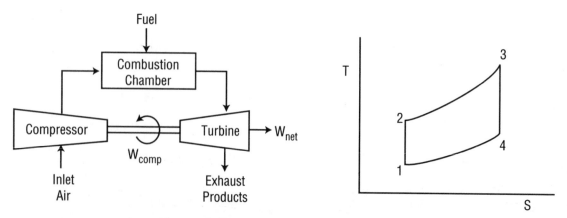

Figure 7. Brayton cycle (gas turbine engine).

Otto Cycle: A Power Cycle

The Otto cycle (Figure 8) is a four-stroke cycle as represented by an idealized internal combustion engine. The four processes involved are:

1-2 Adiabatic compression
2-3 Heat addition at constant volume
3-4 Adiabatic expansion
4-1 Heat rejection at constant volume

The heat flow in and out of the system and the work input and work output terms are:

$q_{in} = c_v (T_3 - T_2)$, $q_{out} = c_v (T_4 - T_1)$

$W_{in} = c_v (T_2 - T_1)$, $W_{out} = c_v (T_3 - T_4)$

The thermal efficiency of the cycle is:

$$\eta_{thermal} = \frac{q_{in} - q_{out}}{q_{in}} = 1 - 1/R^{k-1} = 1 - T_4/T_3 = 1 - T_1/T_2$$

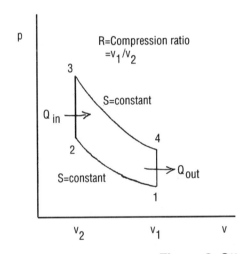
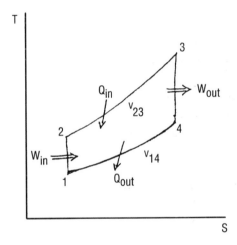

Figure 8. Otto cycle (ideal closed system).

Diesel Cycle: Another Power Cycle

In a diesel engine, only air is compressed; fuel is introduced only at the end of the compression stroke. That is why it is often referred to as a compression-ignition engine. This cycle (Figure 9) uses the heat of the compression process to start the combustion process. The four processes involved are:

1-2 Adiabatic compression
2-3 Heat addition at constant pressure
3-4 Adiabatic expansion (power stroke)
4-1 Heat rejection at constant volume

The heat flow in and out of the system and the work input and work output terms are:

$q_{in} = c_p (T_3 - T_2)$, $q_{out} = c_v (T_4 - T_1)$

$W_{in} = c_v (T_2 - T_1)$, $W_{out} = c_v (T_3 - T_4) + (c_p - c_v) (T_3 - T_2)$

The thermal efficiency of the cycle is:

$$\eta_{thermal} = \frac{q_{in} - q_{out}}{q_{in}} = 1 - \frac{T_4 - T_1}{k (T_3 - T_2)}$$

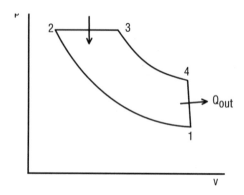

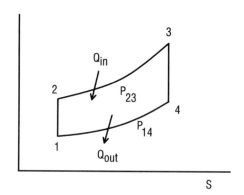

Figure 9. Diesel cycle.

Gas Power Cycles with Regeneration

Use of regeneration is an effective way of increasing the thermal efficiency of the cycle, particularly at low compressor pressure ratios. The Stirling and Ericsson cycles are such attempts to get efficiencies close to that of the ideal Carnot cycle.

Stirling Cycle

This cycle (Figure 10) can come to attain the thermal efficiency very close to that of a Carnot cycle. The isothermal processes can be attained by reheating and intercooling. This cycle is suitable for application in reciprocating machinery. The four processes involved are:

1-2 Heat addition at constant volume (compression)
2-3 Isothermal expansion with heat addition (energy input and power stroke)
3-4 Heat rejection at constant volume
4-1 Isothermal compression with heat rejection

In an ideal regenerator, the quantity of heat rejected during 3-4 is stored in the regenerator and then is restored to the working fluid during the process 1-2. But in reality, there is some loss in between.

The heat flow in and out of the system and the work input and work output terms are:

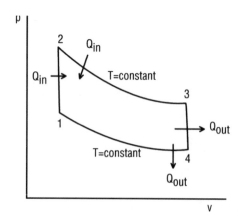

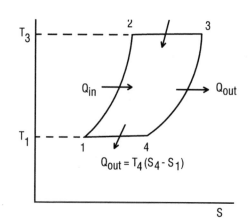

Figure 10. Stirling cycle.

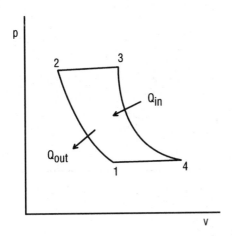 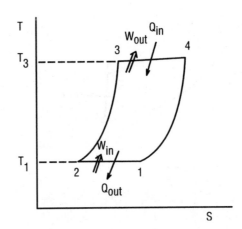

Figure 11. Ericsson cycle.

$$q_{in} = q_{1-2} + q_{2-3}$$

$$W_{in} = W_{4-1} = Q_{out} \qquad W_{out} = W_{2-3} = Q_{in}$$

The thermal efficiency of the cycle is:

$$\eta_{thermal} = \frac{T_3 - T_1}{T_3}$$

Ericsson Cycle

This cycle (Figure 11) can also come very close to attaining the thermal efficiency of a Carnot cycle. The isothermal processes can be attained by reheating and intercooling. Its application is most appropriate to rotating machinery. The four processes involved are:

1-2 Isothermal compression (energy rejection)
2-3 Heat addition at constant pressure
3-4 Isothermal expansion (energy input and power output)
4-1 Heat rejection at constant pressure

The heat flow in and out of the system and the work input and work output terms are:

$$q_{in} = q_{2-3} + q_{3-4}$$

$$W_{in} = W_{1-2} \qquad W_{out} = W_{3-4}$$

The thermal efficiency of the cycle is:

$$\eta_{thermal} = \frac{T_3 - T_1}{T_3}$$

4

Mechanical Seals

Todd R. Monroe, P.E., Houston, Texas
Perry C. Monroe, P.E., Monroe Technical Services, Houston, Texas

BASIC MECHANICAL SEAL COMPONENTS

All mechanical seals are constructed with three basic groups of parts. The first and most important group is the mechanical seal faces, shown in Figure 1. The rotating seal face is attached to the shaft, while the stationary seal face is held fixed to the equipment case via the gland ring.

The next group of seal components is the secondary sealing members. In Figure 1, these members consist of a wedge ring located under the rotating face, an o-ring located on the stationary face, and the gland ring gasket.

The third group of components is the seal hardware, including the spring retainer, springs, and gland ring. The purpose of the spring retainer is to mechanically drive the rotating seal face, as well as house the springs. The springs are a vital component for assuring that the seal faces remain in contact during any axial movement from normal seal face wear, or face misalignment.

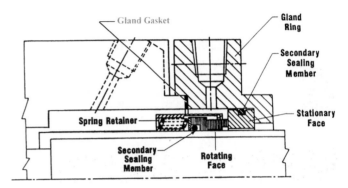

Figure 1. Mechanical seal components. (*Courtesy of John Crane, Inc.*)

SEALING POINTS

There are four main sealing points in a mechanical seal (see Figure 2). The primary sealing point is at the seal faces, Point A. This sealing point is achieved by utilizing two very flat, lapped surfaces, perpendicular to the shaft, that create a very treacherous leakage path. Leakage is further minimized by the rubbing contact between the rotating and stationary faces. In most cases, these two faces are made of one hard material, like tungsten carbide, and a relatively soft material such as carbon-graphite. The carbon seal face generally has the smaller contact area, and is the wearing face. The O.D. and I.D. of the wearing face represent the "seal face dimensions," and are also generically referred to as the "seal face" throughout this chapter.

The second leakage path, Point B, is along the shaft under the rotating seal face. This path is blocked by the secondary o-ring. An additional secondary o-ring, Point C, is used to prevent leakage between the gland ring and the stationary seal face. Point D is the gland ring gasket which prevents leakage between the equipment case and the gland.

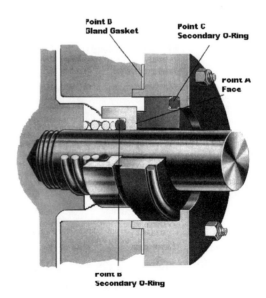

Figure 2. Sealing points. (*Courtesy of Durametallic Corp.*)

MECHANICAL SEAL CLASSIFICATIONS

Mechanical seals can be categorized by certain design characteristics or by the arrangement in which they're used. Figure 3 outlines these classifications. None of these designs, or arrangements, are inherently better than the other. Each has a specific use, and a good understanding of the differences will allow the user to properly apply and maintain each seal type.

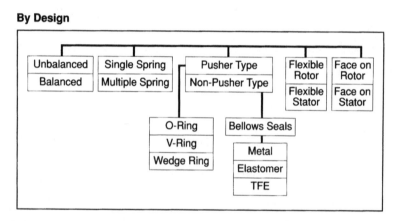

Figure 3. Mechanical seal classifications. (*Courtesy of Durametallic Corp.*)

BASIC SEAL DESIGNS

Pusher Seals

The characteristic design of a pusher seal is the dynamic o-ring at the rotating seal face (see Figure 4). This o-ring must move axially along the shaft or sleeve to compensate for any shaft or seal face misalignment as well as normal face wear. The advantages of a pusher seal are that the design can be used for very high-pressure applications (as much as 3,000 psig), the metal components are robust and

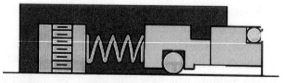

Figure 4. Basic pusher seal. (*Courtesy of Durametallic Corp.*)

come in special alloy materials, and the design is well suited for special applications.

The disadvantages of the pusher design are related to the dynamic o-ring. In a corrosive service, the constant relative motion between the dynamic o-ring and shaft wears away at the protective oxide layer of the shaft or sleeve material, causing fretting corrosion. The fretting will wear a groove in the shaft or sleeve, providing a leakage path for the sealed fluid.

An additional limitation of the dynamic o-ring is the problem of "hang-up," shown in Figure 5. In applications where the sealed fluid can "salt-out" or oxidize, like caustic or hydrocarbons, the normal seal weepage can build under the seal faces and prevent forward movement, thereby creating a leakage path.

Figure 5. Dynamic o-ring "hang-up." (*Courtesy of Durametallic Corp.*)

Another characteristic of the pusher seal design is the use of coil springs for providing mechanical closing force. These springs can be either a single coil spring (Figure 6) or a multiple arrangement of springs (Figure 7). Mechanical seals using a single coil spring are widely used because of their simple design, and the large spring cross-section is good for corrosion resistance. The disadvantages of the single coil are that the applied spring force is very nonuniform

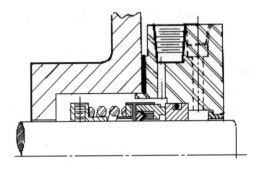

Figure 6. Single coil spring.

and can cause waviness and distortion to the seal face in larger sizes. In addition, the spring can distort at high surface speeds.

Multispring seals (Figure 7) use a series of small coil springs spaced circumferentially around the seal face. This spacing provides a uniform face loading, minimizing the waviness and distortion attributed to spring forces. The multispring arrangement is also less susceptible to high-speed spring distortion under 4,500 fpm.

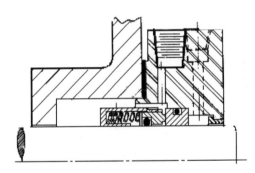

Figure 7. Multi-spring seal. (*Courtesy of John Crane, Inc.*)

Non-Pusher Seals

The characteristic design feature of the non-pusher, or bellows, seal is the lack of a dynamic o-ring. As shown in Figure 8, the non-pusher seal design has a static o-ring in the drive collar of the rotating seal face unit. This is made possible by the bellows, which acts as a pressure containing device, as well as the spring force component. This unique feature provides advantages over the pusher seal in that the static

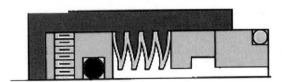

Figure 8. Welded metal-bellows, non-pusher seal. (*Courtesy of Durametallic Corp.*)

Figure 9. Static o-ring, no hang-ups. (*Courtesy of Durametallic Corp.*)

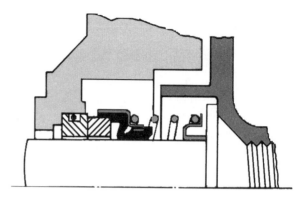

Figure 10. Elastomeric bellows seal.

o-ring virtually eliminates the problem of fretting corrosion and seal face hang-up, as shown in Figure 9.

There are two basic bellows seal designs available: the metal bellows and the elastomeric bellows. Figure 8 depicts a welded metal bellows design, constructed from a series of thin metal leaflets that are laser welded together at the top and bottom to form a pressure-containing spring. The elastomeric bellows (Figure 10) consists of a large rubber bellows, or boot, that is energized by a single coil spring.

While the non-pusher seal has several advantages over the pusher seal, the bellows also provides this seal design with its limitations. In the case of the welded metal bellows, the thin cross-section of the leaflets, .005″ to .009″, limits

the pressure at which the design should be applied to 250 psig. This limitation is due to the pressure acting on the unsupported seal face, causing severe face deflections. The thin leaflets also limit the corrosion allowance to .002″ for chemical services. The elastomeric bellows design is not as pressure limited, but does possess the same limitations as a single coil spring design, in addition to limited chemical resistance of the elastomeric bellows.

Unbalanced Seals

For all mechanical seals, the pressure of the sealed fluid exerts a hydraulic force on the seal face. The axial component of this hydraulic force is known as the hydraulic closing force. Unbalanced seal designs have no provisions for reducing the amount of closing force exerted on the seal face, and for pusher seals, the characteristic trait of this design is for the seal faces to be above the balance diameter, as shown in Figure 11. The balance diameter can be determined by locating the innermost point at which pressure can act on the seal faces. For almost all pusher seal designs, that point is the I.D. of the dynamic o-ring.

A more detailed discussion of seal balancing can be found in the Basic Design Principles section of this chapter. The point to be made here is that the unbalanced seal design is the preferred design for low-pressure applications. Seal face weepage is directly related to the closing force acting on the seal face; the higher the closing force,

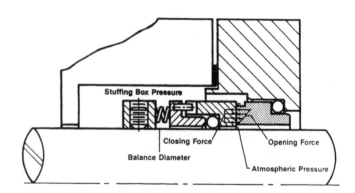

Figure 11. Unbalanced seal. (*Courtesy of Durametallic Corp.*)

the lower the seal face weepage. Unbalanced seal designs inherently have higher closing forces and therefore less seal weepage at lower pressures. Additionally, unbalanced seals are more stable during off-design equipment conditions such as cavitation, high vibration, or misalignment. The only disadvantage to the unbalanced seal design is the pressure limitations. The closing force exerted at the seal face can reach a point where it overcomes the stiffness of the lubricating fluid and literally squeezes the fluid from between the seal faces. This is a destructive condition that should be avoided. Figure 12 lists some recommended pressure limits for unbalanced seals, based on seal size and speed [3].

Seal ID (In)	Shaft Speed (rpm)	Sealing Pressure (psig)
1/2 to 2	Up to 1800	175
	1801 to 3600	100
over 2	Up to 1800	
	1801 to 3600	50

Figure 12. Application limits for unbalanced seals. (*Courtesy of Durametallic Corp.*)

Balanced Seals

It is apparent from our discussion of unbalanced seal designs that the primary purpose of a balanced seal is to reduce the hydraulic closing force acting on the seal face, and therefore provide a seal design suitable for high-pressure applications. For pusher seals, the primary trait of a balanced seal design is the seal faces dropping below the balance diameter, as shown in Figure 13. The shaft or sleeve now has a stepped diameter which allows the seal face dimensions to be reduced. This reduction now exposes more of the front side of the rotating seal face to the seal chamber pressure. Because pressure acts in all directions, this increased exposure opposes or negates a larger portion of the pressure and effectively reduces the closing force acting on the seal face. This reduction of the closing force allows for the maintenance of a good lubrication film between the seal faces at pressures as high as 1,500 psig for a single pusher seal.

While the use of a balanced pusher seal requires a physical step in the shaft or sleeve, a non-pusher or bellows seal design is inherently balanced and therefore requires no stepped sleeve. This can be an important advantage for medium-pressure applications.

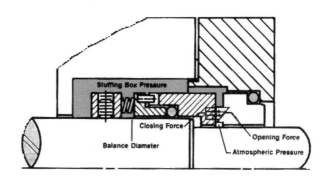

Figure 13. Balanced seal. (*Courtesy of Durametallic Corp.*)

Flexible Rotor Seals

Flexible rotor seal designs (Figure 14) make up the vast majority of the mechanical seals in service. Some of the advantages of the flex rotor design are less cost, less axial space required, and that the springs or bellows are "self-cleaning" due to the effects of centrifugal force. The disadvantages are a speed limitation of 4,500 fpm and the inability to handle severe seal face misalignment. Because the gland ring is bolted directly to the equipment case, the stationary seal face is prone to misalignment due to pipe strain or thermal expansion of the equipment case. For each revolution of the shaft, the flexible rotor must axially compensate for the out-of-perpendicularity of the stationary seal face. As discussed earlier with fretting corrosion, this can be very detrimental to reliable seal performance.

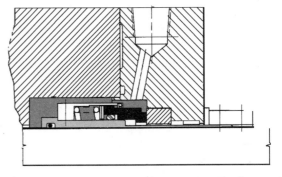

Figure 14. Flexible rotor. (*Courtesy of John Crane, Inc.*)

Flexible Stator Seals

Flexible stator seal designs (Figure 15) make up a smaller, but very necessary, part of the mechanical seals in service. It stands to reason that the disadvantages of the flexible rotor design would be addressed with this design. The primary advantage of the flexible stator design is its ability to handle severe seal face misalignment caused by equipment case distortion by making a one-time adjustment to the rotating seal face. This is of great importance for large, hot equipment in the refining and power markets. The flexible stator is also the design of choice for operating speeds above 4,500 fpm.

The disadvantages of the flexible stator are higher cost, increased axial space requirements, and the limitation of being applied in services with less than 5% solids.

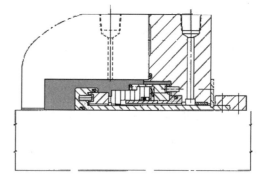

Figure 15. Flexible stator. (*Courtesy of John Crane, Inc.*)

BASIC SEAL ARRANGEMENTS

Single Inside Seals

The single inside seal (Figure 16) is by far the most common seal arrangement used and, for single seals, is the arrangement of choice. The most important consideration for this arrangement is that the seal faces are lubricated by the sealed fluid, and therefore the sealed fluid must be compatible with the environment. Toxic or hazardous fluids should not be handled with a single seal. From an emissions standpoint, volatile organic compounds (VOCs) have been effectively contained with emissions of 500 ppm using a single seal [6]. For corrosive services, the seal must operate in the fluid, so material considerations must be reviewed.

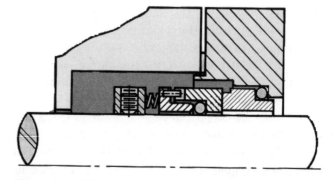

Figure 16. Single inside seal. (*Courtesy of Durametallic Corp.*)

Single Outside Seals

While single outside seals (Figure 17) are not the arrangement of choice, certain situations dictate their usage. Equipment with a very limited seal chamber area is a good candidate for an outside seal. Economics will also impact the use of a single outside seal. To prevent the need for very expensive metallurgies in highly corrosive applications, outside seals are sometimes employed. Because all the metal parts are on the atmospheric side of the seal, only the seal faces and secondary sealing members are exposed to the corrosive product. If outside seals must be used, always use a balanced seal design.

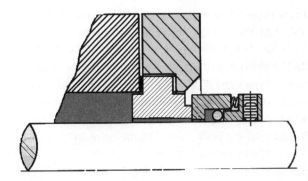

Figure 17. Single outside seal. (*Courtesy of Durametallic Corp.*)

Double Seals

Double seals are used when the product being sealed is incompatible with a single-seal design. As discussed earlier, toxic or hazardous chemicals require special considerations and must be handled with a multiple-seal arrangement. Highly corrosive products can also be safely contained with a double-seal arrangement. The primary purpose of the double seal is to isolate the sealed fluid from the atmosphere, and create an environment in which a mechanical seal can survive. This is accomplished by using two seals that operate in a different fluid, called a barrier fluid. The barrier fluid is there to provide clean, noncorrosive lubrication to the seal faces. To assure that the seals are being lubricated by the barrier fluid, the pressure of the barrier fluid is maintained at 15–25 psig higher than the product in the seal chamber area. This weepage requires that the barrier fluid be chemically compatible with the product.

There are two different arrangements for a double seal; the *back-to-back* arrangement (Figure 18) and the *face-to-face* arrangement (Figure 19). The advantage of the back-to-back arrangement is that none of the metal seal parts are exposed to the product. This is an ideal arrangement for highly corrosive chemicals. The major disadvantage to this arrangement is that it will not take pressure reversals. Under upset conditions, should the barrier pressure be lost, the inboard seal (left seal on Figure 18) would be pushed open, exposing the seal, and possibly the environment, to the product.

Face-to-face double-seal arrangements are designed to accommodate pressure reversals. Should the barrier pressure be lost, the only effect to the seal would be that the seal faces are lubricated by the product instead of the barrier

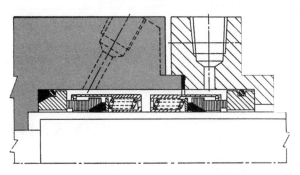

Figure 18. Double seal—back-to-back. (*Courtesy of John Crane, Inc.*)

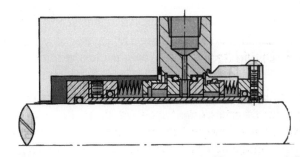

Figure 19. Face-to-face dual seal. (*Courtesy of Durametallic Corp.*)

fluid. While this condition could only be tolerated by the seal for a short time, the potentially major failure of the back-to-back arrangement has been avoided. The disadvantage of the face-to-face arrangement is that one of the seals must operate in the product. This virtually eliminates its use in highly corrosive products.

Tandem Seals

Tandem seals are used when a single seal design is compatible with the product, but emissions to the environment must be severely limited, such as with VOCs. The classical arrangement of a tandem seal is two seals in series, as shown in Figure 20. The primary seal (on the left) functions just like a single seal in that it contains all the pressure and is lubricated by the product. The secondary seal (on the right) serves as a backup seal to the primary and is lubricated by a nonpressurized barrier fluid. The secondary seal also serves as a second "defense" for containing emissions. Under normal conditions, weepage from the primary seal is contained in the nonpressurized barrier fluid and typically vented off to a flare system. In the event that the primary seal should fail, the secondary seal is in place to contain the pressure, and the product, until a controlled shutdown of the equipment can be arranged.

Face-to-face tandem seal arrangements are also available, and are identical to Figure 19. The only difference between the double and tandem seal in this case, with the exception of their purpose, is that the double seal has a pressurized barrier fluid and the tandem seal has a nonpressurized barrier fluid.

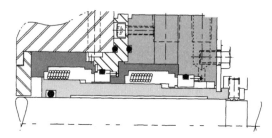

Figure 20. Tandem seal. (*Courtesy of John Crane, Inc.*)

BASIC DESIGN PRINCIPLES

Seal Balance Ratio

As a means of quantifying the amount, or percent, of balance for a mechanical seal, a ratio can be made between the seal face area above the balance diameter versus the total seal face area. This ratio can also be expressed as the area of the seal face exposed to hydraulic closing force versus the total seal face area. In either case, referring to Figure 21, the mathematical expression for the balance ratio of an inside seal design is:

$$\text{Balance Ratio} = \frac{\text{OD}^2 - \text{BD}^2}{\text{OD}^2 - \text{ID}^2}$$

where: OD = seal face outside diameter
       ID = seal face inside diameter
       BD = balance diameter of the seal

As a general rule of thumb, balanced seal designs use a balance ratio of 0.75 for water and nonflashing hydrocar-

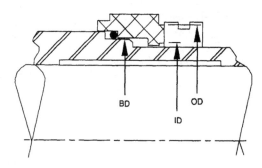

Figure 21. Balance ratio for inside seal. (*API-682. Courtesy of American Petroleum Institute.*)

bons. For flashing hydrocarbons, which are fluids with a vapor pressure greater than atmospheric pressure at the service temperature, the balance ratio is typically 0.80 to 0.85. Unbalanced seal designs typically have a ratio of 1.25 to 1.35.

Seal Hydraulics

As previously discussed, all mechanical seals are affected by hydraulic forces due to the pressure in the seal chamber. Both mechanical and hydraulic forces act on the seal face, and are shown in Figure 22. The total net forces acting on the seal face can be expressed as:

$$F_{Total} = F_c - F_o + F_{sp}$$

where: F_c = closing force
F_o = opening force
F_{sp} = mechanical spring force

The hydraulic closing force can be described mathematically as:

$$F_c = \frac{\pi}{4}(OD^2 - BD^2)(D_p)$$

where: OD = seal face outside diameter (in.)
BD = balance diameter (in.)
$D_p = \Delta P$ across seal face (psi)

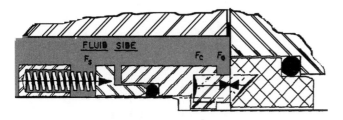

Figure 22. Mechanical seal force diagram. (*Courtesy of Durametallic Corp.*)

The mechanical closing force, or spring force, is expressed as F_{sp}. The amount of spring force is a function of the wire diameter, the number of springs, and the length of displacement. For mechanical seal designs, the range can be from 5 to 15 lbs of force per inch of OD circumference. As a general rule, values of 5 to 7 lbs are typically used. The hydraulic opening force can be expressed as:

$$F_o = \frac{\pi}{4}(OD^2 - ID^2)(D_p)(K)$$

where: OD = seal face outside diameter (in.)
ID = seal face inside diameter (in.)
$D_p = \Delta P$ across seal face (psi)
K = pressure drop factor

The pressure drop factor K shown in the opening force equation can be viewed as a percentage factor for quantifying the amount of differential pressure that is converted to opening force as the fluid migrates across the seal face. Seal faces act as pressure-reducing devices, and their shape has great impact on the K factor. Figure 23 shows various examples of K values and how they affect pressure drop [9]. A value of K = 1 (100%) is known as a *converging seal face*, where all of the differential pressure is used for opening force and the seal relies totally on spring force to remain closed. For a *diverging seal face,* K = 0 (0%), none of the differential pressure is used for opening force. For normal flat seal faces, where K = 0.5 (50%), half the differential pressure is used for opening force. This can also be expressed as a linear pressure drop across the seal face, and is commonly used for hydraulic calculations.

Fluid types also affect the pressure drop factor. For light hydrocarbons, the liquid generally flashes to a gas as it migrates across the seal face. As the liquid expands, it creates higher opening forces, and is expressed as K = 0.5 to 0.8.

As a general rule, K = 0.5 is used for nonflashing liquids, and K = 0.75 is used for flashing liquids.

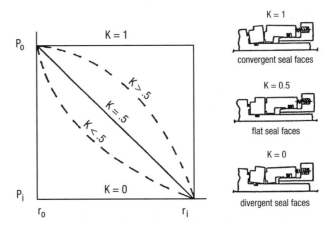

Figure 23. Pressure drop factor K.

Seal Face Pressure

Seal face pressure, or unit loading, is the most common term used when discussing the effects of chamber pressure on the seal faces. The total seal face pressure can be determined by dividing the F_{total} equation by the seal face area:

$$P_{Total} = \frac{F_c}{A} - \frac{F_o}{A} + \frac{F_{sp}}{A}$$

where: $A = \frac{\pi}{4}(OD^2 - ID^2)$

$$P_{Total} = \frac{\frac{\pi}{4}(OD^2 - BD^2)(D_p)}{\frac{\pi}{4}(OD^2 - ID^2)} - \frac{\frac{\pi}{4}(OD^2 - ID^2)(D_p)(K)}{\frac{\pi}{4}(OD^2 - ID^2)}$$
$$+ \frac{F_{sp}}{A}$$

$$P_{Total} = (B)(D_P) - (D_P)(k) + \frac{F_{SP}}{A}$$

$$P_{Total}\,(psi) = (D_p)(B - K) + P_{sp} \qquad [7, 9]$$

where: $D_p = \Delta P$ across seal face (psi)
$\quad$ B = seal balance ratio
$\quad$ K = pressure drop factor
$\quad$ P_{sp} = spring pressure (psi)

Basic Seal Lubrication

Basic seal lubrication occurs as the seal chamber pressure drives the fluid across the seal faces, forcing it through the asperities in the seal face caused by surface roughness, porosity, or waviness. Depending on the seal design and the type of lubricating fluid, one of three basic lubrication modes can be used (see Figure 24).

The most common mode of lubrication is *boundary lubrication*. In this case, the lubricant remains as a fluid all the way across the face, but is not visible on the low-pressure side. This mode is common when sealing oils, acids, and other non-flashing liquids in low- to medium-pressure applications.

The next lubrication mode is called *phase transformation*, and is characteristic of flashing liquids and high-pressure applications. As the fluid migrates across the face, heat generated from fluid shear or rubbing contact elevates the fluid temperature above the vapor pressure, causing a phase change from liquid to vapor. As discussed in the section on pressure drop factors, this vaporization can greatly increase opening forces and must be considered in the design process.

The third mode of lubrication is *full fluid film*, and is characterized by visible leakage on the low pressure side of the seal face. This lubrication mode is used primarily in high performance, noncontacting seals, where seal face wear is virtually eliminated.

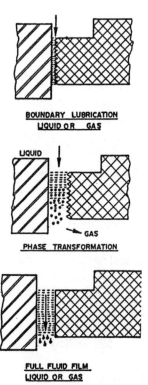

Figure 24. Lubrication modes. (*Courtesy of Durametallic Corp.*)

MATERIALS OF CONSTRUCTION

Hardware material can be made from almost all available metals. These metals are governed by ASTM standards, and will not be discussed in this section.

Seal Face Materials

The old saying goes that the perfect seal face material would have the hardness of a diamond, the strength of alloy steel, the heat transfer abilities of a super conductor, and the self-lubricating properties of teflon. While a compromise to this "standard" is certainly necessary, there are a few seal face materials that come close to these ideal properties. The three best materials available are also the most common face materials used in the petrochemical industry: tungsten carbide, silicon carbide, and carbon-graphite.

Tungsten Carbide

Tungsten carbide is a very popular seal face material due to its good wear characteristic, good corrosion resistance, and durability. The material has a very high modulus of elasticity and is well suited for high-pressure applications where face distortion is a problem. Tungsten carbide typically comes in two grades: nickel bound or cobalt bound. The user must be aware that all tungsten carbides are not created equal. Mechanical seal grade tungsten carbide should have very fine grains, with a maximum of 6%–8% binder material. Depending on the function, some commercial grade tungsten carbides can have as much as 50% binder material.

For mechanical seal applications, tungsten carbide coatings should also be avoided because of problems with coating cracks. The best material configuration for tungsten carbide is a solid homogeneous ring.

Silicon Carbide

Silicon carbide has many of the same good qualities as tungsten carbide, but also contains superior hardness, thermal conductivity, and a very low friction coefficient. If not for the brittle nature of silicon carbide, it would indeed be the ideal seal face material. This material typically comes in two grades: reaction bonded and alpha sintered.

The primary characteristic of reaction-bonded silicon carbide is the 8%–12% free silicon found in the structure. Under conditions of seal face contact, this free silicon can vaporize, leaving behind free carbon atoms at the interface. This reduces the frictional heat at the faces, and promotes good wear characteristic. The disadvantage to reaction-bonded silicon carbide is also the free silicon, which reduces chemical resistance. Chemicals like caustic or hydrofluoric acid will leach out the free silicon and severely limit good seal performance.

Alpha sintered silicon carbide, on the other hand, has no free silicon in the material structure. This provides a seal face material that is virtually inert to any chemicals, but does not offer the superior wear characteristics found in reaction-bonded materials.

Carbon-Graphite

Carbon-graphite is by far the most commonly used seal face material, and is used almost exclusively as the wearing face of a mechanical seal. Carbon-graphite offers outstanding antifriction properties as well as exceptional chemical resistance. This material does have a low modulus of elasticity and is therefore susceptible to distortion in high-pressure applications. There are many hundreds of different carbon grades available, and like tungsten carbide, they are not all created equal. Reputable mechanical seal manufacturers expend great effort in evaluating and testing suitable mechanical seal carbon grades. To assure good reliable seal performance, stick to the grades offered by the seal manufacturers.

Seal Face Compatibility

There are a few general items to keep in mind when applying various seal face combinations:

- One of the seal face materials should be harder than the other, like tungsten carbide versus carbon-graphite.

- For abrasive services, both seal face materials should be harder than the abrasive particles, such as tungsten carbide versus silicon carbide.
- For fluids with a viscosity less than 0.4 cp, one of the seal face materials should be carbon-graphite.

Additionally, not all seal face materials work well together. Figure 25 outlines various seal face combinations for different types of fluids [4]. Figure 26 shows typical temperature limitations for common seal face materials [1].

	Ceramic	Stellite	Tung Car Nickel Bound	Silicon Car. React. Bond.	Silicon Car. Alpha Sint.	Bronze	Carbon
Ceramic	D	X	X	D	D	X	ABCD
Stellite	X	D	D	X	X	D	ACDE
Tungsten Carbide (Nickel Bound)	X	D	BCD	ABD	ABCD	BD	BCDE
Silicon Carbide (Reaction Bonded)	D	X	BD	ABD	ABD	X	ABDE
Silicon Carbide (Alpha Sintered)	D	X	BCD	ABD	ABCD	X	ABCD E
Bronze	X	D	BD	X	X	X	X
Carbon	ABCD	ACDE	BCD	ABDE	ABCD E	X	ABCD

Codes: A – Acids diluted and concentrated
        B – Water and water solution
        C – Caustics diluted and concentrated
        D – Oil and lubricants
        E – Dry running
        X – Not recommended

Figure 25. Seal face compatibility. (*Courtesy of Durametallic Corp.*)

MATERIALS	CONSTRUCTION	MAX TEMP	
		°F	°C
Stellite Face	Welded Stellite Face on Metal Ring	350	177
Tung-Car	Solid tungsten Carbide Ring	750	400
Bronze	Solid Leaded Bronze Ring	350	177
Ceramic	Solid Pure Ceramic Ring	350	177
Carbon	Solid Carbon-Graphic Ring	525	275
Silicon-Carbide	Solid Silicon Carbide Ring	800	427

Figure 26. Temperature limitations of seal face materials. (*Courtesy of Durametallic Corp.*)

Secondary Sealing Materials

Secondary sealing members must be made from materials that are capable of sealing between two different surfaces. Ideally, resilient materials such as elastomeric o-rings should be used whenever temperature and chemical conditions will allow. For more severe conditions, such as high-temperature or corrosive environments, specialized materials like pure graphite or teflon must be used. These materials offer no resilience and must be formed into different shapes, such as wedges or squares, and be mechanically energized. Figure 27 shows temperature limits for the most common secondary sealing materials [1].

Material	Form	Min Temp Limit °F	Max Temp Limit °F
Buna N (Nitrile)	O-Ring	-40	+225
Neoprene	O-Ring	-40	+225
Viton	O-Ring	0	+400
EPR/EPT	O-Ring	-40	+350
Kalrez	O-Ring	0	+500
Teflon	Square Ring	-100	+450
	Wedge/V-Ring	-100	+350
	Gasket	-100	+350
Glass Filled Teflon	Gasket	-175	+450
Pure Graphite	Square Ring	-450	+750
Spiral Wound	Gasket	-450	+1200
Sythetic Organic Fiber With Nitrile	Gasket	-175	+700

Figure 27. Temperature limitations of secondary seal materials. (*Courtesy of Durametallic Corp.*)

DESIRABLE DESIGN FEATURES

Much has been said about the sealing members of a mechanical seal. There are also many design features that should be considered for the mechanical seal hardware.

Gland Rings

The following is a list of design features that should be considered for mechanical seal gland rings:

- Gland rings should be designed to withstand maximum seal chamber design pressure, and should be sufficiently rigid to avoid any distortion that would impair reliable seal performance.
- A minimum pilot length of 0.125″ should be used to properly align the seal to the stationary housing. This consideration would not apply for cartridge seal designs using centering tabs.
- The seal flush should enter through the gland ring and be located directly over the seal faces. The minimum pipe tap diameter should be ½″ NPT.

- Confined gland ring gaskets should be used when space and design limitations permit. Examples would be o-rings or spiral wound gaskets.
- Gland rings should be bolted to the seal chamber with a minimum of four bolts.
- Throttle bushings should be of a nonsparking and nongalling material, with a minimum of 0.025″ diametrical clearance.
- The stationary seal face should be mounted in the gland ring with a circumferential o-ring or other flexible sealing element. Clamp-type arrangements should be avoided.

Sleeves

Metal sleeves are used with mechanical seals to both protect the equipment shaft, in the case of hook type sleeves, and provide a means of installing a seal as a complete unit, as with the cartridge sleeve design. The cartridge seal is the design of choice unless space or other design limitations exist. The following are desirable features for sleeve designs:

- Shaft sleeves should have a minimum wall thickness of 0.100″.
- Materials of construction should be stainless steel as a minimum.
- All shaft sleeves should be "relieved" on the I.D. for ease of installation.

- Shaft sleeves should be manufactured to tolerances suitable for a maximum sleeve run-out of 0.002″ TIR when mounted on the shaft.
- Shaft sleeve sealing members (o-rings, gaskets, etc.) should be located as close to the equipment internals as possible.
- Cartridge sleeve designs should be mechanically secured to the shaft and be capable of maintaining position at maximum discharge pressure. The sleeve should be positively driven with the use of drive keys or dog-point set screws.

EQUIPMENT CONSIDERATIONS

Equipment Checks

One of the most important considerations for reliable seal performance is the operating condition of the equipment. Many times, mechanical seal failures are a direct result of poor equipment maintenance. High vibration, misalignment, pipe strain, and many other detrimental conditions cause poor mechanical seal life. There are also several dimensional checks that are often overlooked.

Because half of the seal is rotating with the shaft, and the other half is fixed to a stationary housing, the dimensional relationships of concentricity and "squareness" are very important. The centrifugal pump is by far the most common piece of rotating equipment utilizing a mechanical seal. For this reason, the dimensional checks will be referenced to the shaft and seal chamber for a centrifugal pump.

Axial Shaft Movement

Axial shaft movement (Figure 28) can be measured by placing a dial indicator at the end of the shaft and gently tapping or pulling the shaft back and forth. The indicator movement should be no more than 0.010″ TIR [1].

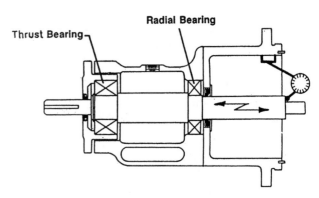

Figure 28. Checking pump for axial shaft movement. (*Courtesy of Durametallic Corp.*)

Radial Shaft Movement

There are two types of radial shaft movement (Figure 29) that need to be inspected. The first type is called *shaft deflection,* and is a good indication of bearing conditions and bearing housing fits. To measure, install the dial indicators as shown, and lift up or push down on the end of the shaft. The indicator movement should not exceed 0.002″ TIR [1].

The second type of radial shaft movement is called *shaft run-out,* and is a good way to check for a bent shaft condition. To measure, install the dial indicators as shown in

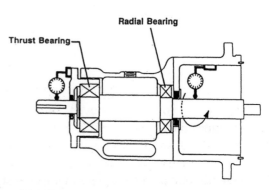

Figure 29. Checking pump for radial shaft movement. (*Courtesy of Durametallic Corp.*)

Seal Chamber Face Run-Out

Figure 29, and slowly turn the shaft. The indicator movement should not exceed 0.003″ TIR [1].

As stated earlier, because the stationary portion of the mechanical seal bolts directly to the pump case, it is very important that the face of the seal chamber be perpendicular to the shaft center-line. To check for "out-of-squareness," mount the dial indicator directly to the shaft, as shown in Figure 30. Sweep the indicator around the face of the seal housing by slowly turning the shaft. The indicator movement should not exceed 0.005″ TIR [1].

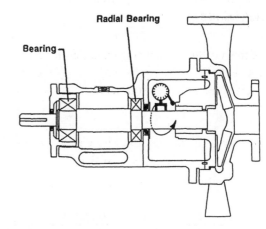

Figure 30. Checking pump for seal chamber face run-out. (*Courtesy of Durametallic Corp.*)

Seal Chamber Bore Concentricity

There are several stationary seal components that have close diametrical clearances to the shaft, such as the throttle bushing. For this reason, it is important for the seal chamber to be concentric with the pump shaft. Additionally, for gland ring designs with O.D. pilots, the outer register must also be concentric. To measure, install the dial indicators as shown in Figure 31. The indicator movement should not exceed 0.005″ TIR [1].

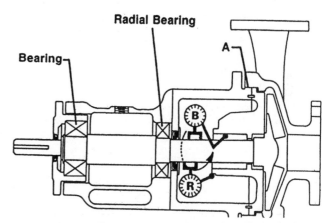

Figure 31. Checking pump for seal chamber bore concentricity. (*Courtesy of Durametallic Corp.*)

CALCULATING SEAL CHAMBER PRESSURE

The seal chamber pressure is a very important data point for selecting both the proper seal design and seal flush scheme. Unfortunately, the seal chamber pressure varies considerably with different pump designs and impeller styles. Some pumps operate with chamber pressures close to suction pressure, while others are near discharge pressure. The easiest and most accurate way to determine the seal chamber pressure on an existing pump is simply to measure it. Install a pressure gauge into a tapped hole in the seal chamber, and record the results with the pump running. The second most accurate method for determining seal chamber pressure is to consult the pump manufacturer. If neither of these two methods is feasible, there are ways of estimating the seal chamber pressure on standard pumps.

Single-Stage Pumps

The majority of overhung process pumps use wear rings and balance holes in the impeller to help reduce the pressure in the seal chamber. The estimated chamber pressure for this arrangement can be calculated with the following equation:

$$P_b = P_s + 0.15 (P_d - P_s)$$

where: P_b = seal chamber pressure (psi)
P_s = pump suction pressure (psi)
P_d = pump discharge pressure

In some special cases, where the suction pressure is very high, pump designers will remove the back wear ring and balance holes in an effort to reduce the loading on the thrust bearing. In this case, the seal chamber pressure (P_b) will be equal to discharge pressure (P_d).

Another common technique for reducing seal chamber pressure is to incorporate pump-out vanes in the back of the impeller. This is used primarily with ANSI-style pumps, and can be estimated with the equation:

$$P_b = P_s + 0.25 (P_d - P_s)$$

The final type of single-stage pump is the double suction pump, and for this pump design, the seal chamber pressure (P_b) is typically equal to the pump suction pressure (P_s).

Multistage Pumps

Horizontal multistage pumps typically are "between bearing" designs, and have two seal chambers. On the low-pressure end of the pump, the seal chamber pressure (P_b) is usually equal to the pump suction pressure (P_S). On the high-pressure end of the pump, a balance piston and pressure balancing line is typically incorporated to reduce both the thrust load and the chamber pressure. Assuming that the balance line is open and clear, the seal chamber pressure is estimated to be:

$$P_b = P_s + 75 \text{ psi}$$

The seal chamber pressure for vertical multistaged pumps can vary greatly with the pump design. The seal chamber can be located either in the suction stream or the discharge stream, and can incorporate a pressure balancing line, with a "breakdown" bushing, on high-pressure applications. Vertical pumps tend to experience more radial movement than horizontal pumps, and for this reason the effectiveness of the balancing line becomes a function of bushing wear. With so many variables, it is difficult to estimate the pressure in the sealing chamber. The best approach is either to measure the pressure directly, or consult the manufacturer.

SEAL FLUSH PLANS

As previously discussed, different seal designs are used in different seal arrangements to handle a vast array of different fluid applications. In every case, the seal must be provided with a clean lubricating fluid to perform properly. This fluid can be the actual service fluid, a barrier fluid, or an injected fluid from an external source. All these options require a different flushing or piping scheme. In an ef-

fort to organize and easily refer to the different seal flush piping plans, the American Petroleum Institute (API) developed a numbering system for centrifugal pumps that is now universally used [7]. The following is a brief discussion of the most commonly used piping schemes, and where they are used.

Single Seals

API Plan 11

The API Plan 11 (Figure 32) is by far the most commonly used seal flush scheme. The seal is lubricated by the pumped fluid, which is recirculated from the pump discharge nozzle through a flow restriction orifice and injected into the seal chamber. In this case, the chamber pressure must be less than the discharge pressure. The Plan 11 also serves as a means of venting gases from the seal chamber area as liquids are introduced in the pump. This is a very important function for preventing dry running conditions, and when at all possible, the piping should connect to the top of the gland. The API Plan 11 is primarily used for clean, cool services.

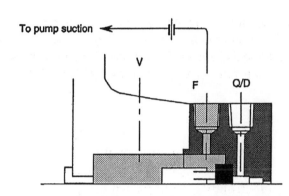

Figure 33. API Plan 13. (*API-682. Courtesy of American Petroleum Institute.*)

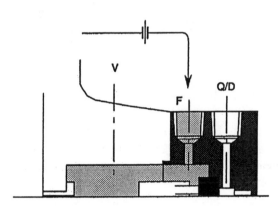

Figure 32. API Plan 11. (*API-682. Courtesy of American Petroleum Institute.*)

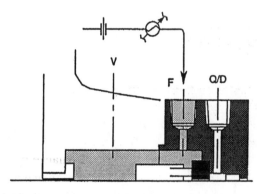

Figure 34. API Plan 21. (*API-682. Courtesy of American Petroleum Institute.*)

API Plan 13

The API Plan 13 (Figure 33) is very similar to the Plan 11, but uses a different recirculation path. For pumps with a seal chamber pressure equal to the discharge pressure, the Plan 13 seal flush is used. Here, the pumped fluid goes across the seal faces, out the top of the gland ring, through a restricting orifice, and into the pump suction. This piping plan is also used primarily in clean, cool applications.

API Plan 21

Figure 34 shows the arrangement for an API Plan 21. This plan is used when the pumpage is to hot to provide good lubrication to the seal faces. A heat exchanger is added in the piping to reduce the fluid temperature before it is in-

troduced into the seal chamber. The heat removal requirement for this plan can be quite high, and is not always the most economical approach.

API Plan 23

The API Plan 23 is also used to cool the seal flush, but utilizes a more economical approach. For the Plan 21, the fluid passes through the heat exchanger one time before it is injected into the seal chamber and then introduced back into the pumping stream. The Plan 23 (Figure 35) recirculates only the fluid that is in the seal chamber. In this case, an internal pumping device is incorporated into the seal design, which circulates a fixed volume of fluid out of the seal chamber through a heat exchanger and back to the gland ring. This greatly reduces the amount of heat removal nec-

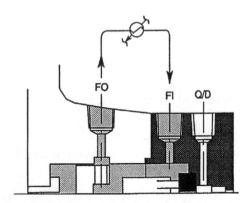

Figure 35. API Plan 23. (*API-682. Courtesy of American Petroleum Institute.*)

essary to achieve a certain flush temperature (and in the process industry, heat is always money). This flush plan is primarily used in boiler feed water applications.

API Plan 32

The last seal flush plan for single seals is API Plan 32, shown in Figure 36. Unlike the previous piping plans, this

arrangement does not use the pumped fluid as a seal flush. In this case, a clean, cool, compatible seal flush is taken from an external source and injected into the seal chamber. This arrangement is used primarily in abrasive slurry applications.

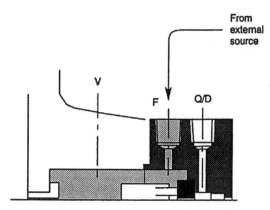

Figure 36. API Plan 32. (*API-682. Courtesy of American Petroleum Institute.*)

Tandem Seals

API Plan 52

Tandem seals consist of two mechanical seals. The primary, or inboard, seal always operates in the pumped fluid, and therefore utilizes the same seal flush plans as the single seals. The secondary, or outboard, seal must operate in a self-contained, nonpressurized barrier fluid. The API Plan 52, shown in Figure 37, illustrates the piping scheme for the barrier fluid. An integral pumping device is used to circulate the barrier fluid from the seal chamber up to the reservoir. Here, the barrier fluid is typically cooled and gravity-fed back to the seal chamber. The reservoir is generally vented to a flare header system to allow the primary seal weepage to exit the reservoir.

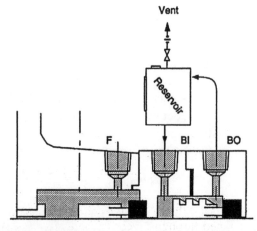

Figure 37. API Plan 52. (*API-682. Courtesy of American Petroleum Institute.*)

Double Seals

API Plan 53

Double seals also consist of two mechanical seals, but in this case, both seals must be lubricated by the barrier fluid. For this reason, the barrier fluid must be pressurized to 15 to 25 psi above the seal chamber pressure. The API Plan 53 (Figure 38) is very similar to Plan 52, with the exception of the external pressure source. This pressure source is typically an inert gas, such as nitrogen.

API Plan 54

The API Plan 54 (Figure 39) uses a pressurized, external barrier fluid to replace the reservoir arrangement. This piping arrangement is typically used for low-pressure applications where local service water can be used for the barrier fluid.

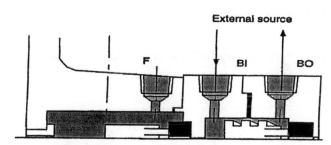

Figure 39. API Plan 54. (*API-682. Courtesy of American Petroleum Institute.*)

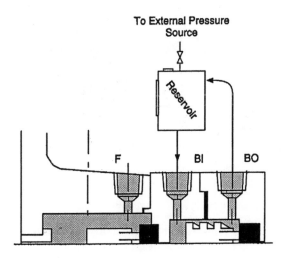

Figure 38. API Plan 53. (*API-682. Courtesy of American Petroleum Institute.*)

INTEGRAL PUMPING FEATURES

Many seal flush piping plans require that the seal lubricant be circulated through a heat exchanger or reservoir. While there are several different ways to accomplish this, the most reliable and cost-effective approach is with an integral pumping feature. There are many different types of integral pumping devices available, but the most common are the radial pumping ring and the axial pumping screw.

Radial Pumping Ring

The radial pumping ring, shown in Figure 40, operates much like a centrifugal pump. The slots in the circumference of the ring carry the fluid as the shaft rotates. When each slot, or volute, passes by the low-pressure area of the discharge tap located in the seal housing, the fluid is pushed out into the seal piping. This design is very dependent on peripheral speed, close radial clearance, and the configuration of the discharge port. A tangential discharge port will produce four times the flow rate, and two times the pressure, of a radial discharge tap. Higher-viscosity fluids also have a negative effect on the output of the radial pumping ring. Fluids with a viscosity higher than 150 SSU, such as oils, will reduce the flow rate by 0.25 and the pressure by 0.5. Figure 41 shows the performance of a typical radial pumping ring [1].

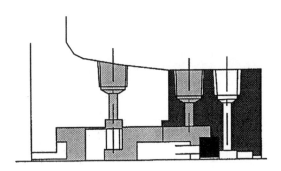

Figure 40. Radial pumping ring. (API-682. Courtesy of American Petroleum Institute.)

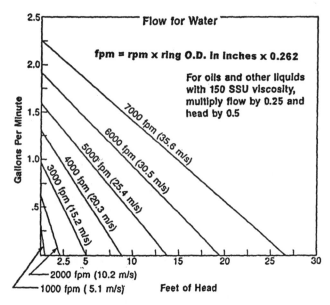

Figure 41. Typical radial pumping ring performance curve. (Courtesy of Durametallic Corp.)

Axial Pumping Screw

The axial pumping screw, shown on the outboard seal of Figure 42, consists of a rotating unit with an O.D. thread and a smooth walled housing. This is called a single-acting pumping screw. Double-acting screws are also available for improved performance and utilize a screw on both the rotating and stationary parts. Unlike the screw thread of a fastener, these screw threads have a square or rectangular cross-section and multiple leads. The axial pumping screw does have better performance characteristics than the radial pumping ring, but while gaining in popularity, the axial pumping screw is still primarily used on high-performance seal designs.

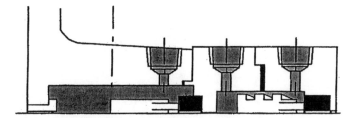

Figure 42. Axial pumping screw. (API-682. Courtesy of American Petroleum Institute.)

Piping Considerations

Integral pumping features are, by their design, very in-efficient flow devices. Consequently, the layout of the seal piping can have a great impact on performance. The following are some general rules for the piping:

• Slope the piping a minimum of ½″ per foot, and eliminate any areas where a vapor pocket could form.
• Provide a minimum of 10 pipe diameters of straight pipe length out of the seal housing before any directional changes are made.

• Minimize the number of fittings used. Eliminate elbows and tees where possible, using long radius bent pipe as a replacement.
• Where possible, utilize piping that is one size larger than the seal chamber pipe connections.

SEAL SYSTEM HEAT BALANCE

Excessive heat is a common enemy for the mechanical seal and to reliable seal performance. Understanding the sources of heat, and how to quantify the amount of heat, is essential for maintaining long seal life. The total heat load (Q_{Total}) can be stated as:

$$Q_{Total} = Q_{sgh} + Q_{hs}$$

where: Q_{total} = total heat load (btu/hr)
$\qquad Q_{sgh}$ = seal generated heat (btu/hr)
$\qquad Q_{hs}$ = heat soak (btu/hr)

Seal-generated heat is produced primarily at the seal faces. This heat can be generated by the shearing of the lubricant between the seal faces, contact between the different asperities in the face materials, or by actual dry running conditions at the face. Any one, or all, of these heat-generating conditions can take place at the same time. A heat value can be obtained from the following equation [2]:

$$Q_{sgh} = 0.077 \times P \times V \times f \times A$$

where $\;Q_{sgh}$ = seal generated heat (btu/hr)
$\qquad P$ = seal face pressure (psi)
$\qquad V$ = mean velocity (ft/min)
$\qquad f$ = face friction factor
$\qquad A$ = seal face contact area (in^2)

and

$$P_{Total} = (B)(D_P) - (D_P)(k) + \frac{F_{SP}}{A}$$

The face friction factor (f) is similar to a coefficient of friction, but is more tailored to the different lubricating conditions and fluids being sealed than to the actual material properties. The following values can be used as a general rule:

f = 0.05 for light hydrocarbons
f = 0.07 for water and medium hydrocarbons
f = 0.10 for oils

For a graphical approach to determining seal-generated heat values, see Figure 43 [2].

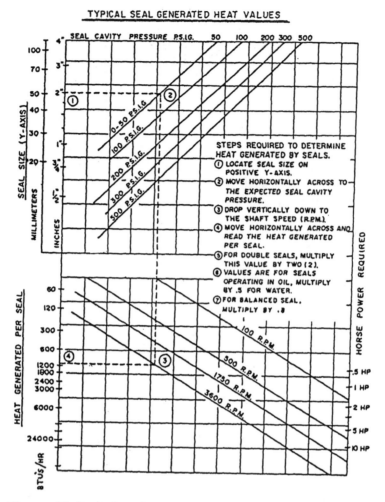

Figure 43. Typical seal-generated heat values. (*Courtesy of Durametallic Corp.*)

Heat soak (Q_{hs}) is the conductive heat flow that results from a temperature differential between the seal chamber and the surrounding environment. For a typical pump application, this would be the temperature differential between the chamber and the back of the pump impeller. Obviously, seals using an API Plan 11 or 13 would have no heat soak. But for Plans 21 or 23, where the seal flush is cooled, there would be a positive heat flow from the pump to the seal chamber. Radiant or convected heat losses from the seal chamber walls to the atmosphere are negligible.

There are many variables that affect heat soak values, such as materials, surface configurations, or film coefficients. In the case of a pump, heat can transfer down the shaft, or through the back plate, and can be constructed from several different materials. To make calculating the heat soak values simpler, a graphical chart, shown in Figure 44, has been provided which is specifically tailored for mechanical seals in centrifugal pump applications [8].

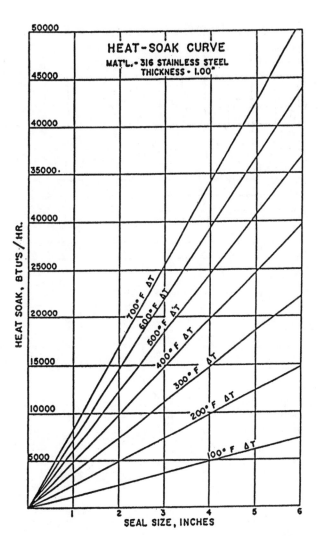

Figure 44. Heat-soak curve for 316 stainless steel. (*Courtesy of Durametallic Corp.*)

FLOW RATE CALCULATION

Once the heat load of the sealing system has been determined, removing the heat becomes an important factor. Seal applications with a high heat soak value will typically require a heat exchanger to help with heat removal. In this case, assistance from the seal manufacturer is required to size the exchanger and determine the proper seal flush flow rate. For simpler applications, such as those using API Plans 11, 13, or 32, heat removal requirements can be determined from a simple flow rate calculation. Using values for seal-generated heat and heat soak, when required, a flow rate value can be obtained from the following equation [2]:

$$\text{FLOW (gpm)} = \frac{Q_{\text{Total}}}{500 \times C_p \times \text{S.G.} \times \Delta T}$$

where: C_p = specific heat (btu/lb-°F)
S.G. = specific gravity
ΔT = allowable temperature rise (°F)

The allowable temperature rise (ΔT) will vary depending on the fluid being sealed. For fluids that are very close to the flashing temperature, the temperature rise should not exceed 5–10°F. For nonflashing fluids, the maximum allowable temperature rise is 20°F. Once the flow is determined, Figure 45 can be used to obtain the proper orifice size [5].

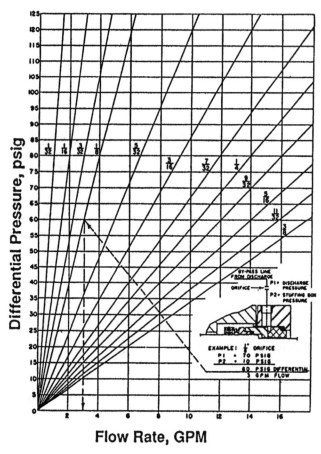

Figure 45. Graph of water flow through a sharp-edged orifice. (*Courtesy of Durametallic Corp.*)

REFERENCES

1. Durametallic Corporation, "Guide to Modern Mechanical Sealing," *Dura Seal Manual,* 8th Ed.
2. Durametallic Corporation, "Sizing and Selecting Sealing Systems," Technical Data SD-1162A.
3. Durametallic Corporation, "Dura Seal Pressure-Velocity Limits," Technical Data SD-1295C.
4. Durametallic Corporation, "Dura Seal Selection Guide," Technical Data SD-634-91.
5. Durametallic Corporation, "Dura Seal Recommendations for Fugitive Emissions Control in Refinery and Chemical Plant Service," Technical Data SD-1475.
6. Durametallic Corporation, "Achieve Fugitive Emissions Compliance with Dura Seal Designs," Technical Data SD-1482B.
7. API Standard 682, "Shaft Sealing Systems for Centrifugal and Rotary Pumps," 1st Ed., October 1994.
8. Adams, Bill, "Applications of Mechanical Seals in High Temperature Services," Mechanical Seal Engineering Seminar, ASME South Texas Section, October 1985.
9. Will, Thomas P., Jr., "Mechanical Seal Application Audit," Mechanical Seal Engineering Seminar, ASME South Texas Section, November 1985.

5

Pumps and Compressors

Bhabani P. Mohanty, Ph.D., Development Engineer, Allison Engine Company
E. W. McAllister, P.E., Houston, Texas

PUMP FUNDAMENTALS AND DESIGN

Pumps convert mechanical energy input into fluid energy. They are just the opposite of turbines. Many of the basic engineering facts regarding fluid mechanics are discussed in a separate chapter. This chapter pertains specifically to pumps from an engineering point of view.

Pump and Head Terminology

Symbol		Variable and Unit
Q	=	flow capacity (gallons/minute or gpm)
cfs	=	flow (ft³/second)
gpm	=	flow (gallons/minute)
P	=	pressure (psi)
bbl	=	barrel (42 gallons)
bpd	=	barrels/day
bph	=	barrels/hour
bhp	=	brake horsepower
whp	=	water horsepower
g	=	acceleration due to gravity (32.16 ft/sec)
T	=	torque (ft. lbs)
t	=	temperature (°F)
s	=	specific gravity of fluid
D	=	impeller diameter (inch)
e	=	pump efficiency (in decimal)
N	=	revolution per minute (rpm)
C	=	specific heat
H	=	total head (ft)
V	=	velocity (ft/sec)
A	=	area (sq. in.)
NPSH	=	net positive suction head (ft of water)
η	=	efficiency
ρ	=	density
γ	=	specific weight of liquid

Pump Design Parameters and Formulas

Following are the pump design parameters in detail:

Flow Capacity: The quantity of fluid discharged in unit time. It can be expressed in one of the following popular units: cfs, gpm, bph, or bpd.

$$gpm = 449 \times cfs = 0.7 \times bph \tag{4}$$

Head: This may also be called the specific energy, i.e., energy supplied to the fluid per unit weight. This quantity may be obtained through Bernoulli's equation. The head is the height to which a unit weight of the fluid may be raised by the energy supplied by the pump.

$$H = 2.31 \times P/s \tag{5}$$

The velocity head is defined as the pressure equivalent of the dynamic energy required to produce the fluid velocity.

Power: Energy consumed by the pump per unit time for supplying liquid energy in the form of pressure.

$$bhp = Q \times H \times s/(3{,}960 \times e) = Q \times P/(1{,}715 \times e) \tag{6}$$

Efficiency: The ratio of useful hydraulic work done to the actual work input. It consists of the product of three components: the volumetric efficiency, the hydraulic efficiency, and the mechanical efficiency.

$$\eta = \eta_v \eta_h \eta_m \tag{7}$$

The overall efficiency varies from 50% for small pumps to 90% for large ones.

Types of Pumps

Pumps fall into two distinct categories: *dynamic pumps* and *positive displacement pumps.*

Dynamic pumps are of two types: centrifugal and axial. They are characterized by the way in which energy is converted from the high liquid velocity at the inlet into pressure head in a diffusing flow passage. Dynamic pumps have a lower efficiency than positive displacement pumps. But their advantages lie in the output of relatively high flow rates compared to their sizes, and their low maintenance costs. They also operate at relatively higher speeds.

Positive displacement pumps are of several types, including reciprocating, rotary screw, and gear pumps. These

pumps operate by forcing a fixed volume of fluid from the inlet pressure section to the discharge section of the pump. In reciprocating pumps, this is done intermittently; and in others this is done continuously. These types of pumps are physically larger than the dynamic pumps for comparable capacity, and they operate at relatively lower speeds.

Table 1 shows major pump types, their characteristics, and their applications.

Source

Cheremisinoff, N. P., *Fluid Flow Pocket Handbook.* Houston: Gulf Publishing Co., 1984.

Table 1
Major Pump Types

Pump Type And Construction Style	Distinguishing Construction Characteristics	Usual Orientation	Usual No. Of Stages	Relative Maintenance Requirement	Comments
Dynamic					Capacity varies with head.
Centrifugal					Low to medium specific speed.
Horizontal					
Single stage overhung, process type	Impeller cantilevered beyond bearings.	Horizontal	1	Low	Most common style used in process services.
Two stage overhung, process type	2 impellers cantilevered beyond bearings.	"	2	Low	For heads above single stage capability.
Single stage impeller between bearings	Impeller between bearings; casing radially or axially split.	"	1	Low	For high flows to 330 m head.
Chemical	Casting patterns designed with thin sections for high cost alloys; small sizes.	"	1	Medium	Low pressure and temperature ratings.
Slurry	Large flow passages, erosion control features.	"	1	High	Low speed; adjustable axial clearance.
Canned	Pump and motor enclosed in pressure shell; no stuffing box.	"	1	Low	Low head-capacity limits for models used in chemical services.
Multistaged, horizontally split casing	Nozzles usually in bottom half of casing.	"	Multi	Low	For moderate temperature-pressure ratings.
Multistage, barrel type	Outer casing confines inner stack of diaphragms.	"	Multi	Low	For high temperature-pressure ratings.
Vertical					
Single stage process type	Vertical orientation.	Vertical	1	Low	Style used primarily to exploit low NPSH requirement.
Multistage, process type	Many stages, low head/stage.	"	Multi	Medium	High head capability, low NPSH requirement.
In-line	Arranged for in-line installation, like a valve	"	*1	Low	Allows low cost installation, simplified piping systems.
High Speed	Speeds to 380 rps, head to 1770 m	"	1	Medium	Attractive cost for high head/low flow.
Sump	Casing immersed in sump for installation convenience and priming ease.	"	1	Low	Low cost installation.
Multistage deep well	Very long shafts	"	Multi	Medium	Water well service with driver at grade.
Axial (Propeller)	Propeller shaped impeller, usually large size.	Vertical	1	Low	A few applications in chemical plants and refineries.
Turbine (Regenerative)	Fluted impeller; flow path like screw around periphery.	Horizontal	1,2	Med. to High	Low flow-high head performance. Capacity virtually independent of head.
Positive Displacement					
Reciprocating					
Piston, plunger	Slow speeds; valves, cylinders, stuffing boxes subject to wear.	Horizontal	1	High	Driven by steam engine cylinders or motors through crankcases.
Metering	Small units with precision flow control system	"	1	Medium	Diaphragm and packed plunger types.
Diaphragm	No stuffing box; can be pneumatically or hydraulically actuated.	"	1	High	Used for chemical slurries; diaphragms prone to failure.
Rotary					
Screw	1, 2 or 3 screw rotors	"	1	Medium	For high viscosity, high flow high pressure.
Gear	Intermeshing gear wheels	"	1	Medium	For high viscosity, moderate pressure, moderate flow.

Centrifugal Pumps

Centrifugal pumps are made in a variety of configurations, such as horizontal, vertical, radial split, and axial split casings. The choice is a function of hydraulic requirements such as the desired pressure and desired flow rate. Other important points to consider are the space limitations at the installation site and the ease of maintenance.

Centrifugal pumps are well-suited either for large volume applications or for large (volume/pressure) ratio applications at smaller volumes. The system variables that dictate the selection are fluid viscosity, fluid specific gravity, head requirement, and the system throughput. These pumps may be used in a series, a parallel, or a series-parallel combination to achieve system objectives.

Power needed to drive the pump is the sum of the power required to overcome all the losses in the system and the power needed to provide the required fluid energy at the system outlet.

Figure 1 shows the vector relationships at the pump inlet and outlet in terms of flow and velocity triangles. The pump head is the difference of total (static and dynamic) heads between the pump's inlet and outlet:

$$H = \frac{P_2 - P_1}{\rho g} + \frac{C_2^2 - C_1^2}{2g} \tag{1}$$

The virtual head (theoretical maximum head for a given set of operating conditions) is given by:

$$H = \frac{u_2 C_2 \cos \alpha_2 - u_1 C_1 \cos \alpha_1}{g} \tag{2}$$

Liquid entering a pump usually moves along the impeller, in which case $\alpha_1 = 90$ degrees and the above relation becomes:

$$H = \frac{u_2 C_2 \cos \alpha_2}{g} = \frac{u_2^2}{g}\left(1 - \frac{w_2}{u_2}\cos\beta_2\right)$$
$$= \frac{u_2^2}{g} - \frac{V}{g(2\pi r_2 b)\tan\beta_2} \tag{3}$$

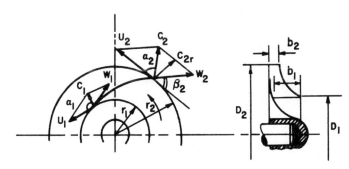

Figure 1. Flow and velocity triangles for centrifugal impellers [6].

where V represents the volumetric rate of flow through the pump and $(2\pi r_2 b)$ represents the cross-section of the flow leaving the impeller.

Source

Cheremisinoff, N. P., *Fluid Flow Pocket Handbook.* Houston: Gulf Publishing Co., 1984.

Net Positive Suction Head (NPSH) and Cavitation

The net positive suction head (NPSH) represents any extra energy added to unit weight of liquid. Mathematically,

$$NPSH = \frac{V_e^2}{2g} = \frac{p_a - p_v - \gamma z_s}{\gamma} \qquad (8)$$

where p_a is the atmospheric pressure, p_v is the vapor pressure of liquid, γ is the specific weight of the liquid, z_s is the suction head, and V is the reference velocity. The NPSH should always be above the vapor pressure of the liquid being pumped. This is important for safe and reliable pump operation. NPSH is given in feet of head above the vapor pressure required at the pump centerline. The NPSH requirement increases as the pump capacity increases. Hence, it is important to consider the range of flow requirements during the pump selection time. However, too much of an operational margin is not good either, because the pump efficiency at the low end of the design range will be lower too. The available NPSH should be about 10 to 15 percent higher than the required NPSH, no more and no less.

When a liquid flows into a space where its pressure is reduced to vapor pressure, it boils and vapor packets develop in it. These bubbles are carried along until they meet a region of higher pressure, where they collapse. This is called *cavitation*, and creates a very high localized pressure that causes pitting of the region. Cavitation lowers efficiency of the fluid machine. It is always accompanied by noise and vibrations. The cavitation parameter σ is defined as:

$$\sigma = \frac{p - p_v}{\rho V^2 / 2} \qquad (9)$$

where p is the absolute pressure at the point, p_v is the vapor pressure of liquid, ρ is the density of the liquid, and V is the reference velocity. This parameter is a nondimensional parameter similar to the pressure coefficient. A hydraulic system is designed to prevent cavitation. The following points should be remembered when addressing cavitation:

1. Avoid low pressures if at all possible. Pressurize the supply tank.
2. Reduce the fluid temperature.
3. Use a larger pipe diameter, and reduce minor losses in the pipe.
4. Use special cavitation-resistant materials or coatings.
5. Small amounts of air entrained into the fluid systems reduce the amount of cavitation damage.
6. The available NPSH should always be more than the required NPSH.

Pumping Hydrocarbons and Other Fluids

It should be remembered that the NPSH specification by a manufacturer is for use with cold water. It does not change much for small changes in water temperature. But for hydrocarbons, these values may be lowered to account for the slower vapor release properties of such complex organic liquids.

The head developed by a pump is independent of the liquid being pumped. (The required horsepower, of course, is dependent on the fluid's specific gravity.) Because of this independence, pump performance curves from water tests are applicable to other Newtonian fluids such as gasoline or alcohol.

The above independence is not true for high-viscosity fluids. Therefore, correction factors have been experimentally established for certain high-viscosity fluids. These correction factors (in limited viscosity and size ranges) may be applied to water curves under the underlying assumptions.

Recirculation

A recirculation problem is just the opposite of a cavitation problem. Cavitation occurs when a pump is forced to operate at a flow rate higher than the intended rate. But if the pump is operated at a rate considerably lower than the one it was designed for, this causes *recirculation*. It results in noise and vibration because the fluid energy is reduced through fluid shear and internal friction.

Pumping Power and Efficiency

A pump runner produces work QγH, where H is the pump head. The energy added by the pump can be determined by writing Bernoulli's equation using the entry and exit points of the pump:

$$h_A = \frac{p_{exit}}{\rho} - \frac{p_{inlet}}{\rho} + \frac{v_{exit}^2}{2g_c} - \frac{v_{inlet}^2}{2g_c} + z_{exit} - z_{inlet} \qquad (10)$$

The pump output is measured in *water horsepower,* or whp; whereas the input horsepower to the pump shaft is called the *brake horsepower,* or bhp. These two are related by:

$$bhp = \frac{whp}{\eta_p} \qquad (11)$$

The difference between the two is called the *friction horsepower,* or fhp:

$$fhp = (bhp - whp)$$

Large horsepower pumps are usually powered by three-phase induction motors, whose synchronous speed is:

$$n = \frac{120f}{N_{Poles}} \qquad (12)$$

Knowing the number of poles N_{Poles} (always an even number), and the frequency f in cycles/second (Hz), one can calculate the synchronous speed. The actual speed is about 4 to 8 percent lower than the synchronous speed. This difference is characterized as *slip*. To obtain the pump operating speed, one could use either gear or belt drives.

Specific Speed of Pumps

The *specific speed* of a homologous unit is widely used as the criterion for selection of a pump for a specific purpose. It can be used to avoid cavitation or to select the most economical pump for a given system layout. The specific speed of a series is defined for the point of best efficiency—one that delivers unit discharge at unit head—and is given by:

$$N_s = \frac{NQ^{1/2}}{H^{3/4}} \qquad (13)$$

Centrifugal pumps have low specific speeds, mixed flow pumps have higher values, and axial flow pumps have even higher values. The specific speed of centrifugal pumps may vary between 1,000 to 15,000, but values above 12,000 are considered impractical. Whenever possible, a value below 8,500 is usually recommended. Note that the relation above is dimensional in nature. Hence the value of N_s depends on the units of discharge and head involved. The above values were given in U.S. customary units, where Q is in *gallons per minute* and H is the NPSH in *feet of fluid*.

Pump Similitude

Two geometrically similar pumps are said to be homologous when

$$\frac{Q}{ND^3} = Constant$$

where the flow rate Q is equal to $C_d A \sqrt{2gH}$, C_d being the discharge coefficient, A is a reference area, and H is the head. Rearranging terms in the equation above, the homologous condition may also be specified as:

$$\frac{H}{N^2D^2} = Constant$$

A systematic Buckingham's Pi Theorem analysis of the functional form for the pump characteristics shows that the nondimensional parameters for a pump may be expressed as:

$$f\left(\frac{Q}{ND^3}, \frac{H}{D}, \frac{g}{N^2D}\right) = 0 \qquad (14)$$

One might draw as many conclusions about pump similitude from the above functional relationship as there are rearrangements that can be made of the terms. As an example, because power is proportional to γQH, the nondimensional power may be derived to be

$$\left(\frac{power}{\rho N^3 D^5}\right)$$

Example 1

A pump delivers 400 gal/min at 3,000 rpm. How much will it deliver at 2,500 rpm?

Solution. Within design limits of the pump, the discharge will be approximately proportional to the speed, $Q_2/Q_1 = N_2/N_1$. Hence $Q_2 = Q_1 \cdot N_2/N_1 = 400 \cdot 2,500/3,000 = 333.33$ gal/min. Similarly, for a given pump design, the head varies as the square of speed, and the power as the cube of speed.

Example 2

Develop a relation for power P in terms of speed N and impeller diameter D.

Solution: From our pump similitude analysis earlier:

$$Q = k_1 ND^3 \text{ and } H = k_2 N^2D^2$$

From our knowledge in fluids:

$$P = \gamma QH = \gamma k_1 k_2 N^3 D^5 = k_3 \gamma N^3 D^5 \qquad (15)$$

Performance Curves

The power required to run a pump depends on two purposes: first, to overcome all the losses in the related flow circuits; and second, to supply the energy to the fluid for the specified task. The first part of the power requirement (to overcome losses) is called the brake horsepower (bhp), and it is the absolute minimal required power even when the volumetric flow rate is zero. The various losses that it accounts for include mechanical friction losses in various components, frictional losses at the impeller, losses due to fluid turbulence, and leakage losses. With increasing flow rate requirements, the bhp increases even for zero head. Figure 2 shows the head-discharge relations.

Figure 3 typifies a manufacturer's performance curve of a pump. In a carpet plot like this one, for any two given values, the rest of the pumping variables may be computed by interpolation and careful extrapolation.

Figure 4 shows the effect of speed change on the pump's performance. Like Figures 2 and 3, this is again a typical sample. For actual curves, one must consult the pump's manufacturer.

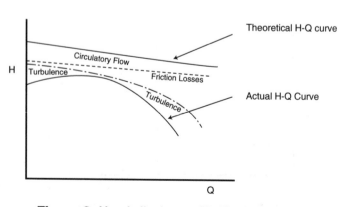

Figure 2. Head-discharge (H-Q) relationship.

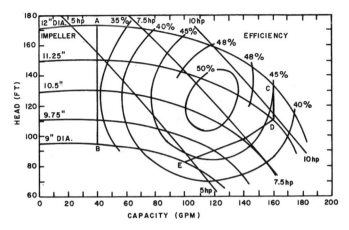

Figure 3. Characteristics of a centrifugal pump [6].

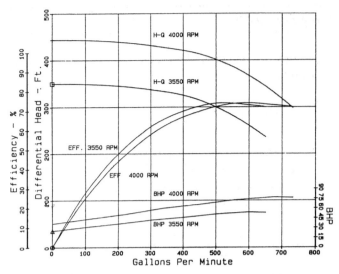

Figure 4. Centrifugal pump performance data for speed change [2].

Sources

Cheremisinoff, N. P., *Fluid Flow Pocket Handbook.* Houston: Gulf Publishing Co., 1984.

McAllister, E. W. (Ed.), *Pipe Line Rules of Thumb Handbook,* 3rd Ed. Houston: Gulf Publishing Co., 1993.

Series and Parallel Pumping

Pumps may be operated in series, in parallel, or in any series-parallel combination. When connected in series, the total available head is the sum total of all heads at a given rate of flow. When connected parallel, the total flow is the sum total of all flows at a given value of head. In other words, series pumping is called *pressure additive;* and parallel pumping is called *flow additive.* (See Figures 5 and 6.)

The same rules are applied to determine the total head or flow calculations for any series-parallel combination

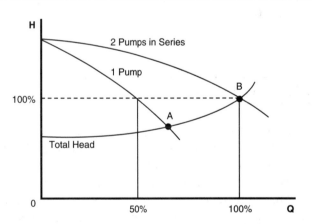

Figure 6. Effect of parallel pumping on H-Q curve.

of pumps. When connected in series, one pump's discharge is connected to another's suction; but pumps in parallel share a common suction and a common discharge line. It is important to match the pumps that are in parallel so that they develop the same head at the same flow rate.

Multiple units are often used to allow a variety of pumping conditions to be met without throttling, and therefore wasting, power.

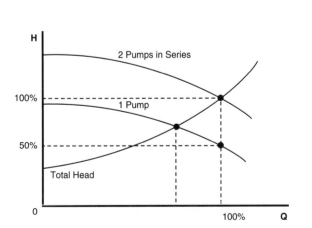

Figure 5. Effect of series pumping on H-Q curve.

Design Guidelines

Pipework

The fluids section in this book should be consulted for pipe calculations, but a general guideline for calculating pipe bore size is:

$$\text{Bore}_{pipe} = \sqrt{\frac{\text{lit}/\text{min}}{0.145}}\ \text{mm} = \sqrt{\frac{\text{gal}/\text{min}}{20}}\ \text{inches} \qquad (16)$$

Table 2 lists common flow velocities in commercial use and Table 3 lists recommended flow velocities based on the fluid's specific gravity SG.

Operating Performance

Table 4 provides a summary of operating limitations of different types of pumps.

Table 2
Common Flow Velocities in Commercial Practice

Medium	Piping		Velocity m/sec	Remarks
Water	Piston pumps	suction	0.5—1.5	
		delivery	1.0—2.0	
	Feed pumps of steam boilers	suction	0.3—0.5	The piping is selected according to its length. A lower velocity is chosen with long piping, a higher velocity for short piping (does not apply to delivery of liquids containing solid particles)
		delivery	2.0—2.5	
	Piping for condensate and sludge	suction	0.3—0.5	
		delivery	1.0—2.0	
	Gravel, sand and other drifted substances	delivery	0.5—2.0	
	Piping for cold water	delivery	1.0—3.0	
	Piping for cold water	suction		
	up to 50mm diameter	max.	1.0	
	up to 100mm diameter	max.	1.3	
	up to 200mm diameter	max.	1.7	
	above 200mm diameter		2.0	
	Piping for cooling water	suction	0.7—1.5	
		delivery	1.0—2.0	
	Pressure water		15.0—30.0	In special cases up to 5m/sec
	Delivery piping in mines		1.0—1.5	
	Supply to water turbines		3.0	Low head
			3.0—7.0	High head
	Municipal water mains, main feed pipings		1.0—2.0	
	municipal water system		0.5—1.2	Normally 0.6 to 0.7m/sec
Petrol Oils	Benzol, gas oil		1.0—2.0	According to viscosity
	Heavy		0.5—2.0	
Hydro-carbons		suction	0.3—0.8	
Air	Low-pressure piping		12—15	
	High-pressure piping		20—25	
Steam	For steam lines up to 4MPa		20—40	Velocities must be chosen economically according to the length of the piping
	High-pressure steam		30—60	
	Superheated steam		39—80	
	Low-pressure heating steam		10—15	
	Exhaust steam		15—40	

Table 3
Recommended Flow Velocities Based on Fluids SG

Pipe Diameter		Power Driven Pumps						Turbine Driven Pumps					
		SG = 1.0		SG = 0.75		SG = 0.5		SG = 1.0		SG = 0.75		SG = 0.5	
inch	mm	ft/sec	m/sec	ft/sec	m/sec	ft/sec	m/sec	ft/sec	m/sec	ft/sec	m/sec	ft/sec	m/sec
2	50	6.00	1.80	7.00	2.10	7.5	2.30	5.00	1.50	5.50	1.70	6.00	1.80
3	75	7.00	2.10	8.00	2.40	8.5	2.60	5.50	1.70	6.00	1.80	6.50	2.00
4	100	8.00	2.40	9.00	2.75	10.0	3.00	6.00	1.80	6.50	2.00	7.00	2.15
6	150	9.00	2.75	10.00	3.00	12.0	3.65	6.50	2.00	7.00	2.15	8.00	2.40
8	200	10.00	3.00	11.25	3.40	13.0	4.00	6.75	2.10	7.50	2.30	8.50	2.60
10	250	11.00	3.25	12.00	3.65	14.0	4.20	7.00	2.15	7.75	2.35	9.00	2.75
12	300	11.50	3.50	12.50	3.80	14.5	4.40	7.00	2.15	8.00	2.40	9.25	2.80
14	350	11.75	3.60	13.00	4.00	15.0	4.50	7.00	2.15	8.00	2.40	9.50	2.90
16 and over	400	12.00	3.65	13.00	4.00	15.0	4.60	7.00	2.15	8.00	2.40	9.50	2.90

Table 4
Summary of Operating Performances of Pumps

Pump Type/Style	Solids Tolerance	Capacity (dm³/s)	Capacity (gph)	Max. Head (m)	Max. Head (ft)	Typical NPSH/Req. (m)	Typical NPSH/Req. (ft)	Max. Kinematic Viscosity (mm²/s)	Max. Kinematic Viscosity (in.²/s)	Efficiency (%)	Max. Pumping Temperature (°C)	Max. Pumping Temperature (°F)
Centrifugal												
Horizontal												
Single-stage overhung	MH	1~320	950~3×10^5	150	492	2~6	6.6~20	650	1.01	20~80	455	851
2-stage overhung	MH	1~75	950~7.1×10^4	425	1394	2~6.7	6.6~22	430	0.67	20~75	455	851
Single-stage impeller between bearings	MH	1~2500	950~2.4×10^6	335	1099	2~7.6	6.6~25	650	1.01	30~90	205~455	401~851
						1.2~6	3.9~20	650	1.01	20~75	205	401
Chemical	MH	65	6.2×10^4	73	239	1.5~7.6	4.9~25	650	1.01	20~80	455	851
Slurry	H	65	6.2×10^4	120	394	2~6	6.6~20	430	0.67	20~70	540	1004
Canned	L	0.1~1250	95~1.2×10^6	1500	4922	2~6	6.6~20	430	0.67	65~90	205~260	401~500
Multi. horiz. split	M	1~700	950~6.7×10^5	1675	5495	2~6	6.6~20	430	0.67	40~75	455	851
Multi., barrel type	M	1~550	950~5.2×10^5	1675	5495	2~6	6.6~20	430	0.67	40~75	455	851
Vertical												
Single-stage process	M	1~650	950~6.2×10^5	245	804	0.3~6	1~20	650	1.01	20~85	345	653
Multistage	M	1~5000	950~4.8×10^6	1830	6004	0.3~6	1~20	430	0.67	25~90	260	500
						2~6	1~20	430	0.67	20~80	260	500
In-line	M	1-750	950~7.1×10^5	215	705	2.4~12	7.9~39.8	109	0.17	10~50	260	500
High-speed	L	0.3~25	285~2.4×10^4	1770	5807	0.3~6.7	1~22	430	0.67	40~75	—	—
Sump	MH	1~45	950~4.3×10^4	60	197	0.3~6	1~20	430	0.67	30–75	205	401
Multi. deep well	M	0.3~25	285~2.4×10^4	1830	6004	~2	6.6	650	1.01	65~85	65	149
Axial (propeller)	H	1~6500	950~6.2×10^6	12	39.4	2~2.5	6.6~8.2	109	0.17	55~85	120	248
Turbine (regenerative)	M	0.1~125	95~1.2×10^5	760	2493	3.7	12	1100	1.71	55~85	290	554
Positive Displacement				(kPa)	(psi)							
Reciprocating												
Piston, plunger	M	1~650	950~6.2×10^5	345000	50038	4.6	15.1	1100	1.71	~20	290	554
Metering	L	0~1	0~950	517000	74985	3.7	12.1	750 (ssu)	1.16 (ssu)	~20	260	500
Diaphragm	L	0.1~6	95~5.7×10^3	34500	5004	~3	~9.8	150×10^6	150×10^6	50~80	260	500
Rotary												
Screw	M	0.1~125	95~1.2×10^5	20700	3002	~3	~9.8	150×10^6	150×10^6	50~80	345	653
Gear	M	0.1~320	95~3.0×10^5	3400	493							

MH—moderately high; H—high; M—medium; L—low

Vacuum Systems

Given two parameters, (a) rate of evacuation and (b) final pressure to be realized, determine the size and number of pumps required for the job. The evacuation time is related to the system parameters by the following equation:

$$\text{Time}_{minutes} = \frac{2.3V}{Q_e} \log_{10} \frac{P_1}{P_2} \qquad (17)$$

where: V = volume to be evacuated (liters), Q_e = effective pump speed (liters/min), and P_1 and P_2 are initial and final pressures (torr).

Note that mechanical pumps are not suitable for producing partial pressures less than about 10^{-3} torr. Highly specialized pumps are used for the purpose.

Noise

The following formula can be used to estimate the general noise level of a centrifugal pump within plus or minus 2 dB.

$$20 \log_{10} \left\{ \frac{P_0\, Q}{10000\, N_s \omega\, a_2^2\, r_o} \right\} \qquad (18)$$

where P_0 is total pressure rise across the impeller (bar), Q is the flow rate (m^3/min), N_s is the specific speed, ω is the angular speed (rad/sec), and a_2 is the impeller outside radius (cm); r_o is width of impeller at outlet (cm).

Noise may also be controlled by changing operating conditions of the pump. The blade frequency noise levels can change about 10 dB by changing the operating point of the pump. Minimum noise levels usually do happen at flow rates around 15% above the design operating point.

Sources

Cheremisinoff, N. P., *Fluid Flow Pocket Handbook.* Houston: Gulf Publishing Co., 1984.

Warring, R. H., *Pumping Manual,* 7th Ed. Houston: Gulf Publishing Co., 1984.

Reciprocating Pumps

A. How a Reciprocating Pump Works

A reciprocating pump is a positive displacement mechanism with liquid discharge pressure being limited only by the strength of the structural parts. Liquid volume or capacity delivered is constant regardless of pressure, and is varied only by speed changes.

Characteristics of a GASO reciprocating pump are 1) positive displacement of liquid, 2) high pulsations caused by the sinusoidal motion of the piston, 3) high volumetric efficiency, and 4) low pump maintenance cost.

B. Plunger or Piston Rod Load

Plunger or piston "rod load" is an important power end design consideration for reciprocating pumps. Rod load is the force caused by the liquid pressure acting on the face of the piston or plunger. This load is transmitted directly to the power frame assembly and is normally the limiting factor in determining maximum discharge pressure ratings. This load is directly proportional to the pump guage discharge pressure and proportional to the square of the plunger or piston diameter.

Occasionally, allowable liquid end pressures limit the allowable rod load to a value below the design rod load. IT IS IMPORTANT THAT LIQUID END PRESSURES DO NOT EXCEED PUBLISHED LIMITS.

C. Calculations of Volumetric Efficiency

Volumetric efficiency (E_v) is defined as the ratio of plunger or piston displacement to liquid displacement. The volumetric efficiency calculation depends upon the internal configuration of each individual liquid body, the piston size, and the compressibility of the liquid being pumped.

D. Tools for Liquid Pulsation Control, Inlet and Discharge

Pulsation Control Tools ("PCT", often referred to as "dampeners" or "stabilizers") are used on the inlet and discharge piping to protect the pumping mechanism and associated piping by reducing the high pulsations within the liquid caused by the motions of the slider-crank mechanism. A properly located and charged pulsation control tool may reduce the length of pipe used in the acceleration head equation to a value of 5 to 15 nominal pipe diameters. Figure 5 is a suggested piping system for power pumps. The pulsation control tools are specially required to compensate for inadequately designed or old/adapted supply and discharge systems.

E. Acceleration Head

Whenever a column of liquid is accelerated or decelerated, pressure surges exist. This condition is found on the inlet side of the pump as well as the discharge side. Not only can the surges cause vibration in the inlet line, but they can restrict and impede the flow of liquid and cause incomplete filling of the inlet valve chamber. The magnitude of the surges and how they will react in the system is impossible to predict without an extremely complex and costly analysis of the system. Since the behavior of the natural frequencies in the system is not easily predictable, as much of the surge as possible must be eliminated at the source. Proper installation of an inlet pulsation control PCT will absorb a large percentage of the surge before it can travel into the system. The function of the PCT is to absorb the "peak" of the surge and feed it back at the low part of the cycle. The best position for the PCT is in the liquid supply line as close to the pump as possible, or attached to the blind flange side of the pump inlet. In either location, the surges will be dampened and harmful vibrations reduced.

RECIPROCATING PUMPS FLOW CHARACTERISTICS

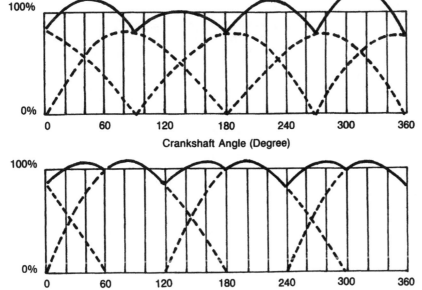

DUPLEX DOUBLE-ACTING

Average Flow — 100%
Maximum Flow — 100% + 24%
Minimum Flow — 100% − 22%
Total Flow Var. — 46%

TRIPLEX SINGLE-ACTING

Average Flow — 100%
Maximum Flow — 100% + 6%
Minimum Flow — 100% − 17%
Total Flow Var. — 23%

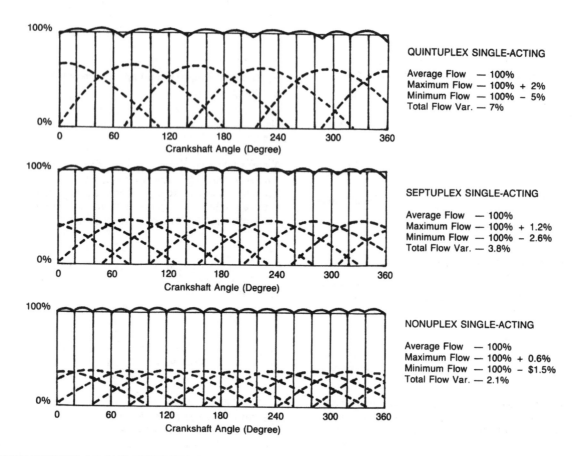

QUINTUPLEX SINGLE-ACTING

Average Flow   — 100%
Maximum Flow — 100% + 2%
Minimum Flow — 100% – 5%
Total Flow Var. — 7%

SEPTUPLEX SINGLE-ACTING

Average Flow   — 100%
Maximum Flow — 100% + 1.2%
Minimum Flow — 100% – 2.6%
Total Flow Var. — 3.8%

NONUPLEX SINGLE-ACTING

Average Flow   — 100%
Maximum Flow — 100% + 0.6%
Minimum Flow — 100% – $1.5%
Total Flow Var. — 2.1%

REQUIRED FORMULAE AND DEFINITIONS

Acceleration Head

$$h_a = \frac{LVNC}{Kg} \qquad V = \frac{GPM}{(2.45)(D)}$$

Where

h_a = Acceleration head (in feet)
L = Length of liquid supply line (in line)
V = Average velocity in liquid supply line (in fps)
N = Pump speed (revolutions per minute)
C = Constant depending on the type of pump
  C = 0.200 for simplex double-acting
    = 0.200 for duplex single-acting
    = 0.115 for duplex double-acting
    = 0.066 for triplex single or double-acting
    = 0.040 for quintuplex single or double-acting
    = 0.028 for septuplex, single or double-acting
    = 0.022 for nonuplex, single or double-acting
K = Liquid compressibility factor
  K = 2.5 For relatively compressible liquids
      (ethane, hot oil)
  K = 2.0 For most other hydrocarbons
  K = 1.5 For amine, glycol and water
  K = 1.4 For liquids with almost no compressibility
      (hot water)
g = Gravitational constant = 32.2 ft/sec²
d = Inside diameter of pipe (inches)

Stroke

One complete uni-directional motion of piston or plunger. Stroke length is expressed in inches.

Pump Capacity (Q)

The capacity of a reciprocating pump is the total volume through-put per unit of time at suction conditions. It includes both liquid and any dissolved or entrained gases at the stated operating conditions. The standard unit of pump capacity is the U.S. gallon per minute.

Pump Displacement (D)

The displacement of a reciprocating pump is the volume swept by all pistons or plungers per unit time. Deduction for piston rod volume is made on double acting piston type pumps when calculating displacement. The standard unit of pump displacement is the U.S. gallon per minute.

For single-acting pumps: $D = \dfrac{Asnm}{231}$

For double-acting piston pumps:

$$D - \frac{(2A - a)\,snm}{231}$$

Where

A = Plunger or piston area, square inch
a = Piston rod cross-sectional area, square inch
    (double-acting pumps)
s = Stroke length, inch
n = RPM of crankshaft
m = Number of pistons or plungers

Plunger or Piston Speed (v)

The plunger or piston speed is the average speed of the plunger of piston. It is expressed in feet per minute.

$$v = \frac{ns}{6}$$

Pressures

The standard unit of pressure is the pound force per square inch.

Discharge Pressure (Pd)—The liquid pressure at the centerline of the pump discharge port.

Suction Pressure (Ps)—The liquid pressure at the centerline of the suction port.

Differential pressure (Ptd)—The difference between the liquid discharge pressure and suction pressure.

Net Positive Suction Head Required (NPSHR)—The amount of suction pressure, over vapor pressure, required by the pump to obtain satisfactory volumetric efficiency and prevent excessive cavitation.

The pump manufacturer determines (by test) the net positive suction head required by the pump at the specified operating conditions.

NPSHR is related to losses in the suction valves of the pump and frictional losses in the pump suction manifold and pumping chambers. Required NPSHR does not include system acceleration head, which is a system-related factor.

Slip (S)

Slip of a reciprocating pump is the loss of capacity, expressed as a fraction or percent of displacement, due to leaks past the valves (including the back-flow through the valves caused by delayed closing) and past double-acting pistons. Slip does not include fluid compressibility or leaks from the liquid end.

Power (P)

Pump Power Input (Pi)—The mechanical power delivered to a pump input shaft, at the specified operating conditions. Input horsepower may be calculated as follows:

$$Pi = \frac{Q \times Ptd}{1714 \times \eta p}$$

Pump Power Output (Po)—The hydraulic power imparted to the liquid by the pump, at the specified operating conditions. Output horsepower may be calculated as follows:

$$Po = \frac{Q \times Ptd}{1714}$$

The standard unit for power is the horsepower.

Efficiencies (n)

Pump Efficiency (ηp) (also called pump mechanical efficiency)—The ratio of the pump power output to the pump power input.

$$\eta p = \frac{Po}{Pi}$$

Volumetric Efficiency (ηv)—The ratio of the pump capacity to displacement.

$$\eta v = \frac{Q}{D}$$

Plunger Load (Single-Acting Pump)

The computed axial hydraulic load, acting upon one plunger during the discharge portion of the stroke is the plunger load. It is the product of plunger area and the guage discharge presssure. It is expressed in pounds force.

Piston Rod Load (Double-Acting Pump)

The computed axial hydraulic load, acting upon one piston rod during the forward stroke (toward head end) is the piston rod load.

It is the product of piston area and discharge pressure, less the product of net piston area (rod area deducted) and suction pressure. It is expressed in pounds force.

Liquid pressure $\left(\frac{pounds}{square\ inch}\right)$ or(psi) =

Cylinder area (square inches) =
(3.1416) x (Radius(inches))² =
$\frac{(3.1416) \times (diameters(inches))^2}{4}$

Cylinder force (pounds)—pressure (psi) x area (square inches)

Cylinder speed or average liquid velocity through piping (feet/second) = $\frac{flow\ rate\ (gpm)}{2.448 \times (inside\ diameter\ (inch))^2}$

Reciprocating pump displacement (gpm) = $\frac{rpm \times displacement\ (cubic\ in/revolution)}{231}$

Pump input horsepower—see horsepower calculations

Shaft torque (foot-pounds) = $\frac{horsepower \times 5252}{shaft\ speed\ (rpm)}$

Electric motor speed (rpm) = $\frac{120 \times frequency\ (Hz)}{number\ of\ poles}$

Three phase motor horsepower (output) = $\frac{1.73 \times ampers \times volts \times efficiency \times power\ factor}{746}$

Static head of liquid (feet) = $\frac{2.31 \times static\ pressure\ (psig)}{specific\ gravity}$

Velocity head of liquid (feet) =

$$\frac{\text{liquid velocity}^2}{g = 32.2 \text{ ft/sec}^2}$$

Absolute viscosity (centipoise) =
specific gravity x Kinematic viscosity (centistrokes)

Kinematic viscosity (centistrokes) =

$$0.22 \times \text{saybolt viscosity (ssu)} = \frac{180}{\text{Saybolt viscosity (ssu)}}$$

Absolute pressure (psia) =
Local atmospheric pressure + gauge pressure (psig)

Gallon per revolution =

$$\frac{\text{Area of plunger (sq in) x length of stroke(in) x number of plungers}}{231}$$

Barrels per day = gal/rev x pump speed (rpm) x 34.3

TABLE 5 Water Compressibility

Compressibility Factor $\beta t \times 10^{-6}$ = Contraction in Unit Volume Per Psi Pressure
Compressibility from 14.7 Psia, 32 F to 212 F and from Saturation Pressure Above 212 F

TEMPERATURE

Pressure Psia	0 C 32 F	20 C 63 F	40 C 104 F	60 C 140 F	80 C 176 F	100 C 212 F	120 C 248 F	140 C 284 F	160 C 320 F	180 C 356 F	200 C 392 F	220 C 428 F	240 C 464 F	260 C 500 F	280 C 536 F	300 C 572 F	320 C 608 F	340 C 644 F	360 C 680 F
200	3.12	3.06	3.06	3.12	3.23	3.40	3.66	4.00	4.47	5.11	6.00	7.27							
400	3.11	3.05	3.05	3.11	3.22	3.39	3.64	3.99	4.45	5.09	5.97	7.21							
600	3.10	3.05	3.05	3.10	3.21	3.39	3.63	3.97	4.44	5.07	5.93	7.15	8.95						
800	3.10	3.04	3.04	3.09	3.21	3.38	3.62	3.96	4.42	5.04	5.90	7.10	8.85	11.6					
1000	3.09	3.03	3.03	3.09	3.20	3.37	3.61	3.95	4.40	5.02	5.87	7.05	8.76	11.4	16.0				
1200	3.08	3.02	3.02	3.08	3.19	3.36	3.60	3.94	4.39	5.00	5.84	7.00	8.68	11.2	15.4				
1400	3.07	3.01	3.01	3.07	3.18	3.35	3.59	3.92	4.37	4.98	5.81	6.95	8.61	11.1	15.1	23.0			
1600	3.06	3.00	3.00	3.06	3.17	3.34	3.58	3.91	4.35	4.96	5.78	6.91	8.53	10.9	14.8	21.9			
1800	3.05	2.99	3.00	3.05	3.16	3.33	3.57	3.90	4.34	4.94	5.75	6.87	8.47	10.8	14.6	21.2	36.9		
2000	3.04	2.99	2.99	3.04	3.15	3.32	3.56	3.88	4.32	4.91	5.72	6.83	8.40	10.7	14.3	20.7	34.7		
2200	3.03	2.98	2.98	3.04	3.14	3.31	3.55	3.87	4.31	4.89	5.69	6.78	8.33	10.6	14.1	20.2	32.9	86.4	
2400	3.02	2.97	2.97	3.03	3.14	3.30	3.54	3.85	4.29	4.87	5.66	6.74	8.26	10.5	13.9	19.8	31.6	69.1	
2600	3.01	2.96	2.96	3.02	3.13	3.29	3.53	3.85	4.28	4.85	5.63	6.70	8.20	10.4	13.7	19.4	30.5	61.7	
2800	3.00	2.95	2.96	3.01	3.12	3.28	3.52	3.83	4.26	4.83	5.61	6.66	8.14	10.3	13.5	19.0	29.6	57.2	238.2
3000	3.00	2.94	2.95	3.00	3.11	3.28	3.51	3.82	4.25	4.81	5.58	6.62	8.08	10.2	13.4	18.6	28.7	53.8	193.4
3200	2.99	2.94	2.94	3.00	3.10	3.27	3.50	3.81	4.23	4.79	5.55	6.58	8.02	10.1	13.2	18.3	27.9	51.0	161.0
3400	2.98	2.93	2.93	2.99	3.09	3.26	3.49	3.80	4.22	4.78	5.53	6.54	7.96	9.98	13.0	17.9	27.1	48.6	138.1
3600	2.97	2.92	2.93	2.98	3.09	3.25	3.48	3.79	4.20	4.76	5.50	6.51	7.90	9.89	12.9	17.6	26.4	45.4	122.4
3800	2.96	2.91	2.92	2.97	3.08	3.24	3.47	3.78	4.19	4.74	5.47	6.47	7.84	9.79	12.7	17.3	25.8	44.5	110.8
4000	2.95	2.90	2.91	2.97	3.07	3.23	3.46	3.76	4.17	4.72	5.45	6.43	7.78	9.70	12.5	17.1	25.2	42.8	101.5
4200	2.95	2.90	2.90	2.96	3.06	3.22	3.45	3.75	4.16	4.70	5.42	6.40	7.73	9.62	12.4	16.8	24.6	41.3	93.9
4400	2.94	2.89	2.90	2.95	3.05	3.21	3.44	3.74	4.14	4.68	5.40	6.36	7.68	9.53	12.2	16.5	24.1	40.0	87.6
4600	2.93	2.83	2.89	2.94	3.05	3.20	3.43	3.73	4.13	4.66	5.37	6.32	7.62	9.44	12.1	16.3	23.6	38.8	82.3
4800	2.92	2.87	2.88	2.94	3.04	3.20	3.42	3.72	4.12	4.64	5.35	6.29	7.57	9.36	12.0	16.0	23.2	37.6	77.7
5000	2.91	2.87	2.87	2.93	3.03	3.10	3.41	3.71	4.10	4.63	5.32	6.25	7.52	9.28	11.8	15.8	22.7	36.6	73.9
5200	2.90	2.85	2.87	2.92	3.02	3.18	3.40	3.69	4.09	4.61	5.30	6.22	7.47	9.19	11.7	15.6	22.3	35.6	70.3
5400	2.90	2.85	2.86	2.91	3.01	3.17	3.39	3.68	4.07	4.59	5.27	6.19	7.41	9.12	11.6	15.3	21.9	34.6	66.9

EXAMPLE: Find the volumetric efficiency of a reciprocating pump with the following conditions:

Type of pump	3 in diam plunger x 5 in stroke triplex
Liquid pumped	Water
Suction pressure	Zero psig
Discharge pressure	1785 psig
Pumping temperature	140 F
c	127.42 cu in
d	35.343 cu in
S	.02

Find βt from Table of Water Compressibility (Table 5).

βt = .00000305 at 140 F and 1800 psia Calculate volumetric efficiency:

Vol. Eff. =

$$\frac{1 - \left[P_{td} \beta t \left(1 + \frac{c}{d} \right) \right]}{1 - P_{td} \beta t} - S =$$

$$\frac{1 - \left[(1785 - 0)(.00000305) \right] \left[1 + \frac{127.42}{35.343} \right]}{1 - (1785 - 0)(.00000305)} - .02$$

= .96026

= 96 per cent

Specific gravity (at 60°F) =
$$\frac{141.5}{131.5 + \text{API gravity (degree)}}$$

Bolt clamp load (lb) =
0.75 x proof strength (psi) x tensile stress area (in²)

Bolt torque (ft-lb) =
$\frac{0.2 \text{ (or } 0.15)}{12}$ x nominal diameter in inches x bolt clamp load (lb)
0.2 for dry
0.15 for lubricated, plating, and hardened washers

Calculating Volumetric Efficiency for Water

The volumetric efficiency of a reciprocating pump, based on capacity at suction conditions, using table of water compressibility, shall be calculated as follows:

Vol. Eff. =

$$1 - \frac{\text{Ptd } \beta t \left(1 + \dfrac{c}{d}\right)}{1 - \text{Ptd } \beta t} - S$$

Where

βt = Compressibility factor at temperature t (degrees Fahrenheit or centigrade). (See Tables 5 and 6).
c = Liquid chamber volume in the passages of chamber between valves when plunger is at the end of discharge stroke in cubic inches

d = Volume displacement per plunger in cubic inches
Ptd = Discharge pressure minus suction pressure in psi
S = Slip, expressed in decimal value

Calculating Volumetric Efficiency For Hydrocarbons

The volumetric efficiency of a reciprocating pump based on capacity at suction conditions, using compressibility factors for hydrocarbons, shall be calculated as follows:

$$\text{Vol. Eff.} = 1 - \left[S - \frac{c}{d} \left(1 - \frac{\rho d}{\rho s} \right) \right]$$

Where

c = Fluid chamber volume in the passages of chamber between valves, when plunger is at the end of discharge strike, in cubic inches
d = Volume displacement per plunger, in cubic inches
P = Pressure in psia (P_s = suction pressure in psia; P_d = discharge pressure in psia)

P_c = Critical pressure of liquid in psia
P_r = Reduced pressure
$$\frac{\text{Actual pressure in psia}}{\text{Critical pressure in psia}} = \frac{P}{P_c}$$

P_{rs} = Reduced suction pressure = $\dfrac{P_s}{P_c}$

P_{rd} = Reduced discharge pressure = $\dfrac{P_d}{P_c}$

TABLE 6 Water Compressibility

Compressibility Factor $\beta t \times 10^{-6}$ = Contraction in Unit Volume Per Psi Pressure
Compressibility from 14.7 Psia at 68 F and 212 F and from Saturation Pressure at 392 F

Pressure Psia	Temperature 20 C 68 F	100 C 212 F	200 C 392 F	Pressure Psia	Temperature 20 C 68 F	100 C 212 F	200 C 392 F
6000	2.84	3.14	5.20	22000	2.61	2.42	3.75
7000	2.82	3.10	5.09	23000	2.59	2.38	3.68
8000	2.80	3.05	4.97	24000	2.58	2.33	3.61
9000	2.78	3.01	4.87	25000	2.57	2.29	3.55
10000	2.76	2.96	4.76	26000	2.56	2.24	3.49
11000	2.75	2.92	4.66	27000	2.55	2.20	3.43
12000	2.73	2.87	4.57	28000	2.55	2.15	3.37
13000	2.71	2.83	4.47	29000	2.54	2.11	3.31
14000	2.70	2.78	4.38	30000	2.53	2.06	3.26
15000	2.69	2.74	4.29	31000	2.52	2.02	3.21
16000	2.67	2.69	4.21	32000	2.51	1.97	3.16
17000	2.66	2.65	4.13	33000	2.50	1.93	3.11
18000	2.65	2.60	4.05	34000	2.49	1.88	3.07
19000	2.64	2.56	3.97	35000	2.49	1.84	3.03
20000	2.63	2.51	3.89	36000	2.48	1.79	2.99
21000	2.62	2.47	3.82				

S = Slip expressed in decimal value
t = Temperature, in degrees Rankine
= Degrees F + 460 (t_s = suction temperature in degrees Rankine; t_d = discharge temperature in degrees Rankine)
T_c = Critical temperature of liquid, in degrees Rankine (See Table 7)
T_r = Reduced temperature
= $\dfrac{\text{actual temp. in degrees Rankine}}{\text{critical temp. in degrees Rankine}}$
= $\dfrac{t}{T_c}$ (See Fig. 7)
T_{rs} = Reduced suction temperature
= $\dfrac{t_s}{T_c}$

T_{rd} = Reduced discharge temperature
= $\dfrac{t_d}{T_c}$
Vol. Eff. = Volumetric efficiency expressed in decimal value.
= $\dfrac{1}{1}$ x   x 62.4 = density of liquid in lb per cu ft
s = Density in lb per cu ft at suction pressure
d = Density in lb per cu ft at discharge pressure
= Expansion factor of liquid
$\dfrac{1}{1}$ = Characteristic constant in grams per cubic centimeter for any one liquid which is established by density measurements and the corresponding values of   (See Table 7).

TABLE 7

Carbon Atoms	Name	T_c Degrees Rankine	P_c Lb Per Sq In	pl/wl Grams Per cc
1	Methane	343	673	3.679
2	Ethane	550	717	4.429
3	Propane	666	642	4.803
4	Butane	766	544	5.002
5	Pentane	847	482	5.128
6	Hexane	915	433	5.216
7	Heptane	972	394	5.285
8	Octane	1025	362	5.340
9	Nonane	1073	332	5.382
10	Decane	1114	308	5.414
12	Dodecane	1185	272	5.459
14	Tetradecane	1248	244	5.483

Example: Find volumetric efficiency of the previous reciprocating pump example with the following new conditions:

Type of Pump	3 inch dia plunger x 5 inch stroke triplex
Liquid pumped	Propane
Suction temperature	70 F
Discharge temperature	80 F
Suction pressure	242 psig
Discharge pressure	1911 psig

Find density at suction pressure:

$$T_{rs} = \frac{t_s}{T_c} = \frac{460 + 70}{666} = .795$$

$$P_{rs} = \frac{P_s}{P_c} = \frac{257}{642} = .4$$

$$\frac{\rho_1}{\omega_1} = 4.803 \text{ (From Table 7, propane)}$$

$$\omega = .1048 \text{ (From Fig. 7)}$$

$$\rho_s = \frac{\rho_1}{\omega_1} \times \omega \times 62.4$$

$$= 4.803 \times .1048 \times 62.4$$
$$= 31.4 \text{ lb per cu ft}$$

Find density at discharge pressure:

$$T_{rd} = \frac{t_d}{T_c} = \frac{460 + 80}{666} = .81$$

$$P_{rd} = \frac{P_d}{P_c} = \frac{1926}{642} = 3.0$$

$$\omega = .1089 \text{ (From Fig. 7)}$$

$$\rho_d = \frac{\rho_1}{\omega_1} \times \omega \times 62.4$$

$$= 4.803 \times .1089 \times 62.4$$
$$= 32.64 \text{ lb per cu ft}$$

Therefore

$$\text{Vol. Eff.} = 1 - \left[S - \frac{c}{d}\left(1 - \frac{\rho_d}{\rho_s}\right)\right]$$

$$= 1 - \left[.02 - \frac{127.42}{35.343}\left(1 - \frac{32.64}{31.4}\right)\right]$$

$$= .8376$$

$$= 83.76 \text{ per cent}$$

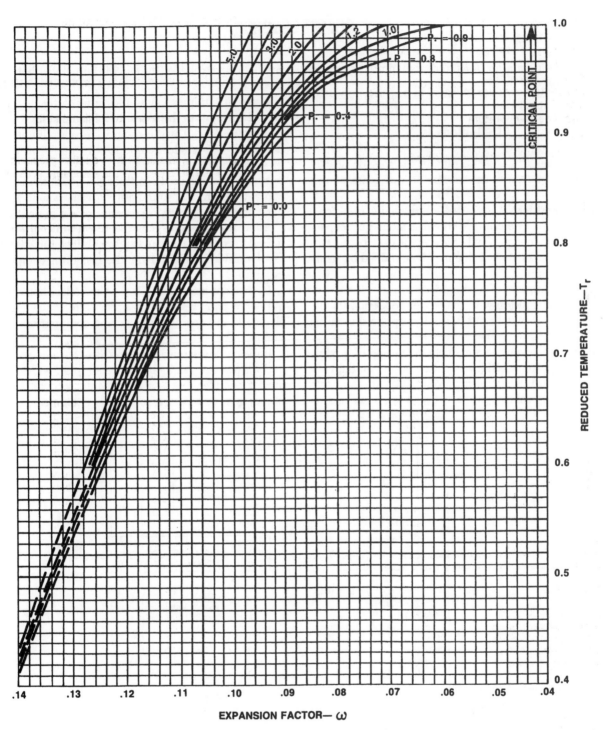

Figure 7. Thermal Expansion and Compressibility of Liquids. (*Reprinted with permission—Goulds Pumps.*)

Source (Reciprocating Pumps)

McAllister, E. W. (Ed.), *Pipe Line Rules of Thumb Hand-book,* 3rd Ed. Houston: Gulf Publishing Co., 1993.

COMPRESSORS*

The following data are for use in the approximation of horsepower needed for compression of gas.

Definitions

The "N value" of gas is the ratio of the specific heat at constant pressure (C_p) to the specific heat at constant volume (C_v).

If the composition of gas is known, the value of N for a gas mixture may be determined from the molal heat capacities of the components. If only the specific gravity of gas is known an approximate N value may be obtained by using the chart in Figure 1.

Piston Displacement of a compressor cylinder is the volume swept by the piston with the proper deduction for the piston rod. The displacement is usually expressed in cubic ft per minute.

Clearance is the volume remaining in one end of a cylinder with the piston positioned at the end of the delivery stroke for this end. The clearance volume is expressed as a percentage of the volume swept by the piston in making its full delivery stroke for the end of the cylinder being considered.

Ratio of compression is the ratio of the absolute discharge pressure to the absolute inlet pressure.

Actual capacity is the quantity of gas compressed and delivered, expressed in cubic ft per minute, at the intake pressure and temperature.

Volumetric efficiency is the ratio of actual capacity, in cubic ft per minute, to the piston displacement, in cubic ft per minute, expressed in percent.

Adiabatic horsepower is the theoretical horsepower required to compress gas in a cycle in which there is no transfer of sensible heat to or from the gas during compression or expansion.

Isothermal horsepower is the theoretical horsepower required to compress gas in a cycle in which there is no change in gas temperature during compression or expansion.

Indicated horsepower is the actual horsepower required to compress gas, taking into account losses within the compressor cylinder, but not taking into account any loss in frame, gear or power transmission equipment.

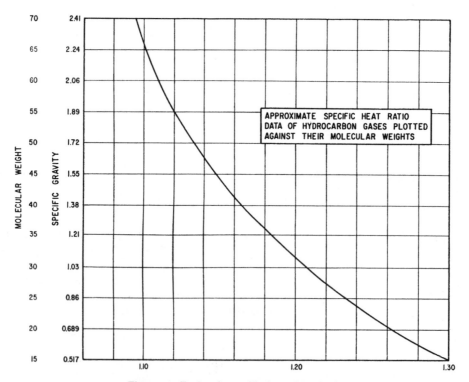

Figure 1. Ratio of specific heat (n-value).

---

*Reprinted from *Pipe Line Rules of Thumb Handbook,* 3rd Ed., E. W. McAllister (Ed.), Gulf Publishing Company, Houston, Texas, 1993.

Compression efficiency is the ratio of the theoretical horsepower to the actual indicated horsepower, required to compress a definite amount of gas. The efficiency, expressed in percent, should be defined in regard to the base at which the theoretical power was calculated, whether adiabatic or isothermal.

Mechanical efficiency is the ratio of the indicated horsepower of the compressor cylinder to the brake horsepower delivered to the shaft in the case of a power driven machine. It is expressed in percent.

Overall efficiency is the product, expressed in percent, of the compression efficiency and the mechanical efficiency. It must be defined according to the base, adiabatic isothermal, which was used in establishing the compression efficiency.

Piston rod gas load is the varying, and usually reversing, load imposed on the piston rod and crosshead during the operation, by different gas pressures existing on the faces of the compressor piston.

The maximum piston rod gas load is determined for each compressor by the manufacturer, to limit the stresses in the frame members and the bearing loads in accordance with mechanical design. The maximum allowed piston rod gas load is affected by the ratio of compression and also by the cylinder design; i.e., whether it is single or double acting.

Performance Calculations for Reciprocating Compressors

Piston Displacement

Single acting compressor:

$$P_d = [S_t \times N \times 3.1416 \times D^2]/[4 \times 1{,}728] \quad (1)$$

Double acting compressor without a tail rod:

$$P_d = [S_t \times N \times 3.1416 \times (2D^2 - d^2)]/[4 \times 1{,}728] \quad (2)$$

Double acting compressor with a tail rod:

$$P_d = [S_t \times N \times 3.1416 \times 2 \times (D^2 - d^2)]/[4 \times 1{,}728] \quad (3)$$

Single acting compressor compressing on frame end only:

$$P_d = [S_t \times N \times 3.1416 \times (D^2 - d^2)]/[4 \times 1{,}728] \quad (4)$$

where P_d = Cylinder displacement, cu ft/min
S_t = Stroke length, in.
N = Compressor speed, number of compression strokes/min
D = Cylinder diameter, in.
d = Piston rod diameter, in.

Volumetric Efficiency

$$E_v = 0.97 - [(1/f)r_p^{1/k} - 1]C - L \quad (5)$$

where E_v = Volumetric efficiency
f = ratio of discharge compressibility to suction compressibility Z_2/Z_1
r_p = pressure ratio
k = isentropic exponent
C = percent clearance
L = gas slippage factor

Let L = 0.3 for lubricated compressors
Let L = 0.07 for non lubricated compressors (6)

These values are approximations and the exact value may vary by as much as an additional 0.02 to 0.03.

Note: A value of 0.97 is used in the volumetric efficiency equation rather than 1.0 since even with 0 clearance, the cylinder will not fill perfectly.

Cylinder inlet capacity

$$Q_1 = E_v \times P_d \quad (7)$$

Piston Speed

$$PS = [2 \times S_t \times N]/12 \quad (8)$$

Discharge Temperature

$$T_2 = T_1(r_p^{(k-1)/k}) \quad (9)$$

where T_2 = Absolute discharge temperature °R
T_1 = Absolute suction temperature °R

Note: Even though this is an adiabatic relationship, cylinder cooling will generally offset the effect of efficiency.

Power

$$W_{cyl} = [144\, P_1 Q_1/33{,}000\, n_{cyl}] \times [k/(k-1)] \times [r_p^{k-1/k} - 1] \quad (10)$$

where n_{cyl} = efficiency
W_{cyl} = Cylinder horsepower

See Figure 2 "Reciprocating compressor efficiencies" for curve of efficiency vs pressure ratio. This curve includes a 95% mechanical efficiency and a valve velocity of 3,000 ft. per minute. Tables 1 and 2 permit a correction to be made to the compressor horsepower for specific gravity and low inlet pressure. While it is recognized that the efficiency is not necessarily the element affected, the desire is to modify the power required per the criteria in these figures. The efficiency correction accom-

plishes this. These corrections become more significant at the lower pressure ratios.

Inlet Valve Velocity

$$V = 288 \times P_d/A \qquad (11)$$

where V = Inlet valve velocity
     A = Product of actual lift and the valve opening periphery and is the total for inlet valves in a cylinder expressed in square in. (This is a compressor vendor furnished number.)

Example. Calculate the following:

Suction capacity
Horsepower
Discharge temperature
Piston speed

Given:

Bore = 6 in.
Stroke = 12 in.
Speed = 300 rpm
Rod diameter = 2.5 in.
Clearance = 12%
Gas = CO_2
Inlet pressure = 1,720 psia
Discharge pressure = 3,440 psia
Inlet temperature = 115°F

Calculate piston displacement using Equation 2.

$$P_d = 12 \times 300 \times 3.1416 \times [2(6)^2 - (2.5)^2]/1,728 \times 4$$
$$= 107.6 \text{ cfm}$$

Calculate volumetric efficiency using Equation 5. It will first be necessary to calculate f, which is the ratio of discharge compressibility to suction compressibility.

$$T_1 = 115 + 460 = 575°R$$

$$T_r = T/T_c$$
$$= 575/548$$
$$= 1.05$$

where T_r = Reduced temperature
    T_c = Critical temperature = 548° for CO_2
    T = Inlet temperature

$$P_r = P/P_c$$
$$= 1,720/1,071$$
$$= 1.61$$

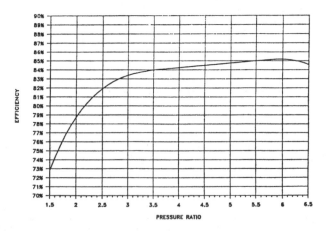

Reciprocating compressor efficiencies plotted against pressure ratio with a valve velocity of 3,000 fpm and a mechanical efficiency of 95 percent.

Figure 2. Reciprocating compressor efficiencies.

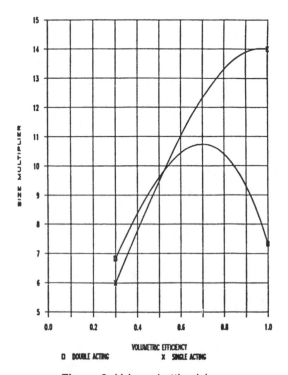

Figure 3. Volume bottle sizing.

where P_r = Reduced pressure
P_c = Critical pressure = 1,071 psia for CO_2
P = Suction pressure

From the generalized compressibility chart (Figure 4):

$Z_1 = 0.312$

Determine discharge compressibility. Calculate discharge temperature by using Equation 9.

$r_p = 3,440/1,720$
$= 2$

$k = cp/cv = 1.3$ for CO_2

$T_2 = 575 \times [2.0^{(1.3-1)/1.3}]$
$= 674.7°R$
$= 674.7 - 460 = 214°F$

$T_r = 674.7/548 = 1.23$

$P_r = 3,440/1,071 = 3.21$

From the generalized compressibility chart (Figure 4):

$Z_2 = 0.575$

$f = 0.575/1.61 = 1.843$

$E_v = 0.97 - [(1/1.843) \times 2.0^{1/1.3} - 1] \times 0.12 - 0.05$
$= 0.929$
$= 93\%$

Note: A value of 0.05 was used for L because of the high differential pressure.

Calculate suction capacity.

$Q_1 = E_v \times P_d$
$= 0.93 \times 107.6$
$= 100.1$ cfm

Calculate piston speed.

$PS = [2 \times S_t \times N]/12$
$= [2 \times 12 \times 300]/12$
$= 600$ ft/min

Table 1
Efficiency Multiplier for Low Pressure

r_p	10	14.7	20	40	60	80	100	150
				Pressure Psia				
3.0	.990	1.00	1.00	1.00	1.00	1.00	1.00	1.00
2.5	.980	.985	.990	.995	1.00	1.00	1.00	1.00
2.0	.960	.965	.970	.980	.990	1.00	1.00	1.00
1.5	.890	.900	.920	.940	.960	.980	.990	1.00

Source: Modified courtesy of the Gas Processors Suppliers Association and Ingersoll-Rand.

Table 2
Efficiency Multiplier for Specific Gravity

r_p	1.5	1.3	1.0	0.8	0.6
			SG		
2.0	0.99	1.0	1.0	1.0	1.01
1.75	0.97	0.99	1.0	1.01	1.02
1.5	0.94	0.97	1.0	1.02	1.04

Source: Modified courtesy of the Gas Processors Suppliers Association.

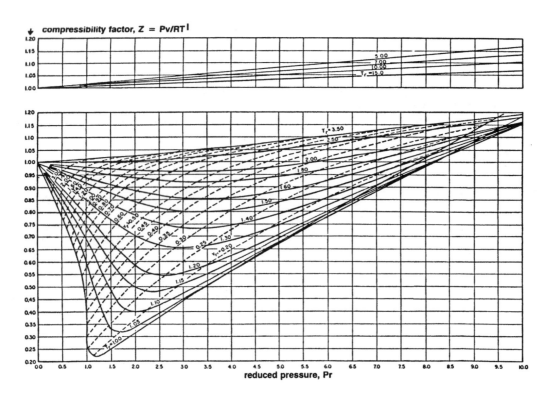

Figure 4. Compressibility chart for low to high values of reduced pressure. Reproduced by permission of *Chemical Engineering*, McGraw Hill Publications Company, July 1954.

Estimating suction and discharge volume bottle sizes for pulsation control for reciprocating compressors

Pressure surges are created as a result of the cessation of flow at the end of the compressor's discharge and suction stroke. As long as the compressor speed is constant, the pressure pulses will also be constant. A low pressure compressor will likely require little if any treatment for pulsation control; however, the same machine with increased gas density, pressure, or other operational changes may develop a problem with pressure pulses. Dealing with pulsa-

tion becomes more complex when multiple cylinders are connected to one header or when multiple stages are used.

API Standard 618 should be reviewed in detail when planning a compressor installation. The pulsation level at the outlet side of any pulsation control device, regardless of type, should be no more than 2% peak-to-peak of the line pressure of the value given by the following equation, whichever is less.

$$P_\% = 10/P_{line}^{1/3} \qquad (12)$$

Where a detailed pulsation analysis is required, several approaches may be followed. An analog analysis may be performed on the Southern Gas Association dynamic compressor simulator, or the analysis may be made a part of the compressor purchase contract. Regardless of who makes the analysis, a detailed drawing of the piping in the compressor area will be needed.

The following equations are intended as an aid in estimating bottle sizes or for checking sizes proposed by a vendor for simple installations—i.e., single cylinder connected to a header without the interaction of multiple cylinders. The bottle type is the simple unbaffled type.

Example. Determine the approximate size of suction and discharge volume bottles for a single-stage, single-acting, lubricated compressor in natural gas service.

Cylinder bore = 9 in.
Cylinder stroke = 5 in.
Rod diameter = 2.25 in.
Suction temp = 80°F
Discharge temp = 141°F
Suction pressure = 514 psia
Discharge pressure = 831 psia
Isentropic exponent, k = 1.28
Specific gravity = 0.6
Percent clearance = 25.7%

Step 1. Determine suction and discharge volumetric efficiencies using Equations 5 and 13.

$$E_{vd} = [E_v \times f]/r_p^{1/k} \qquad (13)$$

$$r_p = 831/514$$
$$= 1.617$$

$Z_1 = 0.93$ (from Figure 5)

$Z_2 = 0.93$ (from Figure 5)

$$f = 0.93/0.93$$
$$= 1.0$$

Calculate suction volumetric efficiency using Equation 5:

$$E_v = 0.97 - [(1/1) \times (1.617)^{1/1.28} - 1] \times 0.257 - 0.03$$
$$= 0.823$$

Calculate discharge volumetric efficiency using Equation 13:

$$E_{vd} = 1 \times [0.823]/[1.617^{1/1.28}]$$
$$= 0.565$$

Calculate volume displaced per revolution using Equation 1:

$$P_d/N = S_t \times 3.1416 \times D^2/[1,728 \times 4]$$
$$= [5 \times 3.1416 \times 9^2]/[1,728 \times 4]$$
$$= 0.184 \text{ cu ft or } 318 \text{ cu in.}$$

Refer to Figure 3, volume bottle sizing, using volumetric efficiencies previously calculated, and determine the multipliers.

Suction multiplier = 13.5
Discharge multiplier = 10.4

Discharge volume = 318 × 13.5
= 3,308 cu in.

Suction volume = 318 × 10.4
= 4,294 cu in.

Calculate bottle dimensions. For elliptical heads, use Equation 14.

$$\text{Bottle diameter } d_b = 0.86 \times volume^{1/3} \qquad (14)$$

Volume = suction or discharge volume

Suction bottle diameter = 0.86 × 4,294^{1/3}
= 13.98 in.

Discharge bottle diameter = 0.86 × 3,308^{1/3}
= 12.81 in.

$$\text{Bottle length} = L_b = 2 \times d_b \qquad (15)$$

Suction bottle length = 2 × 13.98
= 27.96 in.

Discharge bottle length = 2 × 12.81
= 25.62 in.

Source

Brown, R. N., *Compressors—Selection & Sizing,* Houston: Gulf Publishing Company, 1986.

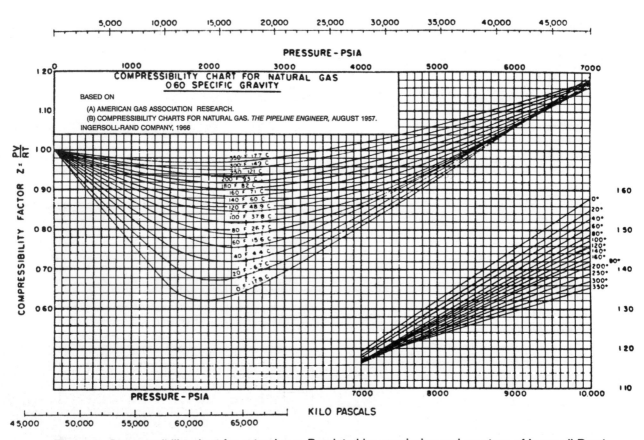

Figure 5. Compressibility chart for natural gas. Reprinted by permission and courtesy of Ingersoll Rand.

Compression horsepower determination

The method outlined below permits determination of approximate horsepower requirements for compression of gas.

1. From Figure 6, determine the atmospheric pressure in psia for the altitude above sea level at which the compressor is to operate.
2. Determine intake pressure (P_s) and discharge pressure (P_d) by adding the atmospheric pressure to the corresponding gage pressure for the conditions of compression.
3. Determine total compression ratio $R = P_d/P_s$. If ratio R is more than 5 to 1, two or more compressor stages will be required. Allow for a pressure loss of approximately 5 psi between stages. Use the same ratio for each stage. The ratio per stage, so that each stage has the same ratio, can be approximated by finding the nth root of the total ratio, when n = number of stages. The exact ratio can be found by trial and error, accounting for the 5 psi interstage pressure losses.
4. Determine the N value of gas from Figure 7, ratio of specific heat.

5. Figure 8 gives horsepower requirements for compression of one million cu ft per day for the compression ratios and N values commonly encountered in oil producing operations.
6. If the suction temperature is not 60°F, correct the curve horsepower figure in proportion to absolute temperature. This is done as follows:

$$HP \times \frac{460° + T_s}{460° + 60°F} = \text{hp (corrected for suction temperature)}$$

where T_s is suction temperature in °F.
7. Add together the horsepower loads determined for each stage to secure the total compression horsepower load. For altitudes greater than 1,500 ft above sea level apply a multiplier derived from the following table to determine the nominal sea level horsepower rating of the internal combustion engine driver.

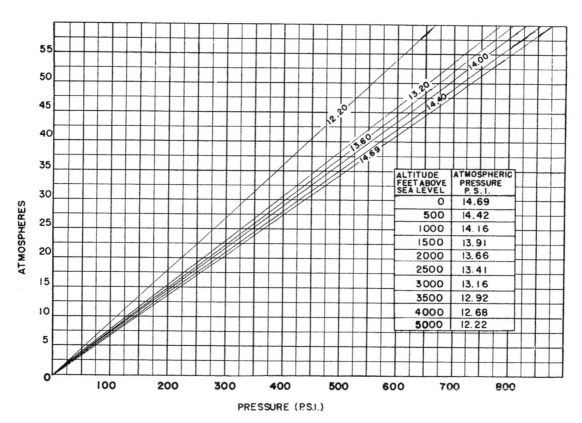

Figure 6. Atmospheres at various atmospheric pressures. *From Modern Gas Lift Practices and Principles,* Merla Tool Corp.

Figure 7. Ratio of specific heat (n-value).

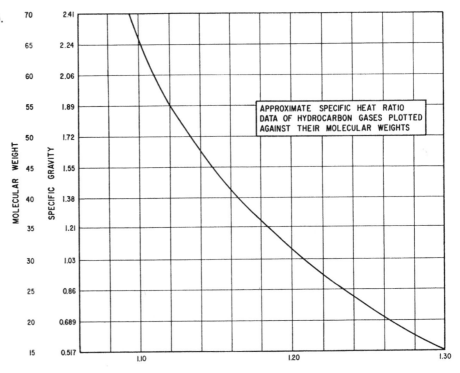

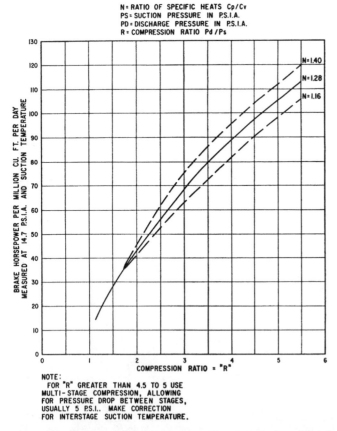

N = RATIO OF SPECIFIC HEATS Cp/Cv
PS = SUCTION PRESSURE IN P.S.I.A.
PD = DISCHARGE PRESSURE IN P.S.I.A.
R = COMPRESSION RATIO Pd/Ps

NOTE:
FOR "R" GREATER THAN 4.5 TO 5 USE
MULTI-STAGE COMPRESSION, ALLOWING
FOR PRESSURE DROP BETWEEN STAGES,
USUALLY 5 P.S.I.. MAKE CORRECTION
FOR INTERSTAGE SUCTION TEMPERATURE.

Figure 8. Brake horsepower required for compressing natural gas.

Altitude—Multiplier		Altitude—Multiplier	
1,500 ft	1.000	4,000 ft	1.12
2,000 ft	1.03	4,500 ft	1.14
2,500 ft	1.05	5,000 ft	1.17
3,000 ft	1.07	5,500 ft	1.20
3,500 ft	1.10	6,000 ft	1.22

8. For a portable unit with a fan cooler and pump driven from the compressor unit, increase the horsepower figure by 7½%.

The resulting figure is sufficiently accurate for all purposes. The nearest commercially available size of compressor is then selected.

The method does not take into consideration the supercompressibility of gas and is applicable for pressures up to 1,000 psi. In the region of high pressures, neglecting the deviation of behavior of gas from that of the perfect gas may lead to substantial errors in calculating the compression horsepower requirements. The enthalpy-entropy charts may be used conveniently in such cases. The procedures are given in sources 1 and 2.

Example. What is the nominal size of a portable compressor unit required for compressing 1,600,000 standard cubic ft of gas per 24 hours at a temperature of 85°F from 40 psig pressure to 600 psig pressure? The altitude above sea level is 2,500 ft. The N value of gas is 1.28. The suction temperature of stages, other than the first stage, is 130°F.

Solution.

P_s = 53.41 psia
P_d = 613.41 psia
$R_t = P_d/P_s = 11.5 = (3.4)^2 = (2.26)^3$

Try solution using 3.44 ratio and 2 stages.

1st stage: 53.41 psia × 3.44 = 183.5 psia discharge
2nd stage: 178.5 psia × 3.44 = 614 psia discharge

Horsepower from curve, Figure 8 = 77 hp for 3.44 ratio

$$\frac{77 \text{ hp} \times 1,600,000}{1,000,000} = 123.1 \text{ (for 60°F suction temp.)}$$

1st stage: $123.1 \text{ hp} \times \dfrac{460 + 85°}{460 + 60°} = 129.1 \text{ hp}$

2nd stage: $123.1 \text{ hp} \times \dfrac{460 + 130°}{460 + 60°} = 139.7 \text{ hp}$

1.05 (129.1 hp + 139.7 hp) = 282 hp

1.075 × 282 hp = 303 hp

Nearest nominal size compressor is 300 hp.

Centrifugal compressors

The centrifugal compressors are inherently high volume machines. They have extensive application in gas transmission systems. Their use in producing operations is very limited.

Sources

1. *Engineering Data Book*, Natural Gasoline Supply Men's Association, 1957.
2. Dr. George Granger Brown: "A Series of Enthalpy-entropy Charts for Natural Gas," *Petroleum Development and Technology*, Petroleum Division AIME, 1945.

Generalized compressibility factor

The nomogram (Figure 9) is based on a generalized compressibility chart.[1] It is based on data for 26 gases, excluding helium, hydrogen, water, and ammonia. The accuracy is about one percent for gases other than those mentioned.

To use the nomogram, the values of the reduced temperature (T/T_c) and reduced pressure (P/P_c) must be calculated first.

where   T = temperature in consistent units
T_c = critical temperature in consistent units
P = pressure in consistent units
P_c = critical pressure in consistent units

Example. P_r = 0.078, T_r = 0.84, what is the compressibility factor, z? Connect P_r with T_r and read z = 0.948.

Source

Davis, D. S., *Petroleum Refiner*, 37, No. 11, (1961).

Reference

1. Nelson, L. C., and Obert, E. F., *Chem. Engr.*, 203 (1954).

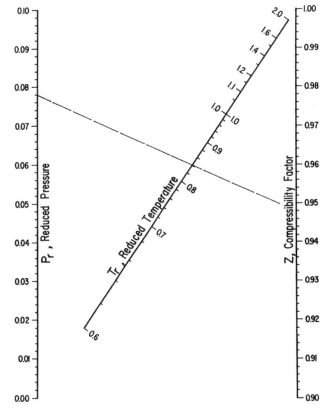

Figure 9. Generalized compressibility factor. (Reproduced by permission *Petroleum Refiner*, Vol. 37, No. 11, copyright 1961, Gulf Publishing Co., Houston.)

Centrifugal Compressor Performance Calculations

Centrifugal compressors are versatile, compact, and generally used in the range of 1,000 to 100,000 inlet cubic ft per minute (ICFM) for process and pipe line compression applications.

Centrifugal compressors can use either a horizontal or a vertical split case. The type of case used will depend on the pressure rating with vertical split casings generally being used for the higher pressure applications. Flow arrangements include straight through, double flow, and side flow configurations.

Centrifugal compressors may be evaluated using either the adiabatic or polytropic process method. An adiabatic process is one in which no heat transfer occurs. This doesn't imply a constant temperature, only that no heat is transferred into or out of the process system. Adiabatic is normally intended to mean adiabatic isentropic. A polytropic process is a variable-entropy process in which heat transfer can take place.

When the compressor is installed in the field, the power required from the driver will be the same whether the process is called adiabatic or polytropic during design. Therefore, the work input will be the same value for either process. It will be necessary to use corresponding values when making the calculations. When using adiabatic head, use adiabatic efficiency and when using polytropic head, use polytropic efficiency. Polytropic calculations are easier to make even though the adiabatic approach appears to be simpler and quicker.

The polytropic approach offers two advantages over the adiabatic approach. The polytropic approach is independent of the thermodynamic state of the gas being compressed, whereas the adiabatic efficiency is a function of the pressure ratio and therefore is dependent upon the thermodynamic state of the gas.

If the design considers all processes to be polytropic, an impeller may be designed, its efficiency curve determined, and it can be applied without correction regardless of pressure, temperature, or molecular weight of the gas being compressed. Another advantage of the polytropic approach is that the sum of the polytropic heads for each stage of compression equals the total polytropic head required to get from state point 1 to state point 2. This is not true for adiabatic heads.

Sample Performance Calculations

Determine the compressor frame size, number of stages, rotational speed, power requirement, and discharge temperature required to compress 5,000 lbm/min of gas from 30 psia at 60°F to 100 psia. The gas mixture molar composition is as follows:

Ethane	5%
Propane	80%
n-Butane	15%

The properties of this mixture are as follows:

$$MW = 45.5$$
$$P_c = 611 \text{ psia}$$
$$T_c = 676°R$$
$$c_p = 17.76$$
$$k_1 = 1.126$$
$$Z_1 = 0.955$$

Before proceeding with the compressor calculations, let's review the merits of using average values of Z and k in calculating the polytropic head.

The inlet compressibility must be used to determine the actual volume entering the compressor to approximate the size of the compressor and to communicate with the vendor via the data sheets. The maximum value of θ is of interest and will be at its maximum at the inlet to the compressor where the inlet compressibility occurs (although using the average compressibility will result in a conservative estimate of θ).

Compressibility will decrease as the gas is compressed. This would imply that using the inlet compressibility would be conservative since as the compressibility decreases, the head requirement also decreases. If the variation in compressibility is drastic, the polytropic head requirement calculated by using the inlet compressibility would be practically useless. Compressor manufacturers calculate the performance for each stage and use the inlet compressibility for each stage. An accurate approximation may be substituted for the stage-by-stage calculation by calculating the polytropic head for the overall section using the average compressibility. This technique results in overestimating the first half of the impellers and underestimating the last half of the impellers, thereby calculating a polytropic head very near that calculated by the stage-by-stage technique.

Determine the inlet flow volume, Q_1:

$$Q_1 = m_1[(Z_1RT_1)/(144P_1)]$$

where m = mass flow
      Z_1 = inlet compressibility factor
      R = gas constant = 1,545/MW
      T_1 = inlet temperature °R
      P_1 = inlet pressure

$$Q_1 = 5,000[(0.955)(1,545)(60 + 460)/(45.5)(144)(30)]$$
$$= 19,517 \text{ ICFM}$$

Refer to Table 3 and select a compressor frame that will handle a flow rate of 19,517 ICFM. A Frame C compressor will handle a range of 13,000 to 31,000 ICFM and would have the following nominal data:

$H_{pnom} = 10,000$ ft-lb/lbm (nominal polytropic head)

$n_p = 77\%$ (polytropic efficiency)

$N_{nom} = 5,900$ rpm

Determine the pressure ratio, r_p.

$$r_p = P_2/P_1 = 100/30 = 3.33$$

Determine the approximate discharge temperature, T_2.

$$n/n - 1 = [k/k - 1]n_p$$
$$= [1.126/(1.126 - 1.000)](0.77)$$
$$= 6.88$$

$$T_2 = T_1(r_p)^{(n-1)/n}$$
$$= (60 + 460)(3.33)^{1/6.88}$$
$$= 619°R = 159°F$$

where T_1 = inlet temp

Determine the average compressibility, Z_a.

$Z_1 = 0.955$ (from gas properties calculation)

where Z_1 = inlet compressibility

$$(P_r)_2 = P_2/P_c$$
$$= 100/611$$
$$= 0.164$$

$$(T_r)_2 = T_2/T_c$$
$$= 619/676$$
$$= 0.916$$

Table 3

Typical Centrifugal Compressor Frame Data*

Frame	Nominal Inlet Volume Flow		Nominal Polytropic Head		Nominal Polytropic Efficiency (%)	Nominal Rotational Speed (RPM)	Nominal Impeller Diameter	
	English (ICFM)	Metric (m³/h)	English (ft-lbf/lbm)	Metric (k·Nm/kg)			English (in)	Metric (mm)
A	1,000–7,000	1,700–12,000	10,000	30	76	11,000	16	406
B	6,000–18,000	10,000–31,000	10,000	30	76	7,700	23	584
C	13,000–31,000	22,000–53,000	10,000	30	77	5,900	30	762
D	23,000–44,000	39,000–75,000	10,000	30	77	4,900	36	914
E	33,000–65,000	56,000–110,000	10,000	30	78	4,000	44	1,120
F	48,000–100,000	82,000–170,000	10,000	30	78	3,300	54	1,370

*While this table is based on a survey of currently available equipment, the instance of any machinery duplicating this table would be purely coincidental.

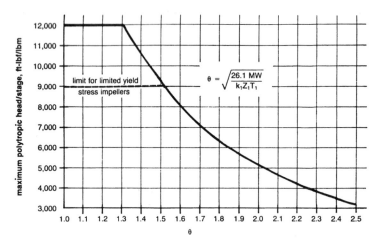

Figure 10. Maximum polytropic head per stage—English system.

Refer to Figure 5 to find Z_2, discharge compressibility.

$$Z_2 = 0.925$$

$$Z_a = (Z_1 + Z_2)/2$$
$$= 0.94$$

Determine average k-value. For simplicity, the inlet value of k will be used for this calculation. The polytropic head equation is insensitive to k-value (and therefore n-value) within the limits that k normally varies during compression. This is because any errors in the $n/(n - 1)$ multiplier in the polytropic head equation tend to balance corresponding errors in the $(n - 1)/n$ exponent. Discharge temperature is very sensitive to k-value. Since the k-value normally decreases during compression, a discharge temperature calculated by using the inlet k-value will be conservative and the actual temperature may be several degrees higher—possibly as much as 25–50°F. Calculating the average k-value can be time-consuming, especially for mixtures containing several gases, since not only must the mol-weighted c_p of the mixture be determined at the inlet temperature but also at the estimated discharge temperature.

The suggested approach is as follows:

1. If the k-value is felt to be highly variable, one pass should be made at estimating discharge temperature based on the inlet k-value; the average k-value should then be calculated using the estimated discharge temperature.
2. If the k-value is felt to be fairly constant, the inlet k-value can be used in the calculations.
3. If the k-value is felt to be highly variable, but sufficient time to calculate the average value is not available, the inlet k-value can be used (but be aware of the potential discrepancy in the calculated discharge temperature).

$$k_1 = k_a = 1.126$$

Determine average $n/(n - 1)$ value from the average k-value. For the same reasons discussed above, use $n/(n - 1) = 6.88$.

Table 4

Approximate Mechanical Losses as a Percentage of Gas Power Requirement*

Gas Power Requirement		Mechanical Losses, L_m (%)
English (hp)	Metric (kW)	
0–3,000	0–2,500	3
3,000–6,000	2,500–5,000	2.5
6,000–10,000	5,000–7,500	2
10,000+	7,500+	1.5

*There is no way to estimate mechanical losses from gas power requirements. This table will, however, ensure that mechanical losses are considered and yield useful values for estimating purposes.

Determine polytropic head, H_p:

$$H_p = Z_a RT_1(n/n - 1)[r_p^{(n-1)/n} - 1]$$
$$= (0.94)(1,545/45.5)(520)(6.88)(3.33)^{1/6.88} - 1]$$
$$= 21,800 \text{ ft-lbf/lbm}$$

Determine the required number of compressor stages, θ:

$$\theta = [(26.1MW)/(k_1 Z_1 T_1)]^{0.5}$$
$$= [(26.1)(45.5)/(1.126)(0.955)(520)]^{0.5}$$
$$= 1.46$$

max H_p/stage from Figure 10 using $\theta = 1.46$

$$\text{Number of stages} = H_p/\text{max. } H_p/\text{stage}$$
$$= 21,800/9,700$$
$$= 2.25$$
$$= 3 \text{ stages}$$

Determine the required rotational speed:

$$N = N_{nom}[H_p/H_{pnom} \times \text{no. stages}]^{0.5}$$
$$= 5,900[21,800/(10,000)(3)]^{0.5}$$
$$= 5,030 \text{ rpm}$$

Determine the required shaft power:

$$PWR_g = mH_p/33,000n_p$$
$$= (5,000)(21,800)/(33,000)(0.77)$$
$$= 4,290 \text{ hp}$$

Mechanical losses (L_m) = 2.5% (from Table 4)
$$L_m = (0.025)(4,290)$$
$$= 107 \text{ hp}$$

$$PWR_s = PWR_g + L_m$$
$$= 4,290 + 107$$
$$= 4,397 \text{ hp}$$

Determine the actual discharge temperature:

$$T_2 = T_1(r_p)^{(n-1)/n}$$
$$= 520(3.33)^{1/6.88}$$
$$= 619°R$$
$$= 159°F$$

The discharge temperature calculated in the last step is the same as that calculated earlier only because of the decision to use the inlet k-value instead of the average k-value. Had the average k-value been used, the actual discharge temperature would have been lower.

Source

Lapina, R. P., *Estimating Centrifugal Compressor Performance*, Houston: Gulf Publishing Company, 1982.

Estimate hp required to compress natural gas

To estimate the horsepower to compress a million cubic ft of gas per day, use the following formula:

$$BHP/MMcfd = \frac{R}{R + RJ}\left(\frac{5.16 + 124 \text{ Log } R}{0.97 - 0.03 R}\right)$$

where R = compression ratio. Absolute discharge pressure divided by absolute suction pressure
J = supercompressibility factor—assumed 0.022 per 100 psia suction pressure

Example. How much horsepower should be installed to raise the pressure of 10 million cubic ft of gas per day from 185.3 psi to 985.3 psi?

This gives absolute pressures of 200 and 1,000.

then $R = \dfrac{1,000}{200} = 5.0$

Substituting in the formula:

$$BHP/MMcfd = \frac{5.0}{5.0 + 5 \times 0.044} \times \frac{5.16 + 124 \times .699}{.97 - .03 \times 5}$$

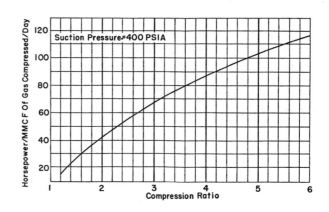

$$= 106.5 \text{ hp} = \text{BHP for 10 MMcfd}$$
$$= 1,065 \text{ hp}$$

Where the suction pressure is about 400 psia, the brake horsepower per MMcfd can be read from the chart.

The above formula may be used to calculate horsepower requirements for various suction pressures and gas physical properties to plot a family of curves.

Estimate engine cooling water requirements

This equation can be used for calculating engine jacket water requirements as well as lube oil cooling water requirements:

$$GPM = \frac{H \times BHP}{500\Delta t}$$

where   H = Heat dissipation in Btu's per BHP/hr. This will vary for different engines; where they are available, the manufacturers' values should be used. Otherwise, you will be safe in substituting the following values in the formula: For engines with water-cooled exhaust manifolds: Engine jacket water = 2,200 Btu's per BHP/hr. Lube oil cooling water = 600 Btu's per BHP/hr.

For engines with dry type manifolds (so far as cooling water is concerned) use 1,500 Btu's/BHP/hr for the engine jackets and 650 Btu's/BHP/hr for lube oil cooling water requirements.

BHP = Brake Horsepower Hour
Δt = Temperature differential across engine. Usually manufacturers recommend this not exceed 15°F; 10°F is preferable.

Example. Find the jacket water requirements for a 2,000 hp gas engine which has no water jacket around the exhaust manifold.

Solution.

$$GPM = \frac{1,500 \times 2,000}{500 \times 10}$$

$$GPM = \frac{3,000,000}{5,000} = 600 \text{ gallons per min}$$

The lube oil cooling water requirements could be calculated in like manner.

Estimate fuel requirements for internal combustion engines

When installing an internal combustion engine at a gathering station, a quick approximation of fuel consumptions could aid in selecting the type fuel used.

Using Natural Gas: Multiply the brake hp at drive by 11.5 to get cubic ft of gas per hour.

Example. Internal combustion engine rated at 50 bhp—3 types of fuel available.

Natural Gas: 50 × 11.5 = 575 cubic ft of gas per hour

Using Butane: Multiply the brake hp at drive by 0.107 to get gallons of butane per hour.
Using Gasoline: Multiply the brake hp at drive by 0.112 to get gallons of gasoline per hour.
These approximations will give reasonably accurate figures under full load conditions.

Butane: 50 × 0.107 = 5.35 gallons of butane per hour

Gasoline: 50 × 0.112 = 5.60 gallons of gasoline per hour

REFERENCES

1. Brown, R. N., *Compressors: Selection and Sizing,* 2nd Ed. Houston: Gulf Publishing Co., 1997.
2. McAllister, E. W. (Ed.), *Pipe Line Rules of Thumb Handbook,* 3rd Ed. Houston: Gulf Publishing Co., 1993.
3. Lapina, R. P., *Estimating Centrifugal Compressor Performance,* Vol. 1. Houston: Gulf Publishing Co., 1982.
4. Warring, R. H., *Pumping Manual,* 7th Ed. Houston: Gulf Publishing Co., 1984.
5. Warring, R. H. (Ed.), *Pumps: Selection, Systems, and Applications,* 2nd Ed. Houston: Gulf Publishing Co., 1984.
6. Cheremisinoff, N. P., *Fluid Flow Pocket Handbook.* Houston: Gulf Publishing Co., 1984.
7. Streeter, V. L. and Wylie, E. B., *Fluid Mechanics.* New York: McGraw-Hill, 1979.

Carl R. Branan, Engineer, El Paso, Texas*

*Reprinted from *Rules of Thumb for Chemical Engineers,* Carl R. Branan (Ed.), Gulf Publishing Company, Houston, Texas, 1994.

Motors: Efficiency

Table 1 from the *GPSA Engineering Data Book* [1] compares standard and high efficiency motors. Table 2 from GPSA compares synchronous and induction motors. Table 3 from Evans [2] shows the effect of a large range of speeds on efficiency.

Table 1
Energy Evaluation Chart
NEMA Frame Size Motors, Induction

HP	Approx. Full Load RPM	Amperes Based on 460V		Efficiency in Percentage at Full Load	
		Standard Efficiency	High Efficiency	Standard Efficiency	High Efficiency
1	1,800	1.9	1.5	72.0	84.0
	1,200	2.0	2.0	68.0	78.5
1½	1,800	2.5	2.2	75.5	84.0
	1,200	2.8	2.6	72.0	84.0
2	1,800	2.9	3.0	75.5	84.0
	1,200	3.5	3.2	75.5	84.0
3	1,800	4.7	3.9	75.5	87.5
	1,200	5.1	4.8	75.5	86.5
5	1,800	7.1	6.3	78.5	89.5
	1,200	7.6	7.4	78.5	87.5
7½	1,800	9.7	9.4	84.0	90.2
	1,200	10.5	9.9	81.5	89.5
10	1,800	12.7	12.4	86.5	91.0
	1,200	13.4	13.9	84.0	89.5
15	1,800	18.8	18.6	86.5	91.0
	1,200	19.7	19.0	84.0	89.5
20	1,800	24.4	25.0	86.5	91.0
	1,200	25.0	24.9	86.5	90.2
25	1,800	31.2	29.5	88.5	91.7
	1,200	29.2	29.1	88.5	91.0
30	1,800	36.2	35.9	88.5	93.0
	1,200	34.8	34.5	88.5	91.0
40	1,800	48.9	47.8	88.5	93.0
	1,200	46.0	46.2	90.2	92.4
50	1,800	59.3	57.7	90.2	93.6
	1,200	58.1	58.0	90.2	91.7
60	1,800	71.6	68.8	90.2	93.6
	1,200	68.5	69.6	90.2	93.0
75	1,800	92.5	85.3	90.2	93.6
	1,200	86.0	86.5	90.2	93.0
100	1,800	112.0	109.0	91.7	94.5
	1,200	114.0	115.0	91.7	93.6
125	1,800	139.0	136.0	91.7	94.1
	1,200	142.0	144.0	91.7	93.6
150	1,800	167.0	164.0	91.7	95.0
	1,200	168.0	174.0	91.7	94.1
200	1,800	217.0	214.0	93.0	94.1
	1,200	222.0	214.0	93.0	95.0

Table 2
Synchronous vs. Induction 3 Phase, 60 Hertz, 2,300 or 4,000 Volts

HP	Speed RPM	Synch. Motor Efficiency Full Load 1.0 PF	Induction Motor Efficiency Full Load	Power Factor
3,000	1,800	96.6	95.4	89.0
	1,200	96.7	95.2	87.0
3,500	1,800	96.6	95.5	89.0
	1,200	96.8	95.4	88.0
4,000	1,800	96.7	95.5	90.0
	1,200	96.8	95.4	88.0
4,500	1,800	96.8	95.5	89.0
	1,200	97.0	95.4	88.0
5,000	1,800	96.8	95.6	89.0
	1,200	97.0	95.4	88.0
5,500	1,800	96.8	95.6	89.0
	1,200	97.0	95.5	89.0
6,000	1,800	96.9	95.6	89.0
	1,200	97.1	95.5	87.0
7,000	1,800	96.9	95.6	89.0
	1,200	97.2	95.6	88.0
8,000	1,800	97.0	95.7	89.0
	1,200	97.3	95.6	89.0
9,000	1,800	97.0	95.7	89.0
	1,200	97.3	95.8	88.0

Table 3
Full Load Efficiencies

hp	3,600 rpm	1,200 rpm	600 rpm	300 rpm
5	80.0	82.5	—	—
	—	—	—	—
20	86.0	86.5	—	—
	—	—	—	82.7*
100	91.0	91.0	93.0	—
	—	—	91.4*	90.3*
250	91.5	92.0	91.0	—
	—	93.9*	93.4*	92.8*
1,000	94.2	93.7	93.5	92.3
	—	95.5*	95.5*	95.5*
5,000	96.0	95.2	—	—
	—	—	97.2*	97.3*

*Synchronous motors, 1.0 PF

Sources

1. *GPSA Engineering Data Book*, Gas Processors Suppliers Association, Vol. I, 10th Ed.
2. Evans, F. L., *Equipment Design Handbook for Refineries and Chemical Plants, Vol. 1, 2nd Ed.* Houston: Gulf Publishing Co., 1979.

Motors: Starter Sizes

Here are motor starter (controller) sizes.

Polyphase Motors

	Maximum Horsepower Full Voltage Starting	
NEMA Size	230 Volts	460–575 Volts
00	1.5	2
0	3	5
1	7.5	10
2	15	25
3	30	50
4	50	100
5	100	200
6	200	400
7	300	600

Single Phase Motors

	Maximum Horsepower Full Voltage Starting (Two Pole Contactor)	
NEMA Size	115 Volts	230 Volts
00	1.3	1
0	1	2
1	2	3
2	3	7.5
3	7.5	15

Source

McAllister, E. W., *Pipe Line Rules of Thumb Handbook*, 3rd Ed. Houston: Gulf Publishing Co., 1993.

Motors: Service Factor

Over the years, oldtimers came to expect a 10–15% service factor for motors. Things are changing, as shown in the following section from Evans.

For many years it was common practice to give standard open motors a 115% service factor rating; that is, the motor would operate at a safe temperature at 15% overload. This has changed for large motors, which are closely tailored to specific applications. Large motors, as used here, include synchronous motors and all induction motors with 16 poles or more (450 rpm at 60 Hz).

New catalogs for large induction motors are based on standard motors with Class B insulation of 80°C rise by resistance, 1.0 service factor. Previously, they were 60°C rise by thermometer, 1.15 service factor.

Service factor is mentioned nowhere in the NEMA standards for large machines; there is no definition of it. There is no standard for temperature rise or other characteristics at the service factor overload. In fact, the standards are being changed to state that the temperature rise tables are for motors with 1.0 service factors. Neither standard synchronous nor enclosed induction motors have included service factor for several years.

Today, almost all large motors are designed specifically for a particular application and for a specific driven machine. In sizing the motor for the load, the hp is usually selected so that additional overload capacity is not required. Customers should not have to pay for capability they do not need. With the elimination of service factor, standard motor base prices have been reduced 4–5% to reflect the savings.

Users should specify standard hp ratings, without service factor for these reasons:

1. All of the larger standard hp are within or close to 15% steps.
2. As stated in NEMA, using the next larger hp avoids exceeding standard temperature rise.
3. The larger hp ratings provide increased pull-out torque, starting torque, and pull-up torque.
4. The practice of using 1.0 service factor induction motors would be consistent with that generally followed in selecting hp requirements of synchronous motors.
5. For loads requiring an occasional overload, such as startup of pumps with cold water followed by continuous operation with hot water at lower hp loads, using a motor with a short time overload rating will probably be appropriate.

Induction motors with a 15% service factor are still available. Large open motors (except splash-proof) are available for an addition of 5% to the base price, with a specified temperature rise of 90° C for Class B insulation by resistance at the overload horsepower. This means the net price will be approximately the same. At nameplate hp the service factor rated motor will usually have less than 80° C rise by resistance.

Motors with a higher service factor rating such as 125% are also still available, but not normally justifiable. Most smaller open induction motors (i.e., 200 hp and below, 514 rpm and above) still have the 115% service factor rating. Motors in this size range with 115% service factor are standard, general purpose, continuous-rated, 60 Hz, design A or B, drip-proof machines. Motors in this size range which normally have a 100% service factor are totally en-

closed motors, intermittent rated motors, high slip design D motors, most multispeed motors, encapsulated motors, and motors other than 60 Hz.

Source

Evans, F. L., *Equipment Design Handbook for Refineries and Chemical Plants, Vol. 1, 2nd Ed.* Houston: Gulf Publishing Co., 1979.

Motors: Useful Equations

The following equations are useful in determining the current, voltage, horsepower, torque, and power factor for AC motors:

$$\text{Full Load } I = [hp(0.746)]/[1.73\,E\,(\text{eff.})\,PF]$$
$$\text{(three phase)}$$
$$= [hp(0.746)]/[E\,(\text{eff.})\,PF]$$
$$\text{(single phase)}$$
$$kVA \text{ input} = IE\,(1.73)/1{,}000 \text{ (three phase)}$$
$$= IE/1{,}000 \text{ (single phase)}$$
$$kW \text{ input} = kVA \text{ input } (PF)$$
$$hp \text{ output} = kW \text{ input } (\text{eff.})/0.746$$
$$= \text{Torque (rpm)}/5{,}250$$
$$\text{Full Load Torque} = hp\,(5{,}250\text{ lb.-ft.})/\text{rpm}$$
$$\text{Power Factor} = kW \text{ input}/kVA \text{ input}$$

where

E = Volts (line-to-line)
I = Current (amps)
PF = Power factor (per unit = percent PF/100)
eff = Efficiency (per unit = percent eff./100)
hp = Horsepower
kW = Kilowatts
kVA = Kilovoltamperes

Source

Evans, F. L., *Equipment Design Handbook for Refineries and Chemical Plants, Vol. 1, 2nd Ed.* Houston: Gulf Publishing Co., 1979.

Motors: Relative Costs

Evans gives handy relative cost tables for motors based on voltages (Table 1), speeds (Table 2), and enclosures (Table 3).

Table 1
Relative Cost at Three Voltage Levels of Drip-Proof 1,200-rpm Motors

	2,300-Volts	4,160-Volts	13,200-Volts
1,500-hp	100%	114%	174%
3,000-hp	100	108	155
5,000-hp	100	104	145
7,000-hp	100	100	133
9,000-hp	100	100	129
10,000-hp	100	100	129

Table 2
Relative Cost at Three Speeds of Drip-Proof 2,300-Volt Motors

	3,600-Rpm	1,800-Rpm	1,200-Rpm
1,500-hp	124%	94%	100%
3,000-hp	132	100	100
5,000-hp	134	107	100
7,000-hp	136	113	100
9,000-hp	136	117	100
10,000-hp	136	120	100

Table 3
Relative Cost of Three Enclosure
Types 2,300-volt, 1,200-rpm Motors

	Drip-proof	Force Ventilated*	Totally-Enclosed Inert Gas or Air Filled**
1,500-hp	100%	115%	183%
3,000-hp	100	113	152
5,000-hp	100	112	136
7,000-hp	100	111	134
9,000-hp	100	111	132
10,000-hp	100	110	125

Does not include blower and duct for external air supply.
**With double tube gas to water heat exchanger. Cooling water within manufacturer's standard conditions of temperature and pressure.*

Source

Evans, F. L., *Equipment Design Handbook for Refineries and Chemical Plants, Vol. 1, 2nd Ed.* Houston: Gulf Publishing Co., 1979.

Motors: Overloading

When a pump has a motor drive, the process engineer must verify that the motor will not overload from extreme process changes. The horsepower for a centrifugal pump increases with flow. If the control valve in the discharge line fully opens or an operator opens the control valve bypass, the pump will tend to "run out on its curve," giving more flow and requiring more horsepower. The motor must have the capacity to handle this.

Source

Branan, C. R., *The Process Engineer's Pocket Handbook, Vol. 2.* Houston: Gulf Publishing Co., 1978.

Steam Turbines: Steam Rate

The theoretical steam rate (sometimes referred to as the water rate) for steam turbines can be determined from Keenan and Keyes [1] or Mollier charts following a constant entropy path. The theoretical steam rate[1] is given as lb/hr/kw which is easily converted to lb/hr/hp. One word of caution—in using Keenan and Keyes, steam pressures are given in *PSIG*. Sea level is the basis. For low steam pressures at high altitudes appropriate corrections must be made. See the section on Pressure Drop Air-Cooled Air Side Heat Exchangers, in this handbook, for the equation to correct atmospheric pressure for altitude.

The theoretical steam rate must then be divided by the efficiency to obtain the actual steam rate. See the section on Steam Turbines: Efficiency.

Sources

1. Keenan, J. H., and Keyes, F. G., "Theoretical Steam Rate Tables," *Trans. A.S.M.E.* (1938).
2. Branan, C. R., *The Process Engineer's Pocket Handbook, Vol. 1.* Houston: Gulf Publishing Co., 1976.

Steam Turbines: Efficiency

Evans [1] provides the following graph of steam turbine efficiencies.

Smaller turbines can vary widely in efficiency depending greatly on speed, horsepower, and pressure conditions.

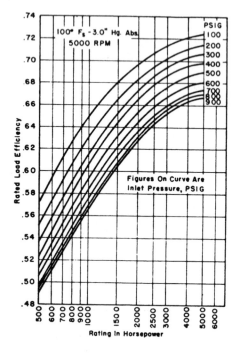

Figure 1. Typical efficiencies for mechanical drive turbines.

Very rough efficiencies to use for initial planning below 500 horsepower at 3,500 rpm are

Horsepower	Efficiency, %
1–10	15
10–50	20
50–300	25
300–350	30
350–500	40

Some designers limit the speed of the cheaper small steam turbines to 3,600 rpm.

Sources

1. Evans, F. L., *Equipment Design Handbook for Refineries and Chemical Plants, Vol. 1, 2nd Ed.* Houston: Gulf Publishing Co., 1979.
2. Branan, C. R., *The Process Engineer's Pocket Handbook, Vol. 1.* Houston: Gulf Publishing Co., 1976.

Gas Turbines: Fuel Rates

Gas turbine fuel rates (heat rates) vary considerably; however, Evans [1] provides the following fuel rate graph for initial estimating. It is based on gaseous fuels.

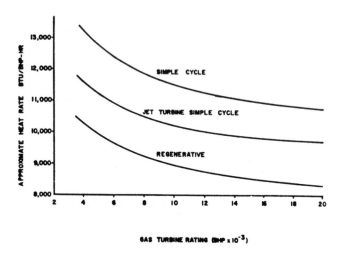

Figure 1. Approximate gas turbine fuel rates.

The *GPSA Engineering Data Book* [2] provides the following four graphs (Figures 2–5) showing the effect of altitude, inlet pressure loss, exhaust pressure loss, and ambient temperature on power and heat rate.

GPSA [2] also provides a table showing 1982 Performance Specifications for a worldwide list of gas turbines, in their Section 15.

Sources

1. Evans, F. L., *Equipment Design Handbook for Refineries and Chemical Plants, Vol. 1, 2nd Ed.* Houston: Gulf Publishing Co., 1979.
2. *GPSA Engineering Data Book*, Gas Processors Suppliers Association, Vol. I, 10th Ed.

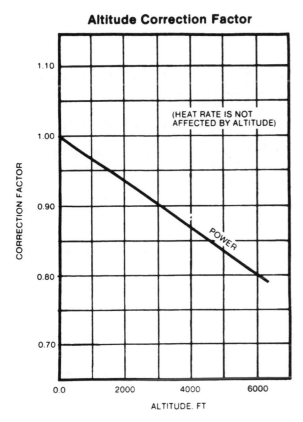

Figure 2. Altitude Correction Factor.

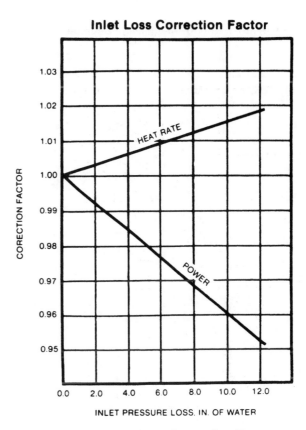

Figure 3. Inlet Loss Correction Factor.

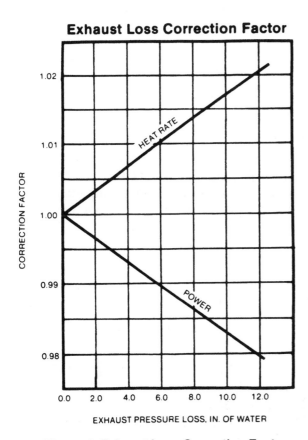

Figure 4. Exhaust Loss Correction Factor.

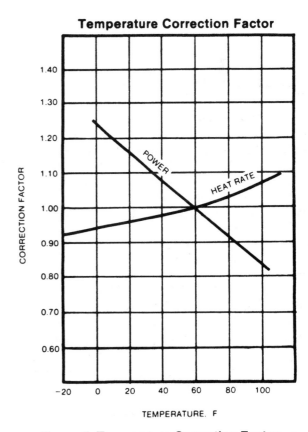

Figure 5. Temperature Correction Factor.

Gas Engines: Fuel Rates

Here are heat rates, for initial estimating, for gas engines.

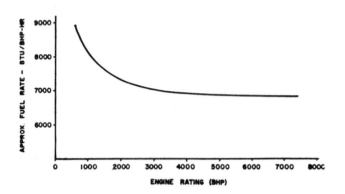

Figure 1. Approximate gas engine fuel rates.

Source

Evans, F. L. *Equipment Design Handbook for Refineries and Chemical Plants, Vol. 1, 2nd Ed.* Houston: Gulf Publishing Co., 1979.

Gas Expanders: Available Energy

With high energy costs, expanders will be used more than ever. A quickie rough estimate of actual expander available energy is

$$\Delta H = C_p T_1 \left[1 - \left(\frac{P_2}{P_1} \right)^{(K-1)/K} \right] 0.5 \tag{1}$$

where

ΔH = Actual available energy, Btu/lb
C_p = Heat capacity (constant pressure), Btu/lb °F
T_1 = Inlet temperature, °R
P_1, P_2 = Inlet, outlet pressures, psia
$K = C_p/C_v$

To get lb/hr-hp divide as follows:

$$\frac{2545}{\Delta H}$$

A rough outlet temperature can be estimated by

$$T_2 = T_1 \left(\frac{P_2}{P_1} \right)^{(K-1)/K} + \left(\frac{\Delta H}{C_p} \right) \tag{2}$$

For large expanders, Equation 1 may be conservative. A full rating using vendor data is required for accurate results. Equation 1 can be used to see if a more accurate rating is worthwhile.

For comparison, the outlet temperature for gas at critical flow across an orifice is given by

$$T_2 = T_1 \left(\frac{P_2}{P_1} \right)^{(K-1)/K} = T_1 \left(\frac{2}{K+1} \right) \tag{3}$$

The proposed expander may cool the working fluid below the dew point. Be sure to check for this.

Source

Branan, C. R., *The Process Engineer's Pocket Handbook, Vol. 1.* Houston: Gulf Publishing Co., 1976.

7

Gears

Leonard L. Haas, Manager, Lift Fan Design, Allison Advanced Development Company

This chapter is intended as a brief guide for the engineer who has an occasional need to consider gear design. The methods presented are for estimating only, and full analysis should be done in accordance with the standards of the American Gear Manufacturers Association (AGMA) or the International Standards Organization (ISO). The engineer should also reference the many good books covering the complete subject of gear design.

Ratios and Nomenclature

Consider the most common gear application of reducing motor speed to machine speed. The necessary gear ratio is the ratio of the motor speed to the machine speed. The magnitude of the required ratio may affect the type of gear or gear arrangement. See Table 1 for the range of ratios typically practical for different types of gears and arrangements, and see also the section on gear types later in the chapter. Nomenclature is given in Table 2.

Table 1
Typical Gear Ratios

Type of Gearset	Min. Ratio	Max. Ratio
External spur gear	1:1	5:1
Internal spur gear	1.5:1	7:1
External helical gear	1:1	10:1
Internal helical gear	1.5:1	10:1
Cylindrical worm	3:1	100:1
Straight bevel gear	1:1	8:1
Spiral bevel gear	1:1	8:1
Epicyclic planetary	3:1	12:1
Epicyclic star	2:1	11:1
Epicyclic solar	1.2:1	1.7:1

Table 2
Nomenclature

d	Pitch diameter of pinion*	D	Pitch diameter of gear
C	Center distance	m_g	Gear ratio
P_{ac}	Allowable based on contact	K_{all}	Allowable K factor
n	Speed of pinion	F	Contacting face width
P	Transmitted power	F_d	Face-to-diameter ratio
J	Tooth form geometry factor	S_{at}	Allowable bending stress
a_c	Chordal addendum for caliper measurement	P_d	Transverse diametral pitch
		C_d	Combined derating factor
t	Arc tooth thickness	C_v	Dynamic factor
C_a	Application factor	t_c	Normal chordal thickness
C_m	Mounting factor	γ	Pitch line velocity
P_n	Normal diametral pitch	D_{bev}	Bevel gear diameter
d_{bev}	Bevel pinion diameter	Γ	Pitch angle
N_T	Total number of teeth	N_G	Number of gear teeth
N_P	Number of pinion teeth		

*The smaller diameter gear in a pair is called the pinion.
In a one-to-one ratio, the definition is meaningless.

Spur and Helical Gear Design

The most common arrangement is to use spur or helical gears on parallel shafts. If the required ratio is greater than the recommended ratio for a single set, then a number of sets in series are used with the total ratio being the product of the individual gear set ratios.

Consider a case in which the required ratio is in the practical range for a single gear set. If the only consideration were to fit gears of the desired ratio on parallel shafts separated by some desired distance, the diameter of the gears could be calculated by Equations 1 and 2.

$$d = \frac{2 \times C}{m_g + 1} \tag{1}$$

$$D = 2 \times C - d \tag{2}$$

The normal requirement is to make the gears large enough to transmit a certain power. The power capacity depends on the diameter, the face width, the size of the teeth (diametral pitch), and the material. Each gear member must have a safe stress margin relative to both contact stress and bending stress. Both stress levels depend on the diameter, the face width, and the material, while the bending stress also depends on the size and form of the teeth. To estimate the size of gear set needed to transmit a required power at a given speed, first determine the size required for a safe level of contact stress, using Equation 3a:

$$d = \sqrt{\frac{126000 \times P_{ac}}{K_{all} \times n \times F \times C_d} \times \frac{(m_g + 1)}{m_g}} \tag{3a}$$

The stress calculation equation can be rearranged to calculate the allowable power that can be transmitted, resulting in equation 3b:

$$P_{ac} = \frac{K_{all} \times n \times F \times d^2 \times C_d}{126,000} \times \left[\frac{m_g}{m_g + 1} \right] \tag{3b}$$

Note: Gear power equations always use the rpm and diameter of the pinion even when the power rating of the gear is being calculated.

The power, speed, and ratio are known values, but K_{all} and C_d are unknown, and F and d are the size of the desired pinion. The allowable K factor (K_{all}) depends on the material to be used for the gears. In general, the harder the material, the higher the allowable K, but there is usually more manufacturing cost. For estimating the size of a gear set, allowable values can be taken from Table 3. The allowable K must be selected for the gear member with the lower value; the capacity of the pair is based on the lesser material. If the materials are through-hardened alloy steel, there should be a hardness differential between the pinion and the gear as shown in the pinion and gear columns in Table 3.

Table 3
Allowable Material Factors

BHN of Pinion	BHN of Gear	K_{all} Factor	Sat
225	180	110	31,000–41,000
350	300	350–450	36,000–47,000
575	575	500–800	45,000–55,000
58 Rc	58 Rc	600	60,000–70,000

The derating factor C_d is used to allow for the manufactured quality, operating conditions, and installation of the gear set. Actual gears have tolerance and are not manufactured perfectly. They are mounted in support structures that are not in perfect alignment, and the shafts, bearings, and supports deflect when the gear reaction loads are applied. Derating factors can be included in the power-to-stress equation to allow for these conditions. There are several significant conditions that should be considered in the derating factor, and many minor conditions. The major considerations are speed, gear accuracy, type of load, reliability requirements, and flexibility of gear mounting (rigidity of shafts, bearings, and gear case). The design and operating conditions should be considered when selecting a value for C_d. C_d is defined from a combination of factors as shown in Equation 4.

$$C_d = C_a + C_v \times C_m \tag{4}$$

The application factor C_a is selected to compensate for the nature of the driving and driven machines. If both machines are smooth, such as a gas turbine and a centrifugal blower, the factor can be 1.00. If a machine runs with dy-

namic load pulses, as in a reciprocating engine, or shock loads, as in a crusher drive, then a larger value of C_a must be used to allow for the dynamic loads that are above the nominal design load. The application factor is also adjusted for use and life requirements. If the gears are needed only for a short life and the use is intermittent, smaller gears at higher load levels can be used and designed with the "short life" factors. Conversely, to achieve long life with high reliability, the gears should be designed to a lower load level, and the "extra life" factors should be used.

Based on the driving/driven machines and the life required, choose a C_a factor from Table 4. "Smooth" machines include:

- gas turbines
- steam turbines
- electric motors
- centrifugal blowers

"Rough" machines include:

- reciprocating engines and compressors
- crushers
- pulverizers
- rolling mills

Table 4
Application Factor, C_a

Load Type	Required Life		
	Short 200 hrs	Normal 1,000 hrs	Extra 10,000 hrs
Smooth/Smooth	0.50	1.00	1.25
Smooth/Rough	0.63	1.25	1.56
Rough/Rough	0.75	1.50	1.88

The dynamic factor C_v allows for the increased load caused by inaccuracies in the gear teeth that result in nonuniform transmission of load from the driving gear to the driven gear. The dynamic load tends to increase with speed, as defined by the pitch line velocity (v, see Equation 5). It will be necessary to assume a diameter, or velocity, to choose the C_v factor and to recalculate later if necessary. The dynamic load is also very dependent on the precision of the gear tooth form manufactured. Select from the columns of Table 5 as follows:

$$v = .262 \times d \times n \tag{5}$$

Table 5
Dynamic Factor, C_v

	Quality			
PLV	Low Quality	Standard Commercial	Precision	Extra Precision
0–2,000	1.35	1.18	1.10	1.05
2,000–5,000	1.53	1.26	1.14	1.05
5,000–10,000	*	1.33	1.20	1.05
> 10,000	*	*	1.25	1.05

Better precision quality is required for these higher speeds.

Gears manufactured by cutting only, with the main priority of low cost, normally fall in the first column—low quality. Most gears that are surface hardened without finishing after hardening also fit into the first column.

Normal commercial-cut-only gears usually fit into the second column.

Finish grinding, or shaving of through-hardened gears, is usually required to qualify for the precision column.

Extra-precision gears require careful finish grinding of the gear teeth as well as precision control of gear blank dimensions and accurate mountings.

The mounting factor C_m allows for the deflection of the shafts, bearings, and housing that will cause the misalignment of the pinion teeth relative to the gear teeth. Misalignment also results from inaccuracy of manufacture of the gear teeth and deviation from theoretical helix angle. The C_m factor should be selected from Table 6 based on the precision of manufacture of the gears and housing as well as the support design of the drive.

When the derating factor has been selected, the pinion diameter and the face width are both unknown. For estimating these, the proportion of face width to diameter F_d can be assumed. There is a wide range of acceptable values, but 0.75 to 1.25 is a good starting value. Equation 2 can be rearranged to calculate the pinion diameter required.

$$d = \left[\frac{126,000 \times P \times C_d}{F_d \times K_{all} \times n} \times \left(\frac{m_g + 1}{m_g} \right) \right]^{1/3} \tag{6}$$

Table 6
Mounting Factor, C_m

Mounting	C_m
Rigid	1.10
Normal	1.30
Overhung	1.60

$$F_d = \frac{F}{d} \tag{7}$$

With the pinion diameter estimated, the face and the gear diameter can be established by Equations 8 and 9.

$$F = d \times F_d \tag{8}$$

$$D = d \times m_g \quad \text{or} \quad D = (2 \times C) - d \tag{9}$$

When the diameters have been established, it is time to determine the number of teeth to put on the gears. The size of gear teeth is defined in terms of *diametral pitch*. Diametral pitch is defined as the number of teeth per inch of pitch diameter. For a helical gear, the size of teeth can be measured in the plane of rotation, called *transverse diametral pitch* (P_d), or in the normal plane, called *normal diametral pitch* (P_n). For a spur gear, the normal plane is the plane of rotation. When gear teeth are cut, the tool works in the normal plane. For most gears, standard cutting tools are used; therefore, it is best to use standard whole number values for the normal diametral pitch. The normal and transverse pitches are related by the helix angle as shown in Equation 10. Obviously, the number of teeth on the gears must be a whole number, so slight adjustments to pitch diameter may be necessary. For helical gears, the diameter is a function of normal diametral pitch and helix angle, so slight changes in helix angle can be used to get a whole number of teeth on a given pitch diameter. The size of the gear teeth, and therefore the number, must be selected to provide a satisfactory bending stress. The required diametral pitch can be estimated using Equation 11.

$$P_n = P_d \times \cos(\psi) \tag{10}$$

$$P_d = \frac{n \times d \times F \times J \times S_{at}}{252,000 \times P \times C_d} \tag{11}$$

S_{at} can be selected from Table 3 for the material to be used for the gears. The geometry factor J is calculated for the number of teeth, the pressure angle, the tooth proportions, and the fillet radius. For approximation, the values in Table 7 can be used. When the P_d value is estimated from Equation 11, it should be rounded down to the nearest whole number to use standard cutters. Because of the different number of teeth and the possible different materials, the pitch requirement should be calculated for both the pinion and the gear, and the smaller value used.

The actual number of teeth can be calculated from Equations 12 and 13. The result will probably be a fractional number. The center distance, or the helix angle, must be adjusted to get a whole number. An alternative is to use Equation 14. Adjust the center distance or helix angle to ob-

Table 7
J Factors

Spur	Number of Teeth				
	16	20	25	30	50 or more
1 to 1	.299 P	.333 P	.359 P	.388 P	.442 P
	.299 G	.333 G	.359 G	.388 G	.442 G
3 to 1	.315 P	.351 P	.379 P	.409 P	.463 P
	.407 G	.428 G	.444 G	.462 G	.495 G
5 to 1	.319 P	.357 P	.385 P	.415 P	.469 P
	.434 G	.451 G	.464 G	.479 G	.507 G
Helical (12-degree)					
1 to 1	.422 P	.465 P	.496 P	.530 P	.589 P
	.422 G	.465 G	.496 G	.530 G	.589 G
3 to 1	.455 P	.496 P	.526 P	.588 P	.611 P
	.549 G	.574 G	.593 G	.613 G	.648 G
5 to 1	.465 P	.506 P	.534 P	.565 P	.616 P
	.581 G	.602 G	.617 G	.633 G	.662 G

tain an even number for the total (N_T), and then divide the total to pinion and gear to give the ratio closest to the original desired ratio. Equation 15 can be used.

$$N_p = d \times P_n \times \cos(\psi) \quad \text{or} \quad N_p = d \times P_d \tag{12}$$

$$N_g = D \times P_n \times \cos \psi \quad \text{or} \quad N_g = D \times P_d \tag{13}$$

$$N_T = 2 \times C \times P_n \times \cos \psi \tag{14}$$

$$N_p = \frac{N_T}{m_g + 1} \tag{15}$$

The final number of teeth should be a reasonable value. The choice of number of teeth can be a function of speed. The best measure of speed is pitch line velocity, which is calculated per Equation 5.

The number of teeth should not be less than 16 unless more detailed analysis is done to determine the optimum tooth form. Above 4,000 ft/min pitch line velocity, at least 20 teeth should be used, and at least 26 teeth above 10,000 ft/min. If the estimating procedure gives a number of pinion teeth less than recommendation, then a larger diametral pitch should be used, and the strength requirement satisfied by using a higher-strength material or an increase in diameter or face width.

Gear Set Sizing

Example. Estimate the size of gear set required for a 500-hp high-speed electric motor running 3,500 rpm to drive a reciprocating compressor at 700 rpm. Assume that there are no dimensional requirements (any center distance will work). Because of the output torque, this will be a relatively large gear set. Therefore, a single helical, through-hardened gear set will be selected. For single helical gears, a good starting point would be to choose a helix angle of 12 degrees; for double helical gears, 28 degrees would be a better choice.

Start with the gear ratio: 3,500/700 = 5.0 to 1.

From Table 3, assume through-hardened alloy steel, 350 BHN for the pinion and 300 BHN for the gear; therefore, $K_{all} = 400$.

From Table 4, select C_a for a motor driving a compresser to give a long-life gear set: $C_a = 1.56$.

To select a value for C_v, an estimate of pitch line velocity is needed. Assume that the pinion will be approximately 10 inches in diameter and use Equation 5 to calculate the velocity:

$$v = .262 \times 10.0 \times 3,500 = 7,336 \text{ ft/min}$$

From Table 5, the value for C_v can be selected as 1.33, and C_m as 1.3.

From Equation 4:

$$C_d = 1.56 \times 1.33 \times 1.30 = 2.70$$

Assume $F_d = 1.0$, and then Equation 6 can be used to determine the pinion diameter:

$$d = \left[\frac{126,000 \times 500 \times 2.70}{1.0 \times 400 \times 3,500} \times \left(\frac{5.0+1}{5.0} \right) \right]^{1/3} = 5.26 \text{ in.}$$

At this point, the pitch line velocity must be checked to determine if the estimate used to select the dynamic factor C_v was valid.

$$v = .262 \times 5.26 \times 3,500 = 4,823 \text{ ft/min}$$

This is lower than the estimated velocity and would give a C_v factor of 1.26 from Table 5. However, this would only reduce the diameter of the pinion by 2%, so there is no need to recalculate.

Per Equations 8 and 9:

$F = 1.0 \times 5.26 = 5.26$ in.

$D = 5.26 \times 5.0 = 26.32$ in.

$$C = \frac{5.26 + 26.32}{2} = 15.79 \text{ in.}$$

Next, the size and number of teeth for the pinion and gear must be determined. From Table 3, the allowable bending stress S_{at} is selected: 45,000 psi for the pinion and 40,000 psi for the gear. The remaining factors needed are the J factors for the gears. If we assume that the number of teeth on the pinion will be approximately 25, then the J factors can be taken from Table 7: $J_P = .534$ and $J_G = .617$.

Now, Equation 11 can be used to solve for the size of the gear teeth, the diametral pitch P_d. Because the J factors and allowable bending stresses are different for the pinion and the gear, it will be necessary to calculate both and use the smaller value, which is the larger size gear teeth (the smaller the diametral pitch P_d, the larger the teeth).

$$P_{dP} = \frac{3,500 \times 5.26 \times 5.26 \times .534 \times 45,000}{252,000 \times 500 \times 2.70} = 6.84$$

$$P_{dG} = \frac{3,500 \times 5.26 \times 5.26 \times .617 \times 40,000}{252,000 \times 500 \times 2.70} = 7.03$$

Continue to determine the actual size and number of teeth based on the smaller P_d, 6.84:

$P_n = 6.84 \times \cos(12) = 6.69$

As stated above, the P_d should be a nominal integer if it is desired to use standard tools. Therefore, the P_n will change to 6.0. Now the number of teeth can be calculated using Equations 12 and 13.

$N_p = 5.26 \times 6.0 \times \cos(12) = 30.87$

Because the number of teeth must be a whole integer number, use 31 teeth for the pinion:

$N_G = m_g \times N_p = 5.0 \times 31 = 155$

In this case, the desired gear ratio has been achieved exactly, but the rounding of the normal diametral pitch requires recalculation of the pitch diameters and the center dis-

tance. Equations 12 and 13 can be reorganized to calculate the diameters based on the nominal diametral pitch:

$$d = \frac{N_P}{P_n \times \cos \psi} = \frac{31}{6.0 \times \cos(12)} = 5.28$$

$D = m_g \times d = 5.0 \times 5.28 = 26.41$

and

$$C = \frac{d + D}{2} = \frac{5.28 + 26.41}{2} = 15.85$$

A more accurate rating of this gear set by the current AGMA standards gives a capacity of 537 hp, limited by pitting resistance with a bending fatigue capacity of 900 hp. The bending capacity should normally be 1.5 to 2.0 times the pitting capacity because it is much more desirable to have the gears suffer surface deterioration than to have tooth breakage.

Drawing Data

When the pinion and gear are designed, it is necessary to put adequate information on manufacturing drawings for the parts to be produced. The drawings must have the normal graphic representation of the part configuration, the dimensions that define the size, and the normal material and specification notes. In addition, there are dimensions required to produce the gear teeth that are usually shown on the drawing in table form. Table 8 shows an example of the the data required. To determine some of the manufacturing dimensions, it is necessary to select a value for backlash, or running clearance. For normal conditions, the backlash

Table 8
Typical Gear Drawing Data Table

Number of teeth	xx
Normal diametral pitch	xx
Transverse diametral pitch	xx
Normal pressure angle	xx°
Transverse pressure angle	xx°
Pitch diameter	x.xxx
Normal circular thickness	.xxx–.xxx
Normal chordal addendum	.xxx
Normal chordal thickness	.xxx–.xxx
Whole depth	.xxx
Helix angle left (or right) hand	xx°
Lead	x.xxx

is a function of diametral pitch and can be selected from Table 9. For very high speeds or unusual temperature ranges, more detailed study is required to determine the proper backlash.

In addition to the physical size and data used in the design, dimensions are required to allow machining the teeth with the precision required for smooth operation. The basic method of measuring the teeth is to use a tooth caliper. There are two calipers, one to measure the tooth thickness and one

Table 9
Backlash

Diametral Pitch	Backlash (B)	Diametral Pitch	Backlash (B)
1	.025–.040	5	.006–.009
1½	.018–.027	6	.005–.008
2	.014–.020	7	.004–.007
2½	.011–.016	8–9	.004–.006
3	.009–.014	10–13	.003–.005
4	.007–.011	14–32	.002–.004

to control the depth point where the measurement is made. The thickness is called the *normal chordal thickness,* and it is measured at the *normal chordal addendum.* Equations 16 and 17 can be used to calculate the thickness and addendum.

$$a_c = a + \frac{d}{2}\left(1 - \cos\frac{180 \times t}{\pi \times d}\right) \qquad (16)$$

$$t_c = d \times \sin\frac{180 \times t}{\pi \times d} \qquad (17)$$

$$t = \frac{\pi \times d}{2 \times N} - \frac{B}{2} \qquad (18)$$

The dimensions are calculated for both the pinion and the gear by using the appropriate number of teeth (N_P or N_G) in Equations 16 and 17. Machining tolerance is provided by calculating with both the minimum and the maximum values of backlash in Equation 18.

Bevel Gear Design

Bevel gear sizes can be estimated with the same methods used for spur and helical gears. First, a few differences in normal practice between bevel gears and parallel shaft gears must be considered. The pitch diameters and tooth size (diametral pitch) for bevel gears are defined at the large end of the gear. To determine load capacity, the load can be considered to be applied at the mid-face where the diameter is called the *mean diameter.* The face width is very consistently sized as one-third of the outer cone distance, based on good design and machine capabilities (see Figure 1).

To estimate size by Equation 6, the face factor F_d must be calculated from Equations 19 and 20. The mean diameter can then be estimated from an equation similar to Equation 6 with the constant adjusted for bevel gears (Equation 21). The face and bevel pinion diameter can then be determined from Equations 22 through 24.

$$\Gamma = \tan^{-1}\frac{1}{m_g} \qquad (19)$$

$$F_d = \frac{0.2}{\sin\Gamma} \qquad (20)$$

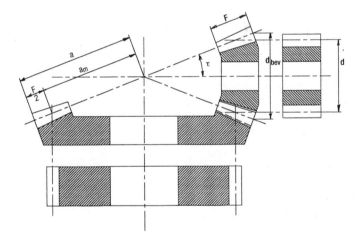

Figure 1. Spur gear equivalent for bevel gears.

Pitch diameter d can now be estimated by modifying Equation 6:

$$d = \left[\frac{100,000 \times P \times C_d}{F_d \times K_{all} \times n} \times \left(\frac{m_g + 1}{m_g}\right)\right]^{1/3} \qquad (21)$$

$$F = d \times F_d \tag{22}$$

$$a = \frac{d}{2.0 \times \sin \Gamma} + \frac{F}{2} \tag{23}$$

$$d_{bev} = 2.0 \times a \times \sin \Gamma \tag{24}$$

The size of bevel gear teeth is calculated the same as with spur and helical gears except a size factor is included. When an average value of size factor is included in Equation 11, the result is Equation 25. Equation 25 can be used to estimate the tooth size. The tapered tooth of the bevel gear changes in size across the face width. The diametral pitch calculated from Equation 25 is the size at the mid-face (mean diameter). The standard nomenclature for bevel gears defines the diametral pitch at the large end. Due to the different methods of manufacturing, it is not necessary to round the calculated pitch value to a whole number, but only to adjust the value to the size at the large end of the gear, per Equation 26.

$$P_d = \frac{n \times d \times F \times J \times S_{at}}{390,000 \times P \times C_d} \tag{25}$$

$$P_{dbev} = P_d \frac{d}{d_{bev}} \tag{26}$$

$$N_{Pbev} = d \cdot P_{dbev} \tag{27}$$

The number of teeth must be adjusted to a whole number, and then the pitch P_{dbev} must be adjusted so the number of teeth will fit the diameter.

$$N_{Gbev} = N_{Pbev} \times m_g \tag{28}$$

After rounding the number of gear teeth to a whole number, the gear diameter can be calculated:

$$D_{bev} = \frac{N_{Gbev}}{P_{dbev}} \tag{29}$$

Bevel Gear Sizing

Example. Estimate the size of bevel gearing for a steam turbine running 4,200 rpm to drive a centrifugal pump requiring 350 hp and running at 2,000 rpm. Most spiral bevel gears are carburized and hardened—this is best for the higher rpm in this application. For normal practice the

pressure angle will be 20 degrees and the spiral angle 30 degrees.

Start with the gear ratio: 4,200/2,000 = 2.1 to 1.

From Table 3, for gears surface-hardened to 58 R_c, the $K_{all} = 600$.

From Table 4, select C_a. Based on smooth driving and driven machines and extra life, choose $C_a = 1.25$.

For an initial value of C_v, assume a diameter d of 4.0 inches. Per Equation 5 calculate the pitch line velocity:

$$v = .262 \times 4.0 \times 4,200 = 4,402 \text{ ft/min}$$

Use standard commercial practice, which would be to cut the gears, then carburize and harden, and then finish by lapping. From Table 5, use $C_v = 1.26$, and from Table 6, use $C_m = 1.30$.

Therefore, from Equation 4:

$$C_d = 1.25 \times 1.26 \times 1.30 = 2.05$$

Now use Equations 19 and 20 to determine the face-to-diameter factor F_d based on the gear ratio of the bevels:

$$\Gamma = \tan^{-1} \frac{1}{2.1} = 25.46$$

$$F_d = \frac{0.2}{\sin 25.46} = 0.47$$

Now the equivalent mean diameter d can be calculated using Equation 21.

$$d = \left[\frac{100,000 \times 50 \times 2.05}{.47 \times 600 \times 4,200} \times \left(\frac{2.1+1}{2.1} \right) \right]^{1/3} = 4.485$$

From Equation 20:

$$F = 4.485 \times 0.47 = 2.11$$

The cone distance a is calculated per Equation 23:

$$a = \frac{4.485}{2 \times \sin 25.46} + \frac{2.109}{2} = 6.271$$

The actual pitch diameter of the bevel pinion is calculated from Equation 24:

$d_{bev} = 2.0 \times 6.271 \times \sin 25.46 = 5.392$

Use Equation 5 to calculate the actual pitch line velocity:

$v = .262 \times 5.392 \times 4,200 = 5,933$ ft/min

This is close enough to 5,000 ft/min to validate the original C_v factor of 1.26. Proceed to Equation 25 to estimate the tooth size at the mean diameter. From Table 3, the value of S_{at} can be selected as 65,000 psi. Assume 22 teeth on the pinion and select J factors from the helical section since this gear set is *spiral* bevel, the equivalent of helical in a parallel shaft gear set.

$J_p = .520$ and $J_g = .560$

Because the material strengths of the pinion and the gear are the same, the pinion with the smaller J factor will dictate the tooth size.

$$P_d = \frac{4,200 \times 4.485 \times 2.109 \times .520 \times 65,000}{390,000 \times 350 \times 2.05} = 4.799$$

From Equation 26 the actual diametral pitch at the large end of the pinion is:

$$P_{dbev} = \frac{4.485 \times 4.799}{5.392} = 3.991$$

From Equation 27:

$N_{Pbev} = 5.392 \times 3.99 = 21.52$

Obviously, the number of teeth must be an integer number, so use $N_{Pbev} = 22$. Working Equation 26 backwards, define the final diametral pitch:

$$P_{dbev} = \frac{22}{5.392} = 4.08$$

From Equation 28:

$N_{Gbev} = 22 \times 2.1 = 46.2$

Use 46 teeth. From Equation 29:

$$D_{bev} = \frac{46}{4.08} = 11.275$$

Thus, the size of the bevel pinion and gear and the numbers of teeth on the gears are estimated. When this gear set is rated by AGMA standards, the results show 423 HP capacity which shows that the estimate is reasonable and conservative.

Cylindrical Worm Gear Design

Worm gears have a number of unique characteristics besides the arrangement of perpendicular shafts offset by the center distance. The input worm is basically a screw thread which makes one revolution to advance the gear wheel one tooth. This makes it possible to have very high gear ratios, especially since the worm can be made with multiple start threads. Due to the sliding nature of the tooth contact, the efficiency can be poor, with typical values being 90% to 50% with the lower values in the high-ratio designs. This characteristic can be used to make a self-locking drive in which the output gear cannot drive the input worm. This is generally the case when the lead angle is 5 degrees or less. Caution must be exercised if this characteristic is desired, because the difference between dynamic and static coefficient of friction can cause a self-locking drive to unlock due to vibration or any slight initiation from the input. Since the efficiency can be low, it is best to think in terms of two different power ratings: the output power to drive the load and the input power which also includes the friction loss load.

Input Power Rating. The rating equation has two parts: the first is the transmitted power and the second is the friction power loss in the mesh.

$$P = \frac{W_t D n}{126,000\, m_g} + \frac{v W_f}{33,000} \qquad (30)$$

The maximum tooth load is:

$$W_t = 900 \times D^{0.8}\, F_c K_r K_v \qquad (31)$$

This is based on a hardened and ground worm running with a centrifugal-cast bronze wheel with a physical face width of 6 inches or less. A wider face, up to 12 inches, would be derated up to 20%. For a chill-cast bronze wheel, derate by 20%, and by 30% for a sand-cast bronze wheel.

F_c is the effective face width, which is the actual face width but not exceeding ⅔ of the mean diameter of the worm. K_r is the ratio correction factor taken from Figure 2. The velocity factor, K_v, is a function of sliding velocity and can be read from Figure 3. The sliding velocity is:

$$v = \frac{\Pi\,nd}{12\cos\lambda} \qquad (32)$$

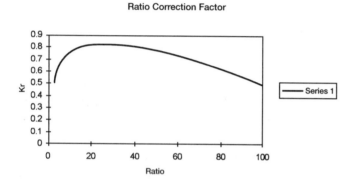

Figure 2. Ratio correction factor.

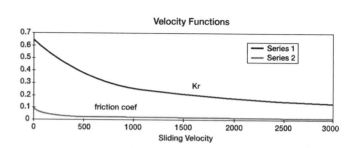

Figure 3. Velocity functions.

where γ = lead angle of the worm thread at the mean diameter

With the tooth load calculated, the friction force can be calculated:

$$W_f = \frac{\mu W_t}{\cos\lambda \cos\phi} \qquad (33)$$

The design of the finished gear drive must allow for cooling the heat from the friction part of the input power. Worm drives frequently have cooling fans on the high-speed input shafts and cooling fins cast onto the housing. It should also be noted that surface finish is critical on the tooth surfaces, and lubricating oil properties are very important.

Materials

Most gears are made of alloy steel. The main criteria for selecting material is the fact that the load capacity of the gear set is proportional to the hardness of the material. There are two major material categories: surface-hardened and through-hardened. Through-hardened alloy steel is normally limited to the range of 38 Rc maximum. One characteristic of through-hardened gear sets that might not be expected is the hardness relationship between the pinion and the gear. For best life and durability, the pinion should be at least 2 Rc points harder than the gear. When both members are the same size—one-to-one ratio—equal hardness works satisfactorily. Surface hardening can increase the surface to as much as 60 Rc while the softer core maintains a ductility and toughness. Of the various methods that can be used to surface-harden gears, three are most common. *Carburizing* is the most common method used to achieve the maximum hardness and gear load capacity. The greatest drawback with carburizing

is the significant geometric distortion introduced during the quenching operation. This requires a finishing operation to restore the dimensional accuracy in almost all designs. As an alternative, *nitriding* can achieve surface hardness in the range of 50 to 60 Rc depending on the steel alloy used. The distortion is usually very low so that finishing is not generally required. The nitriding operation requires a long furnace time—40 to 120 hours in proportion to the case depth—and therefore is normally limited to smaller case depth used for smaller-size teeth and may be impractical for gears with large-size teeth. *Induction* hardening takes a number of forms and can be used with a wide range of case depths. Distortion is usually low. This process requires careful development and, sometimes, tool development to assure consistent quality. Without proper development, the result may give good surface and core hardness but may have problems with ductility and fatigue life.

Summary of Gear Types

With so many types and arrangements of gearing available, a summary is provided below.

Parallel Shaft

This is the most common type of gear and, as the label implies, this type of gear set operates with the axes of rotation parallel to each other. The most common use of parallel shaft gears is to change the speed, and torque, of the driven shaft relative to the driving shaft. The driven shaft also rotates in the opposite direction (unless one of the gears is an internal gear). Unless the two gears are equal in diameter, the smaller diameter member is called a pinion.

Spur. Spur gears are the most basic type of gear. The gear teeth are parallel to the axis of the shaft.

Helical. As the name implies, the teeth on a helical gear have a lead angle relative to the axis of rotation and follow the curve of a helix across the face width of the gear. The tooth load is shared by more pairs of teeth and can be transferred from tooth to tooth more smoothly than the more simple spur gear. Therefore, when all other parameters are equal, a helical gear set can carry more load and run quieter with less vibration than a spur gear set. While the basic cost to manufacture a helical gear is usually no greater than a spur, there is a penalty in the form of a thrust component to the gear reaction loads that must be supported by the shaft and bearings.

Double Helical (Herringbone). A way to counter the thrust loads of the helical gear is the double helical gear. This is accomplished by dividing the face width of the gear into two halves and using the opposite hand of helix for each half. In order to use conventional manufacturing machines, a cutter runout space must be provided between the two halves; this adds to the overall width of the gear and makes it bigger than the equivalent helical gear. With special cutting machines, the space between the two halves can be eliminated, and this type of gear is called *herringbone*. Some of the finishing methods used to improve the capacity and precision of gears cannot be used with herringbone gears. Since nothing in this world is perfect, the gear tooth circle is never perfectly concentric with the shaft axis of rotation, and this eccentricity contributes to vibration and dynamic load. This is of particular importance in the case of double helical gears because the two halves of the face each have specific runouts and combine to create an additional axial runout. While this is not generally a problem, it can require additional manufacturing effort. It is also imperative that the gear shaft bearing and coupling designs allow the two halves of the gear face to share the tooth load equally, as any external thrust loads that react through the gears will cause an overload in one half.

Bevel

Bevel gears have the teeth formed on a cone in place of a cylinder, and the axes of rotation intersect rather than being parallel lines. The most common arrangement has the axes intersecting at 90 degrees; however, other angles can be used, such as seen in Vee drives for boat transmissions.

Straight Tooth. The straight tooth bevel is the spur gear of the bevel family. Being on a cone, the teeth are tapered in thickness from the inner end of the face to the outer end.

Spiral Bevel. The spiral bevel gear is the equivalent of the helical gear on a cone. While the teeth on a straight bevel follow a ray line along the cone from one end of the face to the other, the spiral bevel tooth is modified in two ways. The tooth is set at a spiral angle, similar to the helix angle of the helical gear, and it is curved with the radius of the cutter head used to hold the blades that cut the teeth.

Zerol. The zerol is a special form of the spiral bevel that has the teeth curved with the cutter radius, but with a spiral angle of zero degrees. The curved tooth form gives some of the smoother-action characteristics of the spiral bevel; but with no spiral angle, the thrust reaction is not transmitted to the bearings.

Hypoid. The hypoid gear set is very similar to the spiral bevel set except the input pinion axis is offset so that the axes no longer intersect. More sliding is introduced in the tooth contact which results in a slight reduction in efficiency, but some geometric shaft arrangement problems can be solved.

Worm

Worm gear sets have their input and output axes perpendicular and offset by the center distance. While this arrangement may be an advantage for some applications, the worm gear type is more frequently chosen for other characteristics. Very high gear ratios can be achieved in a single gear stage. However, efficiency goes down as ratio goes up. This is sometimes used to advantage since high-ratio worm sets are the only gears normally designed to be self-locking. A self-locking set acts as a brake, and the gears lock if the output shaft tries to drive the input.

Buying Gears and Gear Drives

One of the most important considerations in purchasing gears is to work with a reliable and experienced vendor. A good source of information on suppliers is the American Gear Manufacturers Association (see References). It is also important to inform the vendor of all possible data about the requirements, application, and use planned for the gears or drive. Keep the specification of detailed gear data to a minimum and allow the vendor to apply his experience to help you get the best possible product. However, the most detailed possible design information should be required to be submitted with the vendor's quotation. The idea is to give the vendor freedom to offer the most appropriate product but to require detailed data with the quotation for evaluation in selecting the best offering. Many times, a second quotation will be in order.

REFERENCES

1. American Gear Manufacturers Association, AGMA and ISO Standards, 1500 King St., Suite 201, Alexandria, VA 22314.
2. Dudley, Darle W., *Practical Gear Design*. New York: McGraw-Hill, Inc., 1984.
3. Drago, Raymond J., *Fundamentals of Gear Design*. Stoneham, MA: Butterworth Publishers, 1988.
4. Townsand, Dennis P., *Dudley's Gear Handbook*. New York: McGraw-Hill, Inc., 1991.

8
Bearings

C. Richard Lenglade, Jr., Development Engineer, Allison Engine Company

TYPES OF BEARINGS

There are two general categories of bearings: rolling element bearings and journal bearings. Most of this chapter is devoted to rolling element bearings because, for most industrial equipment, these are the most common bearings in usage. On the other hand, journal bearings have their place on some types of equipment, and are covered briefly at the end of the chapter.

Rolling element bearings consist of four basic components: the inner ring, the outer ring, the cage or separator or retainer, and the rolling elements, either balls or rollers. The inner ring is mounted on the shaft with the rolling elements between it and the outer ring, which goes in the housing.

Rolling element bearings can be grouped into two basic types: ball bearings and roller bearings. Each type has its advantages and disadvantages which are described below in the discussion for each type of bearing.

Ball Bearings

Ball bearings have a number of advantages over roller bearings, but they also have some disadvantages. Advantages are:

- Low friction
- Low heat generation
- Higher speeds
- Low cost
- Take both radial and thrust loads
- Less sensitive to mounting errors

Disadvantages are:

- Lower life
- Lower load capacity

There are many different types of ball bearings, each designed for a particular type of application. The most common type of ball bearing is the Conrad or deep groove type (Figure 1). It is suitable for radial loads, thrust loads in both directions, or a combination of both. This bearing uses either a two-piece riveted cage or a snap-on polymeric cage. This feature of the bearing tends to limit its top end speed where a one-piece cage is needed, but it is suitable for most industrial machine speeds.

Another common type of ball bearing is the angular contact ball bearing (Figure 2). This bearing is designed primarily for thrust loads but can take limited radial loads if sufficient thrust loads are also present. The thrust load must be in one direction only on single bearings. This bearing has the advantage of higher capacity and longer life than a deep groove bearing because one of the rings is counterbored, allowing more balls to be assembled in the bearing. Another advantage for very high speeds is that a one piece cage can be used, if necessary. Angular contact bearings are available in several different contact angles, depending on how much thrust will be present relative to the radial load.

Figure 1. Deep groove (Conrad) ball bearing. (*Courtesy SKF USA, Inc.*)

Figure 2. Angular contact ball bearing. (*Courtesy SKF USA, Inc.*)

Because single angular contact bearings can take thrust in only one direction, they are often used in pairs. This is sometimes called a duplex bearing, or a duplex set (Figure 3). The two single bearings are mounted with their counterbores in opposite directions, allowing thrust in both directions. Duplex bearings can also be conveniently preloaded as a set to provide very rigid and accurate shaft position control and stiffness.

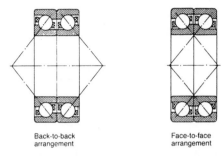

Figure 3. Duplex sets of angular contact ball bearings. (*Courtesy SKF USA, Inc.*)

A variation of the angular contact bearing is the split inner ring bearing (Figure 4). This is a ball bearing with the inner ring split circumferentially, allowing a single row bearing to take thrust in either direction. These bearings are used mostly in the aircraft industry due to their cost.

Figure 4. Split inner ring ball bearing. (*Courtesy SKF USA, Inc.*)

Self-aligning, double-row ball bearings are a somewhat specialized two-row bearing (Figure 5). The outer ring raceway is a portion of a curve with only the inner ring having grooves for the balls to ride in. This allows the bearing to be internally self-aligning, and can compensate for considerable mounting or even dynamic misalignment in the shaft/housing system. Its major disadvantage is that because of the flat outer raceway, the load capacity is not very high.

Figure 5. Self-aligning ball bearing. (*Courtesy SKF USA, Inc.*)

Finally, thrust-type ball bearings are bearings with a 90° contact angle (Figure 6). They cannot take any radial load, but can take considerable thrust load and high speeds. They are somewhat of a specialty bearing due to the special mounting systems required.

Figure 6. Thrust ball bearing. (*Courtesy SKF USA, Inc.*)

Roller Bearings

Roller bearings are usually used for applications requiring greater load carrying capacity than a ball bearing. Roller bearings are generally much stiffer structurally and provide greater fatigue life than do ball bearings of a comparable size. Their advantages and disadvantages tend to be the opposite of ball bearings. Advantages are:

• Greater load capacity
• Greater fatigue life

- Some types take both radial and thrust loads
- Some types less sensitive to mounting errors

Disadvantages are:

- Higher friction
- Higher heat generation
- Moderate speeds
- Higher cost

There are three basic types of roller bearings: cylindrical or straight roller bearings, spherical roller bearings, and tapered roller bearings. As with the ball bearings, each has its strengths and weaknesses.

Cylindrical Roller Bearings

Cylindrical roller bearings have the lowest frictional characteristics of all other roller bearings, which makes them more suitable for high speed operation. They also have the highest radial load carrying capacity. They are not designed for carrying axial loads, although some configurations can handle very small axial loads, such as shaft positioning, when there is no external thrust load. Cylindrical roller bearings are also very sensitive to misalignment. Often, their rollers have a partial or even a full crown to help this situation.

Cylindrical roller bearings are available in a variety of rib configurations (Figure 7). These are illustrated below. In general, there must be at least two ribs on one of the rings. One or two ribs on the other ring allow the bearing to lo-cate the shaft as long as there is no external thrust load. This feature is used in gear trains by using two cylindrical bearings to support the spur gear shaft with no ball bearing. The typical cylindrical roller bearing is free to float axially. It has two roller guiding ribs on one ring and none on the other. Then a ball bearing or other thrust type bearing is used on the other end of the shaft to locate it.

Spherical Roller Bearings

Spherical roller bearings (Figure 8) are so named because the cross-section of one of the raceways, usually the outer raceway, makes up a portion of a sphere. The rollers of this type of bearing are barrel shaped and usually symmetrical but sometimes off-center or asymmetrical. The bearings are available in both single- or double-row configurations, but the double-row design is by far the most common. Spherical roller bearings are capable of carrying high radial loads or, in the double-row versions, a combination of radial and axial loads. The single-row design cannot take any thrust loading.

The great advantage of the spherical roller bearing over the ball bearing or cylindrical roller bearing is its ability to take considerable amounts of misalignment without reduction of capacity. The misalignment can be either static or dynamic, and as much as 3 to 5 degrees depending on the internal geometry of the bearing. It can also take much more thrust load than a ball bearing of the same size. Its biggest disadvantage is that it is the most difficult bearing type to manufacture. It costs several times as much as a

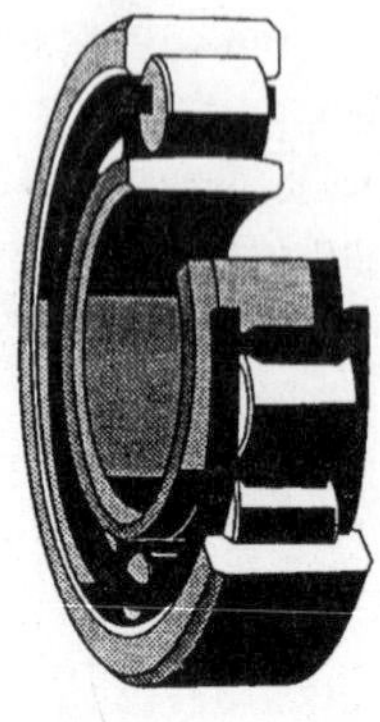

Roller Bearing Types						
		C	D	E	F	G
Bearing Type						
Inner Ring Flanges		Both Sides	Both Sides	None	One Fixed One Separable	One Side
Outer Ring Flanges		None	One Side	Both Sides	Both Sides	Both Sides

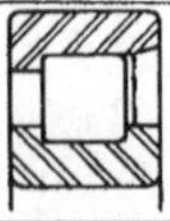

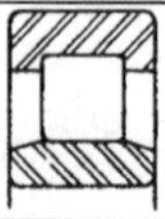

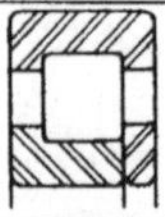

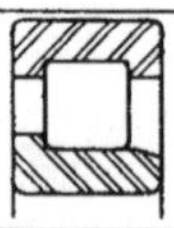

Figure 7. Cylindrical roller bearing rib configurations [16]. (*Courtesy SKF USA, Inc.*)

| Bearing on adapter sleeve | Bearing on withdrawal sleeve | Bearing with cylindrical bore |

Figure 8. Spherical roller bearings. (*Courtesy SKF USA, Inc.*)

cylindrical roller bearing with the same load capacity. The other significant disadvantage is that it has more friction and heat generation than any other type of bearing.

Tapered Roller Bearings

Tapered roller bearings (Figure 9) are similar to cylindrical roller bearings except that the roller is tapered from one end to the other and the raceways are angled to match the roller taper. Unlike cylindrical roller bearings, they can take large thrust loads or a combination of radial and thrust loads. Tapered roller bearings can be mounted on the shaft in pairs, taking thrust in both directions and completely controlling the shaft location. They also have more load capacity than a spherical roller bearing of the same size and

are much less difficult to manufacture, providing a significant cost advantage.

The biggest disadvantage of the tapered roller bearing is its tapered design. In operation, the raceway forces push the roller to one end of the bearing so that there must be a guide flange present to keep the roller in the bearing. This sliding contact causes friction and heat generation and makes the bearing generally unsuitable for high speeds. The other disadvantage of this bearing is that it is sensitive to misalignment, just like a cylindrical roller bearing. In general, tapered roller bearings have the same .001″ per inch requirement for full load capacity.

Because the tapered roller bearing has evolved a little differently than other types of roller bearings, its part terminology is different. Inner rings are frequently called cones and outer rings are called cups.

Figure 9. Tapered roller bearings. (*Courtesy SKF USA, Inc.*)

Standardization

Bearings are one of the earlier manufactured items to have become standardized. Today, almost all bearings are made to a strict standard, for many features, that is the same around the world in many aspects, especially in the areas of boundary plan and tolerances. A standardized set of definitions has been developed by the American Bearing Manufacturers Association (ABMA) for the various bearing components and some of their key dimensions and tolerances. To better understand the discussions that follow and to better communicate with bearing suppliers, some of these definitions as given in ANSI/AFBMA Standard 1-1990 [4] are included here.

Inner ring: A bearing ring incorporating the raceway(s) on its outside surface.
Cone: An inner ring of a tapered roller bearing.

Outer ring: A bearing ring incorporating the raceway(s) on its inside surface.
Cup: An outer ring of a tapered roller bearing.
Cage: A bearing part which partly surrounds all or several of the rolling elements and moves with them. Its purpose is to space the rolling elements and generally also to guide and/or retain them in the bearing.
Separator: Another word for cage.
Retainer: Another word for cage.
Rolling element: A ball or roller which rolls between raceways.
Raceway: A surface of a load supporting part of a rolling bearing, suitably prepared as a rolling track for the rolling elements.
Bearing bore diameter (bore): The bore or I.D. of the inner ring of a rolling bearing.

Bearing outside diameter (O.D.): The outside surface of the outer ring of a rolling bearing.

Bearing width (width): The axial distance between the two ring faces designated to bound the width of a radial bearing. For a single row tapered roller bearing this is the axial distance between the back face of the cup and the opposite face of the cone.

In the United States, this standardization is controlled by the American National Standards Institute (ANSI) together with the American Bearing Manufacturers Association (ABMA), formerly the Anti-Friction Bearing Manufacturers Association (AFBMA). The ABMA has published a large number of standards on bearings, including boundary plans, tolerances, life calculations and load ratings, gauging practices, ball specifications, mounting practices, and packaging. It also works together with the International Standards Organization (ISO) in the development of international standards. The three engineering committees of the ABMA develop these standards for the United States and consist of the Annular Bearing Engineering Committee (ABEC—for ball bearings), the Roller Bearing Engineers Committee (RBEC—for roller bearings), and the Ball Manufacturers Engineers Committee (BMEC—for bearing balls only).

The basic boundary dimension plan consists of the inner ring bore or I.D., the outer ring O.D., and the bearing width. It is important to realize that bearing boundary dimension plans are so standardized that they need to be factored into every machine design. The shaft and housing should be sized to correspond to one of the standard bearing boundary plans. Only the gas turbine aircraft engine industry, due to its special designs and extremely low volume usage, can violate the standard boundary plans, and even then it is usually cost-effective to take them into account. The good news is that there are a tremendous number of dimensional variations in the standards, many of which are commonly produced.

Boundary plans are done in terms of millimeters. This is true both around the world and in the United States.

Some manufacturers make a variety of bearings with the inner ring bore dimension in even fractional inch sizes, but even these bearings are merely variations of a metric boundary plan with an undersized or oversized bore to the nearest fractional inch. The entire range of boundary plan variations is given in the ANSI/ABMA Standard 19-1974 [9] for tapered roller bearings and ANSI/ABMA Standard 20-1987 [10] for ball bearings and cylindrical and spherical roller bearings, and are too extensive to list here. However, Table 1 lists the boundary plans for the most commonly available ball and roller bearings.

The exception to the above comments and the table on boundary plan standardization are the tapered roller bearings. ABMA also publishes standards for tapered roller bearings, but they do not follow the same boundary plan rules as other bearings. Some tapered roller bearings have metric boundary plans, but many more have inch dimension boundary plans. The sizes that are available in general come from the standardization plan developed by The Timken Co., and are not as easily categorized as the other types of bearings. Their standards have effectively been adopted by the rest of the world.

Another important area of standardization by ANSI/ABMA is tolerances. Certain dimensional features of bearings have had the allowable tolerances in manufacture standardized. These features are: bore and O.D. variation (roundness), width variation, bore and O.D. diameter variation (taper), side face runout with bore, raceway radial runout, and raceway axial runout. In general, these tolerances control the running accuracy of a bearing. The tolerances have been grouped into classes and numbered—the higher the number, the higher the bearing precision. For ball bearings they are ABEC-1, -3, -5, -7, and -9, and for roller bearings they are RBEC -1, -3, and -5. In both cases, Class 1 bearings are standard commercial bearings, and Class 5 are standard high precision, or aircraft precision, bearings. A complete listing of bearing tolerances can be found in ANSI/ABMA Standard 4. Some precision bearing manufacturers list some of these tolerances in their catalogs.

Table 1
Common Boundary Dimension Plans

Dimensions in mm

Bore Size	Dimensions Series											
	1 8		1 9		1 0		0 2		0 3		0 4	
	O.D.	Width	O.D.	Width	O.D.	Width	O.D.	Width	O.D.	Width	O.D.	Width
10	19	5	22	6	26	8	30	9	35	11	37	12
12	21	5	24	6	28	8	32	10	37	12	42	13
15	24	5	28	7	32	9	35	11	42	13	52	15
17	26	5	30	7	35	10	40	12	47	14	62	17
20	32	7	37	9	42	12	47	14	52	15	72	19
22	34	7	39	9	44	12	50	14	56	16		
25	37	7	42	9	47	12	52	15	62	17	80	21
28	40	7	45	9	52	12	58	16	68	18		
30	42	7	47	9	55	13	62	16	72	19	90	23
32	44	7	52	10	58	13	65	17	75	20		
35	47	7	55	10	62	14	72	17	80	21	100	25
40	52	7	62	12	68	15	80	18	90	23	110	27
45	58	7	68	12	75	16	85	19	100	25	120	29
50	65	7	72	12	80	16	90	20	110	27	130	31
55	72	9	80	13	90	18	100	21	120	29	140	33
60	78	10	85	13	95	18	110	22	130	31	150	35
65	85	10	90	13	100	18	120	23	140	33	160	37
70	90	10	100	16	110	20	125	24	150	35	180	42
75	95	10	105	16	115	20	130	25	160	37	190	45
80	100	10	110	16	125	22	140	26	170	39	200	48
85	110	13	120	18	130	22	150	28	180	41	210	52
90	115	13	125	18	140	24	160	30	190	43	225	54
95	120	13	130	18	145	24	170	32	200	45	240	55
100	125	13	140	20	150	24	180	34	215	47	250	58
105	130	13	145	20	160	26	190	36	225	49	260	60
110	140	16	150	20	170	28	200	38	240	50	280	65
120	150	16	165	22	180	28	215	40	260	55	310	72
130	165	18	180	24	200	33	230	40	280	58	340	78
140	175	18	190	24	210	33	250	42	300	62	360	82
150	190	20	210	28	225	35	270	45	320	65	380	85

Source: ANSI/AFBMA.

Materials

The types of steel used in bearing inner rings, outer rings, balls, and rollers are made especially for bearings. The material properties are extremely important to the life of any bearing. The requirements for bearing steels are high strength, wear resistance, excellent fatigue resistance, and dimensional stability. In addition, they must be capable of being hardened to a high level, producing a very fine and uniform microstructure, having a high level of cleanliness, and having the proper chemistry.

Table 2 lists most of the steels used in bearings. It also gives their useful, continuous temperature limit. The problem is that for off-the-shelf industrial bearings, there is usually not a choice as to the material. If special requirements are needed, a specific bearing manufacturer

Table 2
Common Bearing Steels and Their Temperature Limits

Thru-hardening Steels	Case-hardening Steels	Temperature Limit
TBS-9	4118	300°F
STROLOY 503-A	5120	
52100	8620	
	4620	.
	4720	
	4320	
52100 TYPE 1	4820	350°F
52100 TYPE 2	9310	
440C	3310	
M50	M50NiL	550°F
T-1 (18-4-1)		

should be consulted. Some bearing materials are available in increasing levels of cleanliness, which will increase the life of any bearing; but again, these are only available on special order.

Bearing cages are also available in a variety of materials. The commonly used materials are shown in Table 3 along with the useful, continuous temperature limit for each. Again, common industrial bearings are usually only available in one cage material, selected by the manufacturer for general use. Other cage materials can often be obtained in the high precision bearings of some manufacturers.

Table 3
Common Bearing Cage Materials and Their Temperature Limits

Material	Temperature, °F
Low carbon steel	400
Bronze/brass	600
Aluminum	400
Alloy steel with/silver plate	600
Nylon 6/6	250–300
Phenolic	300
Polyethersulfone	400
Polyetheretherketone	500
Polyamide	600

RATING AND LIFE

ABMA Definitions

To provide a means of evaluating similar bearings from different manufacturers, the ABMA developed standards for the way in which bearing capacity and life are calculated. The ABMA standard on ratings was adopted as ANSI B3.11. The load rating standards have been published as ANSI/ABMA Standard 9-1990 [7] for ball bearings and Standard 11-1990 [8] for tapered roller bearings, spherical roller bearings, and cylindrical roller bearings. As with the dimensions, there are a number of special terms associated with bearing life, and these are also defined in ANSI/ABMA Standard 1, Terminology. Several of the more important ones are given below.

Basic rating life or L_{10} life: The predicted value of life, based on a basic dynamic radial load rating, associated with 90% reliability.

Basic dynamic radial load rating (capacity): That constant stationary radial load which a rolling bearing can theoretically endure for a basic rating life of one million revolutions. Often referred to as the "basic load rating."

Basic static radial load rating (C_O): Static radial load which corresponds to a calculated contact stress at the center of the most heavily loaded rolling element/raceway contact of 580,000 psi. (NOTE: For this contact stress, a total permanent deformation of rolling element and raceway occurs which is approximately .0001 of the rolling diameter.)

Fatigue life (of an individual bearing): The number of revolutions which one of the bearing rings makes in relation to the other ring before the first evidence of fatigue develops in the material of one of the rings or one of the rolling elements. Life may also be expressed in number of hours of operation at a given constant speed of rotation.

The basic dynamic radial load rating is the one used for calculating the life of a bearing and is more useful than the static load rating. When someone talks about the capacity of a bearing, the basic load rating is what he is referring to. The capacity or basic load rating is the best way to compare various bearings of the same type. The formula for bearing capacity is as follows:

For roller bearings:

$$C = f_c \, (i \, l_{eff} \, \cos\alpha)^{7/9} \, Z^{3/4} \, D^{29/27}$$

For ball bearings:

$$C = f_c \, (i \, \cos\alpha)^{0.7} \, Z^{2/3} \, D^{1.8} \quad \text{(for balls larger than 1″, use } D^{1.4})$$

where: C = basic load rating, in pounds
   f_c = a factor which depends on bearing geometry
   i = number of rows of rolling elements
   l_{eff} = effective length of contact between the roller and raceway
   α = bearing contact angle
   Z = number of rolling elements per row
   D = maximum rolling element diameter

These formulas are not complicated, but they do require the knowledge of bearing geometry not usually disclosed by the bearing manufacturers. Because most manufacturers' catalogs contain listings of the capacities of their bearings based on the ABMA standards, it is preferable to use these values to compare one bearing to another and for calculations. These formulas are given here so that the effect of each variable can be judged by the engineer. For instance, judging by the exponents, it can be seen that roller diameter has a greater effect on capacity than either the number of rollers or the roller length.

According to the ABMA standard, the static load rating is that load which will produce a raceway maximum Hertzian contact stress of 580,000 psi. However, this will not usually produce permanent measurable deformation of the bearing raceways and is not directly related to the calculated fatigue life. This is why the static load rating is seldom used except as a guideline for the maximum load that a bearing can take. The nature of the load should also be considered when dealing with static loading. Impact or shock loads will have a more severe effect as will a repetitive cycle. Stiffness of the support structure must also be considered.

Fatigue Life

There are many misconceptions about bearing life. Bearings operating in the field can fail from a variety of causes. Among these are lack of lubrication, corrosion, dirt, wear, and fatigue, to name just a few. It is possible to keep records of operational bearing life and use these as a predictor for future bearings. However, when it comes to new or redesigned applications, the only life that can be calculated is the fatigue life. While it is known that fatigue life failures represent a small percentage of the actual bearing failures in the field, it is still a good yardstick for predicting the reliability of a bearing application.

Bearing fatigue life is generally discussed in terms of calculated L_{10} life. This is also referred to as rating life and also B_{10} life. This L_{10} life or basic rating life is the one that is calculated from the basic dynamic radial load rating or capacity. This life is a statistical value based on high cycle fatigue of the material and is not an absolute value. It is associated with 90% reliability. This means that for any statistically large group of bearings with the same calculated L_{10} life, 10% of them will fail before they reach the calculated life.

There are a number of assumptions included in the calculation of the L_{10} life. It is assumed that the bearing is properly lubricated and the internal geometry is correct. It is assumed that there is no dirt or water present in the bearing. It is also assumed that the loading applied to the bearing is within the bearing's capability and that there is no misalignment. The steel used to make the bearing is assumed to be clean within acceptable bearing standards.

With all of these assumptions, it is easy to see why bearings do not always last as long as their calculated life. In addition, the L_{10} life is a statistical value that does not guarantee the life of any particular bearing, and in fact predicts that some of them (10%) will fail before the calculated life is reached.

The formula for calculating bearing L_{10} fatigue life is relatively simple and based on empirical data. All that is needed is the bearing capacity or basic dynamic load rating from the manufacturer's catalog and the equivalent radial load (discussed in the next section). The formula is as follows:

$$L_{10} \text{ life} = (C/P)^n \text{ in cycles}$$

where: C = basic dynamic load rating
P = equivalent radial load
n = 3 for ball bearings, 10/3 for roller bearings

To convert this formula to hours, the bearing speed must be factored in. The complete formula for bearing L_{10} fatigue life in hours is as follows:

$$L_{10} \text{ life} = \frac{(C/P)^n (1,000,000)}{N (60)} \text{ in hours}$$

where: N = shaft speed (or housing speed, for outer ring rotation)

Life Adjustment Factors

The L_{10} fatigue life formula and the ones for calculating the basic dynamic load ratings are based on empirical data generated in the 1940s and 1950s in the laboratory where all of the conditions could be controlled, resulting in only fatigue-related failures. Because some applications vary, the ABMA has created three life adjustment factors that are intended to be combined with the L_{10} life to obtain an adjusted life, as shown below. Care must be taken in the use of these factors to be sure that the conditions that justify them exist.

$$L_{10}' = a_1 a_2 a_3 L_{10}$$

where: a_1 = life adjustment factor for reliability
a_2 = life adjustment factor for material
a_3 = life adjustment factor for application conditions

There are times when a level of reliability greater than the 90% calculated by L_{10} life is desired. In these cases a_1 can be used as given in the Table 4 (from ANSI/AFBMA Std. 9-1990 [7]).

Over the years, bearing materials and their processing have improved considerably. This means that the empirical data that created the life and capacity formulas are conservative,

since they are based on material cleanliness at the time. In addition, special processing for some steels used in the precision and aircraft bearing industries have improved their cleanliness even further. A few of the materials/processes are listed in Table 5, with a suggested material factor. The problem with using these factors is that these materials are not available in standard, off-the-shelf bearings.

Table 5
Life Adjustment Factor for Material

Material/Process	Life Adjustment Factor
SAE 52100/CEVM	5
SAE M-50/VIM-VAR	6
9310/CEVM	8
8620/CEVM	8
M50 NiL/VIM-VAR	12

The life adjustment factor for application conditions is most commonly used for adjustments due to lubrication, load distribution, clearance, and misalignment. In many cases, there is not enough information available to accurately use these factors. The operating conditions for which a_3 can be assumed to be 1.0 are given in Table 6.

The most common use for the a_3 life adjustment factor is for lubrication effects. This factor can be either greater or less than 1.0, depending on the ratio of the lubricant film thickness to the bearing raceway surface roughness, Λ. (See the lubrication section for a discussion of Λ.) A chart from which the life adjustment factor for lubrication can be calculated is shown in Figure 10.

The bearing load zone refers to the number of rolling elements that are carrying the load for a given condition. The standard life equation assumes a load zone of 180°, which

Table 4
ABMA Life Adjustment Factor for Reliability

Adjusted rating life, L_n	Reliability, per cent From ABMA Standards (1990)	Life factor, a_1
L_{10}	90	1.00
L_5	95	0.62
L_4	96	0.53
L_3	97	0.44
L_2	98	0.33
L_1	99	0.21

Table 6
Standard Conditions for Valid Life Calculation

Load	From 1% to 30% of C
Speed	Within the manufacturer's catalog rating
Temperature	–40°F to +250°F
Lubrication	Oil viscosity giving 1.1 <Λ> 1.3 (where Λ is defined in the text)
Ring Support	Housing and shaft support must keep ring deflections small when compared to the rolling element contact deflections
Misalignment	Between inner and outer rings as follows:

Misalignment		
	Ball bearings	.0030 in/in
	Cylindrical roller bearings	.0005 in/in
	Spherical roller bearings	.0087 in/in
	Tapered roller bearings	.0010 in/in

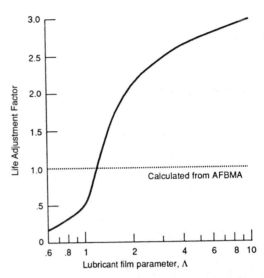

Figure 10. Life adjustment factor as function of lubricant film parameter [12]. (*Reprinted with permission of ASME.*)

implies an operating internal clearance of zero. Because this is not a normally recommended situation, the a_3 factor for load zone is almost always less than 1.0. This type of calculation is usually done in a computer program and combined with all the other application factors. For hand calculations, it is best to ignore this portion of the factor as it is often balanced out by other factors also not used in hand calculations. To a certain extent, this factor has been accounted for in the life equation since it was partly derived from empirical data involving bearings running with some clearance.

Misalignment in excess of the above limits will always result in an a_3 factor less than 1.0. This is also a factor best left to computer programs. However, Figure 11 for ball bearings under radial load illustrates the importance of maintaining good alignment in all situations.

It must be emphasized that care be taken in the use of the three life adjustment factors. A considerable amount of information must be known to use them properly, and often some of the needed information is not available. It is also improper to use one of the adjustment factors to compensate for something like contamination.

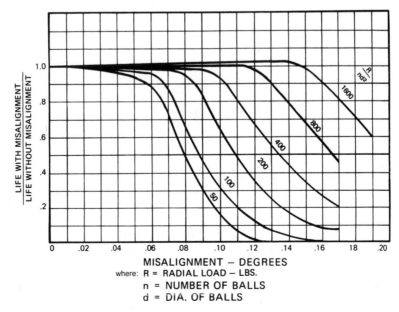

Figure 11. Effect of misalignment on ball bearing fatigue life.

LOAD AND SPEED ANALYSIS

Equivalent Loads

To calculate the L_{10} life of a bearing, the bearing capacity and loading must be known. The previous section discussed the bearing capacity. The loading on a given bearing may be simple or complex, but all of the loads must be converted into a single equivalent load that can be used with the basic dynamic rating discussed above. For radial bearings, the equivalent load is defined by the ABMA:

Equivalent Load: That constant stationary radial load under the influence of which a rolling bearing would have the same life as it will attain under the actual load conditions.

The general formula for the equivalent load is:

$$P = XFr + YFa$$

where: P = equivalent load
   Fr = radial load
   Fa = axial or thrust load
   X = radial factor, taken from bearing manufacturer's catalog
   Y = thrust factor, taken from bearing manufacturer's catalog

For most bearings, the X and Y factors are variable according to the ratio of the thrust to radial load. Table 7a illustrates how the factors are determined for radial deep groove ball bearings, and Table 7b shows how to calculate the factors for radial roller bearings. In both tables, comparing the thrust to radial load ratio to "e" tells which column to use. In Table 7a, the proper row to use is determined by calculating the ratio of thrust load to iZD^2 (where i is the number of rows of balls, Z is the number of balls per row, and D is the ball diameter) and matching the ratio with the values in the table. In Table 7b, the 'α' refers to the bearing contact angle (α = 0 for cylindrical roller bearings). These tables are taken from ANSI/AFBMA Standard 9-1990 [7] and Standard 11-1990 [8]. A more comprehensive table for ball bearings, including thrust bearings and angular contact bearings, can be found in ANSI/AFBMA Standard 9-1990 [7]. Some of these tables are also given in several of the references to varying degrees (10, 12, 13, 15, 16).

Table 7a
Inch Values of X and Y for Radial Ball Bearings[3]

Bearing Type	Relative axial load[1,2]	Single row bearings $\frac{F_a}{F_r} \le e$ X	$\frac{F_a}{F_r} \le e$ Y	$\frac{F_a}{F_r} > e$ X	$\frac{F_a}{F_r} > e$ Y	Double row bearings $\frac{F_a}{F_r} \le e$ X	$\frac{F_a}{F_r} \le e$ Y	$\frac{F_a}{F_r} > e$ X	$\frac{F_a}{F_r} > e$ Y	e
	$\frac{F_a}{iZD^2}$									
	24.92				2.30				2.30	0.19
	50.03				1.99				1.99	0.22
	99.91				1.71				1.71	0.26
Radial contact	149.35				1.55				1.55	0.28
groove ball	200.10	1	0	0.56	1.45	1	0	0.56	1.45	0.30
bearings	300.15				1.31				1.31	0.34
	500.25				1.15				1.15	0.38
	749.65				1.04				1.04	0.42
	999.05				1.00				1.00	0.44

[1]Permissible maximum value depends on bearing design (internal clearance and raceway groove depth). Use first or second column depending on available information.
[2]Values of X, Y, and e for intermediate "relative axial loads" and/or contact angles or obtained by linear interpolation.
[3]Use to obtain P in pounds when D is given in inches.
Source: ANSI/AFBMA Std. 9.

Table 7b
Values of X and Y for Radial Roller Bearings

Bearing Type	$\frac{F_a}{F_r} < e$ X	$\frac{F_a}{F_r} < e$ Y	$\frac{F_a}{F_r} > e$ X	$\frac{F_a}{F_r} > e$ Y	e
Single row, α ≠ 0°	1	0	0.4	0.4cotα	1.5tanα
Double row, α ≠ 0°	1	0.45cotα	0.67	0.67cotα	1.5tanα

Source: ANSI/AFBMA Std. 11.

Cylindrical roller bearings that have opposed integral ribs on inner and outer rings can support limited thrust loads, but these loads do not affect fatigue life. Heavier thrust loads will reduce life due to overturning moments exerted on the rollers, but this effect is difficult to estimate at best.

Due to the design of tapered roller bearings, a radial load will induce a thrust reaction load within the bearing which must be opposed by another bearing somewhere on the shaft, usually another tapered roller bearing mounted in opposition. When only a radial load is applied to a tapered roller bearing, the induced bearing thrust is:

Fa = (.47Fr)/K

where: K = K factor from bearing manufacturer's catalog

There are many variations of mounting arrangements for tapered roller bearings which produce a variety of equations for calculating the total thrust load, including the induced thrust. These are too numerous and rigorous to reproduce here, but the *Bearing Selection Handbook* published by The Timken Company [14] has a complete description.

Loading in the case of a duty cycle can be calculated by means of the formula shown below. When using this formula, the question of what speed to use in the life equation once the load is determined can usually be answered by using the average speed. If the speeds vary widely, the life will not be completely accurate but should be a good estimate.

$$P = \frac{P_1{}^b N_1 t_1 + P_2{}^b N_2 t_2 + \ldots + P_n{}^b N_n t_n}{N_1 t_1 + N_2 t_2 + \ldots + N_n t_n}$$

where: P_n = equivalent load for condition n
  N_n = speed for condition n
  t_n = fraction of time at condition n
  b = 3 for ball bearings; 10/3 for roller bearings

To obtain the most accurate life estimate for a duty cycle, or for a system of bearings, where the life is known for each duty cycle condition or for each bearing, Minor's Rule can be used. The formula for this is as follows:

$$L_{10(weighted)} = \frac{1}{\frac{t_1}{(L_{10})_1} + \frac{t_2}{(L_{10})_2} + \ldots + \frac{t_n}{(L_{10})_n}}$$

where: $(L_{10})_n$ = calculated L_{10} life of condition for bearing n

Contact Stresses

Compared to most mechanical components, bearings operate under high stresses. Even a lightly loaded bearing might have as much as 100,000 psi rolling element/race contact stress. Bearings are designed to take this stress for millions of revolutions. Some applications may only need a very short fatigue life, and the temptation is to use a very small bearing because the life calculations indicate that there is plenty of life. This is often the case when the shaft speed is very low. In this case, the bearing should be designed to a maximum stress level as well as to a life criteria. The maximum stress level for continuous operation is given in Table 8.

**Table 8
Maximum Recommended Stress Level**

| | (All values in ksi) | | |
	Continuous load	Momentary load	Static load
Ball bearing	300	375	610
Roller bearing	320	400	580

Preloading

One of the basic jobs of a bearing is to control the location of the shaft that it is on. Sometimes this control has to be very precise, as in the case of machine tool spindles. The deviation from rotation about the theoretical shaft centerline is called *runout*. There are many factors related to the shaft, housing, and bearings that can affect the runout of the

shaft. One of the common ways to control this runout is by preloading the bearings. Preloading also has several other advantages which are listed below:

• eliminate all radial and axial play
• reduce nonrepetitive runout

- increase rigidity of the shaft system
- impart a known yield to a system
- limit change in contact angle between inner and outer rings at high speed
- prevent undesirable ball dynamics under high acceleration or speed

One of the problems with using preloaded bearings is that they generate more heat when preloaded than when clearance is left in the bearing. If the preload is not carefully designed, bearing operating temperatures can be excessive or even out of control. If the preloaded bearings are not mounted close to each other, they are susceptible to damage due to differential expansion of the shaft or housing system. In addition, the amount of preload adds to the bearing operating load and reduces life.

There are four basic methods of preloading bearings:

1. spring preloading
2. axial adjustment
3. use of duplexed bearings
4. controlled elimination of radial clearance

Spring preloading is a method accomplished by the user. One of the two bearings is mounted with a spring pushing on the outer ring, creating an artificial axial load or preload. The outer ring is able to move axially in the housing so as to maintain a constant spring force. Axial adjustment is done through control of the axial stack dimensions between the two bearings, creating a preload when the complete shaft/housing system is put together. Duplexed bearings are a pair of angular contact ball bearings, or sometimes tapered roller bearings, that are mounted opposed to each other. Their inner rings or outer rings are specially ground to eliminate all axial play in the bearings when they are mounted on the shaft and in the housing.

It is also possible to preload double-row spherical roller bearings not axially, but radially, by very carefully eliminating all of the bearing clearance. This must be done under very closely controlled conditions to avoid overheating of the bearing, and is usually restricted to low or moderate speeds. Controlled clearance bearings are only available from the manufacturer, who should be consulted for application advice. The controlled clearance bearing is not recommended for cylindrical roller bearings.

Special Loads

A shaft is considered indeterminate when it is supported by three or more bearings. The extra bearings will affect the load on the other bearings, inducing leverage forces caused by deflections of the shaft. More than two bearings on a shaft should be avoided, if possible; otherwise, great care must be taken to accurately align the shaft.

Unbalanced loads come in two varieties: ones that are designed to be there, such as a vibratory feeder, and ones that occur during operation, such as dirty fan blades on an air handler. In the second case, if the condition is known to be common, it should be designed for by adding some unbalanced load to the regular operating load and making the bearing selection based on the sum.

Equipment that is designed to have unbalanced load must be analyzed carefully. Although the unbalanced load can be calculated accurately, this type load puts an extra strain on a bearing and is one of the most severe applications for the cage. One way to compensate for this is to multiply the calculated load by an application factor of at least 1.5. It is also suggested that the manufacturer be consulted to determine if the bearing cage is suitable for unbalanced load operation.

Applications involving oscillating motion impose conditions that make bearing selection difficult. Life calculations do not have the same meaning because there are no revolutions. Loading is restricted to only a small portion of the total bearing raceway and on only a few rolling elements. Usually, oscillating motion is so slow that there is no lubricant film built up. The normal mode of failure of this type of application is wear, not fatigue. Therefore, the choice of lubricant is important. Keeping the bearing cavity full of a soft grease with an EP additive or, even better, an EP oil has been found effective in minimizing this wear.

Bearings are often subjected to inertial loading caused by a variety of conditions. Some of the most common are:

- reciprocating motion
- rotary mechanism subject to fluctuating loads
- accelerations
- cyclic or random torsional variations

If any of these are included in the application, whether by design or not, it is important to include them in the load and life calculations. Often, these conditions add to stresses and forces on bearing cages and mounting systems.

There are a number of bearing selection criteria that depend on the bearing loading. Bearing loading is generally expressed in terms of a percentage of the basic dynamic rating or capacity, "C." Table 9 gives the typical groupings for such selections as shaft and housing fits and lubrication practices.

Table 9
Load Ranges for Rolling Element Bearings

Description	Ball	Roller
Very light	<1% C	<%C
Light	1% to 7% C	1% to 8% C
Normal	7% to 15% C	8% to 18% C
Heavy	15% to 30% C	18% to 30% C
Very heavy	>30% C	>30% C

where: C = bearing basic (dynamic) capacity

Effects of Speed

There are a number of considerations for bearing selection associated with speed. Because most applications have the shaft and bearing inner ring turning, shaft speed and bearing speed are often interchanged. Speed is a relative term. Obviously, any given speed will have a greater effect on a larger bearing than on a small one. The most common way of comparing speeds of bearings is by using the term DN. This is calculated by multiplying the bearing bore diameter, in millimeters, by the shaft speed, in rpm. (A 50-mm bore bearing running at 2,000 rpm has a DN level of 100,000.) This factor can be used for all types of bearings, although different types of bearings have different DN levels that are critical.

Most bearing manufacturers specify speed limits for their products in their catalogs. These serve as useful guides for the majority of normal bearing applications. They are based on good lubrication, moderate load, and a reasonable thermal environment. In addition to affecting the life calculation, speed causes heat generation in a bearing. The speed beyond which bearing temperatures exceed a critical value is often the limiting factor. At the other end of the spectrum, the speed may be so low that a good lubricant film is never developed, reducing the life as discussed in the lubrication section. Generally, however, it is the upper limit of speed that is of most concern. Capability of the lubricant, seal requirements, thermal requirements, and even bearing design will affect this limit.

In general, bearings can operate in oil lubrication at higher speeds than in grease. Clearance-type seals can operate at higher speeds than lip seals. One-piece bearing cage designs can operate higher than riveted or assembled cages. Some special bearing steels have higher temperature limits than standard materials, allowing higher speed operation where high speed heat generation would affect the material properties.

Table 10 contains general guidelines for limiting speeds of the different types of bearings. Because lubricant interaction is so important, two conditions are given. To illustrate that these recommendations are not hard and fast, the highest speed attained in each type of bearing is also given. These speeds were usually obtained with very special bearing designs and test machine set-ups and in no way indicate typical performance.

There are several effects of high speed operation that should be considered in bearing selection. Most of these effects do not occur at normal industrial equipment speeds. Gas turbine engines and high speed machine tools are examples of where they can be a problem. These are:

• skidding
• spin-to-roll ratio
• ball excursions
• centrifugal loading

Skidding is a condition in which the bearing loading is so light that it cannot create enough traction for the rolling

Table 10
Bearing Speed Limits

Bearing Type	Approximate Normal Speed Limits, DN		Highest Speeds Attained
	Grease or Oil Bath	Circulating Oil	
Radial or angular contact ball bearings	300,000	500,000	3,500,000
Cylindrical roller bearings	300,000	500,000	3,500,000
Tapered roller bearings	150,000	300,000	3,500,000
Spherical roller bearings	200,000	300,000	1,000,000

elements to roll at the given speed, so they tend to slide along. This can be reduced by increasing the rolling element load by reducing the number of balls or rollers, by reducing the lubricant factor Λ, thereby increasing the traction, and in the case of cylindrical roller bearings, by using a purposely out-of-round outer ring to pinch the rollers and increase the roller load.

Spin-to-roll ratio is a measure of the amount of spinning that a ball does compared to its rolling around the raceway. High spin-to-roll ratio in high speed ball bearings can cause excessive wear. Ball excursions is another ball bearing effect that causes cage wear at high speed when the thrust-to-radial load ratio is too low. Both of these effects are difficult design problems that should be referred to a bearing manufacturer for analysis.

Centrifugal loading is the effect of increased loading of the roller/outer raceway contact due to high speed. This will affect any type of bearing, reducing the life from the standard calculation because of the increased stresses on the outer raceway. This cannot be eliminated but must be considered in the selection of high speed bearings. One situation created by this effect is that, according to the capacity equation (see above), capacity goes up as rolling element diameter increases. At high speeds, a point is reached where a larger rolling element will cause the life to be lower due to centrifugal loading. These situations are best analyzed by special computer programs used by the bearing manufacturers.

LUBRICATION

General

Adequate lubrication of rolling bearings is required for achieving the life calculated for any bearing. In a correctly operating rolling element bearing, a thin film of lubricant separates the rolling elements from the raceways. This film should be of sufficient thickness to actually prevent the rolling element surfaces from touching the inner and outer ring raceways. Contact of the raceway surfaces will result in wear, scoring, and possible seizure. Providing this film is the primary function of the lubricant to four types of internal bearing contact:

- true rolling contact of the rolling element/raceways
- sliding contact between the cage and other bearing components
- partial sliding/rolling contact in some bearing types
- sliding contact between the rollers and guide ribs in roller bearings

In addition, the lubricant has several important secondary functions:

- protection from corrosion
- dissipation of heat
- exclusion of contaminants
- flushing away of wear products and debris

The requirements of a lubricant for rolling element bearings are often more severe than realized. In a rolling element bearing, there are conditions of both rolling and sliding with extremely high contact pressures. The lubricant must withstand high rates of shear and mechanical working not generally prevalent in other mechanical components. For these reasons, proper attention to lubrication is vital for successful bearing operation.

Oils

Oil is a liquid lubricant which can be pumped, circulated, atomized, filtered, cleaned, heated, and cooled, making it more versatile than grease. It is suitable for many severe applications involving extreme speeds and high temperatures. On the other hand, it is more difficult to seal or retain in bearings and housings and, in general, involves a more complicated system than grease.

Viscosity, the measure of an oil's thickness, is the most important property of lubricating oil. The selection of proper viscosity is essential and is based primarily on expected operating temperatures of bearings. Excessive oil viscosity may cause skidding of rolling elements and high friction. Insufficient oil viscosity may result in metal-to-metal contact of the rolling surfaces.

There are two general categories for liquid lubricants: petroleum or mineral oils, and synthetic oils. Mineral oils are lower in cost and have excellent lubricating properties. Synthetic lubricants have been developed to satisfy the need for a wider operating temperature range than is possible with mineral oils. This development has been prompted by the extreme environmental demands of military and aerospace applications. There is a wide range of synthetic types with varying temperature limits. The maximum temperature limits for the common types are given in Table 11. The major disadvantage of synthetics is that they do not have the same load-carrying capacity as do mineral oils at typical industrial equipment operating temperatures. Also, synthetics are rarely compatible with mineral oils, so care must be taken when both types are being used in proximity.

Table 11
Approximate Temperature Limits for Oils, °F

Petroleum or mineral	300
Superrefined petroleum	350
Synthetic hydrocarbon	400
Synthetic esters	400
Silicones	500
Polyphenolether	500
Perfluorinated compounds	600

Greases

Grease is a combination of mineral oil or a synthetic fluid and a suitable thickener (often called *soap*). The percentage of the oil in grease is usually about 80%, but can range from 70% to 97%. Grease consistency or stiffness is determined primarily by the thickener and base oil viscosity. Greases of a given consistency may be formulated from various combinations of thickener and base oil viscosity so that greases of equal stiffness are not necessarily equal in performance.

Greases considered satisfactory lubricants for rolling element bearings are combinations of soap or nonsoap agents, mineral oil, and additives. Soaps such as sodium, calcium, barium, aluminum, lithium, complexes of these soaps, and nonsoaps such as silica and special clays are generally used. Rust and oxidation inhibitors and extreme pressure agents are often added. Lithium and lithium-complex thickeners seem to give the best all-purpose performance, but each type has its advantages.

Performance of a grease depends on several different factors. The lubricating capability of the grease is mainly dependent on the properties of the base oil used. The corrosion protection of the grease is determined by the thickener. While the temperature limit of grease can be restricted by the base oil also, it is usually restricted by the thickener. Table 12 shows the maximum temperature limits for the most common thickener types.

Be careful to avoid mixing greases of different soap bases. The combination will usually be worse than either one by itself, and sometimes worthless. Care should be taken when mixing greases with the same soap bases from different manufacturers, although this is usually not a problem. However, it is not always obvious what the soap base is unless the manufacturer's data sheet is consulted. In no case should mineral oil greases be mixed with greases using a synthetic oil.

Table 12
Approximate Temperature Limits for Grease Thickeners, °F

Calcium	170
Aluminum	180
Barium	225
Sodium	250
Lithium	300
Synthetics	<500

Lubricant Selection

The major criteria for selection of a lubricant is the viscosity. The selection of the proper viscosity oil is especially important for bearings operating in the high load, speed, or temperature ranges. As mentioned in the section on life adjustment factors, it is necessary to have Λ between 1.1 and 1.3. The Λ factor is a measure of the ratio of the oil film thickness to the surface roughness of the raceways in contact. In other words, the oil film needs to be thicker than the raceway roughness so that there is never metal-to-metal contact.

The viscosity of most oils changes dramatically with a change in temperature. When determining the operating temperature, it is the oil temperature that is important. Generally, the oil temperature is 5° to 20°F greater than that of the bearing housing.

The oil film thickness is calculated through the theory of *elastohydrodynamic* lubrication. This involves the elastic deformations of the raceway contacts, and the pressure-viscosity effects and hydrodynamics of the lubricant. This theory is very complex and best left to computer programs. However, a simplified method for determining if the oil film thickness is sufficient involves using Figures 12 and 13. From Figure 12, find the oil viscosity needed based on the

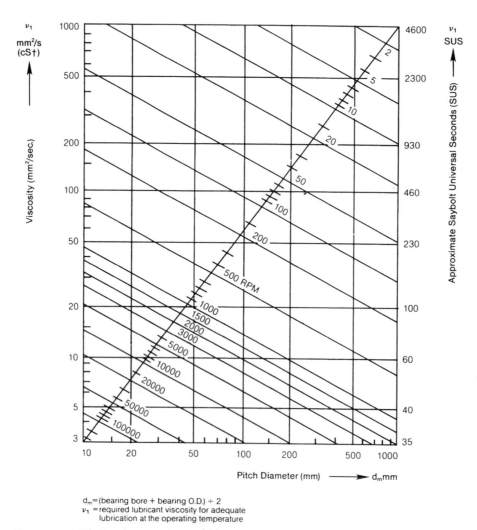

$d_m = $ (bearing bore + bearing O.D.) ÷ 2
$\nu_1 = $ required lubricant viscosity for adequate
lubrication at the operating temperature

Figure 12. Minimum required lubricant viscosity [15]. (*Courtesy SKF USA, Inc.*)

bearing size and the operating speed. Then, using Figure 13, combine that viscosity with the operating temperature of the oil to determine what grade of oil is needed. This should give a ballpark estimate of which oil to use: It assumes the use of a mineral oil with a viscosity index of 95. If a much different oil is to be used (like a synthetic, for example), it is best to consult with a bearing manufacturer or oil supplier to make a more detailed calculation.

If the Λ factor is less than 1.1, the bearing life will be reduced. If Λ is greater than 1.3, the bearing life will be increased. This implies that the highest viscosity possible should be used. However, as the viscosity goes up, so does the operating temperature. As temperature goes up, the oil viscosity goes down and maintenance activity goes up. This all means that there is a practical limit to the life improvement from higher oil viscosity.

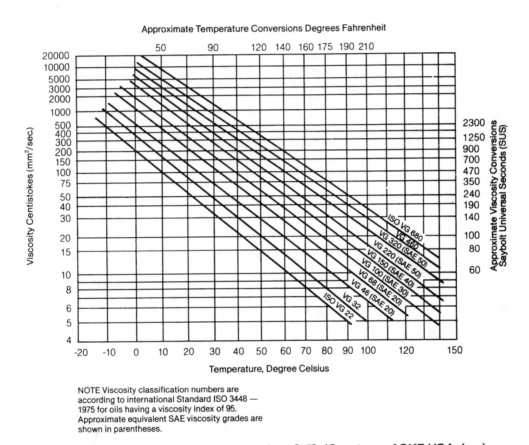

Figure 13. Viscosity-temperature chart [15]. (*Courtesy of SKF USA, Inc.*)

Lubricating Methods

There are a variety of methods to apply the proper amount of oil to a bearing. The most common are as follows:

- Oil bath
- Circulating systems
- Jet lubrication
- Mist lubrication
- Wick feed

The simple oil bath method is satisfactory for low and moderate speeds. The oil level in the housing should not be less than the lip of the outer ring, nor higher than the center of the lowest rolling element. The oil level should only be checked when the bearing is not rotating.

Circulating oil is an excellent way to lubricate a bearing, especially on large machines, and can reduce maintenance and prolong the life of the oil in severe operating conditions.

It can be used with either a wet or dry sump, with the oil usually introduced on one side of the bearing and drained on the other. A system shutoff with loss of pressure is a desired feature. This system is good for all speeds and loads. The main drawback of such a system is the cost.

Jet lubrication is a special type of circulating oil system used on very high speed bearings such as in a gas turbine engine. Most of the oil in this method is used for cooling of the bearing. This method can be used at speed levels up to 1.5 million DN with proper design. The jet should be aimed at the largest space between the cage and the ring lands.

Mist lubrication is of two types, which are distinguished by the method of generating the mist. In some applications, the mist is generated by a flinger that dips into the oil and throws it into the air in the vicinity of the bearing. Sometimes gears substitute for the flinger. Another way to generate the mist is to spray a jet of oil against the side on the inside of the machinery. This method can be very effective for bearings where cooling is not needed.

The second type of oil mist lubrication is when the mist is produced by a special mist generator. The oil mist is formed in an atomizer and supplied to the bearing housing under suitable pressure. This method of lubrication has proven very effective in reducing the operating temperature, not so much by air cooling as by the flow of air, preventing excess oil from accumulating in the bearing. Since the air pressurizes the housing and escapes through the seals, the entrance of moisture and grit is retarded. No drain is needed, as the quantity of oil supplied is very small. The problems with mist oil generators is that the immediate area

may be coated with oil and if the oil generator shuts off, the bearings cannot survive long because of the small amount of oil supplied. Also, if there are air pressures created by other parts of the mechanical system, they should be checked to make sure they are not restricting the flow of the air mist under all operating conditions.

Wick feed is also suitable for high speeds because, again, a small amount of oil is delivered to the bearing. Careful maintenance is needed to make sure the cup never runs dry and that the wick is always in contact with the source.

Grease systems are not as numerous as those for oil. Many bearings come from the manufacturer with a supply of grease already in them. Grease can be added to other bearings by hand, filling the internal volume of the bearing one-third to one-half full. A grease gun with a grease fitting on the housing can be used, but care should be exercised not to overfill the housing and cause overheating of the bearing. One method of gauging the amount of grease to add is to add grease slowly while the bearing is running until some grease is just visible coming out either seal. The grease fitting should then be removed briefly to allow any grease backpressure to relieve itself, and then be reinstalled.

There are automatic grease systems on the market that can relieve a lot of maintenance activity when a number of bearings can be grouped into a system. The bearings and/or their housings need to be packed with grease before the system is operated. The disadvantages of these systems is their initial cost and that all bearings on any one system must be able to use the same grease. Bearings with special needs would need a separate system.

Relubrication

The proper relubrication of bearings is often of equal or greater importance than the initial selection of lubricant. The establishment of proper relubrication procedures is also one of the most difficult aspects of lubrication. This is because bearing requirements vary so much depending on the load, speed, temperature, and environmental conditions of operation. The basic concept is to replace the lubricant at a rate that will compensate for deterioration from all causes. Although recommendations can be made, often the best guide can only be established by experience.

The frequency at which the oil must be changed is mainly dependent on the temperature and quantity of oil used. A temperature rise of the oil of 15° to 20°F can double the

rate of oxidation. In an oil bath system, the frequency of oil change can vary from once a year if the oil temperature does not exceed 120°F to four times a year for oil temperatures of 220°F. This assumes that there is no contamination of the oil and that no oil is lost through the seals. These same rules would hold true for a circulating system, modified slightly by how often the total oil quantity is circulated. There is no relubrication necessary for mist or wick feed lubrication, as all of the oil is lost. Of course, it is imperative that the reservoirs or feed cups are kept supplied with oil.

Relubrication with grease is more complicated than with oil. The major variables to be considered are bearing size and type, speed, operating temperature, and the type of grease used.

Some bearings are lubricated for life and not only do not require relubrication but usually have no provision for it. This typically applies to small bearings, and the life referred to is the life of the lubricant—not necessarily the life of the bearing. For other bearings, when new grease is added to an operating bearing, the used grease condition and the amount of grease added to start purging of the grease can be used as a relubrication guide. As a general guideline, Figure 14 can be used and then modified by experience. The chart is valid for stationary machines where loading conditions are normal. The use of a good quality grease is assumed, and the temperature should not exceed 160°F. The relubrication intervals should be halved for every 30°F increase in temperature above 160°F, but the temperature limit of the grease must not be exceeded. When contamination is known to be a concern, relubrication intervals should be reduced accordingly. The best protection for the bearings is a good maintenance program.

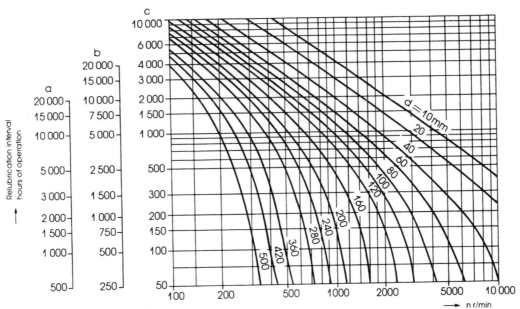

a Radial ball bearings
b Cylindrical roller bearings, needle roller bearings
c Spherical roller bearings, taper roller bearings, thrust ball bearings
d Bearing bore diameter

Figure 14. Relubrication interval [15]. (*Courtesy of SKF USA, Inc.*)

Cleaning, Preservation, and Storage

Bearings come from the manufacturer in a very clean condition. They are usually coated with a preservative oil and wrapped in special corrosion-resistant paper. Bearings should not be cleaned by the user before assembling them on a machine unless something has happened to them after the box has been opened. Cleanliness in bearings is very important. Tests have been done showing decreasing bearing life with increasing oil contamination. New bearings should always be stored in their original packaging whenever possible.

If bearings need to be cleaned, the chemicals used should be consistent with the bearing materials, especially the cage. Many solvents can harm nylon or other types of nonmetal cages. For general cleaning, mineral spirits is recommended. Other solvents will work well on the metal parts, but many have environmental drawbacks. After cleaning, bearings are extremely vulnerable to corrosion and handling damage. They should be dried and coated with oil or preservative fluid as soon as possible. If compressed air is used to dry the bearing, *DO NOT* allow the bearing to spin under the force of the air. This is not only dangerous, but serious damage to the bearing can result. After preserving, the bearings should be wrapped in a neutral grease-proof paper, foil, or plastic film. For a more detailed procedure on cleaning unshielded bearings, recommendations by ABEC have been reproduced in the SKF Bearing Installation and Maintenance Guide [15].

When machinery with rolling element bearings is to be idle for a long period of time, some extra measures should be taken. Bearings should be relubricated before shut-down. No rolling bearing lubricant has been developed which will completely protect a bearing against moisture, but some oils and greases are better than others. Compounded oils and lithium base greases are more water-repellent than others. In severe cases of bearings exposed to the elements for a long storage period, one method of protecting them is to completely fill the housing and bearing with a good water-resistant grease. The only problem with this is that when it comes time to start the machine back up, the excess grease must be removed first, or else some means of allowing the housing and bearing to self-purge without overheating the bearing must be used.

MOUNTING

Rolling element bearings have extremely accurate component dimensions. The inner ring bore and outer ring O.D. are manufactured to very close tolerances to fit their supporting members—the shaft and housing. It follows that the shaft and housing must also be machined to close tolerances. Even with collar-mounted bearings, the quality of the shafting will affect the performance of the bearing.

Shafting

There are several different types of mounting methods available for commercial bearings. These include collar mounting, adapter sleeve, and direct press fit, both straight and tapered bearing seat. Table 13 lists the advantages and disadvantages of each kind.

The aspects of shaft quality that affect bearings are geometric and dimensional accuracy, surface finish deflections, material, and hardness. The geometric accuracy includes not only the bearing seats but the shoulders. Because the inner and outer rings of rolling element bearings are relatively elastic, imperfections in the shafting and the housing can be translated directly to distortion of the bearing raceways. This is especially true of shaft out-of-roundness, taper, and shoulder squareness. A shaft that is not straight can cause dynamic misalignment that severely increases the bearing loading.

Although one of the advantages of using collar-mounted bearings is the use of commercial shafting, some care should be taken. On more critical applications, it is recommended that turned, ground, and polished shafting be used. Other commercial shafting is often quite a bit undersize and can cause the collar to eventually come loose. It is best for the shaft to be no more than .001″ undersize. The best

Table 13 Mounting Methods

Mounting Type	Advantages	Disadvantages
Eccentric locking collar	Quick and easy.	Least reliable, can come loose.
Set screw locking collar	Can use commercial shafting.	Set screw slightly better.
Adapter sleeve	Can use commercial shafting. Positive mounting.	Bearing must have tapered bore. Not all bearings available with a tapered bore. Additional hardware is needed. No accurate axial location.
Press fit—tapered bore	Ease of mounting. Ease of dismounting. Positive mounting.	Requires machined tapered shaft bearing seat. Bearing must have tapered bore. Not all bearings available with a tapered bore.
Press fit—straight bore	Positive mounting. Easier to machine shaft.	Precision machined shaft seat.

mounting for poor quality shafting is adapter sleeve mounted bearings. For this type of mounting, the shaft can be up to .003″ to .004″ undersize and still give a secure fit.

The surface roughness of the shaft may cause loss of press fits and excessive wear and fretting of the bearing seat (if it is too rough). A maximum limit for roughness on the bearing seat of 63 Ra is recommended. If integral seals will also contact the shaft surface, a finish between 10 and 20 Ra is the maximum recommended.

Sometimes a shaft surface itself is used as the inner raceway of the bearing. This is mainly true of cylindrical roller bearings, although occasionally of other types. The most common usage of this concept is for gearbox or transmission applications in which there is a gear on the shaft between two bearings. The advantage is that there is no need to press fit an inner ring on the shaft, and the locknut and lock washer usually associated with keeping the inner ring on the shaft is not needed. The disadvantage is that the shaft raceway surface has to have the same tolerances and finishes that

an actual inner raceway would have. This makes the shaft difficult to make and costly, and can cause a maintenance problem if the inner raceway fails on one end but the gear and other raceway are still good. For this type of application, the shaft raceways should have a surface hardness of Rockwell HRC 59 minimum, a maximum surface roughness of 15 Ra, and freedom from objectionable lobing and waviness. Some manufacturers' catalogs list the raceway diameters needed for different size bearings.

The majority of bearings are mounted on a shaft with a very close or interference fit. The contact pressure and movement of the rollers on the inner or outer ring during operation causes them to fret and even creep around the shaft or housing. The amount of press or interference fit needed varies considerably depending on the application. The factors that must be considered are speed, load magnitude and direction, stiffness of the supporting structure, and the temperature range of the system. Shaft fits recommended for various types of applications are listed in Table 14. In

Table 14
Selection of Shaft Tolerance Classifications for Metric
Radial Ball and Roller Bearings of Tolerance Classes ABEC-1, RBEC-1

Dimensions in inches

DESIGN & OPERATING CONDITIONS			BALL BEARINGS			CYLINDRICAL ROLLER BEARINGS			SPHERICAL ROLLER BEARINGS		
Rotational Conditions	Inner Ring Axial Displaceability	Radial Loading	d Over	Incl.	Tolerance Classification (1)	d Over	Incl.	Tolerance Classification (1)	d Over	Incl.	Tolerance Classification (1)
Inner Ring Rotating in relation to Load Direction		Light	0 0.71	0.71 All	h5 j6 (2)	0 1.57 5.51 12.6 19.7	1.57 5.51 12.6 19.7 All	j6 (2) k6 (2) m6(2) n6 p6	0 1.57 3.94 12.6 19.7	1.57 3.94 12.6 19.7 All	j6 (2) k6 (2) m6(2) n6 p6
or		Normal	0 0.71	0.71 All	j5 k5	0 1.57 3.94 5.51 12.6 19.7	1.57 3.94 5.51 12.6 19.7 All	k5 m5 m6 n6 p6 r6	0 4.72 2.56 3.94 5.51 11.0 19.7	1.57 2.56 3.94 5.51 11.0 19.7 All	k5 m5 m6 n6 p6 r6 r7
Load Direction is Indeterminate		Heavy	0.71 3.94	3.94 All	k5 m5	0 1.57 2.56 5.51 7.87 19.7	1.57 2.56 5.51 7.87 19.7 All	m5 m6 n6 p6 r6 r7	0 1.57 2.56 3.94 5.51 7.87	1.57 2.56 3.94 5.51 7.87 All	m5 m6 n6 p6 r6 r7
Inner Ring Stationary in Relation to Load Direction	Inner Ring must be easily axially displaceable	Light Normal Heavy	All Sizes		g6	All Sizes		g6	All Sizes		g6
	Inner Ring need not be easily axially displaceable	Light Normal Heavy	All Sizes		h6	All Sizes		h6	All Sizes		h6
Pure Thrust (Axial) Load			All Sizes		j6	Consult Bearing Manufacturer					

(1) Tolerance Classifications shown are for solid steel shaft. Numerical values are listed in Table 15.
   For hollow or nonferrous shafts, tighter fits may be needed.
(2) If greater accuracy is needed, substitute j5, k5 and m5 for j6, k6, and m6 respectively.

Source: ANSI/AFBMA Std. 7-1988.

general, the higher the load, the heavier the press fit needed. This same trend is true of speed. An interference fit is generally recommended for the ring, which rotates relative to the major load. A loose or slip fit is recommended for the stationary ring. If the shaft is hollow, a heavier press fit is usually needed. The fits in the table are valid for an operating temperature range between 32° and 250°F and when the speed level is less than 600,000 DN.

The fit classes recommended in Table 14 refer to the bearing/shaft diameter fits given in Table 15. This table gives the bearing bore and tolerance for commercial grade (ABEC 1 and RBEC 1) bearings and the corresponding shaft diameter tolerance from the nominal bearing bore for a range of bearing sizes and fit classes. A complete listing can be found in ANSI/ABMA Standard 7-1996 [6], but many bearing companies reprint portions of the listings in their catalogs.

For unusual applications, it is necessary to calculate the correct fit. These calculations are based on thin wall ring theory. In general, some level of fit pressure must be maintained while at the same time the inner ring hoop stress is within allowable limits. These limits are about 25 ksi for rings made of through hardened material, and 35 ksi for rings made from a carburizing or case-hardened steel.

Table 15
Shaft Fitting Practice for Metric Radial Ball and Roller Bearings of Tolerance Classes ABEC-1, RBEC-1

Part II — Dimensions in Inches — Deviations and Fits in .0001 Inches

TOLERANCE CLASSIFICATIONS

over	incl	Deviation	g6 Shaft Dev	g6 Fit	h6 Shaft Dev	h6 Fit	h5 Shaft Dev	h5 Fit	j5 Shaft Dev	j5 Fit	j6 Shaft Dev	j6 Fit	k5 Shaft Dev	k5 Fit	k6 Shaft Dev	k6 Fit	m5 Shaft Dev	m5 Fit	m6 Shaft Dev	m6 Fit	n6 Shaft Dev	n6 Fit	p6 Shaft Dev	p6 Fit	r6 Shaft Dev	r6 Fit	r7 Shaft Dev	r7 Fit
0.3937	0.7087	0 / -3	-2 / -7	7L / 1T	0 / -4	4L / 3T	0 / -3	3L / 3T	+2 / -1	1L / 5T	+3 / -1	1L / 6T	+4 / -1	0T / 7T			+6 / +3	3T / 9T										
0.7087	1.1811	0 / -4	-3 / -8	8L / 1T	0 / -5	5L / 4T			+2 / -2	2L / 6T	+4 / -2	2L / 8T	+4 / +1	1T / 8T			+7 / +3	3T / 11T										
1.1811	1.9685	0 / -4.5	-4 / -10	10L / 0.5T	0 / -6	6L / 4.5T			+2 / -2	2L / 6.5T	+4 / -2	2L / 8.5T	+5 / +1	1T / 9.5T	+7 / +1	1T / 11.5T	+8 / +4	4T / 12.5T	+10 / +4	4T / 14.5T								
1.9685	3.1496	0 / -6	-4 / -11	11L / 2T	0 / -7	7L / 6T			+2 / -3	3L / 8T	+5 / -3	3L / 11T	+6 / +1	1T / 12T	+8 / +1	1T / 14T	+9 / +4	4T / 15T	+12 / +4	4T / 18T	+15 / +8	8T / 21T						
3.1496	4.7244	0 / -8	-5 / -13	13L / 3T	0 / -9	9L / 8T			+2 / -4	4L / 10T	+5 / -4	4L / 13T	+7 / +1	1T / 15T	+10 / +1	1T / 18T	+11 / +5	5T / 19T	+14 / +5	5T / 22T	+18 / +9	9T / 26T	+23 / +15	15T / 31T				
4.7244	7.0866	0 / -10	-6 / -15	15L / 4T	0 / -10	10L / 10T			+3 / -4	4L / 13T	+6 / -4	4L / 16T	+8 / +1	1T / 18T	+11 / +1	1T / 21T	+13 / +6	6T / 23T	+16 / +6	6T / 26T	+20 / +11	11T / 30T	+27 / +17	17T / 37T	+35 / +26	26T / 45T		
7.0866	7.8740	0 / -12	-6 / -17	17L / 6T	0 / -11	11L / 12T			+3 / -5	5L / 15T	+6 / -5	5L / 18T	+9 / +2	2T / 21T			+15 / +7	7T / 27T	+18 / +7	7T / 30T	+24 / +12	12T / 36T	+31 / +20	20T / 43T	+42 / +30	30T / 54T		
7.8740	8.8583	0 / -12	-6 / -17	17L / 6T	0 / -11	11L / 12T			+3 / -5	5L / 15T	+6 / -5	5L / 18T	+9 / +2	2T / 21T			+15 / +7	7T / 27T	+18 / +7	7T / 30T	+24 / +12	12T / 36T	+31 / +20	20T / 43T	+43 / +31	31T / 55T	+50 / +31	31T / 62T
8.8583	9.8425	0 / -12	-6 / -17	17L / 6T	0 / -11	11L / 12T			+3 / -5	5L / 15T	+6 / -5	5L / 18T	+9 / +2	2T / 21T			+15 / +7	7T / 27T	+18 / +7	7T / 30T	+24 / +12	12T / 36T	+31 / +20	20T / 43T	+44 / +33	33T / 56T	+51 / +33	33T / 63T
9.8425	11.0236	0 / -14	-7 / -19	19L / 7T	0 / -13	13L / 14T			+3 / -6	6L / 17T	+6 / -6	6L / 20T	+11 / +2	2T / 25T			+17 / +8	8T / 31T	+20 / +8	8T / 34T	+26 / +13	13T / 40T	+35 / +22	22T / 49T	+50 / +37	37T / 64T	+57 / +37	37T / 71T
11.0236	12.4016	0 / -14	-7 / -19	19L / 7T	0 / -13	13L / 14T			+3 / -6	6L / 17T	+6 / -6	6L / 20T	+11 / +2	2T / 25T			+17 / +8	8T / 31T	+20 / +8	8T / 34T	+26 / +13	13T / 40T	+35 / +22	22T / 49T	+51 / +39	39T / 65T	+59 / +39	39T / 73T

Housings

The housing should provide a rigid support for the bearing. Housings may be separate components fastened to a machine frame or foundation, or they may be an integral part of the machine. In addition to supporting the load, the housing protects the bearing and often provides other features such as a lubricant reservoir, a lubricant flow system, cooling, and seals.

There are so many things that affect the selection of a housing that it is difficult to make any specific recommendations. Table 16 lists many of the factors that may affect the housing design or selection.

Bearing outside diameters (O.D.) are held to tolerances almost as close as the bores. A system of fits has been developed by the ABMA to provide flexibility in selecting housing fits. Housing fits recommended for various types of applications are listed in Table 17 (from ANSI/AFBMA Standard 7-1996 [6]). The class of fit is determined by the nature of loading, axial movement requirements, temperature conditions, housing materials, and design. In most cases, the outer rings are subjected to stationary loads that permit a loose housing fit when matched with a tight shaft fit. Fits must also account for differential thermal expansion between the bearing and housing so that the bearing O.D. is always able to move axially. However, a loose fit should never be greater than necessary. Excessive looseness results in less accurate shaft centering and additional outer ring deformation under load.

The classes of fits referred to in Table 17 are given in Table 18 for a limited range of bearing sizes. ANSI/ABMA Standard 7-1996 [6] presents a wider range of bearing sizes as well as additional fit classes. The bearing O.D. tolerances shown in the table are for standard commercial (ABEC 1 or RBEC 1) bearings. Precision-class bearings have tighter O.D. tolerances, and therefore, different fit ranges. These are also given in the ABMA standards.

Table 16
Housing Design Considerations

Loading	Accuracy
Magnitude of load: variable or constant	Axial control of shaft
Direction of load: variable or constant	Radial control of shaft
Shock	Bearing/housing fit
Vibration	Squareness and concentricity

Environment	Servicing and Maintenance
Corrosion resistance	Installation problems
Radiation resistance	Removal: frequent, or only at failure
Heat and cold resistance	Relubrication: regreasing or changing oil
Magnetic permeability	

Styling, Appearance, and Cost	Accessories and Auxiliaries
	Lubrication: grease or oil
Housing: solid or two-piece design	Lube method: circulating, bath, mist
Construction: casting or fabrication	Seals and sealing
Weight: massive or light	Controls: thermocouples, switches, sensors

Source: Link-Belt Bearing Technical Journal [11].

Again, many bearing companies include portions of these tables in their catalogs.

In most bearing/shaft/housing systems, it is necessary to have one fixed bearing to locate the shaft, and one expansion bearing. The purpose of the expansion bearing is to prevent preloading of the two bearings against each other. This is often accomplished through housing design. For ball bearings and spherical roller bearings, this is done by using a loose fit of the bearing, in its housing or on the shaft.

Table 17
Selection of Housing Tolerance Classifications for Metric Radial Ball and Roller Bearings of Tolerance Classes ABEC-1, RBEC-1

DESIGN AND OPERATING CONDITIONS				TOLERANCE CLASSIFICATION (1)
Rotational Conditions	Loading	Other Conditions	Outer Ring Axial Displaceability	
Outer Ring Stationary in relation to load direction	Light Normal or Heavy	Heat input through shaft	Outer ring easily axially displaceable	G7 (3)
		Housing split axially		H7 (2)
		Housing not split axially		H6 (2)
	Shock with temporary complete unloading		Transitional range (4)	J6 (2)
Load Direction indeterminate	Light			J6 (2)
	Normal or Heavy	Split housing not recommended		K6 (2)
	Heavy shock			M6 (2)
Outer Ring Rotating in relation to load direction	Light			M6 (2)
	Normal or Heavy			N6 (2)
	Heavy	Thin wall housing not split	Outer ring not easily axially displaceable	P6 (2)

(1) For cast iron or steel housings. Numerical values are listed in Table 18. For housings of non-ferrous alloys tighter fits may be needed.
(2) Where wider tolerances are permissible, use tolerance classifications H8, H7, J7, K7, M7, N7 and P7 in place of H7, H6, J6, K6, M6, N6 and P6 respectively.
(3) For large bearings and temperature differences between outer ring and housings greater than 10 degrees C, F7 may be used instead of G7.
(4) The tolerance zones are such that outer ring may be either tight or loose in the housing.

Source: ANSI/AFBMA Std. 7-1988.

Table 18

Housing Fitting Practice for Metric Radial Ball and Roller Bearings of Tolerance Classes ABEC-1, RBEC-1

Part II

Deviations and Fits in 0.0001 Inches

TOLERANCE CLASSIFICATIONS

Bearing O.D. D — over	incl	Deviation	F7 Housing Deviation	F7 Resultant Fit	G7 Housing Deviation	G7 Resultant Fit	H8 Housing Deviation	H8 Resultant Fit	H7 Housing Deviation	H7 Resultant Fit	H6 Housing Deviation	H6 Resultant Fit	J6 Housing Deviation	J6 Resultant Fit	J7 Housing Deviation	J7 Resultant Fit	K6 Housing Deviation	K6 Resultant Fit	K7 Housing Deviation	K7 Resultant Fit	M6 Housing Deviation	M6 Resultant Fit	M7 Housing Deviation	M7 Resultant Fit	N6 Housing Deviation	N6 Resultant Fit	N7 Housing Deviation	N7 Resultant Fit	P6 Housing Deviation	P6 Resultant Fit	P7 Housing Deviation	P7 Resultant Fit
0.7087	1.1811	0 / −3.5	+8 / +16	19.5L / 8L	+3 / +11	14.5L / 3L	0 / +13	16.5L / 0	0 / +8	11.5L / 0	0 / +5	8.5L / 0	−2 / +3	6.5L / 2T	−4 / +5	8.5L / 4T	−4 / +1	4.5L / 4T	−6 / +2	5.5L / 6T	−7 / −2	1.5L / 7T	−8 / 0	3.5L / 8T	−9 / −4	0.5L / 9T	−11 / −3	0.5L / 11T	−12 / −7	3.5L / 12T	−14 / −6	2.5T / 14T
1.1811	1.9685	0 / −4.5	+10 / +20	24.5L / 10L	+4 / +13	17.5L / 4L	0 / +15	19.5L / 0	0 / +10	14.5L / 0	0 / +6	10.5L / 0	−2 / +4	8.5L / 2T	−4 / +6	10.5L / 4T	−5 / +1	5.5L / 5T	−7 / +3	7.5L / 7T	−8 / −2	2.5L / 8T	−10 / 0	4.5L / 10T	−11 / −5	0.5L / 11T	−13 / −3	1.5L / 13T	−15 / −8	3.5L / 15T	−17 / −7	2.5T / 17T
1.9685	3.1496	0 / −5	+12 / +24	29L / 12L	+4 / +16	21L / 4L	0 / +18	23L / 0	0 / +12	17L / 0	0 / +7	12L / 0	−2 / +5	10L / 2T	−5 / +7	12L / 5T	−6 / +2	7L / 6T	−8 / +4	9L / 8T	−9 / −2	3L / 9T	−12 / 0	5L / 12T	−13 / −6	1T / 13T	−15 / −4	1L / 15T	−18 / −10	5T / 18T	−20 / −8	3T / 20T
3.1496	4.7244	0 / −6	+14 / +28	34L / 14L	+5 / +19	25L / 5L	0 / +21	27L / 0	0 / +14	20L / 0	0 / +9	15L / 0	−2 / +6	12L / 2T	−5 / +9	15L / 5T	−7 / +2	8L / 7T	−10 / +4	10L / 10T	−11 / −2	4L / 11T	−14 / 0	6L / 14T	−15 / −6	0 / 15T	−18 / −4	2L / 18T	−20 / −12	6T / 20T	−23 / −9	3T / 23T
4.7244	5.9055	0 / −7	+17 / +33	40L / 17L	+6 / +21	28L / 6L	0 / +25	32L / 0	0 / +16	23L / 0	0 / +10	17L / 0	−3 / +7	14L / 3T	−6 / +10	17L / 6T	−8 / +2	9L / 8T	−11 / +5	12L / 11T	−13 / −3	4L / 13T	−16 / 0	7L / 16T	−18 / −8	1T / 18T	−20 / −5	2L / 20T	−24 / −14	7T / 24T	−27 / −11	4T / 27T
5.9055	7.0866	0 / −10	+17 / +33	43L / 17L	+6 / +21	31L / 6L	0 / +25	35L / 0	0 / +16	26L / 0	0 / +10	20L / 0	−3 / +7	17L / 3T	−6 / +10	20L / 6T	−8 / +2	12L / 8T	−11 / +5	15L / 11T	−13 / −3	7L / 13T	−16 / 0	10L / 16T	−18 / −8	2L / 18T	−20 / −5	5L / 20T	−24 / −14	4L / 24T	−27 / −11	1T / 27T
7.0866	9.8425	0 / −12	+20 / +38	50L / 20L	+6 / +24	36L / 6L	0 / +28	40L / 0	0 / +18	30L / 0	0 / +11	23L / 0	−3 / +9	21L / 3T	−6 / +12	24L / 6T	−9 / +2	14L / 9T	−13 / +5	17L / 13T	−15 / −3	9L / 15T	−18 / 0	12L / 18T	−20 / −9	3L / 20T	−24 / −6	6L / 24T	−28 / −16	4L / 28T	−31 / −13	1T / 31T
9.8425	12.4016	0 / −14	+22 / +43	57L / 22L	+7 / +27	41L / 7L	0 / +32	46L / 0	0 / +20	34L / 0	0 / +13	27L / 0	−3 / +10	24L / 3T	−6 / +14	28L / 6T	−11 / +2	16L / 11T	−14 / +6	20L / 14T	−16 / −4	10L / 16T	−20 / 0	14L / 20T	−22 / −10	4L / 22T	−26 / −6	8L / 26T	−31 / −19	5L / 31T	−35 / −14	0 / 35T
12.4016	15.7480	0 / −16	+24 / +47	63L / 24L	+7 / +30	46L / 7L	0 / +35	51L / 0	0 / +22	38L / 0	0 / +14	30L / 0	−3 / +11	27L / 3T	−7 / +15	31L / 7T	−11 / +3	19L / 11T	−16 / +7	23L / 16T	−18 / −4	12L / 18T	−22 / 0	16L / 22T	−24 / −10	6L / 24T	−29 / −6	10L / 29T	−34 / −20	4L / 34T	−39 / −16	0 / 39T
15.7480	19.6850	0 / −18	+27 / +52	70L / 27L	+8 / +33	51L / 8L	0 / +38	56L / 0	0 / +25	43L / 0	0 / +16	34L / 0	−3 / +13	31L / 3T	−8 / +17	35L / 8T	−13 / +3	21L / 13T	−18 / +7	25L / 18T	−20 / −4	14L / 20T	−25 / 0	18L / 25T	−26 / −11	7L / 26T	−31 / −7	11L / 31T	−37 / −22	4L / 37T	−43 / −18	0 / 43T

L=Loose T=Tight

Source: ANSI/AFBMA Std. 7-1988.

Bearing Clearance

The establishment of correct bearing clearance is essential for reliable performance of rolling element bearings. Excessive bearing clearance will result in poor load distribution within the bearing, decreased fatigue life, and possible excessive dynamic excursions of the rotating system. Insufficient bearing clearance may result in excessive operating temperature or possible thermal lockup and catastrophic failure.

Most bearings are manufactured with an initial radial internal clearance. This clearance is expressed over the diameter. It is called radial clearance to distinguish it from axial clearance or end play. The terms radial clearance and diametral clearance are used interchangeably in the rolling bearing industry. The radial internal clearance is defined by the outer ring raceway contact diameter minus the inner ring raceway contact diameter minus twice the rolling element diameter. This initial unmounted clearance is changed by the shaft and housing fits, shaft speed, and by the thermal gradients existing in the system and created by operation of the bearing. After all of these factors have been considered, the bearing "operating clearance" should usually be positive. The exception to this occurs with preloaded bearings where the clearance has been carefully selected to provide shaft control. Clearances of only .0001" or .0002" are acceptable, but very small changes in thermal gradients can eliminate such a clearance and cause problems.

Generally, higher speed bearings will need higher operating clearance to allow a margin for unknown thermal gradients. Lower speed bearings, especially those with heavy loads, will perform best with smaller operating clearance. If the housing will remain much cooler than the bearing during operation, extra clearance is often needed to account for the fact that the shaft and inner ring will expand, while the housing and outer ring will not. In general, ball bearings need less operating clearance than do roller bearings. A rule of thumb for minimum operating clearance of a cylindrical roller bearing is .0003" to .0005". Ball bearings can be slightly less, and spherical roller bearings should be slightly more. The above considerations must be used to go from an operating clearance to the unmounted internal radial clearance that must be obtained in the bearing.

After both the shaft and housing fits have been selected, it is absolutely necessary to go back and review the internal radial clearance of the bearings. If a relatively tight fit has been selected, a bearing with more than standard clearance is usually needed. Interference fits always reduce the internal clearance of the bearing. For bearings mounted on solid shafts, the reduction in clearance will be about 80%–90% of the interference fit. For housings, this factor is about 90% of the interference fit. These factors can change significantly for hollow shafts and thin section housings. Again, this can be calculated by using thin ring theory.

The clearance manufactured into the unmounted bearing has been standardized by ANSI/ABMA in Standard 20-1987 [10] for ball and roller bearings (except tapers). For some types of bearings a similar format is used, but the actual values of clearance are selected by the manufacturer. Table 19 gives the radial internal clearance classifications. The internal fit refers to the relative amount of clearance inside the bearing.

Tables 20 and 21 illustrate the radial internal clearance values for ball and roller bearings, respectively, established by ANSI/ABMA. A complete version of these tables can be found in ANSI/AFBMA Standard 20-1987 [10]. Commercial and precision bearings can normally be obtained off the shelf with the clearances listed, although *tight* and *extra loose* bearings are not always stocked in all sizes. For special applications, clearances other than those listed can be obtained on special order. Special clearances are not necessarily more costly to make except that the quantity would be low and delivery much longer. However, if the combination of fits and special circumstances of operation require more clearance than available in the standards, there is no alternative to getting a nonstandard clearance bearing.

Table 19
Radial Internal Clearance Classifications

ANSI/ABMA Identification Code	Internal Fit
2	Tight
0	Standard
3	Loose
4	Extra loose

Table 20
Radial Internal Clearance Values for Radial Contact Ball Bearings

Clearance values in 0.0001 inch

d mm over	incl.	SYMBOL 2* min.	max.	SYMBOL 0* (Normal) min.	max.	SYMBOL 3* min.	max.	SYMBOL 4* min.	max.	SYMBOL 5* min.	max.
2.5	6	0	3	1	5	3	9	–	–	–	–
6	10	0	3	1	5	3	9	6	11	8	15
10	18	0	3.5	1	7	4	10	7	13	10	18
18	24	0	4	2	8	5	11	8	14	11	19
24	30	0.5	4.5	2	8	5	11	9	16	12	21
30	40	0.5	4.5	2	8	6	13	11	18	16	25
40	50	0.5	4.5	2.5	9	7	14	12	20	18	29
50	65	0.5	6	3.5	11	9	17	15	24	22	35
65	80	0.5	6	4	12	10	20	18	28	26	41
80	100	0.5	7	4.5	14	12	23	21	33	30	47
100	120	1	8	6	16	14	26	24	38	35	55
120	140	1	9	7	19	16	32	28	45	41	63
140	160	1	9	7	21	18	36	32	51	47	71
160	180	1	10	8	24	21	40	36	58	53	79
180	200	1	12	10	28	25	46	42	64	59	91

* These symbols relate to the Identification Code.

Source: ANSI/AFBMA Std. 20-1987.

Table 21
Radial Internal Clearance Values for Cylindrical Roller Bearings

Clearance values in 0.0001 inches

d mm Over	Incl.	Tight (2)* low	high	Normal (0)* low	high	Loose (3)* low	high	Extra Loose (4)* low	high
	10	4	8	8	12	14	18	18	22
10	18	4	8	8	12	14	18	18	22
18	24	4	8	8	12	14	18	18	22
24	30	4	10	10	14	16	20	20	24
30	40	5	10	10	16	18	22	22	28
40	50	6	12	12	18	20	26	26	32
50	65	6	14	14	20	22	30	30	35
65	80	8	16	16	24	28	35	35	43
80	100	10	18	18	28	32	41	41	49
100	120	10	20	20	32	37	47	47	57
120	140	12	24	24	35	41	53	53	63
140	160	14	26	26	39	45	59	59	71
160	180	14	30	30	43	49	65	65	79
180	200	16	32	32	47	55	71	71	87
200	225	18	35						
225	250	20	39						
250	280	22	43						
280	315	24	47						
315	355	26	53						

*These symbols relate to the Identification Code.

Source: ANSI/AFBMA Std. 20-1987.

Seals

Bearing seals have two basic functions: to keep contaminants out of the bearing and to keep the lubricant in the bearing. The design of the seal depends heavily on exactly what the seal is supposed to do. The nature of the contaminant, shaft speed, temperature, allowable leakage, and type of lubricant must be considered. Sealing can be an important consideration since in field use more bearings fail from contamination than from fatigue. There are two major categories of seals: contact seals and clearance seals. Each has its advantages and disadvantages for different applications. Contact seals vary widely from a simple felt strip to precision face seals made flat to millionths of an inch. In all cases, there is contact between moving and nonmoving surfaces, which provides a barrier to contaminants and loss of lubricant. There is a tremendous variety of materials and configurations used for contact seals.

The main limitation of contact seals is the sliding friction between the seal and shaft or rubbing surface. Seals for commercial bearing application can use felt seals up to 500 to 1,000 feet per minute surface velocity. Lip seals, probably the most common contact seal, can be used up to 2,000 to 3,000 feet per minute with common materials, and up to 5,000 feet per minute with special materials. Special carbon circumferential seals and face seals can be used at very high speeds, but these types of seals are very special and not suitable for the average industrial application.

Lip seals are excellent for sealing solids, liquids, and gases at reasonable pressures. The most common lip seal material is Buna-N, a synthetic rubber compound. This is the material usually used for bonded lip seals where a thin rubber lip is attached to a metal holder and attached directly to the bearing. It is also used in commercial cartridge-type lip seals where the rubber is held by a metal case and a spring is used to control lip pressure against the shaft. This type of seal can have high torque and heat generation and requires lubrication. For the effective application of lip seals, the rubbing surface roughness should be 10 to 20 Ra. Smoother than this can result in leakage while rougher can cause leakage and premature wear. Bearings with built-in lip seals already have this type surface ground on the bearing. Housing seals usually rub on the shaft itself, which must have a smooth surface with no spiraling.

Labyrinth seals, often called clearance seals, do not have rubbing contact between the seal and rotating member. It is this feature that gives them their principle advantage: no frictional drag or heat generation. Because of this, they are the most commonly used seal for high speeds. Their disadvantage is that they cannot be used to seal against pressure, and they are less effective against liquid and should not be used when even partially submerged. Seal effectiveness often depends on the availability of regular maintenance to keep the area around them clean and to lubricate them where necessary. Grease combined with a labyrinth seal can form a very effective barrier when properly maintained. Seal clearance must be carefully analyzed to keep the seal gap as small as possible but still maintain some gap at all operating points. To retain oil, labyrinth seals may need to be vented and usually must provide an oil return drain within the seal.

For extreme sealing conditions, special seal designs must be created. There is no exact formula for the design of special sealing systems because the conditions are so varied. Engineering experience is the biggest factor, and consulting with one of the bearing manufacturers that offers sealed bearings or with a seal company is recommended. One of the most common considerations is to use a combination of two or more seals at a given location. A good example is the Link-Belt D8 grease-flushable auxiliary seal shown in Figure 15.

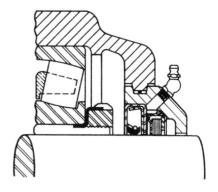

Figure 15. D8 Independently Flushable Seal [11]. (*Courtesy Link-Belt Bearing Div., Rexnord Corp.*)

SLEEVE BEARINGS

A sleeve bearing (also called a journal bearing) is a simple device for providing support and radial positioning while permitting rotation of a shaft. It is the oldest bearing device known to man. In the broad category of sleeve bearings can be included a great variety of materials, shapes, and sizes. Materials used include an infinite number of metallic alloys, sintered metals, plastics, wood, rubber, ceramic, solid lubricants, and composites. Types range from a simple hole in a cast-iron machine frame to some exceedingly complex gas-lubricated high-speed rotor bearings.

Sleeve bearings do have a number of advantages over rolling element bearings, as well as some disadvantages. Advantages are:

1. Inherently quiet operation because there are no moving parts.
2. If properly selected and maintained, they do not fail suddenly.
3. Wear is gradual, allowing scheduling of replacement.
4. Well suited to oscillating movement of the shaft.
5. With proper material selection, excessive moisture and submersion can be tolerated.
6. With proper material selection, extreme temperatures can be accommodated.

Disadvantages are:

1. High coefficient of friction.
2. For the same boundary plan, much less load capacity.
3. Life is not predictable except through experience.

In the application of sleeve bearings, the most important factor is the selection of the actual bearing material. The three most common industrial materials are babbitt, bronze, and cast iron. After these, there is an amazing variety of different bearing materials, often specialized for a particular application. In most cases, the details of selection are unique and assistance should be obtained from the manufacturer of the sleeve material.

Plain bearings made from babbitt are universally accepted as providing reasonable capacity and dependable service, often under adverse conditions. Babbitt is a relatively soft bearing material, which minimizes the danger of scoring or damage to shafts or rotors. It often can be repaired quickly on the spot by, for example, rescraping or pouring of new metal. Ambient temperatures should not exceed 130°F, and the actual bearing operating temperature must not exceed 200°F. Babbitt bearings are usually restricted to applications involving light to moderate loads and mild shock.

Bronze bearings are more suitable than babbitt for heavier loads bearings (75% to 200% higher), depending on specific conditions of load and speed. Bronze withstands higher shock loads and permits somewhat higher speed operation. It is usually restricted to 300°F ambient temperatures if properly lubricated. Bronze is a harder material than babbitt and has a greater tendency to score or damage shafts in the event of malfunction such as lack of relubrication. Field repair of bronze bearings generally requires removing shims and scraping or replacement of bushings. Bronze bushings commonly are available in both cast and sintered forms.

Cast-iron bearings are generally low in cost and suitable for many slow-moving shafts and oscillating or reciprocating arms supporting relatively light loads. The lubricating characteristics of cast iron are attributed to the free graphite flakes present in the material. With the use of cast-iron bearings, higher shaft clearance is usually utilized. Thus, any large wear particles or debris will not join or seize the bearing. This material has been used to temperatures as high as 1000°F (where ordinary lubricants are ineffective), under light loads and slow speed intermittent operations.

Lubrication is just as important in sleeve bearings as it is in rolling element bearings. There are three basic conditions of lubrication for sleeve bearings: full film or hydrodynamic, boundary, and extreme boundary lubrication. In full film lubrication, the mating surfaces of the shaft and bearing material are completely separated by a relatively thick film of lubricant. Boundary lubrication occurs when the separating film becomes very thin. Extreme boundary occurs when mating surfaces are in direct contact at various high points. The first two categories give long bearing life, while the third results in wear and shorter life.

In a full film bearing, the coefficient of friction is from .001 to .020, depending on the mating surfaces, clearances, lubricant type and viscosity, and speed. For a boundary lubricated bronze bearing, it is .08 to .14. Friction in a bearing design is important because temperature and wear are directly related to it. The lower the coefficient of friction, the longer the life of the bearing.

Either oil or grease can be used for lubrication as long as the temperature limitations for the grease or oil are not exceeded. Oil viscosity should be chosen between 100 and 200 SUS at the estimated operating temperature. Grease is the most common lubricant used for sleeve bearings, mainly due to lubricant retention. Grease lubricated bearings usually operate with a boundary film. Many sleeve bearings use grooving to improve lubrication on long sleeves. If the sleeve length-to-diameter ratio is greater than 1.5:1, a groove should be used.

Under certain operating conditions, dry lubrication can be used successfully with sleeve bearings. Graphited cast-bronze bearings are commonly used at elevated temperatures, in low speed or high load applications, or where the bearings are inaccessible for relubrication. Typical operating conditions for graphited bearings are 50 psi load with speeds to 30 sfm or a maximum PV factor of 1,500.

There are a number of factors that combine to determine the type of lubrication a bearing will have. Any of the following changes in the application would result in improved lubrication and longer life:

- A greater supply of lubricant available at the bearing
- Increased shaft speed, which gives increased oil film thickness
- Reducing the load, which will increase the oil film thickness
- Better alignment
- Smoother surface finishes
- Use of a higher-viscosity lubricant

The load carrying ability of a sleeve bearing is usually expressed in pounds per square inch (psi). This is calculated by dividing the applied load in pounds by the projected bearing area in square inches. Projected bearing area is found by multiplying the bearing bore diameter by the effective length of the sleeve. Few industrial bearings are loaded over 3,000 psi, and most are carrying loads under 400 psi. With cast-bronze sleeve bearings, 1,000 psi is acceptable. A usable figure for flat thrust washers is 100 psi. Figure 16 shows the maximum loads for various materials.

Another way of evaluating load capacity is through its maximum PV factor. The PV factor is the bearing load pres-

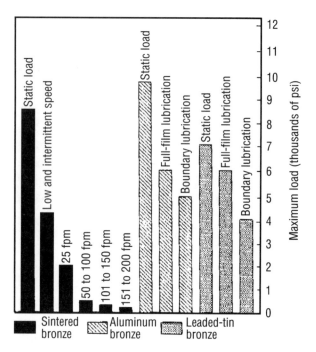

Figure 16. Load rating of three common bronzes. Temperatures should not exceed 300°F with most lubricants. (*From 1996 Power Transmission Design Handbook [18]*).

sure times the surface velocity of the shaft in feet per minute (sfm). For speeds above 200 sfm, use a PV factor of 20,000 for bronze sleeves and 10,000 for babbitt sleeves. Of course, there are maximum load limits and maximum and minimum speed limits that must also be kept in mind when using the PV factors. PV factors for other materials should be obtained from the sleeve manufacturers.

Very careful shaft alignment is necessary during installation. Shaft journals must turn freely without binding in the bearing, otherwise, excessive heat and seizure can result. Sharp edges on the shaft or the bearing surface can act as scrapers to destroy lubricant films. Do not extend shaft keyways into bearing bores. Shafting should be of the proper size and finish. Shaft diameters for rigid sleeve bearing units are usually held to the regular commercial tolerances as shown in Table 22. Standard shaft surface roughness of 32 Ra is acceptable for most applications. Graphited sleeves should have shaft roughness reduced to 12 Ra. When picking the housing style, consider the direction of loading. Avoid loading cast-iron housings in tension, whether one- or two-piece styles. If this cannot be avoided, try to obtain cast-steel housings.

Table 22
Recommended Shaft Tolerances for Journal Bearings

Shaft Diameters	Recommended Tolerance
Through 2″	Nominal to −.003″
2¹⁄₁₆ through 4″	Nominal to −.004″
4¹⁄₁₆ through 6″	Nominal to −.005″
6¹⁄₁₆ through 13″	Nominal to −.006″

From Link-Belt Technical Journal [11].

REFERENCES

1. Lundberg, G. and Palmgren A., "Dynamic Capacity of Rolling Bearings," *Acta Polytechnica,* Mechanical Engineering Series, Vol. 1, No. 3, Royal Swedish Academy of Engineering Sciences, Stockholm, 1947.

2. Lundberg, G. and Palmgren A., "Dynamic Capacity of Roller Bearings," *Acta Polytechnica,* Mechanical Engineering Series, Vol. 2, No. 4, Royal Swedish Academy of Engineering Sciences, Stockholm, 1947.

3. Anderson, W. J., "Bearing Fatigue Life Prediction," National Bureau of Standards, No. 43NANB716211, 1987.

4. American National Standard (ANSI/AFBMA) Std 1-1990, "Terminology for Anti-friction Ball and Roller Bearings and Parts."

5. American National Standard (ANSI/ABMA) Std 4-1984, "Tolerance Definitions and Gaging Practices for Ball and Roller Bearings."

6. American National Standard (ANSI/ABMA) Std. 7-1996, "Shafting and Housing Fits for Metric Radial Ball and Roller Bearings (Except Tapered Roller Bearings) Conforming to Basic Boundary Plans."

7. American National Standard (ANSI/AFBMA) Std 9-1990, "Load Ratings and Fatigue Life for Ball Bearings."

8. American National Standard (ANSI/AFBMA) Std 11-1990, "Load Ratings and Fatigue Life for Roller Bearings."

9. American National Standard (ANSI/AFBMA) Std 19-1974, "Tapered Roller Bearings, Radial, Inch Design."

10. American National Standard (ANSI/AFBMA) Std 20-1987, "Radial Bearings of Ball, Cylindrical Roller, and Spherical Roller Types, Metric Design."

11. *Bearing Technical Journal,* Link-Belt Bearing Div., Rexnord Corporation, 1982.

12. Bamberger, E. N., et al., *Life Adjustment Factors for Ball and Roller Bearings—An Engineering Design Guide,* ASME, New York, 1967.

13. Harris, T. A., *Rolling Bearing Analysis.* New York: John Wiley & Sons, Inc., 1966.

14. *Bearing Selection Handbook Revised—1986,* The Timken Co., 1986.

15. *Bearing Installation and Maintenance Guide,* SKF USA, Inc., 1988.

16. *MRC Aerospace Ball and Roller Bearings, Engineering Data Catalog,* SKF USA, Inc., 1993.

17. Zaretsky, Erwin V. (Editor), *STLE Life Factors for Rolling Bearings.* Society of Tribologists and Lubrication Engineers, 1992.

18. *1996 Power Transmission Design Handbook,* Penton Publishing, Inc., copyrighted Dec. 1995.

9
Piping and Pressure Vessels

R. R. Lee, Vice President—International Sales, Lee's Materials Services, Inc., Houston, Texas[1]
E. W. McAllister, P.E., Houston, Texas[2]
Jesse W. Cotherman, former Chief Engineer, Miller Pipeline Corp., Indianapolis, Ind.[3]
Dennis R. Moss, Supervisor of Vessel Engineering, Fluor Daniel, Inc., Irvine, Calif.[4]

[1]Process Plant Pipe
[2]Transportation Pipe Lines
[2,3]Pipe Line Condition Monitoring
[4]Pressure Vessels

PROCESS PLANT PIPE

Standard pipe is widely used in the process industries and is manufactured to ASTM standards (ANSI B36.10). Pipe charts, such as the one in Table 1, and careful attention to purchase order descriptions when shipping or receiving pipe help achieve accurate results. A description of piping, definitions, and how various types are manufactured follows.

Definitions and Sizing

Pipe Size

In pipe of any given size, the variations in wall thickness do not affect the outside diameter (OD), just the inside diameter (ID). For example, 12-in. nominal pipe has the same OD whether the wall thickness is 0.375 in. or 0.500 in. (Refer to Table 1 for wall thickness of pipe).

Pipe Length

Pipe is supplied and referred to as single random, double random, longer than double random, and cut lengths.

Single random pipe length is usually 18–22 ft threaded and coupled (T&C), and 18–25 ft plain end (PE).

Double random pipe lengths average 38–40 feet.

Cut lengths are made to order within ±⅛-in. Some pipe is available in about 80-ft lengths.

The major manufacturers of pipe offer brochures on their process of manufacturing pipe. The following descriptions are based upon vendor literature and specifications.

Seamless Pipe

This type of pipe is made by heating billets and advancing them over a piercer point. The pipe then passes through a series of rolls where it is formed to a true round and sized to exact requirements.

Electric Weld

Coils or rolls of flat steel are fed to a forming section that transforms the flat strip of steel into a round pipe section. A high-frequency welder heats the edges of the strip to 2,600°F at the fusion point. Pressure rollers then squeeze the heated edges together to form a fusion weld.

Double Submerged Arc Weld

Flat plate is used to make large-diameter pipe (20-in. to 44-in.) in double random lengths. The plate is rolled and pressed into an "O" shape, then welded at the edges both inside and outside. The pipe is then expanded to the final diameter.

Continuous Weld

Coiled skelp (skelp is semi-finished coils of steel plate used specifically for making pipe), is fed into a flattener, and welded to the trailing end of a preceding coil, thus forming a continuous strip of skelp. The skelp travels through a furnace where it is heated to 2,600°F and then bent into an oval by form rollers. It then proceeds through a welding stand where the heat in the skelp and pressure exerted by the rolls forms the weld. The pipe is stretched to a desired OD and ID, and cut to lengths. (Couplings, if ordered for any size pipe, will be hand tight only.)

Source

Lee, R. R., *Pocket Guide to Flanges, Fittings, and Piping Data,* 2nd Ed. Houston: Gulf Publishing Co., 1992.

Table 1
Pipe Chart

NOMINAL PIPE SIZE INCHES	OUTSIDE DIAMETER INCHES	I.P.S.	SCHEDULE	WALL INCHES	INSIDE DIAMETER INCHES	WT/FT POUNDS
⅛	.405		10S	.049	.307	.1863
		40	40S Std.	.068	.269	.2447
		80	80S Ex. Hvy.	.095	.215	.3145
¼	.540		10S	.065	.410	.3297
		40	40S Std.	.088	.364	.4248
		80	80S Ex. Hvy.	.119	.302	.5351
⅜	.675		10S	.065	.545	.4235
		40	40S Std.	.091	.493	.5676
		80	80S Ex. Hvy.	.126	.423	.7388
½	840		5S	.065	.710	.5383
			10S	.083	.674	.6710
		40	40S Std.	.109	.622	.8510
		80	80S Ex. Hvy.	.147	.546	1.088
		160		.188	.466	1.309
			XX Hvy.	.294	.252	1.714
¾	1.050		5S	.065	.920	.6838
			10S	.083	.884	.8572
		40	40S Std.	.113	.824	1.131
		80	80S Ex. Hvy.	.154	.742	1.474
		160		.219	.614	1.944
			XX Hvy.	.308	.434	2.441
1	1.315		5S	.065	1.185	.8678
			10S	.109	1.097	1.404
		40	40S Std.	.133	1.049	1.679
		80	80S Ex. Hvy.	.179	.957	2.172
		160		.250	.815	2.844
			XX Hvy.	.358	.599	3.659

Table 1 (Continued)
Pipe Chart

NOMINAL PIPE SIZE INCHES	OUTSIDE DIAMETER INCHES	I.P.S.	SCHEDULE	WALL INCHES	INSIDE DIAMETER INCHES	WT/FT POUNDS
1¼	1.660		5S	.065	1.530	1.107
			10S	.109	1.442	1.806
		40	40S Std.	.140	1.380	2.273
		80	80S Ex. Hvy.	.191	1.278	2.997
		160		.250	1.160	3.765
			XX Hvy	.382	.896	5.214
1½	1.900		5S	.065	1.770	1.274
			10S	.109	1.682	2.085
		40	40S Std.	.145	1.610	2.718
		80	80S Ex. Hvy.	.200	1.500	3.631
		160		.281	1.338	4.859
			XX Hvy.	.400	1.100	6.408
2	2.375		5S	.065	2.245	1.604
			10S	.109	2.157	2.638
		40	40S Std.	.154	2.067	3.653
		80	80S Ex. Hvy.	.218	1.939	5.022
		160		.344	1.689	7.462
			XX Hvy.	.436	1.503	9.029
2½	2.875		5S	.083	2.709	2.475
			10S	.120	2.635	3.531
		40	40S Std.	.203	2.469	5.793
		80	80S Ex. Hvy.	.276	2.323	7.661
		160		.375	2.125	10.01
			XX Hvy.	.552	1.771	13.69
3	3.500		5S	.083	3.334	3.029
			10S	.120	3.260	4.332
		40	40S Std.	.216	3.068	7.576
		80	80S Ex.Hvy.	.300	2.900	10.25
		160		.438	2.624	14.32
			XX Hvy.	.600	2.300	18.58

(table continued on next page)

Table 1 (Continued)
Pipe Chart

NOMINAL PIPE SIZE INCHES	OUTSIDE DIAMETER INCHES	I.P.S.	SCHEDULE	WALL INCHES	INSIDE DIAMETER INCHES	WT/FT POUNDS
3½	4.000	5	5S	.083	3.834	3.472
		10	10S	.120	3.760	4.973
		40	40S Std.	.226	3.548	9.109
		80	80S Ex. Hvy.	.318	3.364	12.50
			XX Hvy.	.636	2.728	22.85
4	4.500		5S	.083	4.334	3.915
			10S	.120	4.260	5.613
		40	40S Std.	.237	4.026	10.79
		80	80S Ex. Hvy.	.337	3.826	14.98
		120		.438	3.624	19.00
		160		.531	3.438	22.51
			XX Hvy.	.674	3.152	27.54
4½	5.00		40 Std.	.247	4.506	12.53
			80 Ex. Hvy.	.355	4.290	17.61
			XX Hvy.	.710	3.580	32.43
5	5.563		5S	.109	5.345	6.349
			10S	.134	5.295	7.770
		40	40S Std.	.258	5.047	14.62
		80	80S Ex. Hvy.	.375	4.813	20.78
		120		.500	4.563	27.04
		160		.625	4.313	32.96
			XX Hvy.	.750	4.063	38.55
6	6.625		5S	.109	6.407	7.585
			10S	.134	6.357	9.289
		40	40S Std.	.280	6.065	18.97
		80	80S Ex. Hvy.	.432	5.761	28.57
		120		.562	5.491	36.39
		160		.719	5.189	45.35
			XX Hvy.	.864	4.897	53.16

Table 1 (Continued)
Pipe Chart

NOMINAL PIPE SIZE INCHES	OUTSIDE DIAMETER INCHES	I.P.S.	SCHEDULE	WALL INCHES	INSIDE DIAMETER INCHES	WT/FT POUNDS
7	7.625	40	Std.	.301	7.023	23.57
		80	Ex. Hvy.	.500	6.625	38.05
			XX Hvy.	.875	5.875	63.08
8	8.625		5S	.109	8.407	9.914
			10S	.148	8.329	13.40
		20		.250	8.125	22.36
		30		.277	8.071	24.70
		40	40S Std.	.322	7.981	28.55
		60		.406	7.813	35.64
		80	80S Ex. Hvy.	.500	7.625	43.39
		100		.594	7.439	50.95
		120		.719	7.189	60.71
		140		.812	7.001	67.76
			XX Hvy.	.875	6.875	72.42
		160		.906	6.813	74.69
9	9.625	40	Std.	.342	8.941	33.90
		80	Ex. Hvy.	.500	8.625	48.72
			XX Hvy.	.875	7.875	81.77
10	10.750		5S	.134	10.482	15.19
			10S	.165	10.420	18.70
		20		.250	10.250	28.04
		30		.307	10.136	34.24
		40	40S Std.	.365	10.020	40.48
		60	80S Ex. Hvy.	.500	9.750	54.74
		80		.594	9.564	64.43
		100		.719	9.314	77.03
		120		.844	9.064	89.29
		140		1.000	8.750	104.13
		160		1.125	8.500	115.64
11	11.750	40	Std.	.375	11.000	45.55
		80	Ex. Hvy.	.500	10.750	60.07
			XX Hvy.	.875	10.000	101.63

(table continued on next page)

Table 1 (Continued)
Pipe Chart

NOMINAL PIPE SIZE INCHES	OUTSIDE DIAMETER INCHES	I.P.S.	SCHEDULE	WALL INCHES	INSIDE DIAMETER INCHES	WT/FT POUNDS
12	12.750		5S	.165	12.420	22.18
			10S	.180	12.390	24.20
		20		.250	12.250	33.38
		30		.330	12.090	43.77
			40S Std.	.375	12.000	49.56
		40		.406	11.938	53.52
			80S Ex. Hvy.	.500	11.750	65.42
		60		.562	11.626	73.15
		80		.688	11.376	88.63
		100		.844	11.064	107.32
		120		1.000	10.750	125.49
		140		1.125	10.500	139.67
		160		1.312	10.126	160.27
14	14.000	10		.250	13.500	36.71
		20		.312	13.376	45.61
		30	Std.	.375	13.250	54.57
		40		.438	13.124	63.44
			Ex. Hvy.	.500	13.000	72.09
		60		.594	12.814	85.05
		80		.750	12.500	106.13
		100		.938	12.126	130.85
		120		1.094	11.814	150.9
		140		1.250	11.500	170.21
		160		1.406	11.188	189.1
16	16.000	10		.250	15.500	42.05
		20		.312	15.376	52.27
		30	Std.	.375	15.250	62.58
		40	Ex. Hvy.	.500	15.000	82.77
		60		.656	14.688	107.5
		80		.844	14.314	136.61
		100		1.031	13.938	164.82
		120		1.219	13.564	192.43
		140		1.438	13.124	223.64
		160		1.594	12.814	245.25

Table 1 (Continued)
Pipe Chart

NOMINAL PIPE SIZE INCHES	OUTSIDE DIAMETER INCHES	I.P.S.	SCHEDULE	WALL INCHES	INSIDE DIAMETER INCHES	WT/FT POUNDS
18	18.000	10		.250	17.500	47.39
		20		.312	17.376	58.94
			Std.	.375	17.250	70.59
		30		.438	17.124	82.15
			Ex. Hvy.	.500	17.000	93.45
		40		.562	16.876	104.67
		60		.750	16.500	138.17
		80		.938	16.126	170.92
		100		1.156	15.688	207.96
		120		1.375	15.250	244.14
		140		1.562	14.876	274.22
		160		1.781	14.438	308.5
20	20.000	10		.250	19.500	52.73
		20	Std.	.375	19.250	78.60
		30	Ex. Hvy.	.500	19.000	104.13
		40		.594	18.814	123.11
		60		.812	18.376	166.4
		80		1.031	17.938	208.87
		100		1.281	17.438	256.1
		120		1.500	17.000	296.37
		140		1.750	16.500	341.09
		160		1.969	16.064	379.17
22	22.000	10		.250	21.500	58.07
		20	Std.	.375	21.250	86.61
		30	X Hvy.	.500	21.000	114.81
		60		.875	20.250	197.41
		80		1.125	19.750	250.81
		100		1.375	19.250	302.88
		120		1.625	18.750	353.61
		140		1.875	18.250	403.0
		160		2.125	17.750	451.06
24	24.000	10		.250	23.500	63.41
		20	Std.	.375	23.250	94.62
			Ex. Hvy.	.500	23.000	125.49

(table continued on next page)

Table 1 (Continued)
Pipe Chart

NOMINAL PIPE SIZE INCHES	OUTSIDE DIAMETER INCHES	I.P.S.	SCHEDULE	WALL INCHES	INSIDE DIAMETER INCHES	WT/FT POUNDS
		30		.562	22.876	140.68
		40		.688	22.626	171.29
		60		.969	22.064	238.35
		80		1.219	21.564	296.58
		100		1.531	20.938	367.39
		120		1.812	20.376	429.39
		140		2.062	19.876	483.1
		160		2.344	19.314	542.13
26	26.000	10		.312	25.376	85.60
			Std.	.375	25.250	102.63
		20	X Hvy.	.500	25.000	136.17
28	28.000	10		.312	27.376	92.26
			Std.	.375	27.250	110.64
		20		.500	27.000	146.85
		30		.625	26.750	182.73
30	30.000	10		.312	29.376	98.93
			Std.	.375	29.250	118.65
		20	Ex. Hvy.	.500	29.000	157.53
		30		.625	28.750	196.08
32	32.000	10		.312	31.376	105.59
			Std.	.375	31.250	126.66
		20		.500	31.000	168.21
		30		.625	30.750	209.43
		40		.688	30.624	230.08
34	34.000	10		.312	33.376	112.25
			Std.	.375	33.250	134.67
		20		.500	33.000	178.89
		30		.625	32.750	222.78
		40		.688	32.624	244.77
36	36.000	10		.312	35.375	118.92
			Std.	.375	35.250	142.68
			Ex. Hvy.	.500	35.000	189.57

Table 1 (Continued)
Pipe Chart

NOMINAL PIPE SIZE INCHES	OUTSIDE DIAMETER INCHES	I.P.S.	SCHEDULE	WALL INCHES	INSIDE DIAMETER INCHES	WT/FT POUNDS
42	42.000		Std.	.375	41.250	166.71
		20	X Hvy.	.500	41.000	221.61
		30		.625	40.750	276.18
		40		.750	40.500	330.41
48	48.000		Std.	.375	47.250	190.74
			X Hvy.	.500	47.000	253.65

Data in Table 1 courtesy of Tioga Pipe Supply Company.

Pipe Specifications

ASTM A-120

Sizes ⅛-in. to 16-in. standard weight, extra strong, and double extra strong (Std. Wt., XS, XXS). The specification covers black and hot-dipped galvanized welded and seamless average wall pipe for use in steam, gas, and air lines.

Markings. Rolled, stamped or stenciled on each length of pipe: the brand name, ASTM A-120, and the length of the pipe. In case of bundled pipe, markings will appear on a tag attached to each bundle. Table 2 shows a bundling schedule.

ASTM A-53

Sizes ⅛-in. to 26-in., standard weight, extra strong, and double extra strong, ANSI schedules 10 through 160 (see Table 1 for ANSI pipe schedules). The specification covers seamless and welded black and hot-dipped galvanized

Table 2
Bundling Schedule

Nominal Pipe Size (in.)	Number Pieces per Bundle	Standard Weight Pipe Total Length (ft)	Total Weight (lbs)	Extra Strong Pipe Total Length (ft)	Total Weight (lbs)
⅛	30	630	151	630	195
¼	24	504	212	504	272
⅜	18	378	215	378	280
½	12	252	214	252	275
¾	7	147	166	147	216
1	5	105	176	105	228
1¼	3	63	144	63	189
1½	3	63	172	63	229

average wall pipe for conveying oil, water, gas, and petroleum products.

Markings. Rolled, stamped or stenciled with brand name, kind, schedule, length of pipe, and type of steel used. In case of bundles, markings will appear on a bundle tag.

ASTM A-106

Sizes ⅛ to 26-in., ANSI schedules to 160. The specification covers seamless carbon steel average wall pipe for high-temperature service.

Markings. Rolled, stamped or stenciled with brand name, type such as ASTM A-106A, A-106B, A-106C (the A, B, C, indicate tensile strengths and yield point designations), the test pressure, and length of pipe. In case of bundles, the markings will appear on a bundle tag.

API-5L

Sizes ⅛-in. to 48-in., standard weight through double extra strong. The specification covers welded and seamless pipe suitable for use in conveying oil, water, and gas.

Markings. Paint stenciled with brand name, the API monogram, size, grade, steel process, type of steel, length, and weight per foot on pipe 4-in. and larger. In case of bundles, the markings will be on the bundle tag. Couplings, if ordered, will be hand tight.

Source

Lee, R. R., *Pocket Guide to Flanges, Fittings, and Piping Data,* 2nd Ed. Houston: Gulf Publishing Co., 1992.

Storing Pipe

Step 1—Pipe Racks

Figure 1 shows a pipe rack made by using 12 × 12-in. timbers. The rack has been assigned a number for materials accounting purposes. Do not store pipe directly on the ground. If rack materials are not available, then use the pipe itself by preparing a rack from the pipe with a few boards under each end.

Step 2—Layers

Form the first layer of pipe with one end straight, and other joints straight across the rack. Secure the stack by nailing wooden blocks to the sills, against the side of the pipe on the inside edges (see Figure 1).

Step 3—Measure

Tally each joint of pipe in the layer. Use a paint stick or suitable marker to mark each joint according to length, size, schedule, and purchase order item number.

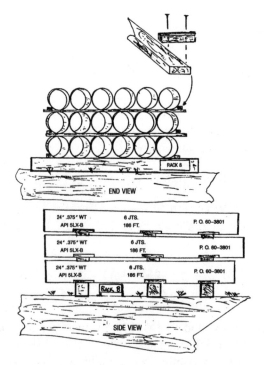

Figure 1. Schematic of rack for storing pipe.

Total the footage on the layer of pipe, and then mark the total footage and number of joints on the outside pipe for future inventory purposes. Apply color codes to pipe at this time if applicable.

Step 4—Dunnage

Apply sufficient dunnage of the same thickness across the pipe with wooden blocks nailed to one side. Stack the next layer of pipe directly over the first layer with the straight ends in line with each other. Then follow steps 2, 3, and 4.

Continue to follow the steps until the rack is considered full by the supervisor.

Rules for Storing Pipe

1. Do not mix pipe sizes and schedules on the same pipe rack.

2. Keep the pipe storage area clean to prevent accidents.
3. Do not crowd the storage areas. Leave room for large trucks and cranes.
4. Make a physical count of the pipe on a weekly or monthly basis to verify your materials accounting records as correct.
5. Always measure pipe within tenths of an inch. Measure the entire length of pipes, including couplings and threads.

Source

Lee, R. R., *Pocket Guide to Flanges, Fittings, and Piping Data,* 2nd Ed. Houston: Gulf Publishing Co., 1992.

Calculations to Use

If the outside diameter (OD) and the wall thickness of a pipe (t) are known, then you may calculate the weight per foot with the following equation:

Weight per foot = $10.68 \times (OD - t) \times t$

Example: What is the weight per foot of a 3-in. pipe with a .216-in. wall thickness and an OD of 3.500 in.? Using the equation,

$$\text{Weight per foot} = 10.68 \times (3.500 - .216) \times .216$$
$$= 7.58 \text{ lbs/ft}$$

Another method to determine weight per foot of pipe where the outside diameter and wall thickness are known is called the Baiamonte plate method. It is based on a square foot of plate 1 inch thick weighing 40.833 lbs, and uses the following equation:

$$\text{Weight per foot} = 40.833 \times \left(\frac{OD - t}{12} \right) \times \pi \times t$$

Example: What is the weight per foot of an 8-in. pipe with a wall thickness of .322 in.? Table 1 shows that an 8-in. pipe has an OD of 8.625 ins. So, using the equation,

$$\text{Weight per foot} = 40.833 \times \left(\frac{8.625 - .322}{12} \right) \times 3.1416 \times t$$
$$= 28.58 \text{ lbs/ft}$$

Source

Lee, R. R., *Pocket Guide to Flanges, Fittings, and Piping Data,* 2nd Ed. Houston: Gulf Publishing Co., 1992.

TRANSPORTATION PIPE LINES*

Steel Pipe Design

The maximum allowable design pressure stress will depend upon the intended service for the pipe line.

Pipe lines to be used for transporting liquid petroleum are covered by ANSI/ASME B31.4—"Liquid Petroleum Transportation Piping Systems." Pipe lines used for transporting gas are covered by ANSI/ASME B31.8—"Gas Transmission and Distribution Systems."

Pipe lines which must be operated in compliance with the Federal Pipeline Safety Regulations will also need to comply with the applicable parts of these regulations: Part 192 for gas transportation systems, and Part 195 for liquid transportation systems.

Gas Pipe Lines—ANSI/ASME B31.8

The maximum allowable pressure is calculated by the following equation:

$$P = (2St/D) \times F \times E \times T$$

where P = Design pressure, lb/in.^{2}
 S = Specified minimum yield strength, lb/in.^{2} (see Table 1)
 t = Nominal wall thickness, in.
 D = Nominal outside diameter, in.
 F = Design factor (see Table 2)
 E = Longitudinal joint factor (see Table 3)
 T = Temperature derating factor (see Table 4)

Class location definitions may be obtained from p. 192.111 of Part 192 of the Federal Pipeline Safety Regulations.

A typical calculation is as follows:

Pipe: 16″ OD × 0.250″ wt API 5LX X52 ERW
Location: Class 1, therefore F = 0.72 (see Table 2)
Temperature: 90°F, Temp. factor T = 1 (see Table 4)
Joint Factor: E = 1.0 (see Table 3)

$P = (2 \times 52,000 \times 0.250/16.0) \times 0.72 \times 1 \times 1$
$P = 1,170$ lb/in.^{2} gauge

Table 1
Specified Minimum Yield Strength for Steel and Iron Pipe Commonly Used in Piping Systems

Specification	Grade	Type[1]	SMYS, psi
API 5L	A25	BW, ERW, S	25,000
API 5L	A	ERW, FW, S, DSA	30,000
API 5L	B	ERW, FW, S, DSA	35,000
API 5LS (Note (2)]	A	ERW, DSA	30,000
API 5LS	B	ERW, DSA	35,000
API 5LS	X42	ERW, DSA	42,000
API 5LS	X46	ERW, DSA	46,000
API 5LS	X52	ERW, DSA	52,000
API 5LS	X56	ERW, DSA	56,000
API 5LS	X60	ERW, DSA	60,000
API 5LS	X65	ERW, DSA	65,000
API 5LS	X70	ERW, DSA	70,000
API 5LX (Note (2)]	X42	ERW, FW, S, DSA	42,000
API 5LX	X46	ERW, FW, S, DSA	46,000
API 5LX	X52	ERW, FW, S, DSA	52,000
API 5LX	X56	ERW, FW, S, DSA	56,000
API 5LX	X60	ERW, FW, S, DSA	60,000
API 5LX	X65	ERW, FW, S, DSA	65,000
API 5LX	X70	ERW, FW, S, DSA	70,000
ASTM A53	Open Hrth. Bas. Oxy., Elec. Furn.	BW	25,000
ASTM A53	Bessemer	BW	30,000

Reproduced from ANSI/ASME Code B31-8-1982, Appendix D. Reprinted courtesy of The American Society of Mechanical Engineers.

*This section reprinted from *Pipe Line Rules of Thumb Handbook,* 3rd Ed., E. W. McAllister (Ed.), Gulf Publishing Company, Houston, Texas, 1993.

Table 1 (Continued)
Specified Minimum Yield Strength for Steel and Iron Pipe Commonly Used in Piping Systems

Specification	Grade	Type[1]	SMYS, psi
ASTM A53	A	ERW, S	30,000
ASTM A53	B	ERW, S	35,000
ASTM A106	A	S	30,000
ASTM A106	B	S	35,000
ASTM A106	C	S	40,000
ASTM A134	–	EFW	[Note (3)]
ASTM A135	A	ERW	30,000
ASTM A135	B	ERW	35,000
ASTM A139	A	ERW	30,000
ASTM A139	B	ERW	35,000
ASTM A333	1	S, ERW	30,000
ASTM A333	3	S, ERW	35,000
ASTM A333	4	S	35,000
ASTM A333	6	S, ERW	35,000
ASTM A333	7	S, ERW	35,000
ASTM A333	8	S, ERW	75,000

Reproduced from ANSI/ASME Code B31-8-1982, Appendix D. Reprinted courtesy of The American Society of Mechanical Engineers.

Table 2
Values of Design Factor F

Construction Type (See 841.151)	Design Factor F
Type A	0.72
Type B	0.60
Type C	0.50
Type D	0.40

Reproduced from ANSI/ASME Code B31-8-1982, Table 841.1A. Reprinted courtesy of The American Society of Mechanical Engineers.

Table 4
Temperature Derating Factor T For Steel Pipe

Temperature, °F	Temperature Derating Factor T
250 or less	1.000
300	0.967
350	0.933
400	0.900
450	0.867

NOTE: For intermediate temperatures, interpolate for derating factor.

Reproduced from ANSI/ASME Code B31-8-1982, Table 841.1C. Reprinted courtesy of The American Society of Mechanical Engineers.

Table 3
Longitudinal Joint Factor E

Spec. Number	Pipe Class	E Factor
ASTM A53	Seamless	1.00
	Electric Resistance Welded	1.00
	Furnace Welded	0.60
ASTM A106	Seamless	1.00
ASTM A134	Electric Fusion Arc Welded	0.80
ASTM A135	Electric Resistance Welded	1.00
ASTM A139	Electric Fusion Welded	0.80
ASTM A211	Spiral Welded Steel Pipe	0.80
ASTM A381	Double Submerged-Arc-Welded	1.00
ASTM A671	Electric Fusion Welded	1.00*
ASTM A672	Electric Fusion Welded	1.00*
API 5L	Seamless	1.00
	Electric Resistance Welded	1.00
	Electric Flash Welded	1.00
	Submerged Arc Welded	1.00
	Furnace Butt Welded	0.60
API 5LX	Seamless	1.00
	Electric Resistance Welded	1.00
	Electric Flash Welded	1.00
	Submerged Arc Welded	1.00
API 5LS	Electric Resistance Welded	1.00
	Submerged Welded	1.00

NOTE: Definitions for the various classes of welded pipe are given in 804.243
*Includes Classes 12, 22, 32, 42, and 52 only.

Reproduced from ANSI/ASME Code B31-8-1982, Table 841.1B. Reprinted courtesy of The American Society of Mechanical Engineers.

Liquid Pipe Lines—ANSI/ASME B31.4

The internal design pressure is determined by using the following formula:

$$P = (2St/D) \times E \times F$$

where P = Internal design pressure, lb/in.^{2} gauge
   S = Specified minimum yield strength, lb/in.^{2} (see Table 5)
   t = Nominal wall thickness, in.
   D = Nominal outside diameter of the pipe, in.
   E = Weld joint factor (see Table 6)
   F = Design factor of 0.72

Note: Refer to p. 195.106 of Part 195 Federal Pipeline Safety Regulations for design factors to be used on offshore risers and platform piping and cold worked pipe.

A typical calculation of the internal design pressure is as follows:

Pipe: 26″ OD × 0.3125″ wt API 5LX X52 ERW
Weld joint factor E = 1.0 (see Table 6)
Design factor F = 0.72

$$P = (2 \times 52{,}000 \times 0.3125/26) \times 1 \times 0.72$$
$$P = 900 \text{ lb/in.}^2 \text{ gauge}$$

Table 5
Tabulation of Examples of Allowable Stresses for
Reference Use in Piping Systems

Allowable stress values (S) shown in this Table are equal to 0.72 × E (weld joint factor) × specified minimum yield strength of the pipe.

Allowable stress values shown are for new pipe of known specification. Allowable stress values for new pipe of unknown specification, ASTM A 120 specification or used (reclaimed) pipe shall be determined in accordance with 402.3.1.

For some Code computations, particularly with regard to branch connections [see 404.3.1 (d) (3)] and expansion, flexibility, structural attachments, supports, and restraints (Chapter II, Part 5), the weld joint factor E need not be considered.

For specified minimum yield strength of other grades in approved specifications, refer to that particular specification.

Allowable stress value for cold worked pipe subsequently heated to 600 F (300 C) or higher (welding excepted) shall be 75 percent of value listed in Table.

Definitions for the various types of pipe are given in 400.2.

(Metric Stress Levels are given in MPa [1 Megapascal = 1 million pascals])

Specification	Grade	Specified Min Yield Strength psi (MPa)	Notes	(E) Weld Joint Factor	(S) Allowable Stress Value −20 F to 250 F (−30 C to 120 C) psi (MPa)
Seamless					
API 5L	A25	25,000 (172)	(1)	1.00	18,000 (124)
API 5L, ASTM A53, ASTM A106	A	30,000 (207)	(1) (2)	1.00	21,600 (149)
API 5L, ASTM A53, ASTM A106	B	35,000 (241)	(1) (2)	1.00	25,200 (174)
ASTM A106	C	40,000 (278)	(1) (2)	1.00	28,800 (199)
ASTM A524	I	35,000 (241)	(1)	1.00	25,200 (174)
ASTM A524	II	30,000 (207)	(1)	1.00	21,600 (149)
API 5LU	U80	80,000 (551)	(1) (4)	1.00	57,600 (397)
API 5LU	U100	100,000 (689)	(1) (4)	1.00	72,000 (496)
API 5LX	X42	42,000 (289)	(1) (2) (4)	1.00	30,250 (208)
API 5LX	X46	46,000 (317)	(1) (2) (4)	1.00	33,100 (228)
API 5LX	X52	52,000 (358)	(1) (2) (4)	1.00	37,450 (258)
API 5LX	X56	56,000 (386)	(1) (4)	1.00	40,300 (278)
API 5LX	X60	60,000 (413)	(1) (4)	1.00	43,200 (298)
API 5LX	X65	65,000 (448)	(1) (4)	1.00	46,800 (323)
API 5LX	X70	70,000 (482)	(1) (4)	1.00	50,400 (347)
Furnace Welded-Butt Welded					
ASTM A53		25,000 (172)	(1) (2)	0.60	10,800 (74)
API 5L Class I & Class II	A25	25,000 (172)	(1) (2) (3)	0.60	10,800 (74)
API 5L (Bessemer), ASTM A53 (Bessemer)		30,000 (207)	(1) (2) (5)	0.60	12,950 (89)

Table 5 (Continued)

Specification	Grade	Specified Min Yield Strength psi (MPa)	Notes	(E) Weld Joint Factor	(S) Allowable Stress Value −20F to 250 F (−30 C to 120 C) psi (MPa)
Furnace Welded-Lap Welded					
API 5L Class I		25,000 (172)	(1) (2) (6)	0.80	14,400 (99)
API 5L Class II		28,000 (193)	(1) (2) (6)	0.80	16,150 (111)
API 5L (Bessemer)		30,000 (207)	(1) (2) (6)	0.80	17,300 (119)
API 5L Electric Furnace		25,000 (172)	(1) (2) (6)	0.80	14,400 (99)
Electric Resistance Welded and Electric Flash Welded					
API 5L	A25	25,000 (172)	(1) (7)	1.00	18,000 (124)
API 5L, ASTM A53, ASTM A135	A	30,000 (207)	(2)	0.85	18,360 (127)
API 5L, API 5LS, ASTM A53, ASTM A135	A	30,000 (207)	(1)	1.00	21,600 (149)
API 5L, ASTM A53, ASTM A135	B	35,000 (241)	(2)	0.85	21,420 (148)
API 5L, API 5LS, ASTM A53, ASTM A135	B	35,000 (241)	(1)	1.00	25,200 (174)
API 5LS, API 5LX	X42	42,000 (289)	(1) (2) (4)	1.00	30,250 (208)
API 5LS, API 5LX	X46	46,000 (317)	(1) (2) (4)	1.00	33,100 (228)
API 5LS, API 5LX	X52	52,000 (358)	(1) (2) (4)	1.00	37,450 (258)
API 5LS, API 5LX	X56	56,000 (386)	(1) (4)	1.00	40,300 (279)
API 5LS, API 5LX	X60	60,000 (413)	(1) (4)	1.00	43,200 (297)
API 5LS, API 5LX	X65	65,000 (448)	(1) (4)	1.00	46,800 (323)
API 5LS, API 5LX	X70	70,000 (482)	(1) (4)	1.00	50,400 (347)
API 5LU	U80	80,000 (551)	(1) (4)	1.00	57,600 (397)
API 5LU	U100	100,000 (689)	(1) (4)	1.00	72,000 (496)
Electric Fusion Welded					
ASTM A 134	–	–		0.80	–
ASTM A 139	A	30,000 (207)	(1) (2)	0.80	17,300 (119)
ASTM A 139	B	35,000 (241)	(1) (2)	0.80	20,150 (139)
ASTM A 155	–	–	(2) (8)	0.90	–
ASTM A 155	–	–	(1) (8)	1.00	–
Submerged Arc Welded					
API 5L, API 5LS	A	30,000 (207)	(1)	1.00	21,600 (149)
API 5L, API 5LS	B	35,000 (241)	(1)	1.00	25,200 (174)
API 5LS, API 5LX	X42	42,000 (289)	(1) (2) (4)	1.00	30,250 (208)
API 5LS, API 5LX	X46	46,000 (317)	(1) (2) (4)	1.00	33,100 (228)
API 5LS, API 5LX	X52	52,000 (358)	(1) (2) (4)	1.00	37,450 (258)
API 5LS, API 5LX	X56	56,000 (386)	(1) (4)	1.00	40,300 (278)
API 5LS, API 5LX	X60	60,000 (413)	(1) (4)	1.00	43,200 (298)
API 5LS, API 5LX	X65	65,000 (448)	(1) (4)	1.00	46,800 (323)
API 5LS, API 5LX	X70	70,000 (482)	(1) (4)	1.00	50,400 (347)
API 5LU	U80	80,000 (551)	(1) (4)	1.00	57,600 (397)
API 5LU	U100	100,000 (689)	(1) (4)	1.00	72,000 (496)
ASTM A 381	Y35	35,000 (241)	(1) (2)	1.00	25,200 (174)
ASTM A 381	Y42	42,000 (290)	(1) (2)	1.00	30,250 (209)
ASTM A 381	Y46	46,000 (317)	(1) (2)	1.00	33,100 (228)
ASTM A 381	Y48	48,000 (331)	(1) (2)	1.00	34,550 (238)
ASTM A 381	Y50	50,000 (345)	(1)	1.00	36,000 (248)
ASTM A 381	Y52	52,000 (358)	(1)	1.00	37,450 (258)
ASTM A 381	Y60	60,000 (413)	(1)	1.00	43,200 (298)
ASTM A 381	Y65	65,000 (448)	(1)	1.00	46,800 (323)

NOTES (1) Weld joint factor E (see Table 402.4.3) and allowable stress value are applicable to pipe manufactured after 1958.

(2) Weld joint factor E (see Table 402.4.3) and allowable stress value are applicable to pipe manufactured before 1959.

(3) Class II produced under API 5L 23rd Edition, 1968, or earlier has a specified minimum yield strength of 28,000 psi (193 MPa).

(4) Other grades provided for in API 5LS, API 5LU, and API 5LX not precluded.

(5) Manufacture was discontinued and process deleted from API 5L in 1969.

(6) Manufacture was discontinued and process deleted from API 5L in 1962.

(7) A25 is not produced in electric flash weld.

(8) See applicable plate specification for yield point and refer to 402.3.1 for calculation of (S).

Table 6
Weld Joint Factor

Specification Number	Pipe Type (1)	Weld Joint Factor E	
		Pipe Mfrd. Before 1959	Pipe Mfrd. After 1958
ASTM A53	Seamless	1.00	1.00
	Electric-Resistance-Welded	0.85 (2)	1.00
	Furnace Lap-Welded	0.80	0.80
	Furnace Butt-Welded	0.60	0.60
ASTM A106	Seamless	1.00	1.00
ASTM A134	Electric-Fusion (Arc)-Welded single or double pass	0.80	0.80
ASTM A135	Electric-Resistance-Welded	0.85 (2)	1.00
ASTM A139	Electric-Fusion-Welded single or double pass	0.80	0.80
ASTM A155	Electric-Fusion-Welded	0.90	1.00
ASTM A381	Electric-Fusion-Welded, Double Submerged Arc-Welded	–	1.00
API 5L	Seamless	1.00	1.00
	Electric-Resistance-Welded	0.85 (2)	1.00
	Electric-Flash-Welded	0.85 (2)	1.00
	Electric-Induction-Welded	–	1.00
	Submerged Arc-Welded	–	1.00
	Furnace Lap-Welded	0.80	0.80 (3)
	Furnace Butt-Welded	0.60	0.60
API 5LS	Electric-Resistance-Welded	–	1.00
	Submerged Arc-Welded	–	1.00
API 5LX	Seamless	1.00	1.00
	Electric-Resistance-Welded	1.00	1.00
	Electric-Flash-Welded	1.00	1.00
	Electric-Induction-Welded	–	1.00
	Submerged Arc-Welded	1.00	1.00
API 5LU	Seamless	–	1.00
	Electric-Resistance-Welded	–	1.00
	Electric-Flash-Welded	–	1.00
	Electric-Induction-Welded	–	1.00
	Submerged Arc-Welded	–	1.00
Known	Known	(4)	(5)
Unknown	Seamless	1.00 (6)	1.00 (6)
Unknown	Electric-Resistance or Flash-Welded	0.85 (6)	1.00 (6)
Unknown	Electric-Fusion-Welded	0.80 (6)	0.80 (6)
Unknown	Furnace Lap-Welded or over NPS 4	0.80 (7)	0.80 (7)
Unknown	Furnace Butt-Welded or NPS 4 and smaller	0.60 (8)	0.60 (8)

NOTES: (1) Definitions for the various pipe types (weld joints) are given in 400.2.
(2) A weld joint factor of 1.0 may be used for electric-resistance-welded or electric-flash-welded pipe manufactured prior to 1959 where (a) pipe furnished under this classification has been subjected to supplemental tests and/or heat treatments as agreed to by the supplier and the purchaser, and such supplemental tests and/or heat treatment demonstrate the strength characteristics of the weld to be equal to the minimum tensile strength specified for the pipe, or (b) pipe has been tested as required for a new pipeline in accordance with 437.4.1.
(3) Manufacture was discontinued and process deleted from API 5L in 1962.
(4) Factors shown above for pipe manufactured before 1959 apply for new or used (reclaimed) pipe if pipe specification and pipe type are known and it is known that pipe was manufactured before 1959 or not known whether manufactured after 1958.
(5) Factors shown above for pipe manufactured after 1958 apply for new or used (reclaimed) pipe if pipe specification and pipe type are known and it is known that pipe was manufactured after 1958.
(6) Factor applies for new or used pipe of unknown specification and ASTM A120 if type of weld joint is known.
(7) Factor applies for new or used pipe of unknown specification and ASTM A120 if type of weld joint is known to be furnace lap-welded, or for pipe over NPS 4 if type of joint is unknown.
(8) Factor applies for new or used pipe of unknown specification and ASTM A120 if type of weld joint is known to be furnace butt-welded, or for pipe NPS 4 and smaller if type of joint is unknown.

PIPE LINE CONDITION MONITORING

Pig-based Monitoring Systems

Pipe line pigs are frequently used for pipe line commissioning, cleaning, filling, dewaxing, batching, and more recently pipe line monitoring. This last type of pig can be designed to carry a wide range of surveillance and monitoring equipment and can be used at regular intervals to check internal conditions rather than continuously monitoring the line. Data, however, can be built up over a period of time to provide a history of the line. This information can be used to predict or estimate when maintenance, line cleaning, or repairs are required. If a leak is detected, for example, by flow meter imbalance, the location can be found by using a pig with acoustic equipment on board. This will alarm when the detection equipment output reaches a maximum and the precise location of the pig can be confirmed by radio transmitters also mounted on board.

Pigs require tracking because they may become stuck, at a point of debris build-up, for example. Pigging should be carried out at a steady speed, but occasionally the pig may stop and start, particularly in smaller lines. Information on when and where the pig stops is therefore important in interpreting the inspection records. Pig tracking is not new and many such proprietary systems exist. In the best systems, however, a picture of the line is often programmed in so that outputs from junctions, valves, cross-overs, and other geometries act as an aid to location. Pig tracking can make use of the acoustic methods discussed earlier. When the sealing cups at the front of the pig encounter a weld, vibrational or acoustic signals are generated. Each pipe line therefore has its characteristic sound pattern. When a crack occurs this pattern changes from the no-leak case and the location can be found from direct comparison. The technology has become so advanced that information on dents, buckles, ovality, weld penetration, expansion, and pipe line footage can be generated.

The equipment is often simple, consisting of sensor, conditioning, and amplifier circuits and suitable output and recording devices. Such a device developed by British Gas is shown in Figure 1. The range of detection is dependent on the pipe line diameter and the type of pig. Operational data have shown that light pigs in a 200 mm line can be detected at a range of 8 km, increasing to 80 km for a heavy pig in a 900 mm line. As the signals travel at acoustic velocity this means a signal from a pig at 80 km range will take 190 seconds to be picked up. Such technology is now becoming routine in both offshore gas and onshore liquid lines.

Source

McAllister, E. W. (Ed.), *Pipe Line Rules of Thumb Handbook,* 3rd Ed. Houston: Gulf Publishing Co., 1993.

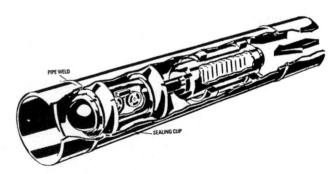

Figure 1. "Intelligent" pipe line monitoring pig. (*Courtesy British Gas*)

Coupons

Coupon samples give an excellent indication of pipe condition, at least at the point at which the coupon is taken. With the application of current techniques, sampling can be a relatively easy procedure on certain types of lines. Coupon sampling involves taking a small section out of the wall of a pipe to check the condition of the line. This is done with a section-retrieving hole saw. The hole saw can be incorporated into a pipeline-attachable pressure box if line pressures and local ordinances prohibit a blowing hole. Once the section is taken, a plug or screw can be driven into the hole to act as a stop. The pressure box, if it was required, can then be removed and the plug sealed with techniques applied to small leaks, such as thermoset two-part plastics.

Source

Jesse W. Cotherman, former Chief Engineer, Miller Pipeline Corp., Indianapolis, IN.

Manual Investigation

Manual investigation of the pipe from inside is the most definitive way to check line condition. Visual indications can immediately be followed up with ultrasonic wall thickness measurements, magnetic field measurement of possible cracks, x-ray inspection, and possibly even fluorescent penetrant inspection. Pipe joints can be leak-checked with equipment that straddles the joint.

Manual investigation is not, however, without disadvantages. Perhaps the biggest disadvantages is that the line must be shut down. This means that the customer must either be able to do without the services of the line or be supplied by other means while the line is down.

If this obstacle can be overcome, the next hurdle is the human technician. This may seem a little unbelievable to someone with claustrophobia, but a definite mental make-up is required for underground work. The knowledge that one is a long way underground, even if the space is not confining, can be a disconcerting experience. If the technician is mentally suited for underground work, physical access comes next.

Physical access can be a constraint on both small and large pipes. On the small side, a 16-inch diameter line is about the smallest size pipe a small adult technician can get into and still do any kind of functional work. At the opposite extreme, special scaffolding may be required to reach the top and sides of larger pipes.

Source

Jesse W. Cotherman, former Chief Engineer, Miller Pipeline Corp., Indianapolis, IN.

Cathodic protection for pipe lines

Estimate the rectifier size required for an infinite line. Refer to Figure 1. If coating conductance tests have not been performed, pick a value from Table 1.

Table 1
Typical Values of Coating Conductance

Micromhos/sq ft	Coating Condition
1–10	Excellent coating—high resistivity soil
10–50	Good coating—high resistivity soil
50–100	Excellent coating—low resistivity soil
100–250	Good coating—low resistivity soil
250–500	Average coating—low resistivity soil
500–1,000	Poor coating—low resistivity soil

Use a value of -0.3 volts for ΔE_x. This is usually enough to raise the potential of coated steel to about -0.85 volts.

Use a value of 1.5 volts for ΔE at the drain point. Higher values may be used in some circumstances; however, there may be a risk of some coating disbondment at higher voltages.

Calculate I at the drain point.

$$\Delta I_A = \text{amp./in.} \times \Delta E_x/0.3 \times D$$

where D = pipe OD, in.

Example. 30-in. OD line with coating conductivity = 100 micromhos/sq ft. What is the current change at

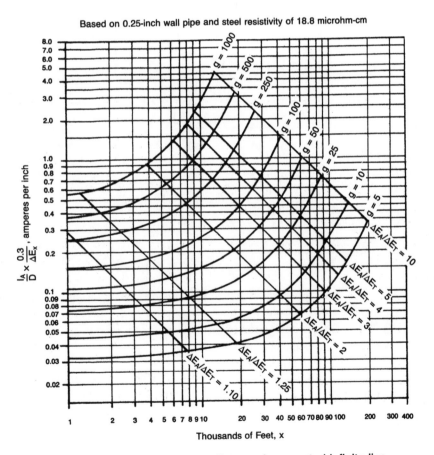

Based on 0.25-inch wall pipe and steel resistivity of 18.8 microhm-cm

Figure 1. Drainage current vs. distance for a coated infinite line.

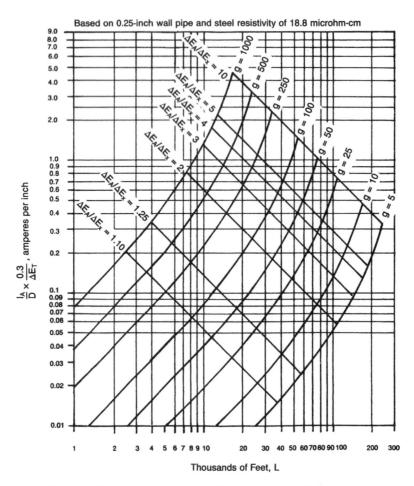

Based on 0.25-inch wall pipe and steel resistivity of 18.8 microhm-cm

Thousands of Feet, L

Figure 2. Drainage current vs. distance for a coated finite line.

the drain point, and how far will this current protect an infinitely long pipe line?

$$\Delta E_A = 1.5$$

$$\Delta E_x = 0.3$$

$$\Delta E_A/\Delta E_x = 5.0$$

Refer to Figure 1 and read

$$L = 30,000 \text{ ft}$$

$$\Delta I_A/D \times 0.3/\Delta E_x = 0.7 \text{ (from Figure 1)}$$

$$\Delta I_A = 0.7 \times 30 = 21 \text{ amp.}$$

21 amps will protect the line for 30,000 ft in either direction from the drain point.

Estimate rectifier size for a finite line. Refer to Figure 2. A 3½-in. OD line 20,000 ft long is protected by insulating flanges at both ends and has a coating conductivity of 500 micromhos/sq ft. What is $\Delta E_A/E_T$ and what ΔI_A is required? Assume that

$$E_T = 0.3$$

From Figure 2 read

$$\Delta E_A/E_T = 5.8$$

$$\Delta E_A = 1.74$$

$$\Delta I_A/D \times 0.3/E_T = 1.8 \text{ amp./in.}$$

$$\Delta I_A = 1.8 \times 3.5 = 6.3 \text{ amp.}$$

Estimate ground bed resistance for a rectifier installation.

Example. A ground bed for a rectifier is to be installed in 1,000 ohm-cm soil. Seven 3-in. × 60-in. vertical graphite anodes with backfill and a spacing of 10 ft will be used. What is the resistance of the ground bed?

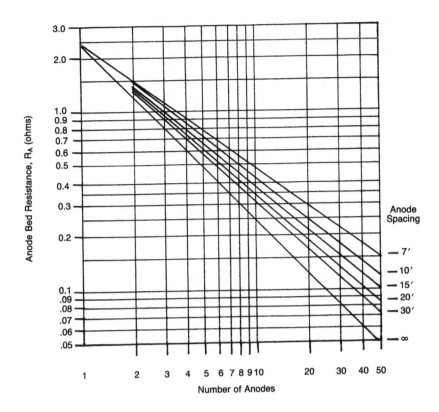

Figure 3. Anode bed resistance vs. number of anodes. 3-in. × 60-in. vertical graphite anodes in backfill.

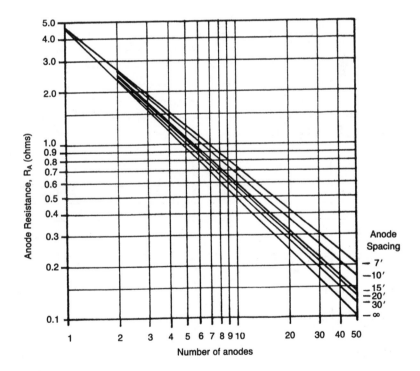

Figure 4. Anode bed resistance vs. number of anodes. 2-in. × 60-in. vertical bare anodes.

Refer to Figure 3. The resistance is 0.56 ohms.

Figure 3 is based on a soil resistivity of 1,000 ohms-cm. If the soil resistivity is different, use a ratio of the actual soil resistivity divided by 1,000 and multiply this by the reading obtained from Figure 3.

A rectifier ground bed is to be composed of 10 − 2 × 60-in. bare "Duriron" silicon anodes spaced 20 ft apart. The soil resistivity is 3,000 ohms-cm. What is the ground bed resistance? Refer to Figure 4. The resistivity from the chart is 0.55 ohms. Since the chart is based on soil resistivity of

1,000 ohms-cm, the ground bed resistivity is 3,000/1,000 × 0.55 or 1.65 ohms.

The resistance of multiple anodes installed vertically and connected in parallel may be calculated with the following equation:

$$R = 0.00521P/NL \times (2.3\text{Log } 8L/d - 1 + 2L/S\text{Log } 0.656N) \qquad (1)$$

where   R = ground bed resistance, ohms
        P = soil resistivity, ohm-cm
        N = number of anodes
        d = diameter of anode, ft
        L = length of anode, ft
        S = anode spacing, ft

If the anode is installed with backfill such as coke breeze, use the diameter and length of the hole in which the anode is installed. If the anode is installed bare, use the actual dimensions of the anode.

Figure 5 is based on Equation 1 and does not include the internal resistivity of the anode. The resistivity of a single vertical anode may be calculated with Equation 2.

$$R = 0.00521P/L \times (2.3\text{Log } 8L/d - 1) \qquad (2)$$

If the anode is installed with backfill, calculate the resistivity using the length and diameter of the hole in which the anode is installed. Calculate the resistivity using the actual anode dimensions. The difference between these two values is the internal resistance of the anode. Use the value of P, typically about 50 ohm-cm, for the backfill medium.

Figure 5 is based on 1,000 ohm-cm soil and a 7-ft × 8-in. hole with a 2-in. × 60-in. anode.

Example. Determine the resistivity of 20 anodes installed vertically in 1,500 ohm-cm soil with a spacing of 20 ft.

Read the ground bed resistivity from Figure 5.

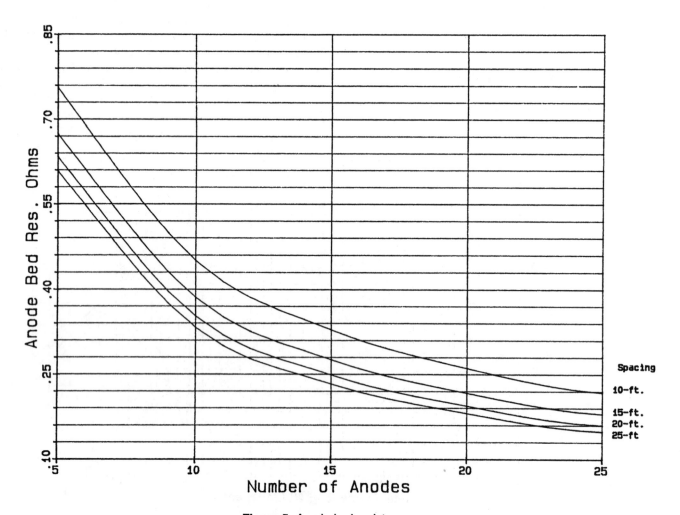

Figure 5. Anode bed resistance.

R = 0.202 ohm

Since the anodes are to be installed in 1,500 ohm-cm soil and Figure 5 is based on 1,000 ohm-cm soil, multiply R by the ratio of the actual soil resistivity to 1,000 ohm-cm.

R = 0.202 × 1,500/1,000

R = 0.303 ohm

The internal resistivity for a single 2-in. × 60-in. vertical anode installed in 50 ohm-cm backfill (7 ft × 8-in. hole) is 0.106 ohm.

Since 20 anodes will be installed in parallel, divide the resitivity for one anode by the number of anodes to obtain the internal resistivity of the anode bank.

0.106/20 = 0.005 ohm

The total resistivity of the 20 anodes installed vertically will therefore be 0.308 ohm (0.303 + 0.005).

Galvanic Anodes

Zinc and magnesium are the most commonly used materials for galvanic anodes. Magnesium is available either in standard alloy or high purity alloy. Galvanic anodes are usually pre-packaged with backfill to facilitate their installation. They may also be ordered bare if desired. Galvanic anodes offer the advantage of more uniformly distributing the cathodic protection current along the pipe line and it may be possible to protect the pipe line with a smaller amount of current than would be required with an impressed current system but not necessarily at a lower cost. Another advantage is that interference with other structures is minimized when galvanic anodes are used.

Galvanic anodes are not an economical source of cathodic protection current in areas of high soil resistivity. Their use is generally limited to soils of 3,000 ohm-cm except where small amounts of current are needed.

Magnesium is the most-used material for galvanic anodes for pipe line protection. Magnesium offers a higher solution potential than zinc and may therefore be used in areas of higher soil resistivity. A smaller amount of magnesium will generally be required for a comparable amount of current. Refer to Figure 6 for typical magnesium anode performance data. These curves are based on driving potentials of −0.70 volts for H-1 alloy and −0.90 volts for Galvomag working against a structure potential of −0.85 volts referenced to copper sulfate.

The driving potential with respect to steel for zinc is less than for magnesium. The efficiency of zinc at low current levels does not decrease as rapidly as the efficiency for magnesium. The solution potential for zinc referenced to a cop-

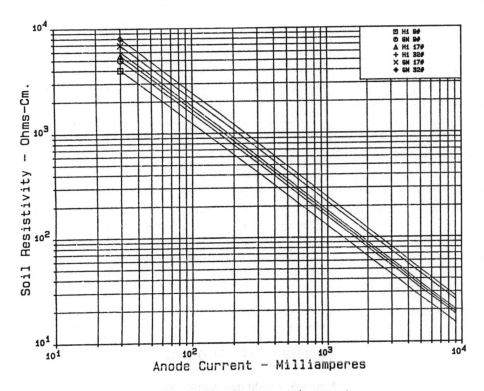

Figure 6. Magnesium anode current.

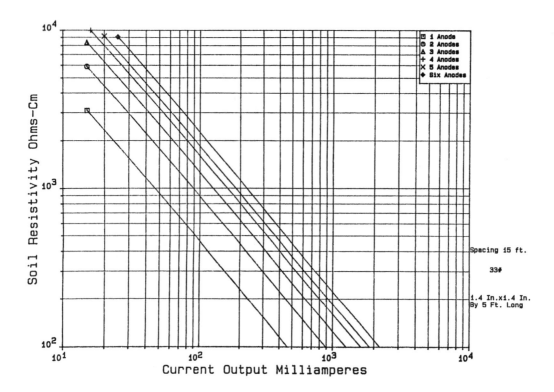

Figure 7a. Current output zinc anodes.

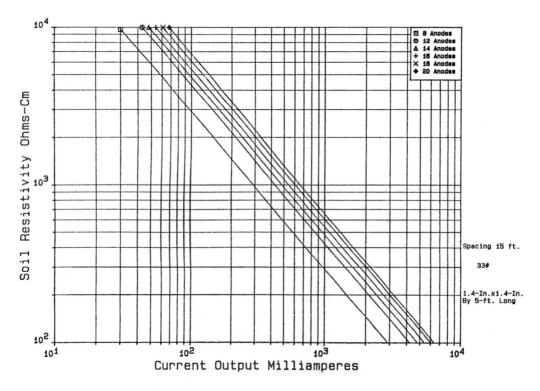

Figure 7b. Current output zinc anodes.

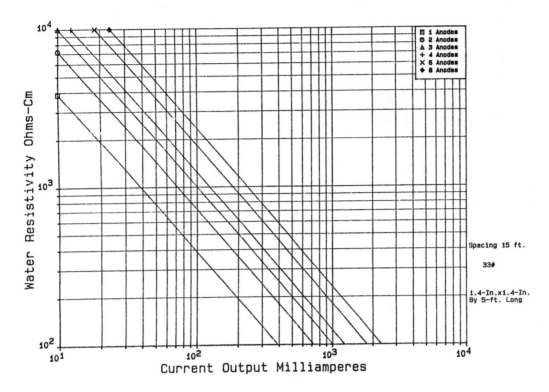

Figure 8a. Current output zinc anodes.

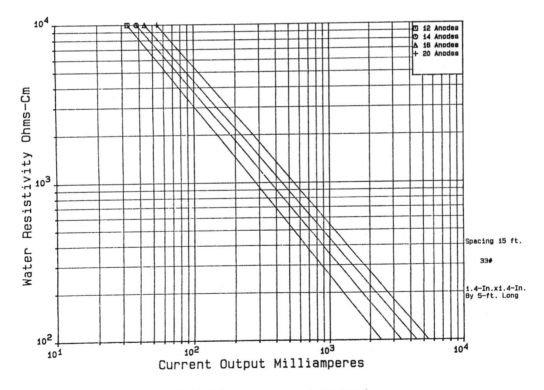

Figure 8b. Current output zinc anodes.

per sulfate cell is −1.1 volts; standard magnesium has a solution potential of −1.55 volts; and high purity magnesium has a solution potential of −1.8 volts.

If, for example, a pipe line is protected with zinc anodes at a polarization potential of −0.9 volts, the driving potential will be −1.1 − (−0.9) or −0.2 volts. If standard magnesium is used, the driving potential will be −1.55 − (−0.9) or −0.65 volts. The circuit resistance for magnesium will be approximately three times as great as for zinc. This would be handled by using fewer magnesium anodes, smaller anodes, or using series resistors.

If the current demands for the system are increased due to coating deterioration, contact with foreign structures, or by oxygen reaching the pipe and causing depolarization, the potential drop will be less for zinc than for magnesium anodes. With zinc anodes, the current needs could increase by as much as 50% and the pipe polarization potential would still be about 0.8 volts. The polarization potential would drop to about 0.8 volts with only a 15% increase in current needs if magnesium were used.

The current efficiency for zinc is 90% and this value holds over a wide range of current densities. Magnesium anodes have an efficiency of 50% at normal current densi-

ties. Magnesium anodes may be consumed by self corrosion if operated at very low current densities. Refer to Figures 7a, 7b, 8a, and 8b for zinc anode performance data. The data in Figures 7a and 7b are based on the anodes being installed in a gypsum-clay backfill and having a driving potential of −0.2 volts. Figures 8a and 8b are based on the anodes being installed in water and having a driving potential of −0.2 volts. [from data prepared for the American Zinc Institute].

Example. Estimate the number of packaged anodes required to protect a pipe line.

What is the anode resistance of a packaged magnesium anode installation consisting of nine 32 lb anodes spaced 7 ft apart in 2,000 ohm-cm soil?

Refer to Figure 9. This chart is based on 17# packaged anodes in 1,000 ohm-cm soil. For nine 32 lb anodes, the resistivity will be

$$1 \times 2{,}000/1{,}000 \times 0.9 = 1.8 \text{ ohm}$$

See Figure 10 for a table of multiplying factors for other size anodes.

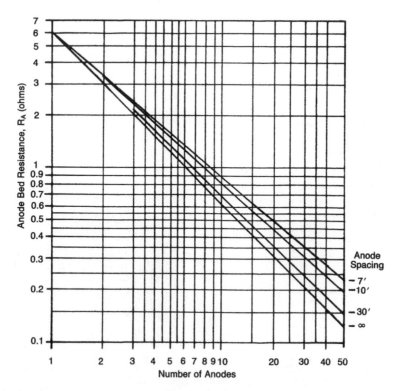

Figure 9. Anode bed resistance vs. number of anodes. 17# packaged magnesium anodes.

Chart based on 17-lb. magnesium anodes installed in 1000 ohm-cm soil in groups of 10 spaced on 10-ft. centers.

For other conditions multiply number of anodes by the following multiplying factors:

For soil resistivity: MF = $\frac{p}{1000}$

For conventional magnesium: MF = 1.3

For 9-lb. anodes: MF = 1.25

For 32-lb. anodes: MF = 0.9

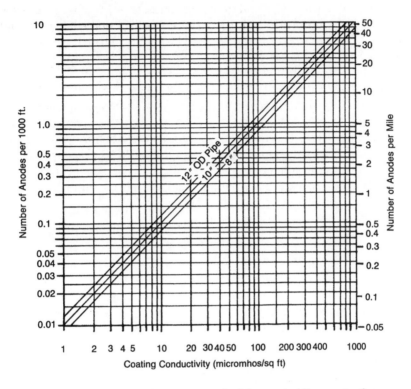

Figure 10. Number of anodes required for coated line protection.

Example. A coated pipe line has a coating conductivity of 100 micromhos/sq ft and is 10,000 ft long, and the diameter is 10¾-in. How many 17 lb magnesium anodes will be required to protect 1,000 ft? Refer to Figure 7 and read 2 anodes per 1,000 ft. A total of twenty 17# anodes will be required for the entire line.

Sources

1. Parker, M. E. and Peattie, E. G., *Pipe Line Corrosion and Cathodic Protection,* 3rd Ed. Houston: Gulf Publishing Co., 1984.
2. McAllister, E. W. (Ed.), *Pipe Line Rules of Thumb Handbook,* 3rd Ed. Houston: Gulf Publishing Co., 1993.

PRESSURE VESSELS

Stress Analysis

Stress analysis is the determination of the relationship between external forces applied to a vessel and the corresponding stress. The emphasis of this discussion is not how to do stress analysis in particular, but rather how to analyze vessels and their component parts in an effort to arrive at an economical and safe design—the difference being that we analyze stresses where necessary to determine thickness of material and sizes of members. We are not so concerned with building mathematical models as with providing a step-by-step approach to the design of ASME Code vessels. It is not necessary to find every stress but rather to know the governing stresses and how they relate to the vessel or its respective parts, attachments, and supports.

The starting place for stress analysis is to determine all the design conditions for a given problem and then determine all the related external forces. We must then relate these external forces to the vessel parts which must resist them to find the corresponding stresses. By isolating the causes (loadings), the effects (stress) can be more accurately determined.

The designer must also be keenly aware of the types of loads and how they relate to the vessel as a whole. Are the effects long or short term? Do they apply to a localized portion of the vessel or are they uniform throughout?

How these stresses are interpreted and combined, what significance they have to the overall safety of the vessel, and what allowable stresses are applied will be determined by three things:

1. The strength/failure theory utilized.
2. The types and categories of loadings.
3. The hazard the stress represents to the vessel.

Membrane Stress Analysis

Pressure vessels commonly have the form of spheres, cylinders, cones, ellipsoids, tori, or composites of these. When the thickness is small in comparison with other dimensions ($R_m/t > 10$), vessels are referred to as membranes and the associated stresses resulting from the contained pressure are called membrane stresses. These membrane stresses are average tension or compression stresses. They are assumed to be uniform across the ves-

sel wall and act tangentially to its surface. The membrane or wall is assumed to offer no resistance to bending. When the wall offers resistance to bending, bending stresses occur in addition to membrane stresses.

In a vessel of complicated shape subjected to internal pressure, the simple membrane-stress concepts do not suffice to give an adequate idea of the true stress situation. The types of heads closing the vessel, effects of supports, variation in thickness and cross section, nozzles, external attachments, and overall bending due to weight, wind, and seismic all cause varying stress distributions in the vessel. Deviations from a true membrane shape set up bending in the vessel wall and cause the direct loading to vary from point to point. The direct loading is diverted from the more flexible to the more rigid portions of the vessel. This effect is called "stress redistribution."

In any pressure vessel subjected to internal or external pressure, stresses are set up in the shell wall. The state of stress is triaxial and the three principal stresses are:

σ_x = longitudinal/meridional stress
σ_ϕ = circumferential/latitudinal stress
σ_r = radial stress

In addition, there may be bending and shear stresses. The radial stress is a direct stress, which is a result of the pressure acting directly on the wall, and causes a compressive stress equal to the pressure. In thin-walled vessels this stress is so small compared to the other "principal" stresses that it is generally ignored. Thus we assume for purposes of analysis that the state of stress is biaxial. This greatly simplifies the method of combining stresses in comparison to triaxial stress states. For thick-walled vessels ($R_m/t < 10$), the radial stress cannot be ignored and formulas are quite different from those used in finding "membrane stresses" in thin shells.

Since ASME Code, Section VIII, Division 1, is basically for design by rules, a higher factor of safety is used to allow for the "unknown" stresses in the vessel. This higher safety factor, which allows for these unknown stresses, can impose a penalty on design but requires much less analysis. The design techniques outlined in this text are a compromise between finding all stresses and utilizing minimum code formulas. This additional knowledge of stresses warrants the use of higher allowable stresses in some cases, while meeting the requirements that all loadings be considered.

In conclusion, "membrane stress analysis" is not completely accurate but allows certain simplifying assumptions to be made while maintaining a fair degree of accuracy. The main simplifying assumptions are that the stress is biaxial and that the stresses are uniform across the shell wall. For thin-walled vessels these assumptions have proven themselves to be reliable. No vessel meets the criteria of being a true membrane, but we can use this tool within a reasonable degree of accuracy.

Failures in Pressure Vessels

Vessel failures can be grouped into four major categories, which describe *why* a vessel failure occurs. Failures can also be grouped into types of failures, which describe *how* the failure occurs. Each failure has a why and how to its history. It may have failed *through* corrosion fatigue because the wrong material was selected! The designer must be as familiar with categories and types of failure as with categories and types of stress and loadings. Ultimately they are all related.

Categories of Failures

1. Material—Improper selection of material; defects in material.
2. Design—Incorrect design data; inaccurate or incorrect design methods; inadequate shop testing.
3. Fabrication—Poor quality control; improper or insufficient fabrication procedures including welding; heat treatment or forming methods.
4. Service—Change of service condition by the user; inexperienced operations or maintenance personnel; upset conditions. Some types of service which require special attention both for selection of material, design details, and fabrication methods are as follows:

   a. Lethal
   b. Fatigue (cyclic)
   c. Brittle (low temperature)
   d. High temperature
   e. High shock or vibration
   f. Vessel contents
   • Hydrogen
   • Ammonia
   • Compressed air
   • Caustic
   • Chlorides

Types of Failures

1. Elastic deformation—Elastic instability or elastic buckling, vessel geometry, and stiffness as well as properties of materials are protection against buckling.
2. Brittle fracture—Can occur at low or intermediate temperatures. Brittle fractures have occurred in vessels made of low carbon steel in the 40°–50°F range during hydrotest where minor flaws exist.
3. Excessive plastic deformation—The primary and secondary stress limits as outlined in ASME Section VIII, Division 2, are intended to prevent excessive plastic deformation and incremental collapse.
4. Stress rupture—Creep deformation as a result of fatigue or cyclic loading, i.e., progressive fracture. Creep is a time-dependent phenomenon, whereas fatigue is a cycle-dependent phenomenon.
5. Plastic instability—Incremental collapse; incremental collapse is cyclic strain accumulation or cumulative cyclic deformation. Cumulative damage leads to instability of vessel by plastic deformation.
6. High strain—Low cycle fatigue is strain-governed and occurs mainly in lower-strength/high-ductile materials.
7. Stress corrosion—It is well known that chlorides cause stress corrosion cracking in stainless steels, likewise caustic service can cause stress corrosion cracking in carbon steels. Material selection is critical in these services.
8. Corrosion fatigue—Occurs when corrosive and fatigue effects occur simultaneously. Corrosion can reduce fatigue life by pitting the surface and propagating cracks. Material selection and fatigue properties are the major considerations.

In dealing with these various modes of failure, the designer must have at his disposal a picture of the state of stress in the various parts. It is against these failure modes that the designer must compare and interpret stress values. But setting allowable stresses is not enough! For elastic instability one must consider geometry, stiffness, and the properties of the material. Material selection is a major consideration when related to the type of service. Design details and fabrication methods are as important as "allowable stress" in design of vessels for cyclic service. The designer and all those persons who ultimately affect the design must have a clear picture of the conditions under which the vessel will operate.

Loadings

Loadings or forces are the "causes" of stresses in pressure vessels. These forces and moments must be isolated both to determine where they apply to the vessel and when they apply to a vessel. Categories of loadings define where these forces are applied. Loadings may be applied over a large portion (general area) of the vessel or over a local area of the vessel. Remember both general and local loads can produce membrane and bending stresses. These stresses are additive and define the overall state of stress in the vessel or component. Stresses from local loads must be added to stresses from general loadings. These combined stresses are then compared to an allowable stress.

Consider a pressurized, vertical vessel bending due to wind, which has an inward radial force applied locally. The effects of the pressure loading are longitudinal and circumferential tension. The effects of the wind loading are longitudinal tension on the windward side and longitudinal compression on the leeward side. The effect of the local inward radial load is some local membrane stresses and local bending stresses. The local stresses would be both circumferential and longitudinal, tension on the inside surface of the vessel, and compressive on the outside. Of course the steel at any given point only sees a certain level of stress or the combined effect. It is the designer's job to combine the stresses from the various loadings to arrive at the worst probable combination of stresses, combine them using some failure theory, and compare the results to an acceptable stress level to obtain an economical and safe design.

This hypothetical problem serves to illustrate how categories and types of loadings are related to the stresses they produce. The stresses applied more or less continuously and uniformly across an entire section of the vessel are primary stresses.

The stresses due to pressure and wind are primary membrane stresses. These stresses should be limited to the Code allowable. These stresses would cause the bursting or collapse of the vessel if allowed to reach an unacceptably high level.

On the other hand, the stresses from the inward radial load could be either a primary local stress or secondary stress. It is a primary local stress if it is produced from an unrelenting load or a secondary stress if produced by a relenting load. Either stress may cause local deformation but will not in and of itself cause the vessel to fail. If it is a primary stress, the stress will be redistributed; if it is a secondary stress, the load will relax once slight deformation occurs.

Also be aware that this is only true for ductile materials. In brittle materials, there would be no difference between primary and secondary stresses. If the material cannot yield to reduce the load, then the definition of secondary stress does not apply! Fortunately current pressure vessel codes require the use of ductile materials.

This should make it obvious that the type and category of loading will determine the type and category of stress. This will be expanded upon later, but basically each combination of stresses (stress categories) will have different allowables, i.e.:

- Primary stress: $P_m < SE$
- Primary membrane local (P_L):
  $P_L = P_m + P_L < 1.5\ SE$
  $P_L = P_m + Q_m < 1.5\ SE$
- Primary membrane + secondary (Q):
  $P_m + Q < 3\ SE$

But what if the loading was of relatively short duration? This describes the "type" of loading. Whether a loading is steady, more or less continuous, or nonsteady, variable, or temporary will also have an effect on what level of stress will be acceptable. If in our hypothetical problem the loading had been pressure + seismic + local load, we would have a different case. Due to the relatively short duration of seismic loading, a higher "temporary" allowable stress would be acceptable. The vessel doesn't have to operate in an earthquake all the time. On the other hand, it also shouldn't fall down in the event of an earthquake! Structural designs allow a one-third increase in allowable stress for seismic loadings for this reason.

For *steady* loads, the vessel must support these loads more or less continuously during its useful life. As a result, the stresses produced from these loads must be maintained to an acceptable level.

For *nonsteady* loads, the vessel may experience some or all of these loadings at various times but not all at once and not more or less continuously. Therefore a temporarily higher stress is acceptable.

For *general* loads that apply more or less uniformly across an entire section, the corresponding stresses must be lower, since the entire vessel must support that loading.

For *local loads,* the corresponding stresses are confined to a small portion of the vessel and normally fall off rapidly in distance from the applied load. As discussed previously, pressurizing a vessel causes bending in certain components. But it doesn't cause the entire vessel to bend. The results are not as significant (except in cyclic service) as those caused by general loadings. Therefore a slightly higher allowable stress would be in order.

Loadings can be outlined as follows:

A. Categories of loadings
  1. General loads—Applied more or less continuously across a vessel section.
    a. Pressure loads—Internal or external pressure (design, operating, hydrotest, and hydrostatic head of liquid).
    b. Moment loads—Due to wind, seismic, erection, transportation.
    c. Compressive/tensile loads—Due to dead weight, installed equipment, ladders, platforms, piping, and vessel contents.
    d. Thermal loads—Hot box design of skirt-head attachment.
  2. Local loads—Due to reactions from supports, internals, attached piping, attached equipment, i.e., platforms, mixers, etc.
    a. Radial load—Inward or outward.
    b. Shear load—Longitudinal or circumferential.
    c. Torsional load.
    d. Tangential load.
    e. Moment load—Longitudinal or circumferential.
    f. Thermal loads.
B. Types of loadings
  1. Steady loads—Long-term duration, continuous.
    a. Internal/external pressure.
    b. Dead weight.
    c. Vessel contents.
    d. Loadings due to attached piping and equipment.
    e. Loadings to and from vessel supports.
    f. Thermal loads.
    g. Wind loads.
  2. Nonsteady loads—Short-term duration; variable.
    a. Shop and field hydrotests.
    b. Earthquake.
    c. Erection.
    d. Transportation.
    e. Upset, emergency.
    f. Thermal loads.
    g. Start up, shut down.

Stress

ASME Code, Section VIII, Division 1 vs. Division 2

ASME Code, Section VIII, Division 1 does not explicitly consider the effects of combined stress. Neither does it give detailed methods on how stresses are combined. ASME Code, Section VIII, Division 2, on the other hand, provides specific guidelines for stresses, how they are combined, and allowable stresses for categories of combined stresses. Division 2 is design by analysis whereas Division 1 is designed by rules. Although stress analysis as utilized by Division 2 is beyond the scope of this discussion, the use of stress categories, definitions of stress, and allowable stresses is applicable.

Division 2 stress analysis considers all stresses in a triaxial state combined in accordance with the maximum shear stress theory. Division 1 and the procedures outlined in this section consider a biaxial state of stress combined in accordance with the maximum stress theory. Just as you would not design a nuclear reactor to the rules of Division 1, you would not design an air receiver by the techniques of Division 2. Each has its place and applications. The following discussion on categories of stress and allowables will utilize information from Division 2, which can be applied in general to all vessels.

Types, Classes, and Categories of Stress

The shell thickness as computed by Code formulas for internal or external pressure alone is often not sufficient to withstand the combined effects of all other loadings. Detailed calculations consider the effects of each loading separately and then must be combined to give the total state of stress in that part. The stresses that are present in pressure vessels are separated into various classes in accordance with the types of loads that produced them, and the hazard they represent to the vessel. Each class of stress must be maintained at an acceptable level and the combined total stress must be kept at another acceptable level. The combined stresses due to a combination of loads acting simultaneously are called stress categories. Please note that this terminology differs from that given in Division 2, but is clearer for the purposes intended here.

Classes of stress, categories of stress, and allowable stresses are based on the type of loading that produced them and on the hazard they represent to the structure. Unrelenting loads produce primary stresses. Relenting loads (self limiting) produce secondary stresses. General loadings produce primary membrane and bending stresses. Local loads produce local membrane and bending stresses. Primary stresses must be kept lower than secondary stresses. Primary plus secondary stresses are allowed to be higher and so on. Before considering the combination of stresses (categories), we must first define the various types and classes of stress.

Types of Stress

There are many names to describe types of stress. Enough in fact to provide a confusing picture even to the experienced designer. As these stresses apply to pressure vessels, we group all types of stress into three major classes of stress, and subdivision of each of the groups is arranged according to their effect on the vessel. The following list of stresses describes types of stress without regard to their effect on the vessel or component. They define a direction of stress or relate to the application of the load.

1. Tensile	10. Thermal
2. Compressive	11. Tangential
3. Shear	12. Load induced
4. Bending	13. Strain induced
5. Bearing	14. Circumferential
6. Axial	15. Longitudinal
7. Discontinuity	16. Radial
8. Membrane	17. Normal
9. Principal	

Classes of Stress

The foregoing list provides examples of types of stress. It is, however, too general to provide a basis with which to combine stresses or apply allowable stresses. For this purpose, new groupings called classes of stress must be used. Classes of stress group stresses according to the type of loading which produced them and the hazard they represent to the vessel.

1. Primary stress
   a. General:
      • Primary general membrane stress, P_m
      • Primary general bending stress, P_b
   b. Primary local stress, P_L
2. Secondary stress
   a. Secondary membrane stress, Q_m
   b. Secondary bending stress, Q_b
3. Peak stress, F

Definitions and examples of these stresses are as follows:

Primary general stress. These stresses act over a full cross-section of the vessel. They are produced by mechanical loads (load induced) and are the most hazardous of all types of stress. The basic characteristic of a primary stress is that it is not self limiting. Primary stresses are generally due to internal or external pressure or produced by sustained external forces and moments. Thermal stresses are never classified as primary stresses.

Primary general stresses are divided into membrane and bending stresses. The need for dividing primary general stress into membrane and bending is that the calculated value of a primary bending stress may be allowed to go higher than that of a primary membrane stress. Primary stresses that exceed the yield strength of the material can cause failure or gross distortion. Typical calculations of primary stress are:

$$\frac{PR}{t}, \frac{F}{A}, \frac{MC}{I}, \text{ and } \frac{TC}{J}$$

Primary general membrane stress, P_m. This stress occurs across the entire cross section of the vessel. It is remote from discontinuities such as head-shell intersections, cone-cylinder intersections, nozzles, and supports. Examples are:

a. Circumferential and longitudinal stress due to pressure.
b. Compressive and tensile axial stresses due to wind.
c. Longitudinal stress due to the bending of the horizontal vessel over the saddles.
d. Membrane stress in the center of the flat head.
e. Membrane stress in the nozzle wall within the area of reinforcement due to pressure or external loads.
f. Axial compression due to weight.

Primary general bending stress, P_b. Primary bending stresses are due to sustained loads and are capable of causing collapse of the vessel. There are relatively few areas where primary bending occurs:

a. Bending stress in the center of a flat head or crown of a dished head.
b. Bending stress in a shallow conical head.
c. Bending stress in the ligaments of closely spaced openings.

Local Primary Membrane Stress, P_L. Local primary membrane stress is not technically a classification of stress but a stress category, since it is a combination of two stresses. The combination it represents is primary membrane stress, P_m, plus secondary membrane stress produced from sustained loads. These have been grouped together in order to limit the allowable stress for this particular combination to a level lower than allowed for other primary and secondary stress applications. It was felt that local stress from sustained (unrelenting) loads presented a great enough hazard for the combination to be "classified" as a primary stress.

A local primary stress is produced either by design pressure alone or by other mechanical loads. Local primary stresses have some self-limiting characteristics like secondary stresses. Since they are localized, once the yield strength of the material is reached, the load is redistributed to stiffer portions of the vessel. However, since any deformation associated with yielding would be unacceptable, an allowable stress lower than secondary stresses is assigned. The basic difference between a primary local stress and a secondary stress is that a primary local stress is produced by a load that is unrelenting; the stress is just redistributed. In a secondary stress, yielding relaxes the load and is truly self limiting. The ability of primary local stresses to redistribute themselves after the yield strength is attained locally provides a safety-valve effect. Thus, the higher allowable stress applies only to a local area.

Primary local membrane stresses are a combination of membrane stresses only. Thus only the "membrane" stress-es from a local load are combined with primary general membrane stresses, not the bending stresses. The bending stresses associated with a local loading are secondary stresses. Therefore, the membrane stresses from a WRC-107-type analysis must be broken out separately and combined with primary general stresses. The same is true for discontinuity membrane stresses at head-shell junctures, cone-cylinder junctures, and nozzle-shell junctures. The bending stresses would be secondary stresses.

Therefore, $P_L = P_m + Q_m$ where Q_m is a local stress from a sustained or unrelenting load. Examples of primary local membrane stresses are:

a. P_m + membrane stresses at local discontinuities:
   1. Head-shell juncture
   2. Cone-cylinder juncture
   3. Nozzle-shell juncture
   4. Shell-flange juncture
   5. Head-skirt juncture
   6. Shell-stiffening ring juncture
b. P_m + membrane stresses from local sustained loads:
   1. Support lugs
   2. Nozzle loads
   3. Beam supports
   4. Major attachments

Secondary stress. The basic characteristic of a secondary stress is that it is self limiting. As defined earlier, this means that local yielding and minor distortions can satisfy the conditions which caused the stress to occur. Application of a secondary stress cannot cause structural failure due to the restraints offered by the body to which the part is attached. Secondary mean stresses are developed at the junctions of major components of a pressure vessel. Secondary mean stresses are also produced by sustained loads other than internal or external pressure. Radial loads on nozzles produce secondary mean stresses in the shell at the junction of the nozzle. Secondary stresses are strain-induced stresses.

Discontinuity stresses are only considered as secondary stresses if their extent along the length of the shell is limited. Division 2 imposes the restriction that the length over which the stress is secondary is $\sqrt{R_m t}$. Beyond this distance, the stresses are considered as primary mean stresses. In a cylindrical vessel, the length $\sqrt{R_m t}$ represents the length over which the shell behaves as a ring.

A further restriction on secondary stresses is that they may not be closer to another gross structural discontinuity than a distance of $2.5 \sqrt{R_m t}$. This restriction is to eliminate the additive effects of edge moments and forces.

Secondary stresses are divided into two additional groups, membrane and bending. Examples of each are as follows: Secondary membrane stress, Q_m.

a. Axial stress at the juncture of a flange and the hub of the flange.
b. Thermal stresses.
c. Membrane stress in the knuckle area of the head.
d. Membrane stress due to local relenting loads.

Secondary bending stress, Q_b.

a. Bending stress at a gross structural discontinuity: nozzles, lugs, etc. (relenting loadings only).
b. The nonuniform portion of the stress distribution in a thick-walled vessel due to internal pressure.
c. The stress variation of the radial stress due to internal pressure in thick-walled vessels.
d. Discontinuity stresses at stiffening or support rings.

Note: For b and c it is necessary to subtract out the average stress which is the primary stress. Only the varying part of the stress distribution is a seondary stress.

Peak stress, F. Peak stresses are the additional stresses due to stress intensification in highly localized areas. They apply to both sustained loads and self-limiting loads. There are no significant distortions associated with peak stresses. Peak stresses are additive to primary and secondary stresses present at the point of the stress concentration. Peak stresses are only significant in fatigue conditions or brittle materials. Peak stresses are sources of fatigue cracks and apply to membrane, bending, and shear stresses. Examples are:

a. Stress at the corner of a discontinuity.
b. Thermal stresses in a wall caused by a sudden change in the surface temperature.
c. Thermal stresses in cladding or weld overlay.
d. Stress due to notch effect (stress concentration).

Categories of Stress

Once the various stresses of a component are calculated, they must be combined and this final result compared to an allowable stress (see Table 1). The combined classes of stress due to a combination of loads acting at the same time are stress categories. Each category has assigned limits of stress based on the hazard it represents to the vessel. The following is derived basically from ASME Code, Section VIII, Division 2, simplified for application to Division 1 vessels and allowable stresses. It should be used as a guideline only because Division 1 recognizes only two categories of stress—primary membrane stress and primary bending stress. Since the calculations of most secondary (thermal and discontinuities) and peak stresses are not included in this text, these categories can be considered for reference only. In addition, Division 2 utilizes a factor K multiplied by the allowable stress for increase due to short term loads due to seismic or upset conditions. It also sets allowable limits of combined stress for fatigue loading where secondary and peak stresses are major considerations. Table 1 sets allowable stresses for both stress classifications and stress categories.

Table 1
Allowable Stresses for Stress Classifications and Categories

Stress Classification or Category	Allowable Stress
General primary membrane, P_m	SE
General primary bending, P_b	$1.5\,SE < .9\,F_y$
Local primary membrane, P_L ($P_L = P_m + Q_{ms}$)	$1.5\,SE < .9\,F_y$
Secondary membrane, Q_m	$1.5\,SE < .9\,F_y$
Secondary bending, Q_b	$3\,SE < 2\,F_y < UTS$
Peak, F	$2\,S_a$
$P_m + P_b + Q_m{}^* + Q_b$	$3\,SE < 2\,F_y < UTS$
$P_L + P_b$	$1.5\,SE < .9\,F_y$
$P_L + P_b + Q_m{}^* + Q_b$	$3\,SE < 2\,F_y < UTS$
$P_L + P_b + Q_m{}^* + Q_b + F$	$2\,S_a$

Notes:
Q_{ms} = membrane stresses from sustained loads
$Q_m{}^*$ = membrane stresses from relenting, self limiting loads
S = allowable stress per ASME Code, Section VIII, Division 1, at design temperature
F_y = minimum specified yield strength at design temperature
$U.T.S.$ = minimum specified tensile strength
S_a = allowable stress for any given number of cycles from design fatigue curves.
Be aware that at certain temperatures for certain materials, $1.5\,SE$ is greater than $.9\,F_y$.

Procedure 1
General Vessel Formulas [1, 3]

Notation

P = internal pressure, psi
D_i, D_o = inside/outside diameter, in.
S = allowable or calculated stress, psi
E = joint efficiency
L = crown radius, in.
R_i, R_o = inside/outside radius, in.
K, M = coefficients
σ_x = longitudinal stress, psi
σ_ϕ = circumferential stress, psi
R_m = mean radius of shell, in.
t = thickness or thickness required of shell, head, or cone, in.
r = knuckle radius, in.

Notes

1. Formulas are valid for:
   a. Pressures < 3,000 psi.
   b. Cylindrical shells where $t \le .5R_i$ or $P \le .385\,SE$. For thicker shells see Reference 1, Para. 1-2.
   c. Spherical shells and hemispherical heads where $t \le .356\,R_i$ or $P \le .665\,SE$. For thicker shells see Reference 1, Para. 1-3.
2. All ellipsoidal and torispherical heads having a minimum specified tensile strength greater than 80,000 psi shall be designed using $S = 20,000$ psi at ambient temperature and reduced by the ratio of the allowable stresses at design temperature and ambient temperature where required.
3. Formulas for factors:

$$K = .167\left[2 + \left(\frac{D}{2h}\right)^2\right]$$

$$M = .25\left(3 + \sqrt{\frac{L}{r}}\right)$$

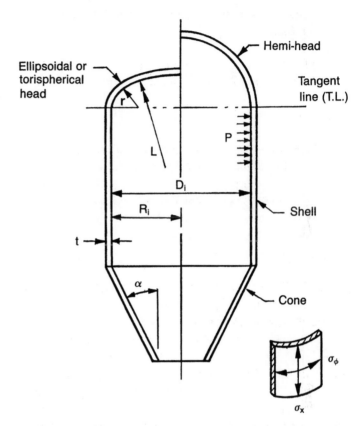

Figure 1. General configuration and dimensional data for vessel shells and heads.

Table 1
General Vessel Formulas

Part	Stress Formula	Thickness, t I.D.	Thickness, t O.D.	Pressure, P I.D.	Pressure, P O.D.	Stress, S I.D.	Stress, S O.D.
Shell							
Longitudinal [Section UG-27(c)(2)]*	$\sigma_x = \dfrac{PR_m}{2t}$	$\dfrac{PR_i}{2SE + .4P}$	$\dfrac{PR_o}{2SE + 1.4P}$	$\dfrac{2SEt}{R_i - .4t}$	$\dfrac{2SEt}{R_o - 1.4t}$	$\dfrac{P(R_i + .4t)}{2Et}$	$\dfrac{P(R_o - 1.4t)}{2Et}$
Circumferential [Section UG-27(c)(1); Section 1-1(a)(1)]*	$\sigma_\phi = \dfrac{PR_m}{t}$	$\dfrac{PR_i}{SE - .6P}$	$\dfrac{PR_o}{SE + .4P}$	$\dfrac{SEt}{R_i + .6t}$	$\dfrac{SEt}{R_o - .4t}$	$\dfrac{P(R_i + .6t)}{Et}$	$\dfrac{P(R_o - .4t)}{Et}$
Heads							
Hemisphere [Section 1-1(a)(2); Section UG-27(d)]*	$\sigma_x = \sigma_\phi = \dfrac{PR_m}{2t}$	$\dfrac{PR_i}{2SE - .2P}$	$\dfrac{PR_o}{2SE + .8P}$	$\dfrac{2SEt}{R_i + .2t}$	$\dfrac{2SEt}{R_o - .8t}$	$\dfrac{P(R_i + .2t)}{2Et}$	$\dfrac{P(R_o - .8t)}{2Et}$
Ellipsoidal [Section 1-4(c)]*	See PROCEDURE 2	$\dfrac{PD_iK}{2SE - .2P}$	$\dfrac{PD_oK}{2SE + 2P(K - .1)}$	$\dfrac{2SEt}{KD_i + .2t}$	$\dfrac{2SEt}{KD_o - 2t(K - .1)}$	See PROCEDURE 2	
2:1 S.E. [Section UG-32d]*	"	$\dfrac{PD_i}{2SE - .2P}$	$\dfrac{PD_o}{2SE + 1.8P}$	$\dfrac{2SEt}{D_i + .2t}$	$\dfrac{2SEt}{D_o - 1.8t}$	"	"
100%-6% Torispherical [Section UG-32(e)]*	"	$\dfrac{.885PL_i}{SE - .1P}$	$\dfrac{.885PL_o}{SE + .8P}$	$\dfrac{SEt}{.885L_i + .1t}$	$\dfrac{SEt}{.885L_o - .8t}$	"	"
Torispherical $L/r < 16.66$ [Section 1-4(d)]*	"	$\dfrac{PL_iM}{2SE - .2P}$	$\dfrac{PL_oM}{2SE + P(M - .2)}$	$\dfrac{2SEt}{L_iM + .2t}$	$\dfrac{2SEt}{L_oM - t(M - .2)}$	"	"
Cone							
Longitudinal	$\sigma_x = \dfrac{PR_m}{2t\cos\alpha}$	$\dfrac{PD_i}{4\cos\alpha(SE + .4P)}$	$\dfrac{PD_o}{4\cos\alpha(SE + 1.4P)}$	$\dfrac{4SEt\cos\alpha}{D_i - .8t\cos\alpha}$	$\dfrac{4SEt\cos\alpha}{D_o - 2.8t\cos\alpha}$	$\dfrac{P(D_i - .8t\cos\alpha)}{4Et\cos\alpha}$	$\dfrac{P(D_o - 2.8t\cos\alpha)}{4Et\cos\alpha}$
Circumferential [Section 1-4(e); Section UG-32(g)]*	$\sigma_\phi = \dfrac{PR_m}{t\cos\alpha}$	$\dfrac{PD_i}{2\cos\alpha(SE - .6P)}$	$\dfrac{PD_o}{2\cos\alpha(SE + .4P)}$	$\dfrac{2SEt\cos\alpha}{D_i + 1.2t\cos\alpha}$	$\dfrac{2SEt\cos\alpha}{D_o - .8t\cos\alpha}$	$\dfrac{P(D_i + 1.2t\cos\alpha)}{2Et\cos\alpha}$	$\dfrac{P(D_o - .8t\cos\alpha)}{2Et\cos\alpha}$

*ASME Boiler and Pressure Vessel Code, Section VIII, Division 1, 1983 Edition, American Society of Mechanical Engineers.

Procedure 2
Stresses in Heads Due to Internal Pressure [3, 4]

Notation

L = crown radius, in.
r = knuckle radius, in.
h = depth of head, in.
R_L = latitudinal radius of curvature, in.
R_m = meridional radius of curvature, in.
σ_ϕ = latitudinal stress, psi
σ_x = meridional stress, psi
P = internal pressure, psi

Formulas

Lengths of R_L and R_m for ellipsoidal heads:

• At equator:

$$R_m = \frac{h^2}{R}$$

$$R_L = R$$

• At center of head:

$$R_m = R_L = \frac{R^2}{h}$$

• At any point X:

$$R_L = \sqrt{\frac{R^4}{h^2} + X^2 \left(1 - \frac{R^2}{h^2}\right)}$$

$$R_m = \frac{R_L^3 \, h^2}{R^4}$$

Notes

1. Latitudinal (hoop) stresses in the knuckle become compressive when the R/h ratio exceeds 1.42. These heads will fail by either elastic or plastic buckling, depending on the R/t ratio.
2. Head types fall into one of three general categories: hemispherical, torispherical, and ellipsoidal. Hemispherical heads are analyzed as spheres and were covered in the previous section. Torispherical (also known as flanged and dished heads) and ellipsoidal head formulas for stress are outlined in the following form.

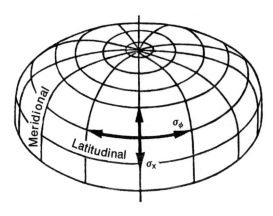

Figure 1. Direction of stresses in a vessel head.

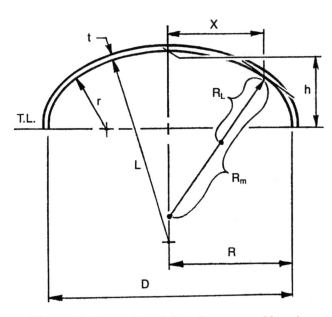

Figure 2. Dimensional data for a vessel head.

TORISPHERICAL HEADS			
σ_x		σ_ϕ	
At Junction of Crown and Knuckle			
$\sigma_x = \dfrac{PL}{2t}$		$\sigma_\phi = \dfrac{PL}{4t}\left(3 - \dfrac{L}{R}\right)$	
In Crown			
$\sigma_x = \dfrac{PL}{2t}$		$\sigma_\phi = \sigma_x$	
In Knuckle			
$\sigma_x = \dfrac{PL}{2t}$		$\sigma_\phi = \dfrac{PL}{t}\left(1 - \dfrac{L}{2r}\right)$	
At Tangent Line			
$\sigma_x = \dfrac{PR}{2t}$		$\sigma_\phi = \dfrac{PR}{t}$	

ELLIPSOIDAL HEADS			
σ_x		σ_ϕ	
At Any Point X			
$\sigma_x = \dfrac{PR_L}{2t}$		$\sigma_\phi = \dfrac{PR_L}{t}\left(1 - \dfrac{R_L}{2R_m}\right)$	
At Center of Head			
$\sigma_x = \dfrac{PR^2}{2th}$		$\sigma_\phi = \sigma_x$	
At Tangent Line			
$\sigma_x = \dfrac{PR}{2t}$		$\sigma_\phi = \dfrac{PR}{t}\left(1 - \dfrac{R^2}{2h^2}\right)$	

Joint Efficiencies (ASME Code) [5]

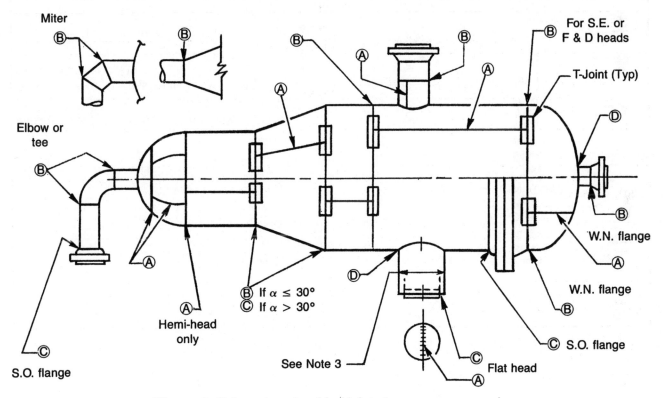

Figure 1. Categories of welded joints in a pressure vessel.

Table 1
Values of Joint Efficiency, E, and Allowable Stress, S*

Extent of Radiography	Case 1								Case 2								Case 3								Case 4							
	Seamless Head		Seamless Shell		Seamless Head		Welded Shell		Welded Head		Seamless Shell		Welded Head		Welded Shell	
	E	S	E	S	E	S	E	S	E	S	E	S	E	S	E	S
Full (RT-1)	1.0	100%	1.0	100%	1.0	100%	1.0	100%	1.0	100%	1.0	100%	1.0	100%	1.0	100%
Spot (RT-3)	1.0	85%	1.0	85%	1.0	85%	.85	100%	.85	100%	1.0	85%	.85	100%	.85	100%
Combination†	1.0	100%	1.0	100%	1.0	100%	.85	100%	1.0	100%	1.0	100%	1.0	100%	.85	100%
None	1.0	80%	1.0	80%	1.0	80%	.7	100%	.7	100%	1.0	80%	.7	100%	.7	100%

*See Note 1.
†See Note 2.

Table 2
Joint Efficiencies

Types of Joints		Full	X-Ray Spot	None
	Single and double butt joints	1.0	.85	.7
	Single butt joint with backing strip	.9	.8	.65
	Single butt joint without backing strip	~	~	.6
	Double full fillet lap joint	~	~	.55
	Single full fillet lap with plugs	~	~	.5
	Single full fillet lap joint	~	~	.45

Notes

1. In Table 1 joint efficiencies and allowable stresses for shells are for longitudinal seams only! All joints are assumed as Type 1 only! Where combination radiography is shown it is assumed that all requirements for full radiography have been met for head, and shell is spot R. T.
2. Combination radiography: Applies to vessels not fully radiographed where the designer wishes to apply a joint efficiency of 1.0 per ASME Code, Table UW-12, for only a specific part of a vessel. Specifically for any part to meet this requirement, you must perform the following:
   - (ASME Code, Section UW-11(5)): Fully x-ray any Cat. A or D butt welds
   - (ASME Code, Section UW-11(5)(b)): Spot x-ray any Category B or C butt welds attaching the part
   - (ASME Code, Section UW-11(5)(a)): All butt joints must be Type 1
3. Any Category B or C butt weld in a nozzle or communicating chamber of a vessel or vessel part which is to have a joint efficiency of 1.0 and exceeds either 10 in. nominal pipe size or 1⅛ in. in wall thickness shall be fully radiographed. See ASME Code, Section UW-11(a)(4).

Properties of Heads

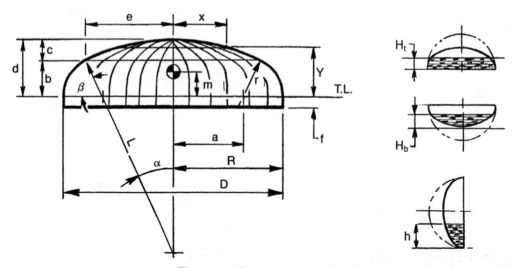

Figure 1. Dimensions of heads.

Formulas

$$a = \frac{D - 2r}{2}$$

$$\alpha = \underset{\sin}{\text{arc}}\left(\frac{a}{L - r}\right)$$

$$\beta = 90 - \alpha$$

$$b = \cos \alpha \, r$$

$$c = L - \cos \alpha \, L$$

$$e = \sin \alpha \, L$$

$$\phi = \frac{\beta}{2}$$

Volume

$$V_1 = \text{(frustum)} = .333b \, \pi(e^2 + ea + a^2)$$

$$V_2 = \text{(spherical segment)} = \pi c^2(L - c/3)$$

$$V_3 = \text{(solid of revolution)}$$

$$= \frac{120r^3\pi \sin \phi \cos \phi + a\phi\pi^2r^2}{90}$$

Total volume: $V_1 + V_2 + V_3$

Table 1
Partial Volumes

Type	Volume to H_t	Volume to H_b	Volume to h
Hemi	$\frac{\pi D^2 H_t}{4}\left[1 - \frac{4H_t^2}{3D^2}\right]$	$\frac{\pi DH_b^2}{2}\left[1 - \frac{2H_b}{3D}\right]$	$\frac{\pi h^2(1.5D - h)}{6}$
2:1 S.E.	$\frac{\pi D^2 H_t}{4}\left[1 - \frac{16H_t^2}{3D^2}\right]$	$\pi DH_b^2\left[1 - \frac{4H_b}{3D}\right]$	$\frac{\pi h^2(1.5D - h)}{12}$
100%-6% F & D	$\frac{3VH_t}{2d}\left[1 - \frac{H_t^2}{3d^2}\right]$	$\frac{3VH_b^2}{2d^2}\left[1 - \frac{H_b}{3d}\right]$	$\frac{3Vh^2}{D^2}\left[1 - \frac{2h}{3D}\right]$

D is in ft.

Table 2
General Data

Type	Surface Area	Volume	C.G. – m Empty	C.G. – m Full	Depth of Head, d	Points on Heads X =	Points on Heads Y =
Hemi	$\pi D^2/2$	$\pi D^3/12$	.2878D	.375D	.5D	$\sqrt{R^2 - Y^2}$	$\sqrt{R^2 - X^2}$
2:1 S.E.	$1.084 \, D^2$	$\pi D^3/24$	.1439D	.1875D	.25D	$.5 \sqrt{D^2 - 16Y^2}$	$.25 \sqrt{D^2 - 4X^2}$
100%-6% F & D	$.9286 \, D^2$	$.0847D^3$	.100D		.169 D		

D is in ft.

Notes

1. Developed length of flat plate (diameter)

$$\text{D.L.} = 2\left(\frac{\beta}{180}\right)\pi r + 2\left(\frac{\alpha}{180}\right)\pi L + 2f$$

2. For 2:1 S.E. heads the crown and knuckle radius may be approximated as follows:

$$L = .9045 \, D$$

$$r = .1727 \, D$$

3. Conversion factors

- Multiply $ft^3 \times 7.48$ to get gallons
- Multiply $ft^3 \times 62.39$ to get lb-water
- Multiply gallons $\times 8.33$ to get lb-water

4. Depth of head

$$A = L - r$$
$$B = R - r$$

$$d = L - \sqrt{A^2 - B^2}$$

Volumes and Surface Areas of Vessel Sections

Notation

l = height of cone, depth of head, or length of cylinder
α = one-half apex angle of cone
D = large diameter of cone, diameter of head or cylinder
R = radius
r = knuckle radius of F & D head
L = crown radius of F & D head
h = partial depth of horizontal cylinder
K, C = coefficients
d = small diameter of truncated cone
V = volume

$$K = \frac{L}{R} - \sqrt{\left(\frac{L}{R} - 1\right)\left(\frac{L}{R} + 1 - \frac{2r}{R}\right)}$$

$$e = \sqrt{1 - \frac{l^2}{R^2}}$$

Table 1
Volumes and Surface Areas of Vessel Sections

Section	Volume	Surface Area
Sphere	$\dfrac{\pi D^3}{6}$	πD^2
Hemi-head	$\dfrac{\pi D^3}{12}$	$\dfrac{\pi D^2}{2}$
2:1 S.E. head	$\dfrac{\pi D^3}{24}$	$1.084\, D^2$
Ellipsoidal head	$\dfrac{\pi D^2 l}{6}$	$2\,\pi R^2 + \dfrac{\pi l^2}{e} \ln \dfrac{1+e}{1-e}$
100–6% F & D head	$.08467\, D^3$	$.9286\, D^2$
F & D head	$\dfrac{2\,\pi R^3 K}{3}$	$\pi R^2 \left[1 + \dfrac{l^2}{R^2}\left(2 - \dfrac{l}{R}\right)\right]$
Cone	$\dfrac{\pi D^2 l}{12}$	$\dfrac{\pi D l}{2 \cos \alpha}$
Truncated cone	$\dfrac{\pi\, l(D^2 + Dd + d^2)}{12}$	$\pi\left(\dfrac{D+d}{2}\right)\sqrt{l^2 + \left(\dfrac{D-d}{2}\right)}$
30° Truncated cone	$.227(D^3 - d^3)$	$1.57\,(D^2 - d^2)$

Table 2
Values of c for Partial Volumes of a Horizontal Cylinder

h/D	c
.1	.0524
.15	.0941
.2	.1424
.25	.1955
.3	.2523
.35	.3119
.4	.3735
.45	.4364
.5	.5
.55	.5636
.6	.6265
.65	.6881
.7	.7477
.75	.8045
.8	.8576
.85	.9059
.9	.9480
.95	.9813

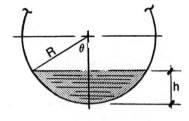

$$\theta = \text{arc cos}\, \frac{R - h}{R}$$

$$V = R^2 l \left[\left(\frac{\pi \theta°}{180}\right) - \sin\theta\,\cos\theta\right]$$

or

$$V = \pi R^2 l c$$

Figure 1. Formulas for partial volumes of a horizontal cylinder.

Maximum Length of Unstiffened Shells

Thickness (in.)

Diameter (in.)	1/4	5/16	3/8	7/16	1/2	9/16	5/8	11/16	3/4	13/16	7/8	15/16	1	1 1/16	1 1/8	1 3/16
36	204															
	∞															
42	168	280														
	313	∞														
48	142	235	358													
	264	437	∞													
54	122	203	306	437												
	228	377	∞													
60	104	178	268	381												
	200	330	499	∞												
66	91	157	238	336	458											
	174	293	442	626	∞											
72	79	138	213	302	408	537										
	152	263	396	561	∞											
78	70	124	193	273	369	483	616									
	136	237	359	508	686	∞										
84	63	110	175	249	336	438	559									
	123	212	327	462	625	816	∞									
90	57	99	157	228	308	402	510	637								
	112	190	300	424	573	748	∞									
96	52	90	143	210	284	370	470	585	715							
	103	173	274	391	528	689	875	∞								
102	48	82	130	190	263	343	435	540	661	795						
	94	160	249	363	490	639	810	1,005	∞							
108	44	76	118	176	245	320	405	502	613	738	875					
	87	148	228	337	456	594	754	935	∞							
114	42	70	109	162	223	299	379	469	571	687	816					
	79	138	211	311	426	555	705	874	1,064	∞						
120	39	65	101	149	209	280	355	440	536	642	762	894				
	74	128	197	287	400	521	660	819	997	∞						
126	37	61	95	138	195	263	334	414	504	603	715	839	974			
	69	120	184	266	374	490	621	770	938	1,124	∞					
132	35	57	88	129	181	242	315	391	475	569	673	789	916	1,053		
	65	113	173	248	348	462	586	727	884	1,060	1,253	∞				
138	33	54	83	121	169	228	297	369	449	538	636	744	864	994		
	62	106	163	234	325	437	555	687	836	1,002	1,185	∞				
144	31	51	78	114	158	214	275	350	426	510	603	705	817	940	1,073	
	59	98	154	221	304	411	526	652	793	950	1,123	1,312	∞			
150		49	74	107	148	201	261	332	405	485	573	669	774	891	1,017	1,152
		92	146	209	286	385	499	619	753	902	1,066	1,246	1,442	∞		
156		46	70	101	140	189	248	309	385	462	546	637	737	846	966	1,095
		87	138	199	271	363	475	590	717	859	1,015	1,186	1,373	∞		
162		44	67	96	133	178	233	294	367	440	520	608	703	806	919	1,042
		83	131	189	258	342	448	562	684	819	968	1,131	1,309	1,509	∞	
	1/4	5/16	3/8	7/16	1/2	9/16	5/8	11/16	3/4	13/16	7/8	15/16	1	1 1/16	1 1/8	1 3/16

Notes:
1. All values are in in.
2. Values are for temperatures up to 500°F.
3. Top value is for full vacuum, lower value is half vacuum.
4. Values are for carbon or low alloy steel ($F_y > 30,000$ psi) based on Figure UCS 28.2 of ASME Code, Section VIII, Div. 1.

Useful Formulas for Vessels [2, 6]

1. Properties of a circle. (See Figure 1.)

   • C. G. of area

$$e_1 = \frac{C^3}{12A_1}$$

$$e_2 = \frac{120C}{\alpha\pi}$$

$$e_3 = \frac{38.197\,(R^3 - r^3)\sin\phi/2}{(R^2 - r^2)\phi/2}$$

   • Chord, C.

$$C = 2R\sin\theta/2$$

$$C = 2\sqrt{2bR - b^2}$$

   • Rise, b.

$$b = .5C\tan\theta/4$$

$$b = R - .5\sqrt{4R^2 - C^2}$$

   • Angle, θ.

$$\theta = 2\arcsin\frac{C}{2R}$$

   • Area of sections

$$A_1 = \frac{\theta\pi R^2 - 180C(R - b)}{360}$$

$$A_2 = \frac{\pi R^2\alpha}{360}$$

$$A_3 = \frac{(R^2 - r^2)\pi\phi}{360}$$

2. Properties of a cylinder.

   • Cross-sectional metal area, A

$$A = 2\pi R_m t$$

   • Section modulus, Z.

$$Z = \pi R_m^2 t$$

$$= \frac{\pi D_m^2 t}{4}$$

$$= \frac{\pi(D^4 - d^4)}{32d}$$

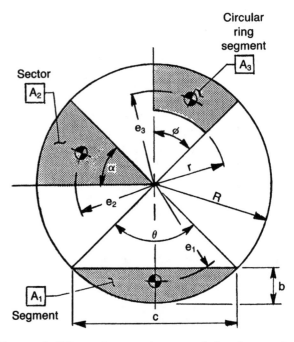

Figure 1. Dimensions and areas of circular sections.

   • Moment of inertia, I.

$$I = \pi R_m^3 t$$

$$= \frac{\pi D_m^3 t}{8}$$

$$= \frac{\pi(D^4 - d^4)}{64}$$

3. Radial displacements due to internal pressure.

   • Cylinder.

$$\delta = \frac{PR^2}{Et}(1 - .5v)$$

   • Cone.

$$\delta = \frac{PR^2}{Et\cos\alpha}(1 - .5v)$$

- Sphere/hemisphere.

$$\delta = \frac{PR^2}{2Et}(1 - v)$$

- Torispherical/ellipsoidal.

$$\delta = \frac{R}{E}(\sigma_\phi - v\sigma_x)$$

where P = internal pressure, psi
R = inside radius, in.
t = thickness, in.
v = Poisson's ratio (.3 for steel)
E = modulus of elasticity, psi
α = ½ apex angle of cone, degrees
σ_ϕ = circumferential stress, psi
σ_x = meridional stress, psi

4. Longitudinal stress in a cylinder due to longitudinal bending moment, M_L.
- Tension.

$$\sigma_x = \frac{M_L}{\pi R^2 tE}$$

- Compression.

$$\sigma_x = (-)\frac{M_L}{\pi R^2 t}$$

where E = joint efficiency
R = inside radius, in.
M_L = bending moment, in.-lb
t = thickness, in.

5. Thickness required for heads due to external pressure.

$$t_h = \frac{L}{\sqrt{\dfrac{E}{16P_e}}}$$

where L = crown radius, in.
P_e = external pressure, psi
E = modulus of elasticity, psi

6. Equivalent pressure of flanged connection under external loads.

$$P_e = \frac{16M}{\pi G^3} + \frac{4F}{\pi G^2} + P$$

where P = internal pressure, psi
F = radial load, lb
M = bending moment, in.-lb
G = gasket reaction diameter, in.

7. Bending ratio of formed plates.

$$\% = \frac{100t}{R_f}\left(1 - \frac{R_f}{R_o}\right)$$

where R_f = finished radius, in.
R_o = starting radius, in. (∞ for flat plates)
t = thickness, in.

8. Stress in nozzle neck subjected to external loads.

$$\sigma_x = \frac{PR_m}{2t_n} + \frac{F}{A} + \frac{MR_m}{I}$$

where R_m = nozzle mean radius, in.
t_n = nozzle neck thickness, in.
A = metal cross-sectional area, in.^{2}
I = moment of inertia, in.^{4}
F = radial load, lb
M = moment, in.-lb
P = internal pressure, psi

9. Circumferential bending stress for out of round shells [2].

$$D_1 - D_2 > 1\%D_{nom}$$

$$R_1 = \frac{D_1 + D_2}{2}$$

$$R_a = \frac{D_1 + D_2}{4} + \frac{t}{2}$$

$$\sigma_b = \frac{1.5PR_1t(D_1 - D_2)}{t^3 + 3\left(\dfrac{P}{E}\right)R_1R_a^2}$$

where D_1 = maximum inside diameter, in.
D_2 = minimum inside diameter, in.
P = internal pressure, psi
E = modulus of elasticity, psi
t = thickness, in.

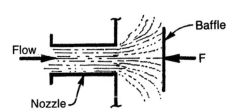

Figure 2. Typical nozzle configuration with internal baffle.

10. Equivalent static force from dynamic flow.

$$F = \frac{V^2 A d}{g}$$

where F = equivalent static force, lb
V = velocity, ft/sec
A = cross-sectional area of nozzle, ft^2
d = density, lb/ft^3
g = acceleration due to gravity, 32.2 ft/sec^2

11. Allowable compressive stress in cylinders [1].

If $\frac{t}{R} \leq .015$, $X = \frac{10^6 t}{R}\left(2 - \frac{200t}{3R}\right)$

If $\frac{t}{R} > .015$, $X = 15,000$

If $\frac{L}{R} \leq 60$, $Y = 1$

If $\frac{L}{R} > 60$, $Y = \dfrac{21,600}{18,000 + \left(\dfrac{L}{R}\right)^2}$

$$F_a = \frac{Q}{A} = XY$$

where t = thickness, in.
R = outside radius, in.
L = length of column, in.
Q = allowable load, lb
A = metal cross-sectional area, in.^{2}
F_a = allowable compressive stress, psi

Material Selection Guide

	Design Temperature, °F	Material	Plate	Pipe	Forgings	Fittings	Bolting
Cryogenic	−425 to −321	Stainless steel	SA-240-304, 304L, 347, 316, 316L	SA-312-304, 304L, 347, 316, 316L	SA-182-304, 304L, 347, 316, 316L	SA-403 304, 304L, 347, 316, 316L	SA-320-B8 with SA-194-8
Cryogenic	−320 to −151	9 nickel	SA-353	SA-333-8	SA-522-I	SA-420-WPL8	
Low temperature	−150 to −76	3½ nickel	SA-203-D	SA-333-3	SA-350-LF63	SA-420-WPL3	SA-320-L7 with SA-194-4
Low temperature	−75 to −51	2½ nickel	SA-203-A				
Low temperature	−50 to −21	Carbon steel	SA-516-55, 60 to SA-20	SA-333-6	SA-350-LF2	SA-420-WPL6	
Low temperature	−20 to 4	Carbon steel	SA-516-All	SA-333-1 or 6			
Low temperature	5 to 32	Carbon steel	SA-285-C				SA-193-B7 with SA-194-2H
Intermediate	33 to 60	Carbon steel	SA-516-All SA-515-All SA-455-II	SA-53-B SA-106-B	SA-105 SA-181-60,70	SA-234-WPB	SA-193-B7 with SA-194-2H
Intermediate	61 to 775	Carbon steel					
Temperature	776 to 875	C-½Mo	SA-204-B	SA-335-P1	SA-182-F1	SA-234-WP1	
Temperature	876 to 1000	1Cr-½Mo	SA-387-12-1	SA-335-P12	SA-182-F12	SA-234-WP12	
Temperature	876 to 1000	1¼Cr-½Mo	SA-387-11-2	SA-335-P11	SA-182-F11	SA-234-WP11	
Temperature	1001 to 1100	2¼Cr-1Mo	SA-387-22-1	SA-335-P22	SA-182-F22	SA-234-WP22	with SA-193-B5 SA-194-3
Elevated	1101 to 1500	Stainless steel	SA-240-347H	SA-312-347H	SA-182-347H	SA-403-347H	
Elevated	1101 to 1500	Incoloy	SB-424	SB-423	SB-425	SB-366	SA-193-B8 with SA-194-8
Elevated	Above 1500	Inconel	SB-443	SB-444	SB-446	SB-366	

From Bednar, H. H., *Pressure Vessel Design Handbook,* Van Nostrand Reinhold Co., 1981.

References

1. ASME Boiler and Pressure Vessel Code, Section VIII, Division 1, 1983 Edition, American Society of Mechanical Engineers.
2. ASME Boiler and Pressure Vessel Code, Section VIII, Division 2, 1983 Edition, American Society of Mechanical Engineers
3. Harvey, J. F., *Theory and Design of Modern Pressure Vessels,* 2nd Ed. New York: Van Nostrand Reinhold Co., 1974.
4. Bednar, H. H., *Pressure Vessel Design Handbook.* New York: Van Nostrand Reinhold Co., 1981.
5. National Board Bulletin, Vol. 32, No. 4, April 1975.
6. Roark, R. J., *Formulas for Stress and Strain,* 5th Ed. New York: McGraw-Hill, 1975.

Source

Moss, Dennis R., *Pressure Vessel Design Manual,* 2nd Ed. Houston: Gulf Publishing Co., 1997.

10

Tribology

Thomas N. Farris, Ph.D., Professor of Aeronautics and Astronautics, Purdue University

INTRODUCTION

Tribology is the science and technology of interacting surfaces in relative motion and the practices related thereto. The word was officially coined and defined by Jost [11]. It is derived from the Greek root *tribos* which means rubbing. Tribology includes *friction, wear,* and *lubrication.*

Tribology has several consequences in mechanical components and modern-day life. Most consequences of friction and wear are considered negative, such as power consumption and the cause of mechanical failure. However, there are also some positive benefits of friction and wear.

It is estimated that 20% of the power consumed in automobiles is used in overcoming friction, while friction accounts for 10% of the power consumption in airplane piston engines and 1.5–2% in modern turbojets. Friction also leads to heat build-up which can cause the deterioration of components due to thermo-mechanical fatigue. Understanding friction is the first step towards reducing friction through clever design, the use of low-friction materials, and the proper use of lubricating oils and greases.

Friction has many benefits, such as the interaction between the tire and the road and the shoe and the floor without which we would not be able to travel. Friction serves as the inherent connecting mechanism in knots, nails, and the nut and bolt assembly. It has some secondary benefits, such as the interaction between the fiber and matrix in composites and damping which may reduce deleterious effects due to resonance.

This chapter begins with a description of contact mechanics and surface topography in sufficient detail to discuss friction, wear, and lubrication in the latter sections. Tribology is a rich subject that cannot be given justice in the space permitted, and the interested reader is encouraged to pursue the subject in greater depth in any of the following books: Bowden and Tabor [3], Rabinowicz [13], Halling [7], Suh [14], Bhushan and Gupta [1], and Hutchings [9]. In addition, an enormous array of material properties is available in tribology handbooks such as Peterson and Winer [12] and Blau [2].

Nomenclature

a	contact half-width or radius
c	approximate radius of elliptical contacts
E	modulus of elasticity
H	hardness
h	surface separation
k	wear coefficient
P	load or load per length
p	contact pressure
p_0	maximum contact pressure
q	shear traction
R_a	roughness
R_i	surface radius
r, x, y, z	position
u, w, δ	displacements
Z_I, Z_{II}	Westergaard functions
$\hat{z}$	= x + iz with i$_M$−$\bar{1}$
μ	coefficient of friction
σ	RMS roughness
σ_i, T	stress components

CONTACT MECHANICS

The solution of contact problems can be reduced to the solution of integral equations in which the known right-hand sides relate to surface geometry and the unknown underneath the integral sign relates to the unknown pressure distribution. Details of the derivation and solution of these equations that highlight the necessary assumptions can be found in Johnson (1985).

Two-dimensional (Line) Hertz Contact of Cylinders

In this section, the contact stress distribution for the frictionless contact of two long cylinders along a line parallel to their axes is derived (Figure 1). This is a special case of the contact of two ellipsoidal bodies first solved by Hertz in 1881.

The origin is placed at the point where the cylinders first come into contact. At first contact the separation of the cylinders is:

$$h = z_1 + z_2$$

$$z_1 = R_1 - \sqrt{R_1^2 - x^2}$$

$$z_2 = R_2 - \sqrt{R_2^2 - x^2} \tag{1}$$

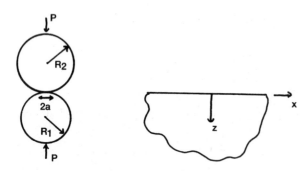

Figure 1. Long cylinders brought into contact by a load per unit length P.

If the contact length is small compared to the size of the cylinders, $a \ll R$, then Equation 1 can be approximated as:

$$h = x^2 \left(\frac{1}{2R_1} + \frac{1}{2R_2} \right) = \frac{x^2}{2R}$$

where $1/R = 1/R_1 + 1/R_2$. The loads cause the cylinders to approach each other a distance $\delta = \delta_1 + \delta_2$. The cylinders must deform to cancel the interpenetration. This is written in equation form as:

$$\overline{w}_1 + \overline{w}_2 = \delta - h$$

or

$$\overline{w}_1 + \overline{w}_2 = \delta - \frac{x^2}{2R}$$

Differentiating gives:

$$\frac{\partial \overline{w}_1}{\partial x} + \frac{\partial \overline{w}_2}{\partial x} = -\frac{x}{R}$$

Using the equations of elasticity, the previous equation can be written as:

$$\int_{-a}^{a} \frac{p(s)ds}{x - s} = \frac{\pi E^*}{2R} x$$

$$-a < x < a$$

where $1/E^* = (1 - v_1^2)/E_1 + (1 - v_2^2)/E_2$

This equation can be inverted for $p(x)$ and the constant of integration is used to assure that the stress is continuous at the edge of contact resulting in:

$$p(x) = \frac{E^*}{2R} \sqrt{a^2 - x^2}$$

The remaining unknown is the contact length (a) which is used to ensure global equilibrium so that:

$$\int_{-a}^{a} p(x)dx = P \qquad \Rightarrow \qquad a = 2\sqrt{\frac{PR}{\pi E^*}}$$

and the maximum contact pressure is:

$$p_0 = p(0)$$

$$p_0 = \sqrt{\frac{PE^*}{\pi R}} \qquad (2)$$

It is interesting to note that the maximum pressure varies as the square root of the load rather than linearly with the load. This is because the contact length increases with the applied load, resulting in an increased area that bears the load. From this perspective, hertzian contacts are very forgiving in the sense that overloads of a factor of two only increase the resulting contact pressure and subsurface stresses by a factor of $\sqrt{2}$.

Ductile materials will be subject to plastic deformation once the maximum shear stress in the material reaches the shear stress at yield in a tensile test. The symmetry of the problem requires that initial yielding occurs along the z-axis, which is a line of symmetry. Because $\tau_{xz} = 0$ along the z-axis, the stresses σ_x and σ_z are principal stresses and the maximum shear stress is one-half of their difference. These stresses along $x = 0$ are [10]:

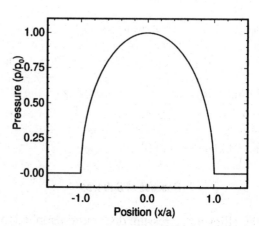

Figure 2. Hertz line contact pressure distribution.

$$\sigma_x(0,z) = -\frac{p_0}{a}\left[\frac{a^2 + 2z^2}{\sqrt{a^2 + z^2}} - 2z\right]$$

$$\sigma_z(0,z) = -\frac{p_0 a}{\sqrt{a^2 + z^2}}$$

$$\tau_{max} = \frac{1}{2}(\sigma_x - \sigma_z) = -p_0\left[\frac{z^2}{a\sqrt{a^2 + z^2}} - z\right] \quad (3)$$

The maximum shear stress occurs below the surface at a depth of $z \approx 0.78a$ and has a value of $\tau_{max} = 0.3p_0$. These stress distributions are shown in Figure 3.

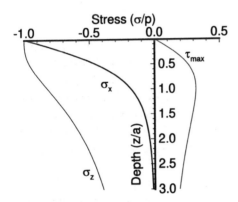

Figure 3. Stress below the surface for Hertz contact.

The Westergaard stress functions [17] can be used in conjunction with a simple FORTRAN program to evaluate the subsurface stress field induced by Hertz contact. For fric-

tionless Hertz contact, the appropriate Westergaard stress function is

$$Z_1 = -\frac{2P}{\pi a^2}\left(\sqrt{a^2 - \hat{z}^2} + i\hat{z}\right) \quad (4)$$

where P is the contact force per unit length, a is the half contact length, and $\hat{z} = x + iz$ with $i = \sqrt{-1}$. This stress function satisfies all of the traction boundary conditions along $z = 0$. The branch cut on the radical in Equation 4 is chosen so that $Z_I \to 0$ as $\hat{z} \to 0$. Contours of the in-plane maximum shear stress:

$$\tau_{max} = \sqrt{\left(\frac{\sigma_x - \sigma_z}{2}\right)^2 + \tau_{xz}^2} \quad (5)$$

are shown in Figure 4. The maximum in-plane shear stress is about $0.3\,p_0$, where p_0 is the maximum contact pressure, and it occurs on the z-axis at a depth of about $z \sim 0.78a$.

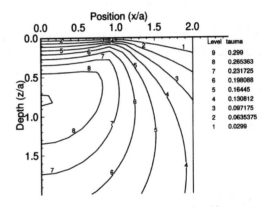

Level	tauma
9	0.299
8	0.265363
7	0.231725
6	0.198088
5	0.16445
4	0.130812
3	0.097175
2	0.0635375
1	0.0299

Figure 4. Stress contours of τ_{max}/p_0 for Hertz contact.

Three-dimensional (Point) Hertz Contact

In this section, the equations for three-dimensional or point contacts are derived. The approach taken is very similar to that used in two-dimensional problems where the point force is superposed to yield the distributed load solutions. The point load solution is derived using Love's axisymmetric stress function, and more details can be found in Chapter 12 of Timoshenko and Goodier [16] and Chapters 3 and 4 of Johnson [10].

Contact of Spheres

For an ellipsoidal pressure distribution applied to a circle of radius, a, such that $p(r) = p_0\sqrt{1 - r^2/a^2}$:

$$\bar{u}_z = \frac{\pi}{4}\frac{1-v^2}{E}\frac{p_0}{a}(2a^2 - r^2), \quad r < a \quad (6)$$

$$\bar{u}_z = \frac{1-v^2}{E}\frac{p_0}{2a}[(2a^2 - r^2)\sin^{-1}(a/r) + r^2(a/r)(1 - a^2/r^2)^{1/2}], \quad r > a$$

Consider the spheres being brought into contact in Figure 5. The load is P, the total approach is δ, and the radius of contact is a. Geometric considerations very similar to those for the contacting cylinders reveal that the sum of the displacements for the two spheres should satisfy:

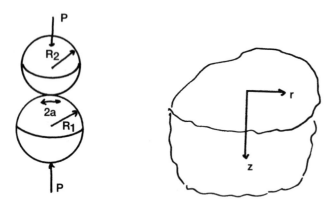

Figure 5. Hertz contact of spheres.

$$\bar{u}_{z1} + \bar{u}_{z2} = \delta - \frac{1}{2R}r^2 \qquad (7)$$

where $1/R = 1/R_1 + 1/R_2$. That is, a pressure distribution which gives a constant plus r^2 term is needed to cancel the potential interpenetration of the spheres. Comparing Equations 6 and 7 reveals that the contacting spheres induce an ellipsoidal pressure distribution and

$$\frac{\pi}{4}\frac{p_0}{aE^*}(2a^2 - r^2) = \delta - \frac{1}{2R}r^2$$

This equation must be valid for any $r < a$ requiring:

$$a = \frac{\pi}{2}\frac{p_0 R}{E^*}$$

and

$$\delta = \frac{\pi}{2}\frac{ap_0}{E^*}$$

Global equilibrium requires:

$$P = \int_0^a p(r)2\pi r dr = \frac{2}{3}p_0\pi a^2$$

Finally:

$$a = \left(\frac{3PR}{4E^*}\right)^{1/3} \qquad (8)$$

$$\delta = \left(\frac{9P^2}{16RE^{*2}}\right)^{1/3} \qquad (9)$$

$$p_0 = \left(\frac{6PE^{*2}}{\pi^3 R^2}\right)^{1/3} \qquad (10)$$

These equations describe Hertz contact for spheres. Notice that these results are nonlinear and that the maximum pressure increases as the load is raised to the ⅓ power.

The corresponding surface stresses can be calculated as:

$$\bar{\sigma}_r = p_0\left\{\frac{1-2v}{3}\frac{a^2}{r^2}[1-(1-r^2/a^2)^{3/2}] \right.$$
$$\left. -(1-r^2/a^2)^{1/2}\right\} \qquad (11)$$

$$\bar{\sigma}_\theta = -p_0\left\{\frac{1-2v}{3}\frac{a^2}{r^2}[1-(1-r^2/a^2)^{3/2}] \right.$$
$$\left. + 2v(1-r^2/a^2)^{1/2}\right\} \qquad (12)$$

$$\bar{\sigma}_z = -p_0(1-r^2/a^2)^{1/2} \qquad (13)$$

inside the contact patch ($r < a$) and

$$\bar{\sigma}_r = -\bar{\sigma}_\theta = p_0\frac{(1-2v)a^2}{3r^2} \qquad (14)$$

outside the contact patch ($r > a$). These stresses are shown in Figure 6. Notice that the radial stress is tensile outside the circle and that it reaches its maximum value at $r = a$. This is the maximum tensile stress in the whole body.

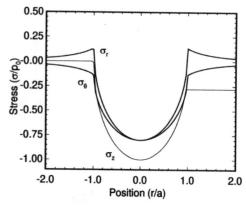

Figure 6. Surface stresses induced by circular point contact ($v = 0.3$).

The stresses along the z-axis (r = 0) can be calculated by first evaluating the stress due to a ring of point force along r = r and integrating from r = 0 to r = a. For example:

$$\sigma_z(0, z) = -\frac{3p_0 z^3}{a} \int_0^a \frac{r\sqrt{a^2 - r^2}\ dr}{(z^2 + r^2)^{5/2}}$$

Making the substitution $a^2 - r^2 = u^2$ and using integration leads to:

$$\sigma_z(0, z) = -p_0 \frac{a^2}{z^2 + a^2} \qquad (15)$$

$$\sigma_r(0, z) = \sigma_\theta(0, z) = p_0 \left\{ -(1 + v)\left[1 - \frac{z}{a}\tan^{-1}\frac{a}{z}\right] \right.$$
$$\left. + \frac{1}{2}\frac{a^2}{z^2 + a^2} \right\} \qquad (16)$$

Symmetry dictates that the maximum shear stress in the body occurs along r = 0. Manipulation of the above equations leads to:

$$\tau_{max} = \frac{1}{2}(\sigma_z(0, z) - \sigma_r(0, z))$$
$$= p_0 \left[(1 + v)\left(1 - \frac{z}{a}\tan^{-1}\frac{a}{z}\right) - \frac{3}{2}\frac{a^2}{z^2 + a^2} \right]$$

For v = 0.3, the maximum shear stress is about $0.31 p_0$ and occurs at a depth of approximately $z \approx 0.48a$. The stresses are plotted for v = 0.3 in Figure 7.

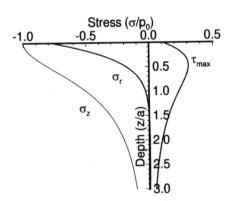

Figure 7. Subsurface stresses induced by circular point contact (v = 0.3).

Elliptical Contact

When solids having unequal curvatures in two directions are brought into contact, the contours of constant separation are ellipses. When the principal curvatures of the two bodies are aligned, the axes of the ellipse correspond to these directions. Placing the x and y axes in the directions of principal curvature, the equation comparable to Equation 7 is:

$$\bar{u}_{z1} + \bar{u}_{z2} = \delta - \frac{1}{2R'}x^2 - \frac{1}{2R''}y^2 \qquad (17)$$

where

$$\frac{1}{R'} = \frac{1}{R'_1} + \frac{1}{R'_2}$$

$$\frac{1}{R''} = \frac{1}{R''_1} + \frac{1}{R''_2},$$

and R'_i are the curvatures in the x direction and R''_i are the curvatures in the y direction.

The contact area is an ellipse, and the resulting pressure distribution is semi-ellipsoidal given by:

$$p(x, y) = p_0 \sqrt{1 - \frac{x^2}{a^2} - \frac{y^2}{b^2}}$$

The actual calculation of a and b is cumbersome. However, for mildly elliptical contacts (Greenwood [5]), the contact can be approximated as circular with:

$$c \equiv \sqrt{ab} \approx \left(\frac{3PR_e}{4E*}\right)^{1/3}$$

$$R_e \equiv \sqrt{R'R''} \qquad (18)$$

with δ and p_0 given by Equations 9 and 10.

Note that for contact of crossed cylinders of radii R_1 and R_2, respectively, the effective radii are given by:

$$\frac{1}{R'} = \frac{1}{R_1} + \frac{1}{\infty}$$

$$\frac{1}{R''} = \frac{1}{\infty} + \frac{1}{R_2}$$

so that if $R_1 = R_2$, the contact patch is circular and the equations of the previous section ("Contact of Spheres") hold.

Effect of Friction on Contact Stress

If contacting cylinders are loaded tangentially as well as normally and caused to slide over each other, then a shear stress will exist at the surface. This shear stress is equal in magnitude to the coefficient of friction μ multiplied by the normal contact pressure. The shear stress acts to oppose the tangential motion of each cylinder. Thus, if the top cylinder moves from the left to the right relative to the bottom cylinder, then the tangential traction on the bottom cylinder is given by:

$$q(x) = \mu \frac{2P}{\pi a^2} \sqrt{a^2 - x^2} \tag{19}$$

The Westergaard stress function that yields these surface tractions is:

$$Z_{II} = -\mu \frac{2P}{\pi a^2} \left(\sqrt{a^2 - \hat{z}^2} + i\hat{z} \right) \tag{20}$$

as can be verified using the equations of Westergaard [17].

Contours of the in-plane maximum shear stress for the combined shear and normal tractions are shown in Figures 8 and 9 for $\mu = 0.1$ and $\mu = 0.4$, respectively. The indenter is sliding over the surface from left to right. Notice that increasing the coefficient of friction increases the maximum shear stress while changing its location to be nearer the surface and off the z-axis towards the leading edge of contact. For the higher value of μ, τ_{max} occurs on the surface.

Tangentially loading spheres so that they slide with respect to each other has similar effects. The subsurface maximum shear stress is increased and moved closer to the surface toward the leading edge of contact. In addition, the surface tensile stress is decreased at the leading edge of contact and increased at the trailing edge of contact.

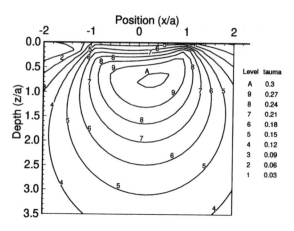

Figure 8. Stress contours of τ_{max}/p_0 for frictional Hertz contact with $\mu = 0.1$. The load is sliding from left to right.

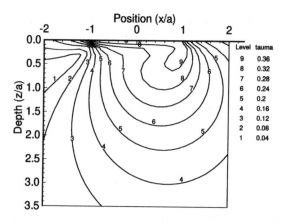

Figure 9. Stress contours of τ_{max}/p_0 for frictional Hertz contact with $\mu = 0.4$. The load is sliding from left to right.

Yield and Shakedown Criteria for Contacts

The maximum shear stress values illustrated above can be used as initial yield criteria for contacts. However, rolling contacts that are loaded above the elastic limit can sometimes develop residual stresses in such a way that the body reaches a state of elastic shakedown. Elastic shakedown implies that there is no repeated plastic deformation and the resulting deleterious fatigue effects. Shake-

down occurs when the sum of the residual stresses and the live stresses do not violate yield anywhere.

Whenever the loads are such that it is possible for such a residual stress state to be developed, then the body does shakedown. This idea makes it possible for shakedown limits to be calculated. As an example, for two-dimensional contacts without friction, the maximum shear stress is about $\tau_{max} = 0.3\,p_0$, implying that initial yield occurs when $0.3p_0 = k$ or $p_0 = 3.3k$ where k is the yield stress in shear. However, it can be shown that for $p_0 < 4k$, elastic shakedown is reached so that subsurface plastic deformation does not continue throughout life. Recalling that the maximum contact pressure increases with the square root of load makes this shakedown effect very important. More details on the concept of residual stress-induced shakedown and additional effects that can lead to shakedown such as strain hardening can be found in Johnson [10].

TOPOGRAPHY OF ENGINEERING SURFACES

The discussion of surface roughness and the techniques used to quantify it are discussed here. Some general work in this area includes Thomas [15] and Greenwood [6].

The length scale of interest is smaller than that considered in the Hertz contact calculations in which contact stresses are calculated to discover what is happening inside the body such as the location of first yield or cracking. Now we are going to focus on the surface of the bodies. One convenient manner of characterizing this surface is to measure its surface roughness. However, it is important to note that mechanical properties are also different near the surface than they are in the bulk of the material.

Table 1
Surface Roughness for Various Finishing Processes

Process	RMS Roughness (Microns)
Grinding	0.8 ~ 0.4
Fine grinding	0.25
Polishing	0.1
Super finishing	0.025 ~ 0.01

Definition of Surface Roughness

Consider a surface profile whose height is given as a function of position $z(x)$. The datum is chosen so that:

$$\int_0^L z(x)dx = 0$$

The average roughness R_a is defined as:

$$R_a = \frac{1}{L}\int_0^L |z|\,dx$$

where the absolute value implies that peaks and valleys have the same contribution.

The root-mean-square roughness or standard deviation is defined by:

$$\sigma^2 = \frac{1}{L}\int_0^L z^2\,dx$$

The RMS roughness is always greater than the average roughness so that:

$$\sigma > R_a$$

and

$\sigma \approx 1.2R_a$

for most surfaces. Typical RMS values for finishing processes are given in Table 1.

Of course, widely different surfaces could give the same R_a and RMS values. The type of statistical quantity needed will depend on the application. One quantity that is used in practice is the bearing area curve which is a plot of the surface area of the surface as a function of height. If the surface does not deform during contact, then the bearing area curve is the relationship between actual area of contact and approach of the two surfaces. This concept leads to discussion of contact of actual rough surface contacts. (See also Figure 10.)

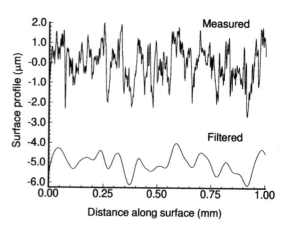

Figure 10. A typical rough surface.

Contact of Rough Surfaces

Much can be gleaned from consideration of the contact of rough surfaces in which it is found that the real area of contact is much less than the apparent area of contact as illustrated by the contact pressure distributions shown in Figure 11. The smooth solid line is the contact pressure for contact with an equivalent smooth surface, while the line showing pressure peaks is the contact pressure for contact with a model periodic rough surface. The dashed line is the moving average of the rough surface contact pressure, and it is very similar to that of the Hertz contact with the smooth surface. Thus, the subsurface stresses are similar for the smooth and rough surface contacts, and yield and plastic flow beneath the surface is not strongly dependent on the surface roughness. This conclusion is the reason that many hertzian contact designs are based on calculated smooth surface pressure distributions with an accompanying call-out on surface roughness.

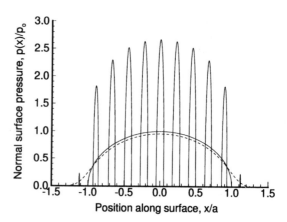

Figure 11. Line contact pressure distribution for periodic rough surface.

Life Factors

The fact that some of the information contained in the rough surface stress field can be inferred from the corresponding Hertz stresses and the surface profile has lead to the development of life factors for rolling element bearings. In these life factor equations, there are terms that account for near-surface metallurgy, surface roughness, lubrica-tion, as well as additional effects. These life factors were summarized recently by longtime practitioners in the bearing design field. This summary can be found in Zaretsky [18] and is written in a format that can be applied easily by the practicing engineer.

FRICTION

Consider a block of weight W resting on an inclined plane. As the plane is tilted to an angle with the horizontal θ, the weight can be resolved into force components perpendicular to the plane N and parallel to the plane F. If θ is less than a certain value, say, θ_s, the block does not move. It is inferred that the plane resists the motion of the block that is driven by the component of the weight parallel to the plane. The force resisting the motion is due to *friction* between the plane and block. As θ is increased past θ_s the block begins to move because the tangential force due to the weight F overcomes the frictional force. It can be shown experimentally that θ_s is approximately independent of the size, shape, and weight of the block. The ratio of F to N at sliding is $\tan \theta_s \equiv \mu$, which is called the *coefficient of friction*. The frictional resistance to motion is equal to the coefficient of friction times the compressive normal force between two bodies. The coefficient of friction depends on the two materials and, in general, $0.05 < \mu < \infty$.

The idea that the resistance to motion caused by friction is $F = \mu N$ is called the Amontons-Coulomb Law of sliding friction. This law is not like Newton's Laws such as $F = ma$. The more we study the friction law, the more complex it becomes. In fact, we cannot fully explain the friction law as evidenced by the fact that it is difficult or even impossible to estimate μ for two materials without performing an experiment. As a point of reference, μ is tabulated for several everyday circumstances in Table 2.

As noted in the rough surface contact section, the real area of contact is invariably smaller than the apparent area of contact. The most common model of friction is based on assuming that the patches of real contact area form junctions in which the two bodies adhere to each other. The resistance to sliding, or friction, is due to these junctions.

Assuming that the real area of contact is equal to the applied load divided by the hardness of the softer material, and that the shear stress required to break the junctions is the yield stress in shear of the weaker material, leads to:

$$\mu = \frac{\tau_y}{H}$$

where μ is the coefficient of friction, τ_y is the yield stress in shear, and H is the hardness. For most materials, the hardness is about three times the yield stress in tension and the Tresca yield conditions assume that the yield stress in shear is about half of the yield stress in tension. Substituting leads to $\mu \approx 1/6$.

The simple adhesive law of friction is attractive in that it is independent of the shape of the bodies and leads to the force acting opposite the direction of motion, resulting in energy dissipation. Its weakness is that it incorrectly predicts that μ is always equal to 1/6. This can be explained in part by the effect of large localized contact pressure on material properties and the contribution of plowing and thermal effects.

Table 2
Typical Coefficients of Friction

Physical Situation	μ
Rubber on cardboard (try it)	0.5–0.8
Brake material on brake drum	1.2
Dry tire on dry road	1
Wet tire on wet road	0.2
Copper on steel, dry	0.7
Ice on wood	0.05

Source: Bowden [3]

WEAR

The rubbing together of two bodies can cause material removal or weight loss from one or both of the bodies. This phenomenon is called *wear*. Wear is a very complex process. It is much more complex even than friction. The complexity of wear is exemplified in Table 3 where wear rate is shown with μ for several material combinations. Radical differences in wear rate occur over relatively small ranges of the coefficient of friction. Note that wear is calculated from wear rate by multiplying by the distance traveled.

There are several standard wear configurations that can be used to obtain wear coefficients and compare material choices for a particular design. A primary source for this information is the *Wear Control Handbook* by Peterson and Winer [12].

Table 3
Friction and Wear from Pin on Ring Tests

Materials	μ	Wear rate $\frac{cm^3}{cm} \times 10^{-12}$
1 Mild steel on mild steel	0.62	157,000
2 60/40 leaded brass	0.24	24,000
3 PTFE (Teflon)	0.18	2,000
4 Stellite	0.60	320
5 Ferritic stainless steel	0.53	270
6 Polyethylene	0.65	30
7 Tungsten carbide on itself	0.35	2

Rings are hardened tool steel except in tests 1 and 7 (Halling [7]).
The load is 400 g and the speed is 180 cm/sec.

As in friction, the most prominent wear mechanism is due to adhesion. Assuming that some of the real contact area junctions fail just below the surface leading to a wear particle results in:

$$V = k \frac{Ws}{H}$$

where V is the volume of material removed, W is the normal load, s is the horizontal distance traveled, H is the hardness of the softer material, and k is the dimensionless wear coefficient representing the probability that a junction will form a wear particle. In this form, wear coefficients vary from $\sim 10^{-8}$ to $\sim 10^{-2}$. There is a wealth of information on wear coefficients published biannually in the proceedings of the International Conference on Wear of Materials, which is sponsored by the American Society of Mechanical Engineers.

LUBRICATION

Lubrication is the effect of a third body on the contacting bodies. The third body may be a lubricating oil or a chemically formed layer from one or both of the contacting bodies (oxides). In general, the coefficient of friction in the presence of lubrication is reduced so that $0.001 < \mu < 0.1$.

Lubrication is understood to fall into three regimes dependent on the component configuration, load, and speed. Under relatively modest loads in conformal contacts such as a journal bearings, moderate pressures exist and the deformation of the solid components does not have a large effect on the lubricant pressure distribution. The bodies are far apart and wear is insignificant. This regime is known as hydrodynamic lubrication.

As loads are increased and the geometry is nonconforming, such as in roller bearings, the lubricant pressure greatly increases and the elastic deformation of the solid components plays a role in lubricant pressure. This regime is known as elastohydrodynamic lubrication, provided the lubrication film thickness is greater than about three times the surface roughness. Once the film thickness gets smaller than this, the solid bodies touch at isolated patches in a mechanism known as *boundary lubrication*. Here, the intense pressures and temperatures make the chemistry of the lubricant surface interaction important.

The lubricant film thickness is strongly dependent on lubricant viscosity at both high and low temperatures. Nondimensional formulas are available for designers to use in distinguishing the regimes of lubrication. Once the regime of lubrication is determined, additional formulas can be used to estimate the maximum contact pressure as well as the minimum film thickness. The maximum contact pressure can then be used in the life factors of Zaretsky [18], and the minimum film thickness can be used in the consideration of lubricant film breakdown. While these formulas are too numerous to summarize here, a primary source is Hamrock [8] in which all of the requisite formulas are defined.

REFERENCES

1. Bhushan, B. and Gupta, B. K., *Handbook of Tribology: Materials, Coatings, and Surface Treatments.* New York: McGraw-Hill, 1991.
2. Blau, P. J. (Ed.), *Friction, Lubrication, and Wear Technology.* ASM Handbook, Vol. 18, ASM, 1992.
3. Bowden, F. P. and Tabor, D., *The Friction and Lubrication of Solids: Part I.* Oxford: Clarendon Press, 1958.
4. Bowden, F. P. and Tabor, D., *Friction: An Introduction to Tribology.* Melbourne, FL: Krieger, 1982.
5. Greenwood, J. A., "A Unified Theory of Surface Roughness," Proceedings of the Royal Society, A393, 1984, pp. 133-157.
6. Greenwood, J. A., "Formulas for Moderately Elliptical Hertzian Contact," *Journal of Tribology,* 107(4), 1985, pp. 501-504.
7. Halling, J. (Ed.), *Principles of Tribology.* MacMillan Press, Ltd., 1983.
8. Hamrock, B. J., *Fundamentals of Fluid Film Lubrication.* New York: McGraw-Hill, 1994.
9. Hutchings, I. M., *Tribology: Friction and Wear of Engineering Materials.* Boca Raton: CRC Press, 1992.
10. Johnson, K. L., *Contact Mechanics.* Cambridge: Cambridge, 1985.
11. Jost, P., "Lubrication (Tribology) Education and Research," Technical report, H.M.S.O., 1966.
12. Peterson, M. B. and Winer, W. O. (Eds.), *Wear Control Handbook.* ASME, 1980.
13. Rabinowicz, E., *Friction and Wear of Materials.* New York: Wiley, 1965.
14. Suh, N. P., *Tribophysics.* Englewood Cliffs: Prentice-Hall, 1986.
15. Thomas, T. R. (Ed.), *Rough Surfaces.* London: Longman, 1982.
16. Timoshenko, S. P. and Goodier, J. N., *Theory of Elasticity,* 3rd Ed. New York: McGraw-Hill, 1970.
17. Westergaard, H. M., "Bearing Pressures and Cracks," *Journal of Applied Mechanics,* 6(2), A49-A53, 1939.
18. Zaretsky, E. V. (Ed.), *STLE Life Factors for Roller Bearings.* Society of Tribologists and Lubrication Engineers, 1992.

11
Vibration

Lawrence D. Norris, Senior Technical Marketing Engineer—Large Commercial Engines, Allison Engine Company, Rolls-Royce Aerospace Group

This chapter presents a brief discussion of mechanical vibrations and its associated terminology. Its main emphasis is to provide practical "rules of thumb" to help calculate, measure, and analyze vibration frequencies of mechanical systems. Tables are provided with useful formulas for computing the vibration frequencies of common mechanical systems. Additional tables are provided for use with vibration measurements and instrumentation. A number of well-known references are also listed at the end of the chapter, and can be referred to when additional information is required.

Vibration Definitions, Terminology, and Symbols

Beating: A vibration (and acoustic) phenomenon that occurs when two harmonic motions (x_1 and x_2) of the same amplitude (X), but of slightly different frequencies are applied to a mechanical system:

$$x_1 = X \cos \omega t$$

$$x_2 = X \cos (\omega + \Delta \omega) t$$

The resultant motion of the mechanical system will be the superposition of the two input vibrations x_1 and x_2, which simplifies to:

$$x = 2X \cos \left(\frac{\Delta \omega}{2} \right) t \cos \left(\omega + \frac{\Delta \omega}{2} \right) t$$

This vibration is called the *beating phenomenon,* and is illustrated in Figure 1. The frequency and period of the *beats* will be, respectively:

$$f_b = \frac{\Delta \omega}{2 \pi} \text{ cycles / sec}$$

$$T_b = \frac{2 \pi}{\Delta \omega} \text{ sec}$$

A common example of beating vibration occurs in a twin engine aircraft. Whenever the speed of one engine varies slightly from the other, a person can easily feel the beating in the aircraft's structure, and hear the vibration acoustically.

Critical speeds: A term used to describe resonance points (speeds) for rotating shafts, rotors, or disks. For example, the critical speed of a turbine rotor occurs when the rotational speed coincides with one of the rotor's natural frequencies.

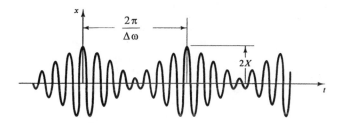

Figure 1. Beating phenomenon.

Damped natural frequency (ω_d or f_d): The inherent frequency of a mechanical system with viscous damping (friction) under free, unforced vibration. Damping decreases the system's natural frequency and causes vibratory motion to decay over time. A system's damped and undamped natural frequencies are related by:

$$\omega_d = \omega_n \sqrt{1 - \zeta^2}$$

Damping (c): Damping dissipates energy and thereby "damps" the response of a mechanical system, causing vibratory motion to decay over time. Damping can result from fluid or air resistance to a system's motion, or from friction between sliding surfaces. Damping force is usually proportional to the velocity of the system: $F = c\dot{x}$, where c is the *damping coefficient,* and typically has units of lb-sec/in or N-sec/m.

Damping ratio (ζ): The damped natural frequency is related to a system's undamped natural frequency by the following formula:

$$\omega_d = \omega_n \sqrt{1 - \zeta^2}$$

The damping ratio (ζ) determines the rate of decay in the vibration of the mechanical system. No vibratory oscillation will exist in a system that is *overdamped* ($\zeta > 1.0$) or *critically damped* ($\zeta = 1.0$). The length of time required for vibratory oscillations to die out in the *underdamped system* ($\zeta < 1.0$) increases as the damping ratio decreases. As ζ decreases to 0, ω_d equals ω_n, and vibratory oscillations will continue indefinitely due to the absence of friction.

Degrees of freedom (DOF): The minimum number of independent coordinates required to describe a system's motion. A single "lumped mass" which is constrained to move in only one linear direction (or angular plane) is said to be a "single DOF" system, and has only one discrete natural frequency. Conversely, continuous media (such as beams, bars, plates, shells, etc.) have their mass evenly distributed and cannot be modeled as "lumped" mass systems. Continuous media have an infinite number of small masses, and therefore have an infinite number of DOF and natural frequencies. Figure 2 shows examples of one- and two-DOF systems.

Equation of motion: A differential equation which describes a mechanical system's motion as a function of time. The number of equations of motion for each mechanical system is equal to its DOF. For example, a system with two-DOF will also have two equations of motion. The two natural frequencies of this system can be determined by finding a solution to these equations of motion.

Forced vibration: When a continuous external force is applied to a mechanical system, the system will vibrate at the frequency of the input force, and initially at its own natural frequency. However, if damping is present, the vibration at ω_d will eventually die out so that only vibration at the forcing frequency remains. This is called the *steady state response* of the system to the input force. (See Figure 3.)

Free vibration: When a system is displaced from its equilibrium position, released, and then allowed to vibrate without any further input force, it will vibrate at its natural frequency (w_n or w_d). Free vibration, with and without damping, is illustrated in Figure 3.

Frequency (ω or f): The *rate* of vibration, which can be expressed either as a circular frequency (ω) with units of radians per second, or as the frequency (f) of the periodic motion in cycles per second (Hz). The periodic frequency (f) is the reciprocal of the period ($f = 1/T$). Since there are 2π radians per cycle, $\omega = 2\pi f$.

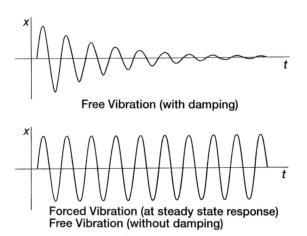

Free Vibration (with damping)

Forced Vibration (at steady state response)
Free Vibration (without damping)

Figure 3. Free vibration, with and without damping.

Harmonic/spectral analysis (Fourier series): Any complex periodic motion or random vibration signal can be represented by a series (called a *Fourier series*) of individual sine and cosine functions which are related harmonically. The summation of these individual sine and cosine waveforms equal the original, complex waveform of the periodic motion in question. When the *Fourier spectrum* of a vibration signal is plotted (vibration amplitude vs. frequency), one can see which discrete vibration frequencies over the entire frequency spectrum contribute the most to the overall vibration signal. Thus, spectral analysis is very useful for troubleshooting vibration problems in mechanical systems. Spectral analysis allows one to pinpoint, via its operational frequency, which component of the system is causing a vibration problem. Modern vibration analyzers

Examples of Single-Degree-of-Freedom Systems

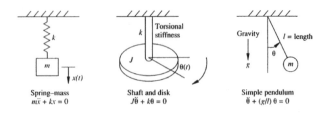

Spring–mass
$m\ddot{x} + kx = 0$

Shaft and disk
$J\ddot{\theta} + k\theta = 0$

Simple pendulum
$\ddot{\theta} + (g/l)\,\theta = 0$

Examples of Two-Degree of Freedom Systems

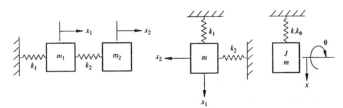

Figure 2. One and two degree of freedom mechanical systems [1]. (*Reprinted by permission of Prentice-Hall, Inc., Upper Saddle River, NJ.*)

with digital microprocessors use an algorithm known as *Fast Fourier Transform* (FFT) which can perform spectral analysis of vibration signals of even the highest frequency.

Harmonic frequencies: Integer multiples of the natural frequency of a mechanical system. Complex systems frequently vibrate not only at their natural frequency, but also at harmonics of this frequency. The richness and fullness of the sound of a piano or guitar string is the result of harmonics. When a frequency varies by a 2:1 ratio relative to another, the frequencies are said to differ by one *octave.*

Mode shapes: Multiple DOF systems have multiple natural frequencies and the physical response of the system at each natural frequency is called its *mode shape.* The actual physical deflection of the mechanical system under vibration is different at each mode, as illustrated by the cantilevered beam in Figure 4.

Natural frequency (ω_n or f_n): The inherent frequency of a mechanical system without damping under free, unforced vibration. For a simple mechanical system with mass m and stiffness k, the natural frequency of the system is:

$$\omega_n = \sqrt{\frac{k}{m}}$$

Node point: *Node points* are points on a mechanical system where no vibration exists, and hence no deflection from the equilibrium position occurs. Node points occur with multiple DOF systems. Figure 4 illustrates the node points for the 2nd and 3rd modes of vibration of a cantilever beam. *Antinodes,* conversely, are the positions where the vibratory displacement is the greatest.

Phase angle (ϕ): Since vibration is repetitive, its periodic motion can be defined using a sine function and phase angle. The displacement as a function of time for a single DOF system in SHM can be described by the function:

$$x(t) = A \sin (\omega t + \phi)$$

where A is the amplitude of the vibration, ω is the vibration frequency, and ϕ is the phase angle. The phase angle sets the initial value of the sine function. Its units can either be radians or degrees. Phase angle can also be used to describe the *time lag* between a forcing function applied to a system and the system's response to the force. The phase relationship between the displacement, velocity, and acceleration of a mechanical system in steady state vibration is illustrated in Figure 5. Since acceleration is the first derivative of velocity and second derivative of displacement, its phase angle "leads" velocity by 90 degrees and displacement by 180 degrees.

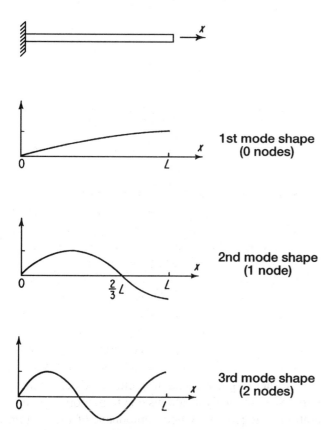

Figure 4. Mode shapes and nodes of the cantilever beam.

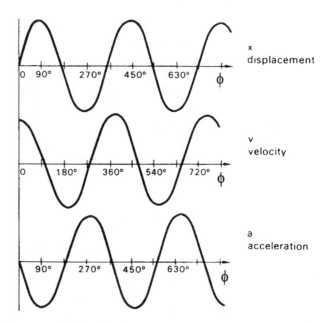

Figure 5. Interrelationship between the phase angle of displacement, velocity, and acceleration [9]. (*Reprinted by permission of the Institution of Diagnostic Engineers.*)

Resonance: When the frequency of the excitation force (forcing function) is equal to or very close to the natural frequency of a mechanical system, the system is excited into *resonance*. During resonance, vibration amplitude increases dramatically, and is limited only by the amount of inherent damping in the system. Excessive vibration during resonance can result in serious damage to a mechanical system. Thus, when designing mechanical systems, it is extremely important to be able to calculate the system's natural frequencies, and then ensure that the system only operate at speed ranges outside of these frequencies, to ensure that problems due to resonance are avoided. Figure 6 illustrates how much vibration can increase at resonance for various amounts of damping.

Rotating unbalance: When the center of gravity of a rotating part does not coincide with the axis of rotation, *unbalance* and its corresponding vibration will result. The unbalance force can be expressed as:

$$F = me\omega^2$$

where m is an equivalent eccentric mass at a distance *e* from the center of rotation, rotating at an angular speed of ω.

Simple harmonic motion (SHM): The simplest form of undamped periodic motion. When plotted against time, the displacement of a system in SHM is a pure sine curve. In SHM, the acceleration of the system is inversely proportional (180 degrees out of phase) with the linear (or angular) displacement from the origin. Examples of SHM are simple one-DOF systems such as the pendulum or a single mass-spring combination.

Spring rate or stiffness (k): The elasticity of the mechanical system, which enables the system to store and release kinetic energy, thereby resulting in vibration. The input force (F) required to displace the system by an amount (x) is proportionate to this spring rate: F = kx. The spring rate will typically have units of lb/in. or N/m.

Vibration: A periodic motion of a mechanical system which occurs repetitively with a time period (cycle time) of T seconds.

Vibration transmissibility: An important goal in the installation of machinery is frequently to isolate the vibration and motion of a machine from its foundation, or vice versa. Vibration isolators (sometimes called *elastomers*) are used to achieve this goal and reduce vibration transmitted through them via their damping properties. *Transmissibility* (TR) is a measure of the extent of isolation achieved, and is the amplitude ratio of the force being transmitted across the vibration isolator (F_t) to the imposing force (F_0). If the frequency of the imposing force is ω, and the natural frequency of the system (composed of the machinery mounted on its vibration isolators) is ω_n, the transmissibility is calculated by:

$$TR = \frac{F_t}{F_0} = \frac{\sqrt{1 + (2\,\zeta\,r)^2}}{\sqrt{(1 - r^2)^2 + (2\,\zeta\,r)^2}}$$

where ζ = damping ratio

$$r = \text{frequency ratio} = \left(\frac{\omega}{\omega_n}\right)$$

Figure 6 shows that for a given input force with frequency (ω), flexible mounting (low ω_n, high r) with very light damping provides the best isolation.

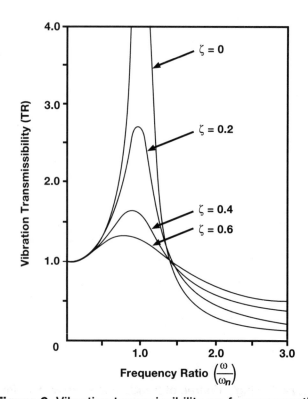

Figure 6. Vibration transmissibility vs. frequency ratio.

Solving the One Degree of Freedom System

A simple, one degree of freedom mechanical system with damped, linear motion can be modeled as a mass, spring, and damper (dashpot), which represent the inertia, elasticity, and the friction of the system, respectively. A drawing of this system is shown in Figure 7, along with a *free body diagram* of the forces acting upon this mass when it is displaced from its equilibrium position. The *equation of motion* for the system can be obtained by summing the forces acting upon the mass. From Newton's laws of motion, the sum of the forces acting upon the body equals its mass times acceleration:

$$\Sigma F = ma = m\ddot{x} = F - kx - c\dot{x}$$

where $F = F(t)$
$x = x(t)$

Simplifying the equation:

$$m\ddot{x} + c\ddot{x} + kx = F$$

This equation of motion describes the displacement (x) of the system as a function of time, and can be solved to determine the system's *natural frequency*. Since damping is present, this frequency is the system's *damped natural frequency*. The equation of motion is a second order differential equation, and can be solved for a given set of initial conditions. Initial conditions describe any force that is being applied to the system, as well as any initial displacement, velocity, or acceleration of the system at time zero. Solutions to this equation of motion are now presented for two different cases of vibration: free (unforced) vibration, and forced harmonic vibration.

Solution for Free Vibration

For the case of free vibration, the mass is put into motion following an initial displacement and/or initial velocity. No external force is applied to the mass other than that force required to produce the initial displacement. The mass is released from its initial displacement at time $t = 0$ and allowed to vibrate freely. The equation of motion, initial conditions, and solution are:

$$m\ddot{x} + c\dot{x} + kx = 0$$

Initial conditons (at time $t = 0$):

$F = 0$ (no force applied)
x_0 = initial displacement
$\dot{x}_0$ = initial velocity

Solution to the Equation of Motion:

$$x(t) = \bar{e}^{\zeta\omega_n t}\left[x_0 \cos \omega_d t + \left(\frac{x_0 \zeta\omega_n + \dot{x}_0}{\omega_d}\right)\sin \omega_d t\right]$$

where : $\omega_n = \sqrt{\dfrac{k}{m}}$
 = undamped natural frequency (rad/sec.)

$\omega_d = \omega_n\sqrt{1 - \zeta^2}$ = damped natural frequency

$\zeta = \dfrac{c}{c_{cr}}$ = damping factor

$c_{cr} = 2\sqrt{km}$ = critical damping

The response of the system under free vibration is illustrated in Figure 8 for the three separate cases of underdamped, overdamped, and critically damped motion. The damping factor (ζ) and damping coefficient (c) for the underdamped system may be determined experimentally, if they are not already known, using the *logarithmic decrement* (ζ) method. The logarithmic decrement is the natural logarithm of the ratio of any two consecutive amplitudes (x) of free vibration, and is related to the damping factor by the following equation:

$$\delta = \ln\frac{x_n}{x_{n+1}} = \frac{2\pi\zeta}{\sqrt{1 - \zeta^2}}$$

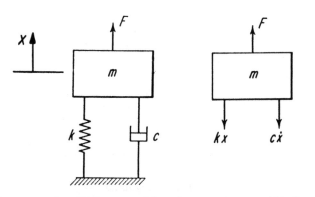

Figure 7. Single degree of freedom system and its free body diagram.

Given the magnitude of two successive amplitudes of vibration, this equation can be solved for ζ and then the damping coefficient (c) can be calculated with the equations listed previously. When ζ is small, as in most mechanical systems, the log-decrement can be approximated by:

$$\delta \cong 2 \pi \zeta$$

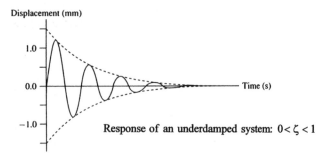

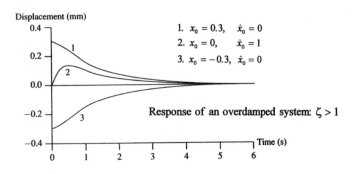

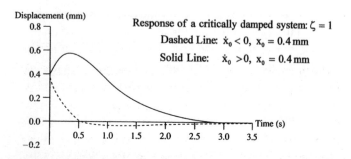

Figure 8. Response of underdamped, overdamped, and critically damped systems to free vibration [1]. *(Reprinted by permission of Prentice-Hall, Inc., Upper Saddle River, NJ.)*

Solution for Forced Harmonic Vibration

We now consider the case where the single degree of freedom system has a constant harmonic force ($F = F_0 \sin \omega t$) acting upon it. The equation of motion for this system will be:

$$m\ddot{x} + c\dot{x} + kx = F_0 \sin \omega t$$

The solution to this equation consists of two parts: free vibration and forced vibration. The solution for the free vibration component is the same solution in the preceding paragraph for the free vibration problem. This free vibration will dampen out at a rate proportional to the system's damping ratio (ζ), after which only the steady state response to the forced vibration will remain. This steady state response, illustrated previously in Figure 3, is a harmonic vibration at the same frequency (ω) as the forcing function (F). Therefore, the steady state solution to the equation of motion will be of the form:

$$x = X \sin (\omega t + \phi)$$

X is the amplitude of the vibration, ω is its frequency, and ϕ is the phase angle of the displacement (x) of the system relative to the input force. When this expression is substituted into the equation of motion, the equation of motion may then be solved to give the following expressions for amplitude and phase:

$$X = \frac{F_0}{\sqrt{(k - m\omega^2)^2 + (c\omega)^2}}$$

$$\phi = \tan^{-1} \frac{c\omega}{k - m\omega^2}$$

These equations may be further reduced by substituting with the following known quantities:

$$\omega = \sqrt{\frac{k}{m}} = \text{undamped natural frequency}$$

$$c_{cr} = 2\sqrt{km} = \text{critical damping}$$

$$\zeta = \frac{c}{c_{cr}} = \text{damping factor}$$

Following this substitution, the amplitude and phase of the steady state response are now expressed in the following nondimensional form:

$$\frac{X\,k}{F_0} = \frac{1}{\sqrt{\left[1-\left(\dfrac{\omega}{\omega_n}\right)^2\right]^2 + \left[2\zeta\left(\dfrac{\omega}{\omega_n}\right)\right]^2}}$$

$$\tan\phi = \frac{2\,\zeta\left(\dfrac{\omega}{\omega_n}\right)}{1-\left(\dfrac{\omega}{\omega_n}\right)^2}$$

These equations are now functions of only the frequency ratio (ω/ω_n) and the damping factor (ζ). Figure 9 plots this nondimensional amplitude and phase angle versus the frequency ratio. As illustrated by this figure, the system goes into *resonance* as the input frequency (ω) approaches the system's natural frequency, and the vibration amplitude at resonance is constrained only by the amount of damping (ζ) in the system. In the theoretical case with no damping, the vibration amplitude reaches infinity. A *phase shift* of 180 degrees also occurs above and below the resonance point, and the rate that this phase shift occurs (relative to a change in frequency) is inversely proportional to the amount of damping in the system. This phase shift occurs instantaneously at $\omega = \omega_n$ for the theoretical case with no damping. Note also that the vibration amplitude is very low when the forcing frequency is well above the resonance point. If sufficient damping is designed into a mechanical system, it is possible to accelerate the system quickly through its resonance point, and then operate with low vibration in speed ranges above this frequency. In rotating machinery, this is commonly referred to as "operating above the critical speed" of the rotor/shaft/disk.

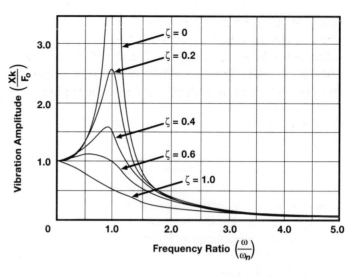

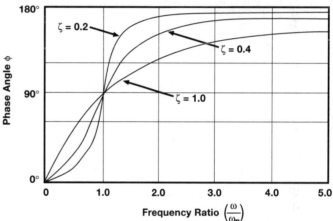

Figure 9. Vibration amplitude and phase of the steady state response of the damped 1 DOF system to a harmonic input force at various frequencies.

Solving Multiple Degree of Freedom Systems

There are a number of different methods used to derive and solve the equations of motion for multiple degree of freedom systems. However, the length of this chapter allows only a brief description of a few of these techniques. Any of the books referenced by this chapter, or any engineering vibration textbook, can be referenced if more information is required about these techniques. Tables at the end of this chapter have listed equations, derived via these techniques, to calculate the natural frequencies of various mechanical systems.

Energy methods: For complex mechanical systems, an energy approach is often easier than trying to determine, analyze, and sum all the forces and torques acting upon a system. The principle of conservation of energy states that the total energy of a mechanical system remains the same over time, and therefore the following equations are true:

Kinetic Energy (KE) + Potential Energy (PE) = constant

$$\frac{d}{dt}(KE + PE) = 0$$

$$KE_{max} = PE_{max}$$

Writing expressions for KE and PE (as functions of displacement) and substituting them into the first equation yields the equation of motion for a system. The natural frequency of the system can also be obtained by equating expressions for the maximum kinetic energy with the maximum potential energy.

Lagrange's equations: Lagrange's equations are another energy method which will yield a number of equations of motion equal to the number of degrees of freedom in a mechanical system. These equations can then be solved to determine the natural frequencies and motion of the system. Lagrange's equations are written in terms of independent, *generalized coordinates* (q_i):

$$\frac{d}{dt}\frac{\partial\,KE}{\partial\,\dot{q}_i} - \frac{\partial\,KE}{\partial\,q_i} + \frac{\partial\,PE}{\partial\,q_i} + \frac{\partial\,DE}{\partial\,\dot{q}_i} = Q$$

where: DE = dissipative energy of the system
Q = generalized applied external force

When the principle of conservation of energy applies, this equation reduces to:

$$\frac{d}{dt}\frac{\partial\,L}{\partial\,\dot{q}_i} - \frac{\partial\,L}{\partial\,q_i} = 0$$

where: L = KE − PE

"L" is called the Lagrangian.

Principle of orthogonality: Principal (normal) modes of vibration for mechanical systems with more than one DOF occur along mutually perpendicular straight lines. This *orthogonality principle* can be very useful for the calculation of a system's natural frequencies. For a two-DOF system, this principle may be written as:

$$m_1\,A_1\,A_2 + m_2\,B_1\,B_2 = 0$$

where A_1, A_2, B_1, and B_2 are the vibration amplitudes of the two coordinates for the first and second vibration modes.

Laplace transform: The *Laplace Transform* method transforms (via integration) a differential equation of motion into a function of an alternative (complex) variable. This function can then be manipulated algebraically to determine the Laplace Transform of the system's response. Laplace Transform pairs have been tabulated in many textbooks, so one can look up in these tables the solution to the response of many complex mechanical systems.

Finite element method (FEM): A powerful method of modeling and solving (via digital computer) complex structures by approximating the structure and dividing it into a number of small, simple, symmetrical parts. These parts are called *finite elements,* and each element has its own equation of motion, which can be easily solved. Each element also has boundary points (nodes) which connect it to adjacent elements. A finite element grid (model) is the complete collection of all elements and nodes for the entire structure. The solutions to the equations of motion for the individual elements are made compatible with the solutions to their adjacent elements at their common node points (boundary conditions). The solutions to all of the elements are then assembled by the computer into global mass and stiffness matrices, which in turn describe the vibration response and motion of the entire structure. Thus, a finite element model is really a miniature lumped mass approximation of an entire structure. As the number of elements (lumped masses) in the model is increased towards infinity, the response predicted by the F.E. model approaches the exact response of the complex structure. Up until recent years, large F.E. models required mainframe computers to be able to solve simultaneously the enormous number of equations of a large F.E. model. However, increases in processing power of digital computers have now made it possible to solve all but the most complex F.E. models with specialized FEM software on personal computers.

Vibration Measurements and Instrumentation

Analytical methods are not always adequate to predict or solve every vibration problem during the design and operation of mechanical systems. Therefore, it is often necessary to experimentally measure and analyze both the vibration frequencies and physical motion of mechanical systems. Sensors can be used to measure vibration, and are called *transducers* because they change the mechanical motion of a system into a signal (usually electrical voltage) that can be measured, recorded, and processed. A number of different types of transducers for vibration measurements are available. Some transducers directly measure the displacement, velocity, or acceleration of the vibrating system, while other transducers measure vibration indirectly by sensing the mechanical strain induced in another object, such as a cantilever beam. Examples of modern spring-mass and strain gage accelerometers, along with schematics of their internal components, are shown in Figure 10. The following paragraphs describe how several of the most widely used vibration transducers work, as well as their advantages and disadvantages.

Spring-Mass Accelerometer Mounted on a Structure

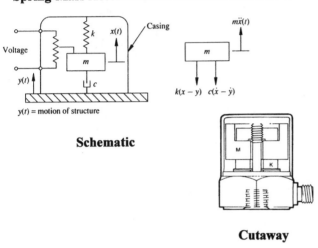

Schematic

Cutaway

Strain Gage Accelerometer Made of a Small Beam

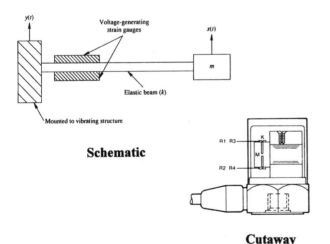

Schematic

Cutaway

Figure 10. Schematics and cutaway drawings of modern spring-mass and strain gage accelerometers. *(Schematics from Inman [1], reprinted by permission of Prentice-Hall, Inc., Upper Saddle River, NJ. Cutaways courtesy of Endevco Corp.)*

Accelerometers

Accelerometers, as their name implies, measure the acceleration of a mechanical system. Accelerometers are *contact* transducers; they are physically mounted to the surface of the mechanism being measured. Most modern accelerometers are *piezoelectric* accelerometers, and contain a spring-mass combination which generates a force proportional to the amplitude and frequency of the mechanical system the accelerometer is mounted upon. This force is applied to an internal piezoelectric crystal, which produces a proportional charge at the accelerometer's terminals. Piezoelectric accelerometers are rated in terms of their *charge sensitivity*, usually expressed as pico-coulombs (electrical charge) per "g" of acceleration. Piezoelectric accelerometers are self-generating and do not require an external power source. However, an external *charge amplifier* is used to convert the electrical charge from the transducer to a voltage signal. Therefore, piezoelectric accelerometers are also rated in terms of their *voltage sensitivity* (usually mV/g) for a given external capacitance supplied by a charge amplifier. The charge produced by the transducer is converted into a voltage by the charge amplifier by electronically dividing the charge by the capacitance in the accelerometer/cable/charge-amplifier system:

$$V = \frac{Q}{C}$$

Accelerometers are calibrated by the manufacturer and a copy of their frequency response curve is included with the transducer. Frequency response is very flat (usually less than ±5%) up to approximately 20% of the resonant (natural) frequency of the accelerometer. The response becomes increasingly nonlinear above this level. The accelerometer should not be used to measure frequencies exceeding 20% ω_n unless the manufacturer's specifications state otherwise.

Advantages:

- Available in numerous designs (such as compression, shear, and strain gage), sizes, weights, and mounting arrangements (such as center, stud, screw mounted, and glue-on). Ease of installation is also an advantage.
- Accelerometers are available for high-temperature environments (up to 1,200°F).
- Wide band frequency and amplitude response.
- Accelerometers are available for high frequency and low frequency (down to DC) measuring capabilities.
- Durable, robust construction and long-term reliability.

Disadvantages:

- Require contact with (mounting upon) the object being measured. Therefore, the mass of the accelerometer must be small relative to this object (generally should be less than 5% of mass of vibrating component being measured).
- Sensitive to mounting (must be mounted securely).
- Sensitive to cable noise and "whip" (change in cable capacitance caused by dynamic bending of the cable).
- Results are not particularly reliable when displacement is calculated by double integrating (electronically) the acceleration signal.

Displacement Sensors and Proximity Probes

There are a number of different types of displacement sensors. The linear variable (voltage) differential transducer (LVDT) is a contact transducer which uses a magnet and coil system to produce a voltage proportional to displacement. One end of the LVDT is mounted to the vibrating object while the other is attached to a fixed reference. In contrast, capacitance, inductance, and proximity sensors are all *noncontact* displacement transducers which do not physically contact the vibrating object. These sensors measure the change in capacitance or magnetic field currents caused by the displacement and vibration of the object being measured. Of all the various noncontact displacement transducers available, the *eddy current proximity sensor* is the most widely used.

The eddy current proximity sensor is supplied with a high-frequency carrier signal to a coil in the tip of this sensor, which generates an eddy current field to any conducting surface within the measurement range of the sensor. Any reduction in the gap between the sensor tip and the object being measured, whether by displacement or vibration of this object, reduces the output voltage of the proximity sensor. The proximity sensor measures only the gap between it and the object in question. Therefore, this sensor is not useful for balancing rotors, since it can only measure the "high" point of the rotor (smallest gap) rather than the "heavy spot" of the rotor. However, proximity probes are very useful when accurate displacement monitoring is required. The orbital motion of shafts is one example where two proximity sensors set at right angles to each other can produce a useful X-Y plot of the orbital motion of a shaft. Orbital motion provides an effective basis for malfunction monitoring of rotating machinery and for dynamic evaluation of relative clearances between bearings and shafts.

Advantages:

- No contact with the object being measured (except for LVDT).
- Small sensor size and weight.
- Can be mounted to a fixed reference surface to give absolute displacement of an object, or to a moving reference surface to give relative displacement of an object.
- Wide frequency response from DC (0 Hz) to over 5,000 Hz.
- Measures displacement directly (no need to integrate velocity and acceleration signals).
- Displacement measurements are extremely useful for analyses such as shaft orbits and run-outs.

Disadvantages:

- LVDT has limited frequency response due to its inertia. It also is a contact transducer, and therefore must be attached to the object being measured.
- Proximity probes are accurate only for a limited measurement range ("gap").
- Not self-generating (requires external power source).
- Limited temperature environment range.
- Proximity sensors are susceptible to induced voltage from other conductors, such as nearby 50 or 60 Hz alternating current power sources.

Useful Relationships Between Dynamic Measurement Values and Units

Dynamic measurements, like vibration, can be expressed in any number of units and values. For example, the vibration of a rotating shaft could be expressed in terms of its displacement (inches), velocity (in/sec), or acceleration (in/sec^2 or "g"). Additionally, since vibration is a periodic, sinusoidal motion, these displacement, velocity, and acceleration units can be expressed in a number of different values (peak, peak-to-peak, rms, or average values). Figure 11 gives a visual illustration of the difference between these values for a pure sine wave, and provides the conversion constants needed to convert one value to another. Figure 12 gives the relationships for sinusoidal motion between displacement, velocity, and acceleration. These relationships can also be plotted graphically on a *nomograph,* as illustrated in Figure 13. The nomograph provides a quick graphical method to convert any vibration measurement, for a given frequency, between displacement, velocity, and acceleration units.

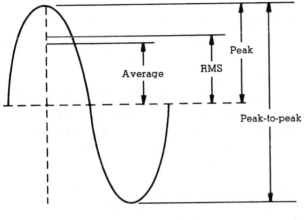

rms value	= 0.707	x peak value
rms value	= 1.11	x average value
peak value	= 1.414	x rms value
peak value	= 1.57	x average value
average value	= 0.637	x peak value
average value	= 0.90	x rms value
peak-to-peak	= 2	x peak value

$$crest\ factor = \frac{peak\ value}{rms\ value} \quad (applies\ to\ any\ varying\ quantity)$$

Figure 11. Dynamic measurement value relationships for sinusoidal motion. (*Courtesy of Endevco Corp.*)

$$d = d_0 \sin 2\pi ft$$
$$v = d_0\ 2\pi f \cos 2\pi ft$$
$$a = -d_0 (2\pi f)^2 \sin 2\pi ft$$
$$G = \frac{acceleration}{g}$$

$$v_0 = 6.28\ f\ d_0 = 3.14\ f\ D$$
$$v_0 = 61.42\ \frac{G}{f}\ in./s\ pk$$
$$= 1.560\ \frac{G}{f}\ m/s\ pk$$
$$d_0 = 9.780\ \frac{G}{f^2}\ inches\ pk$$
$$= 0.2484\ \frac{G}{f^2}\ meters\ pk$$

where:
d_0 = peak displacement
D = pk-pk displacement
G = acceleration in g units
f = frequency in Hz
T = period in seconds
g = 9.806 65 m/s^2
= 386.09 in./s^2
= 32.174 ft/s^2

$$G = 0.0511\ f^2 D$$
(where: D = inches peak-to-peak)

$$G = 2.013\ f^2 D$$
(where: D = meters peak-to-peak)

$$T = \frac{1}{f}\ seconds$$

Figure 12. Displacement, velocity, and acceleration relationships for sinusoidal motion. (*Courtesy of Endevco Corp.*)

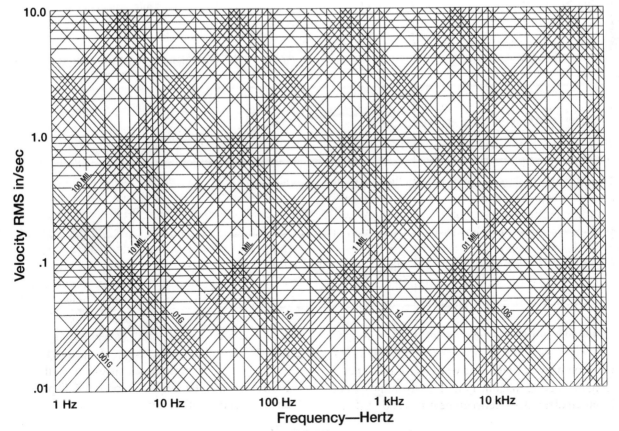

Figure 13. Vibration nomograph.

Table A: Spring Stiffness

Description	Sketch	Stiffness Equation
Springs in parallel		$k = k_1 + k_2$
Springs in series		$k = \dfrac{1}{1/k_1 + 1/k_2}$
Torsional spring		$k = \dfrac{E\,I}{L}$ I = moment of inertia for cross-section L = total length of spring
Rod in torsion		$k = \dfrac{G\,J}{L}$ J = torsion constant for cross-section L = total length of rod
Rod in tension (or compression)		$k = \dfrac{E\,A}{L}$ A = cross-sectional area L = total length of rod
Stiffness of coiled wire spring		$k = \dfrac{G\,d^4}{8\,n\,D^3}$ n = number of coils; d = wire thickness; D = O.D. of coil
Cantilevered beam with load at end		$k = \dfrac{3\,E\,I}{L^3}$ I = moment of inertia for cross-section L = total length of beam
Beam with both ends fixed, load in middle of beam		$k = \dfrac{192\,E\,I}{L^3}$ I = moment of inertia for cross-section L = total length of beam
Beam with both ends pinned, load in middle of beam		$k = \dfrac{48\,E\,I}{L^3}$ I = moment of inertia for cross-section L = total length of beam
Beam with both ends pinned, with off-center load		$k = \dfrac{3\,E\,I\,L}{a^2\,b^2}$ I = moment of inertia for cross-section L = total length of beam = a + b
Beam with one end fixed, one end pinned, load in middle of beam		$k = \dfrac{768\,E\,I}{7\,L^3}$ I = moment of inertia for cross-section L = total length

Table B: Natural Frequencies of Simple Systems

	End mass M; spring mass m, spring stiffness k	$\omega_n = \sqrt{k/(M + m_{,}}$
	End inertia I; shaft inertia I_s, shaft stiffness k	$\omega_n = \sqrt{k/(I + I_s/\varepsilon}$
	Two disks on a shaft	$\omega_n = \sqrt{\dfrac{k(I_1 + I_2)}{I_1 I_2}}$
	Cantilever; end mass M; beam mass m	$\omega_n = \sqrt{\dfrac{k}{M + 0.23m}}$
	Simply supported beam; central mass M; beam mass m	$\omega_n = \sqrt{\dfrac{k}{M + 0.5m}}$
	Massless gears, speed of I_2 n times as large as speed of I_1	$\omega_n = \sqrt{\dfrac{1}{\dfrac{1}{k_1} + \dfrac{1}{n^2 k_2}}}$

$$\omega_n^2 = \frac{1}{2}\left(\frac{k_1}{I_1} + \frac{k_3}{I_3} + \frac{k_1 + k_3}{I_2}\right)$$
$$\pm \frac{1}{2}\sqrt{\left(\frac{k_1}{I_1} + \frac{k_3}{I_3} + \frac{k_1 + k_3}{I_2}\right)^2 - 4\frac{k_1 k_3}{I_1 I_2 I_3}(I_1 + I_2 + I_3)}$$

Note: Torsional Shaft Stiffness (k) above is frequently referred to as torsional shaft rigidity (τ):

$$\tau = \frac{GJ}{L}$$

where: G = modulus of rigidity of the shaft
   J = polar second moment of area of the cross-section
   L = shaft length

For a solid, circular shaft:

$$J = \frac{\pi r^4}{2}$$

Source: from Den Hartog [2], with permission of Dover Publications, Inc.

Table C: Longitudinal and Torsional Vibration of Uniform Beams

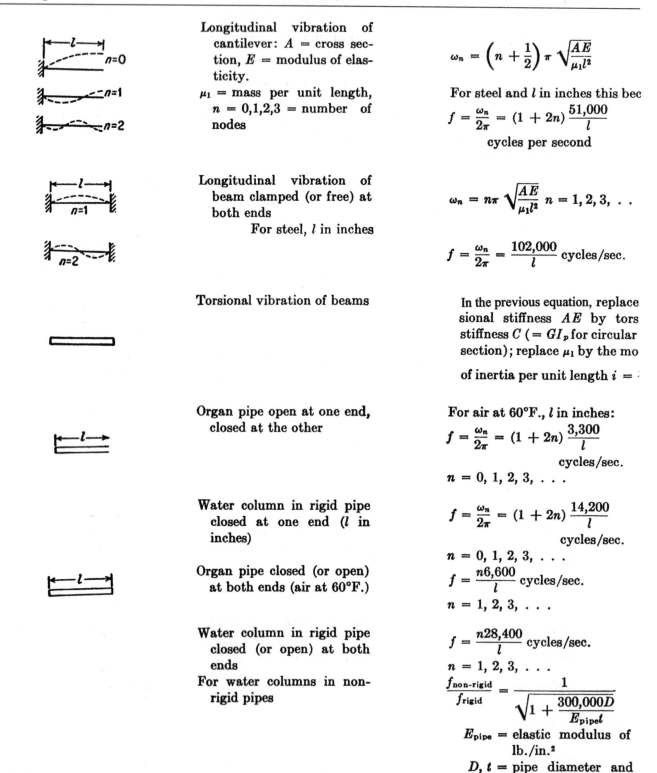

Longitudinal vibration of cantilever: A = cross section, E = modulus of elasticity.
μ_1 = mass per unit length, $n = 0,1,2,3$ = number of nodes

$$\omega_n = \left(n + \frac{1}{2}\right)\pi\sqrt{\frac{AE}{\mu_1 l^2}}$$

For steel and l in inches this bec
$$f = \frac{\omega_n}{2\pi} = (1 + 2n)\frac{51{,}000}{l}$$
cycles per second

Longitudinal vibration of beam clamped (or free) at both ends
For steel, l in inches

$$\omega_n = n\pi\sqrt{\frac{AE}{\mu_1 l^2}} \quad n = 1, 2, 3, \ldots$$

$$f = \frac{\omega_n}{2\pi} = \frac{102{,}000}{l} \text{ cycles/sec.}$$

Torsional vibration of beams

In the previous equation, replace sional stiffness AE by tors stiffness C ($= GI_p$ for circular section); replace μ_1 by the mo
of inertia per unit length $i = $

Organ pipe open at one end, closed at the other

For air at 60°F., l in inches:
$$f = \frac{\omega_n}{2\pi} = (1 + 2n)\frac{3{,}300}{l}$$
cycles/sec.
$n = 0, 1, 2, 3, \ldots$

Water column in rigid pipe closed at one end (l in inches)

$$f = \frac{\omega_n}{2\pi} = (1 + 2n)\frac{14{,}200}{l}$$
cycles/sec.
$n = 0, 1, 2, 3, \ldots$

Organ pipe closed (or open) at both ends (air at 60°F.)

$$f = \frac{n6{,}600}{l} \text{ cycles/sec.}$$
$n = 1, 2, 3, \ldots$

Water column in rigid pipe closed (or open) at both ends

$$f = \frac{n28{,}400}{l} \text{ cycles/sec.}$$
$n = 1, 2, 3, \ldots$

For water columns in non-rigid pipes

$$\frac{f_{\text{non-rigid}}}{f_{\text{rigid}}} = \frac{1}{\sqrt{1 + \dfrac{300{,}000D}{E_{\text{pipe}}t}}}$$

E_{pipe} = elastic modulus of lb./in.²
D, t = pipe diameter and thickness, same ur

Source: from Den Hartog [2], with permission of Dover Publications, Inc.

Table D: Bending (Transverse) Vibration of Uniform Beams

The same general formula holds for all the following cases,

$$\omega_n = a_n \sqrt{\frac{EI}{\mu_1 l^4}}$$

where EI is the bending stiffness of the section, l is the length of the beam, μ_1 is the mass per unit length $= W/gl$, and a_n is a numerical constant, different for each case and listed below

Cantilever or "clamped-free" beam

$a_1 = 3.52$
$a_2 = 22.0$
$a_3 = 61.7$
$a_4 = 121.0$
$a_5 = 200.0$

Simply supported or "hinged-hinged" beam

$a_1 = \pi^2 = 9.87$
$a_2 = 4\pi^2 = 39.5$
$a_3 = 9\pi^2 = 88.9$
$a_4 = 16\pi^2 = 158.$
$a_5 = 25\pi^2 = 247.$

"Free-free" beam or floating ship

$a_1 = 22.0$
$a_2 = 61.7$
$a_3 = 121.0$
$a_4 = 200.0$
$a_5 = 298.2$

"Clamped-clamped" beam has same frequencies as "free-free"

$a_1 = 22.0$
$a_2 = 61.7$
$a_3 = 121.0$
$a_4 = 200.0$
$a_5 = 298.2$

"Clamped-hinged" beam may be considered as half a "clamped-clamped" beam for even a-numbers

$a_1 = 15.4$
$a_2 = 50.0$
$a_3 = 104.$
$a_4 = 178.$
$a_5 = 272.$

"Hinged-free" beam or wing of autogyro may be considered as half a "free-free" beam for even a-numbers

$a_1 = 0$
$a_2 = 15.4$
$a_3 = 50.0$
$a_4 = 104.$
$a_5 = 178.$

Source: from Den Hartog [2], with permission of Dover Publications, Inc.

Table E: Natural Frequencies of Multiple DOF Systems

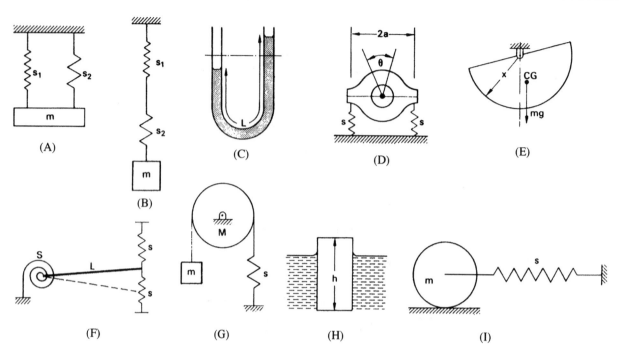

System	Frequency (Hz.)	Reference in Dwg.
Springs in parallel	$\dfrac{1}{2\pi}\sqrt{\dfrac{(s_1 + s_2)}{m}}$	(A)
Springs in series	$\dfrac{1}{2\pi}\sqrt{\left(\dfrac{1}{(1/s_1 + 1/s_1)m}\right)}$	(B)
Manometer column	$\dfrac{1}{2\pi}\sqrt{\left(\dfrac{2g}{L}\right)}$	(C)
Torsion of flexibly supported machine	$\dfrac{2a}{2\pi}\sqrt{\left(\dfrac{s}{J_0}\right)}$	(D)
Pivoted semicircular disc	$\dfrac{1}{2\pi}\sqrt{\left(\dfrac{8g}{3\pi r}\right)}$	(E)
Linear and torsional spring restraint	$\dfrac{1}{2\pi}\sqrt{\left(\dfrac{3S + 6sL^2}{mL^2}\right)}$	(F)
Spring–mass pulley	$\dfrac{1}{2\pi}\sqrt{\left(\dfrac{s}{(M/2) + m}\right)}$	(G)
Floating body	$\dfrac{1}{2\pi}\sqrt{\left(\dfrac{g}{\rho h}\right)}$	(H)
Rolling disc with spring restraint	$\dfrac{1}{2\pi}\sqrt{\left(\dfrac{2s}{3m}\right)}$	(I)

where g = gravitational acceleration
   h = height of floating body
   J_0 = mass moment of inertia of machine about axis of oscillation
   L = length of fluid in manometer or length of uniform rigid rod
   r = radius of pivoted semicircular disc
   s = spring stiffness
   S = torsion spring stiffness
   ρ = density of floating body

Source: from Collacott [9], with permission of Institution of Diagnostic Engineers.

Table F: Planetary Gear Mesh Frequencies

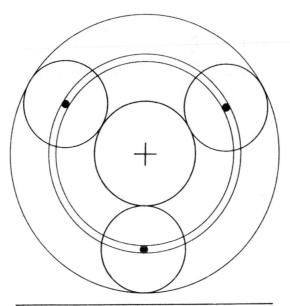

Fixation	Frequency	Equation
Cage	Tooth-meshing frequency	$N_1 t_1$
	High spot on sun	$n N_1$
	High spot on planet	$2 N_2$
	High spot on annulus	$n N_3$
Sun	Tooth-meshing frequency	$N_4 t_1$
	High spot on sun	$n N_4$
	High spot on planet	$\dfrac{2 t_1}{t_2} N_4$
	High spot on annulus	$\dfrac{n t_1}{t_3} N_4$
Annulus	Tooth-meshing frequency	$N_4 t_3$
	High spot on sun	$\dfrac{n t_3}{t_1} N_4$
	High spot on planet	$\dfrac{2 t_3}{t_2} N_4$
	High spot on annulus	N_4

where: n = *number of planet gears*

t_1 = *number of teeth (sun gear)*

t_2 = *number of teeth (planet gears)*

t_3 = *number of teeth (annulus)*

N = *gear speed (rpm or Hz.)*

N_1 = *sun gear speed*

N_2 = *planet gear speed*

N_3 = *annulus speed*

N_4 = *cage speed*

Source: from Collacott [9], with permission of Institution of Diagnostic Engineers.

Table G: Rolling Element Bearing Frequencies and Bearing Defect Frequencies

Rolling Element Bearing Frequencies

Component	Rotational/pass frequency
Outer race element pass	$\dfrac{n}{2}\dfrac{N}{60}\left(1 - \dfrac{d}{D}\cos\beta\right)$
Inner race element pass	$\dfrac{n}{2}\dfrac{N}{60}\left(1 + \dfrac{d}{D}\cos\beta\right)$
Rolling element	$\dfrac{D}{d}\dfrac{N}{60}\left[1 - \left(\dfrac{d}{D}\right)^2\cos\beta\right]$
Cage frequency	$\dfrac{1}{2}\cdot\dfrac{N}{60}\left(1 - \dfrac{d}{D}\cos\beta\right)$

where d = rolling-element diameter
D = bearing pitch diameter
β = contact angle
n = number of rolling elements
N = shaft speed

Rolling Element Bearing Defect Frequencies

Component	Frequency	Significance
Rolling element train	$\dfrac{1}{2}\dfrac{N}{60}\left(1 - \dfrac{d}{D}\cos\beta\right)$	Caused by an irregularity of a rolling element on the cage
Rolling element defect	$2 \times$ rolling element spin $= 2\cdot\dfrac{D}{d}\cdot\dfrac{N}{60}\left[1 - \left(\dfrac{d}{D}\right)^2\cos\beta\right]$	Irregularity strikes the inner and outer cases alternately
Inner race defect	Inner race pass frequency $= \dfrac{n}{2}\cdot\dfrac{N}{60}\left(1 + \dfrac{d}{D}\cdot\cos\beta\right)$	
Outer race defect	Outer race pass frequency $= \dfrac{n}{2}\cdot\dfrac{N}{60}\left(1 - \dfrac{d}{D}\cdot\cos\beta\right)$	Likely to arise if there is a variation in stiffness around bearing housing

Source: from Collacott [9], with permission of Institution of Diagnostic Engineers.

Table H: General Vibration Diagnostic Frequencies

Defect cause	Frequency
Aerodynamic	$N, 2N, 3N \ldots nN$
Bearing assembly loose	$\frac{1}{2}N$
Bearing element distorted	N
Bearing misalignment (plain)	$N, 2N, 3N \ldots nN$
Belt drive faults	$N, 2N, 3N \ldots nN$
Combustion faulty, diesel engines (four-stroke)	$\frac{1}{2}N$
Coupling misalignment	$N, 2N, 3N \ldots nN$
Electrical	NSF or $2SF, 3N, 6N$
Electrical machines, DC (armature slots/commutator segments)	sN
Electrical machines, induction (magnetic field)	$2F_s$
Electrical machines, induction (rotor slots)	either sN or $sN \pm 2F_s$
Electrical machines, synchronous (magnetic field)	$2F_s$
Fan blades	$N, 2N, 3N \ldots nN$
Forces, reciprocating	$N, 2N, 3N \ldots nN$
Journals, eccentric	N
Mechanical looseness	$2N$
Oil whirl	$0.45N$
Pump impellers	nN
Rotor, bent	$N, 2N, 3N \ldots nN$
Shaft, bent	N
Unbalance	N
Whirl, friction-induced	$<0.4N$
Whirl, oil	$0.45N$

where: N = shaft rotation frequency
SF = synchronous frequency
F_S = supply frequency
s = number of slots or segments

Source: from Collacott [9], with permission of Institution of Diagnostic Engineers.

References

Vibration Theory and Fundamentals

1. Inman, Daniel J., *Engineering Vibration.* Upper Saddle River, NJ: Prentice-Hall, 1994.
2. Den Hartog, J. P., *Mechanical Vibrations.* New York: Dover Publications, Inc., 1985.
3. Dimarogonas, A. D. and Haddad, S., *Vibration for Engineers.* Upper Saddle River, NJ: Prentice-Hall, 1992.
4. Thomson, W. T., *Theory of Vibration with Applications.* Upper Saddle River, NJ: Prentice-Hall, 1972.
5. Thomson, W. T., "Vibration" in *Standard Handbook for Mechanical Engineers,* T. Baumeister and L.S. Marks (Editors). New York: McGraw-Hill, 1967.
6. Meirovich, L., *Analytical Methods in Vibration.* New York: The Macmillan Co., 1967.

Solution Methods to Numerous Practical Vibration Problems

7. Seto, W. W., *Theory and Problems of Mechanical Vibrations* (Schaum's Outline Series). New York: McGraw-Hill, 1964.

Vibration Measurement and Analysis Techniques

8. Jackson, C., *The Practical Vibration Primer.* Houston: Gulf Publishing Co., 1979.
9. Collacott, R. A., *Vibration Monitoring and Diagnosis.* London: George Godwin Ltd. (Institution of Diagnostic Engineers), 1979.
10. *Endevco Dynamic Test Handbook,* Rancho Viejo Road, San Juan Capistrano, CA.

12

Materials

Paul S. Korinko, Ph.D., Senior Experimental Metallurgist, Allison Engine Company

In this chapter, material properties, a few definitions, and some typical applications will be presented as guidelines for material selection. The most important rule for material selection is that the operating conditions must be well defined. These conditions include temperature, environment, impurities, stress, strain, cost, and any limitations for life cycle, such as creep or fatigue.

CLASSES OF MATERIALS

Materials can be broken down into three major classes, each having specific characteristics. Metals, ceramics, and polymers constitute these classes, and some typical properties and applications will be described. In addition, fabrication methods for the materials will be described.

Metals are characterized by metallic bonding in which the electrons are shared in a veritable sea of electrons. Each nucleus is surrounded by electrons, but the electrons are not specifically attached to any particular nucleus. Metals are further characterized by their surface appearance and generally have a "metallic" sheen and can be polished to a mirror finish. Metals are also capable of sustaining large loads, they are ductile, and they have reversible elastic properties to a point. They can be alloyed to alter their physical and chemical properties.

Metals and alloys are defined by the major alloying element present. Common engineering alloys consist of iron, nickel, cobalt, aluminum, magnesium, titanium, and copper. These alloys will be discussed in some detail after some definitions are presented.

Polymers are long chains of carbon and hydrogen arranged in specific bonding orientations. The bonds are generally covalent bonds. Noncrystalline or amorphous polymers have a random arrangement of polymer chains. Crystalline polymers have specific polymer chain arrangements. The unit cells (smallest repeating arrangement of atoms to make the full structure) are large (10 nm) relative to metal unit cells (0.3 nm). Polymer properties can be elastic, viscous, or viscoelastic. The actual behavior depends on the temperature, composition, orientation, and degree of crystallinity.

Ceramics are characterized by either ionic bonding in which electrons are lost and there is an electronic force that holds the ions together, or covalent bonding in which the electrons are shared. Because of the electronic nature of the bonding, the net charge on the material must be zero. This requirement makes deformation difficult when like charged atoms must pass closely together. Ceramics can sustain large compressive loads but are very surface defect sensitive for tensile loads. They tend to behave elastically to failure with little or no plastic deformation prior to fracture.

DEFINITIONS

Grain size is a microstructural property of a material that indicates how large the crystals constituting the structure are. The grain size is important for a number of mechanical and physical properties. For example, room temperature strength is increased by having a small grain size. The correlation between grain size and strength is known as the Hall-Petch relationship and is shown in Equation 1:

$$\sigma = \sigma_o + k_y d^{-1/2} \qquad (1)$$

where σ_o is the strength of a single crystal in MPa or ksi and k_y is the slope of the line with units of MPA*mm$^{1/2}$ or ksi*in$^{1/2}$. High-temperature, long-term properties, such as creep, are improved by a large grain size, thus, a balance between the required properties is necessary. A number of methods are used to measure the grain size, with the most common being ASTM E112. In this method, the number of grains per square inch is measured. The larger the number, the finer the grain size, and vice versa.

Alloying is the intentional addition of one or more elements to a parent metal. Most of the metals that are used are alloys; alloys typically have higher strength and other more desirable properties than pure metals. The chemical resistance and oxidation resistance of some alloys is better than the pure elements—nickel alloyed with chromium and aluminum, for example. Alloys can be one of three types: interstitial alloys, in which the added element is much smaller than the parent element and the atoms reside at normally unoccupied positions in the lattice, i.e., interstitial sites; substitutional alloys, in which the addition displaces an atom of the parent metal; or precipitation alloys, in which a cluster of atoms forms a second phase in the parent metal.

Precipitation hardening is a strengthening mechanism for alloys that have specific chemical interactions which can be seen in a type of phase diagram. The solid solubility increases with increasing temperature, and only a certain range of alloying additions will work. The second phase should be stronger than the parent (matrix) metal, is generally brittle, and can interact with the crystal defects (dislocations) that control the deformation of the alloys. Most alloys are strengthened by a combination of the three methods mentioned.

Composites are formed by the addition of discrete particles or fibers to a metal matrix. The strength increase depends on the strength and modulus of both the matrix and the reinforcement addition. For a composite that is strengthened due to isostress, the composite strength σ_c is given by Equation 2:

$$\sigma_c = (1 - V_f)\,\sigma_m + V_f\,\sigma_f \qquad (2)$$

where V_f is the volume fraction of the reinforcing phase, and σ_f and σ_m are the reinforcing phase and matrix strengths.

The modulus of a fiber-reinforced composite tested perpendicular to the fiber axis is given by Equation 3. The symbols are the same as those used above, with E used for the modulus:

$$\frac{1}{E_c} = \frac{V_f}{E_f} + \frac{(1 - V_f)}{E_m} \qquad (3)$$

The following definitions regarding mechanical properties are usually based on a tensile test—a mechanical test in which a standard specimen is pulled uniaxially until failure. The displacement and load are recorded. These data are then converted into stress and strain. In engineering stress and strain, the stress is load divided by the original cross-sectional area; the strain is the change in length divided by the original length. These terms are different than the true stress and strain which are generally ignored in practice but are defined as the load divided by the instantaneous area and increase in length divided by the instantaneous length. The difference does not amount to much in practical applications, but it does change the nature of the stress-strain diagram.

Yield strength refers to the stress at which a certain permanent strain, typically 0.02% or 0.2%, has occurred. Figure 1 shows a typical stress-strain curve with the yield strength determined as in the inset.

Tensile strength is the maximum stress that the specimen withstands, and occurs when the strain is no longer uniform and has become centralized at a band called a "neck." It is often referred to as the onset of necking.

Failure strength is the point at which the specimen separates into two pieces. This stress is not of any practical value.

Elastic modulus, also referred to as Young's modulus, occurs in the linear portion of the stress-strain curve. It is a measure of a material's stiffness, much like a spring constant. The modulus is loosely correlated with the melting point of a material and increases as the melting point increases. Table 1 lists some melting points of metals and their respective elastic modulus. Figure 2 more clearly shows the general trend of increased modulus with increased melting point.

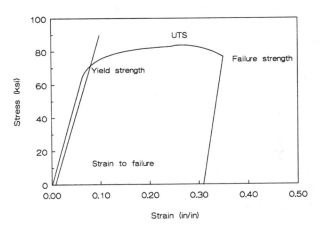

Figure 1. Typical tensile curve showing yield, tensile, and ultimate strengths and elastic and plastic strains.

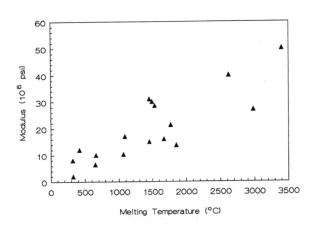

Figure 2. Elastic modulus as a function of melting point. (Data from Table 1.)

Table 1
Modulus and Melting Temperatures of Important Engineering Elements

Element	Crystal Structure	Melting Point (°C)	Modulus (10^6 psi)
Cd	HCP	321	8
Pb	FCC	327	2
Zn	HCP	419	12
Mg	HCP	650	6.4
Al	FCC	660	10
Au	FCC	1,062	10.8
Cu	FCC	1,084	17
Ni	FCC	1,453	31
Nb	BCC	1,453	15
Co	HCP	1,495	30
Fe	BCC	1,536	28.5
Ti	HCP	1,670	16
Pt	FCC	1,770	21.3
Zr	HCP	1,852	13.7
Cr	BCC	1,860	36
Mo	BCC	2,620	40
Ta	BCC	2,980	27
W	BCC	3,400	50

Adapted from the CRC Handbook of Tables for Applied Engineering Science, *2nd Ed. CRC Press, 1984, with permission.*

Strain to failure provides an indication of the amount of energy required to break a specimen. It is measured as indicated in Figure 1. The total strain is increased by the elastic component of the strain.

METALS

Steels

Steels are one of the most commonly used construction materials. Steels are typically termed *plain carbon* or *low alloy,* depending on the type of additions made to the iron base; a new class of steels is the low alloy, high strength variety. A plain carbon steel consists of iron and an alloying addition of carbon. The iron and carbon combine to form a compound phase, known as iron carbide (or cementite), which has the composition of Fe_3C. Iron undergoes an allotropic transformation (a change in crystal structure) of body centered cubic (BCC), also known as alpha iron, to face centered cubic (FCC), also known as gamma iron, at 885°C, and another transformation of FCC to BCC, also known as delta iron, at 1,395°C. The first of these transformations can be either useful or detrimental.

The composition of a steel is indicated by its SAE number. The SAE number has four to five digits. The first two digits indicate the alloying additions, and the last two or three indicate the carbon content; for instance, a 1020 steel is a plain carbon steel with nominally 0.2% carbon, and a 4340 steel has nickel and chromium with 0.4% carbon. Table 2 lists a number of common steels and their SAE numbers. Typical applications and yield strengths are listed in Table 3. In some cases, the composition is not specified, rather, several key properties such as hardness, strength, and ductility are specified and the supplier is free, within reason, to adjust the chemistry to have it meet the mechanical or physical properties. Further, it is apparent that even with the same composition, a number of properties can be developed.

Table 2
Typical Compositions of Steels

AISI-SAE	C	Mn	P	S	Si	Cr	Ni	Mo
1018	0.14–0.20	0.60–0.90	—	—	—	—	—	—
1040	0.36–0.44	0.60–0.90	—	—	—	—	—	—
1095	0.90–1.04	0.30–0.50	—	—	—	—	—	—
4023	0.20–0.25	0.70–0.90	0.035	0.040	0.15–0.30	—	—	0.20–0.30
4037	0.35–0.40	0.70–0.90	0.035	0.040	0.15–0.30	—	—	0.20–0.30
4118	0.18–0.23	0.70–0.90	0.035	0.040	0.15–0.30	0.40–0.60	—	0.08–0.15
4140	0.38–0.43	0.75–1.00	0.035	0.040	0.15–0.30	0.80–1.10	—	0.15–0.25
4161	0.56–0.64	0.75–1.10	0.035	0.040	0.15–0.30	0.80–1.10	—	0.15–0.25
4340	0.38–0.43	0.60–0.80	0.035	0.040	0.15–0.30	0.70–0.90	1.65–2.00	0.20–0.30
5120	0.17–0.22	0.70–0.90	0.035	0.040	0.15–0.30	0.70–0.90	—	—
5140	0.38–0.43	0.70–0.90	0.035	0.040	0.15–0.30	0.70–0.90	—	—
51100	0.98–1.10	0.25–0.45	0.025	0.025	0.15–0.30	0.40–0.60	—	—
8620	0.18–0.23	0.70–0.90	0.035	0.040	0.15–0.30	0.40–0.60	0.40–0.70	0.15–0.25
8640	0.38–0.43	0.75–1.00	0.035	0.040	0.15–0.30	0.40–0.60	0.40–0.70	0.15–0.25
8660	0.56–0.64	0.75–1.00	0.035	0.040	0.15–0.30	0.40–0.60	0.40–0.70	0.15–0.25
9310*	0.08–0.13	0.45–0.65	0.035	0.040	0.15–0.30	1.00–1.40	3.00–3.50	0.08–0.15

* Also contains 0.10–0.15 V.
Typical compositions for steels. Actual compositions depend on class and grade specified.
Adapted from ASM Metals Handbook, Vol. 1, 9th Ed. [2].

Table 3
Typical Mechanical Property Ranges and Applications for Oil Quenched and Tempered Plain Carbon and Alloy Steels

Type	Mechanical Property Range			Applications
	Tensile Strength (ksi)	Yield Strength (ksi)	Ductility (%elongation in 2″)	
	Plain carbon steels			
1040	88–113	62–85	33–19	Crankshafts, bolts
1080	116–190	70–142	24–13	Chisels, hammers
1095	110–186	74–120	26–10	Knives, hacksaw blades
	Alloy steels			
4063	114–345	103–257	24–4	Springs, handtools
4340	142–284	130–228	21–11	Bushings, aircraft tubing
6150	118–315	108–270	22–7	Shafts, pistons, gears

Source: Callister [3]. Reprinted by permission of John Wiley & Sons, Inc.

Steels can be selectively hardened through an appropriate thermal treatment. Surface hardness can be increased by locally increasing the carbon content. Shafts are particularly useful if surface or case hardened. In a case hardened steel, the surface contains a substantially higher carbon content than the core. This provides better wear resistance at the surface for applications such as gears, where there is potentially significant wear at the surface but some impact loading of the core. The surface is hard and somewhat brittle but wear resistant, and the core is tough and more ductile. A nitride layer can also be introduced to increase the surface hardness. Ammonia gas is dissociated, and aluminum in the steel reacts to form aluminum nitrides which impart wear-resistant surfaces to the steel.

Various methods can be used to strengthen steels. The first is to heat treat them. In the process of heat treating, a steel is first heated to the single phase region (γFe) shown in Figure 3 (austenitize). It can then be rapidly cooled (quenched) to form martensite. The martensitic steel is subsequently toughened by tempering. This step occurs at a slightly elevated temperature, but not one too high to prevent overtempering and losing the martensitic structure; Table 4 shows the effect of increasing tempering temperature on 4140 and 4340 steels. It is clear that the strength decreases and the ductility increases with increasing tempering temperature.

A second heat treatment is to austenitize and furnace cool (normalize). This produces a structure that consists of fairly coarse pearlite and either ferrite or cementite depending on the alloy composition, which will have low strength and high ductility.

Yet another treatment is to austenitize and air cool. This cooling rate typically results in finer pearlite and ferrite or cementite structure, properties between quenched and normalized.

Other treatments include heating to below the eutectoid temperature to intentionally coarsen the pearlite (spheroidizing). There are other treatments such as ausforming and ausquenching. For a more detailed description of these

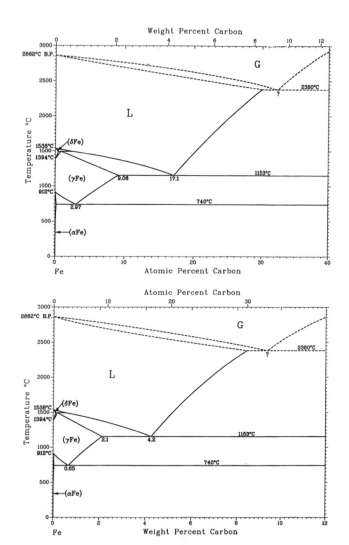

Figure 3. Iron-carbon phase diagram [23]. (With permission, ASM International.)

Table 4
Typical Mechanical Properties of Heat-Treated 4140 and 4340 Steels Oil Quenched from 1,550°F

Tempering Temperature (°F)	Tensile Strength (ksi)	Yield Strength (ksi)	Elongation in 50 mm (%)	Reduction in Area (%)	Hardness HB
4140 Steel					
400	285	252	11.0	42	578
500	270	240	11.0	44	534
600	250	228	11.5	46	495
700	231	212	12.5	48	461
800	210	195	15.0	50	429
900	188	175	16.0	52	388
1,000	167	152	17.5	55	341
1,100	148	132	19.0	58	311
1,200	130	114	21.0	61	277
1,300	117	100	23.0	65	235
4340 Steel					
400	287	270	11	39	520
600	255	235	12	44	490
800	217	198	14	48	440
1,000	180	168	17	53	360
1,200	148	125	20	60	290
1,300	125	108	23	63	250

Adapted from ASM Metals Handbook, Vol 1, 9th Ed. [2].

thermal treatments, almost any introduction to materials science course [6, 7] will be adequate.

The mechanical properties vary significantly for these treatments. The highest yield and tensile strengths will be obtained for the martensitic structure, and the weakest for the spheroidized. The fine pearlite will be stronger than the coarse pearlite.

Tool Steels

Tool steels are characterized by higher carbon contents than conventional steels, and quench and temper heat treatments. They are used as cutting tools, dies, and in other applications where a combination of high strength, hardness, toughness, and high temperature capability are important.

Some typical compositions are shown in Table 5. Typical properties are listed in Table 6. Tool steels can be machined in the annealed condition and then hardened, although distortion from heat treatment can occur.

Table 5
Nominal Composition of Classes of Tool Steels

AISI	USN	C	Mn	Si	Cr	Ni	Mo	W	V
Air–hardening medium alloy cold work steels									
A3	T30103	1.20–1.30	0.40–0.60	0.50 max	4.75–5.50	0.30 max	0.90–1.40	—	0.80–1.40
Shock resistant steels									
S1	T41901	0.40–0.55	0.10–0.40	0.15–1.20	1.00–1.80	0.30 max	0.50 max	1.50–3.00	0.15–0.30
S5	T41905	0.50–0.65	0.60–1.00	1.75–2.25	0.35 max	—	0.50 max	—	0.35 max
Low alloy special purpose tool steels									
L2	T61202	0.45–1.00	0.10–0.90	0.50 max	0.70–1.20	—	0.25 max	—	0.10–0.30
L6	T61206	0.65–0.75	0.25–0.80	0.50 max	0.60–1.20	1.25–2.00	0.50 max	—	0.20–0.30

Adapted from ASM Metals Handbook, *Vol. 1, 9th Ed.* [2].

Table 6
Typical Properties of Tool Steels After Indicated Heat Treatment

Heat Treat Condition	Tensile Strength (ksi)	Yield Strength (ksi)	Elongation in 50 mm (%)	Reduction in Area (%)	Hardness HRC
L2 Annealed	103	74	25	50	96 HRB
Oil quenched from 1,575°F and single tempered at					
400°F	290	260	5	15	54
600°F	260	240	10	30	52
L6 Annealed	95	55	25	55	93 HRB
Oil quenched from 1,550°F and single tempered at					
600°F	290	260	4	9	54
800°F	230	200	8	20	46
S1 Annealed	100	60	24	52	96 HRB
Oil quenched from 1,700°F and single tempered at					
400°F	300	275	—	—	57.5
600°F	294	270	44	12	54
S5 Annealed	105	64	25	50	96 HRB
Oil quenched from 1,600°F and single tempered at					
400°F	340	280	5	20	58
600°F	325	270	7	24	58

Adapted from ASM Metals Handbook, *Vol. 1, 9th Ed.* [2].

Cast Iron

Cast iron is a higher carbon containing iron-based alloy. Cast irons contain more than 2.1% C by weight. They can be cast with a number of different microstructures. The most common is gray cast iron which has graphite flakes in a continuous three-dimensional structure which looks rather like potato chips. This structure promotes acoustic damping and low wear rates because of the graphite.

A second structure involves heat-treating the gray cast iron to form spherodized cast iron. In this structure, the damping capacity is lost but the corrosion resistance is improved.

White iron is very brittle and is formed during cool-down from the melt. It can be used as a wear-resistant surface if the rest of the casting can be ductilized by perhaps forming gray cast iron.

Stainless steels

A special class of iron-based alloys have been developed for resistance to tarnishing and are known as stainless steels. These alloys may be *martensitic* (body centered tetragonal), *austenitic* (FCC), or *ferritic* (BCC) depending on the alloying additions that have been made to the iron.

Use of stainless steels should be considered carefully. The use of some classes should be limited to oxidizing environments in which the alloy has the chance to form a protective oxide scale. Use of alloys requiring the oxide scale for protection in reducing environments, such as carbon monoxide which can electrochemically or thermodynamically convert oxides to metals, can be disastrous. Tables 7 and 8 contain a partial list of common stainless steel compositions and acceptable use environments.

A thin oxide scale forms on the stainless steel and protects it from further oxidation and corrosion. Chromium is typically the element responsible for stainless steel's "stainless" appearance.

Ferritic stainless steels have typically up to 30% Cr and 0.12% C and are moderately strong, solid solution and strain hardened, and low cost. The strengths can be increased by increasing the Cr and C; unfortunately, these actions result in carbide precipitation and subsequent embrittlement. Excessive Cr additions can also promote the precipitation of a brittle second phase known as sigma phase.

Martensitic stainless steels contain up to 17% Cr and from 0.1–1.0% C. These alloys are strengthened by the formation of martensite on cooling from a single-phase austenite field. With the range of carbon contents available, martensite of varying hardness can be produced. Martensitic stainless steels have good hardness, strength, and corrosion resistance. Typical uses are in knives, ball bearings, and valves. They soften at temperatures above 500°C.

Austenitic stainless steels have high chromium and high nickel content. The generic term is 18-8 stainless, which refers to 18% Cr and 8% Ni. The nickel is required to stabilize the gamma or face centered cubic (FCC) phase of the iron, and the Cr imparts the corrosion resistance. These alloys can be used to 1,000°C. Above this temperature, the chromium oxide that forms can vaporize and will not protect the substrate, so rapid oxidation can occur.

Table 7
Composition of Standard Stainless Steels
Composition (%)

Type	UNS Number	C	Mn	Si	Cr	Ni	P	S	Other
Austenitic types									
201	S20100	0.15	5.5–7.5	1.00	16.0–18.0	3.5–5.5	0.06	0.03	0.25 N
304	S30400	0.08	2.00	1.00	18.0–20.0	8.0–10.5	0.045	0.03	—
304L	S30403	0.03	2.00	1.00	18.0–20.0	8.0–12.0	0.045	0.03	—
310	S31000	0.25	2.00	1.50	24.0–26.0	19.0–22.0	0.045	0.03	—
316	S31600	0.08	2.00	1.00	16.0–18.0	10.0–14.0	0.045	0.03	2.0–3.0 Mo
347	S34700	0.08	2.00	1.00	17.0–19.0	9.0–13.0	0.045	0.03	10X%c min Nb+Ta
Ferritic types									
450	S40500	0.045	1.00	1.00	11.5–14.5	—	0.04	0.03	0.1–0.3 Al
430	S43000	0.12	1.25	1.00	16.0–18.0	—	0.04	0.03	—
Martensitic types									
410	S41000	0.15	1.00	1.00	11.5–13.0	—	0.04	0.03	—
420	S42000	0.15	1.00	1.00	12.0–14.0	—	0.04	0.03	—
431	S43100	0.20	1.00	1.00	15.0–17.0	1.25–2.50	0.04	0.03	—
Precipitation– hardening types									
17–4PH	S17400	0.07	1.00	1.00	15.5–17.5	3.0–5.0	0.04	0.03	3.0–5.0 Cu; 0.15–0.45 (Nb+Ta)
17–7PH	S17700	0.09	1.00	1.00	16.0–18.0	6.5–7.75	0.04	0.03	0.75–1.5Al

Adapted from ASM Metals Handbook, *Vol. 3, 9th Ed. [40].*

Table 8
Resistance of Standard Types of Stainless Steel to Various Classes of Environments

Type	Mild Atmospheric and Fresh Water	Atmospheric		Salt water	Mild	Chemical Oxidizing	Reducing
		Industrial	Marine				
Austenitic stainless steels							
201	x	x	x		x	x	
304	x	x	x		x	x	
310	x	x	x		x	x	
316	x	x	x	x	x	x	x
347	x	x	x		x	x	
Ferritic stainless steels							
405	x				x		
430	x	x			x	x	
Martensitic stainless steels							
410	x				x		
420	x						
431	x	x	x		x		
Precipitation hardening stainless steels							
17-4PH	x	x	x		x	x	
17-7PH	x	x	x		x	x	

An "x" notation indicates that the specific type is resistant to the corrosive environment.
Adapted from ASM Metals Handbook, Vol. 3, 9th Ed. [40].

Since austenitic stainless steels are FCC, they tend not to be magnetic. Thus an easy test to separate austenitic stainless steel from ferritic or martensitic alloys is to use a magnet.

Austenitic stainless steels are not as strong as martensitic stainless steels, but can be cold worked to higher strengths than ferritic stainless steels since they are strengthened via solid solution hardening in addition to the cold work. They are more formable and weldable than the other two types of stainless steel. They are also more expensive due to the high nickel content.

The amount of carbon in an austenitic stainless steel is important; if it exceeds 0.03% C, the Cr can form chromium carbides which locally decrease the Cr content of the stainless steel and can sensitize it. A sensitized alloy forms when slowly cooled from below about 870°C to about 500°C. It is prone to corrosion along the grain boundaries where the local Cr content drops below 12%. Figure 4 shows a schematic of a sensitized alloy. A rapid quench through this temperature range should prevent the formation of the chrome carbides. Elements such as Ti or Nb, which are strong carbide formers, can be added to the alloy to form carbides and stabilize the alloy, for example, types 347 and 321.

Austenitic stainless steels also have good low temperature properties. Since they are FCC, they do not undergo a ductile to brittle transition like body centered cubic metals (BCC). Austenitic stainless steels can be used at cryogenic temperatures.

The precipitation hardening alloys are strengthened by the formation of martensite and precipitates of copper-niobium-tantalum.

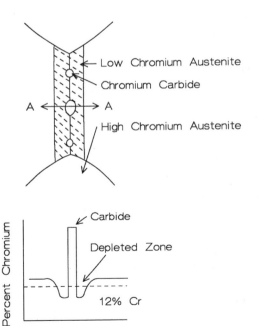

Figure 4. Sensitized stainless steel. Cr content near grain boundary is too low for corrosion protection.

Superalloys

Iron-based superalloys have high nickel contents to stabilize the austenite, chromium for corrosion protection, and niobium, titanium, and aluminum for precipitation hardening. Refractory elements are introduced for solid solution hardening. They also confer some creep resistance. Creep resistance is further enhanced by the presence of small coherent precipitates. Unfortunately, the fine precipitates that improve the creep strength the most are also the most likely to dissolve or coalesce and grow.

Nickel- and cobalt-based superalloys have higher temperature capabilities than iron-based superalloys. The strengthening mechanisms for nickel-based alloys are similar to those for iron-based alloys. The nickel matrix is precipitation hardened with coherent precipitates of niobium, aluminum, and titanium. Carbides and borides are used as grain boundary strengtheners, and refractory elements are added as solid solution strengtheners. The gamma prime (Ni_3Al,Ti) is a very potent strengthener that is a coherent precipitate. These precipitates are present up to 70% in modern, advanced nickel-based alloys. They permit the use of nickel-based alloys to approximately 0.75 times the melting point. Nickel-based alloys are also cast as single crystals which provide significant strength and creep improvements over polycrystalline alloys of the same composition. Some typical compositions and applications are listed in Tables 9 and 10.

Cobalt alloys are not strengthened by a coherent phase like Ni_3Al, rather, they are solid solution hardened and carbide strengthened. Cobalt alloys have higher melting points and flatter stress rupture curves which often allow these alloys to be used at higher absolute temperatures than nickel- or iron-based alloys. Their use includes vanes, combustor liners, and other applications which require high temperature strength and corrosion resistance. Most cobalt-based superalloys have better hot corrosion resistance than nickel-based superalloys. They also have better fabricability, weldability, and thermal fatigue resistance than nickel-based alloys.

Table 9
Nominal Compositions of Typically Used Iron-, Nickel-, and Cobalt-based Superalloys

Alloy	Co	Ni	Fe	Cr	Al	Ti	Mo	W	Nb	Cu	Other
Wrought Alloys											
HASTELLOY® C-4*		Bal		16			16				
HASTELLOY® C-22™*		Bal	3	22			13	3			
HASTELLOY® C-276*		Bal	5	16			16	4			
HASTELLOY® D-205™*		Bal	6	20			2.5			20	5 Si
HASTELLOY® S*		Bal		16			15				La
HASTELLOY® W*		Bal	6	5			24				
HASTELLOY® X*	1.5	Bal	18	22			9	0.6			
HAYNES 188*	Bal	22		22				14			La
HAYNES 214™*		Bal	3	16	4.5						Y
HAYNES 230™*		Bal		22			2	14			La
Alloy 625*		Bal		21			9		3.5		
Alloy 718*		Bal	19	18	0.5	1			5		
Waspaloy*	14	Bal		19	1.5	3	4				
INCONEL® MA 754†		Bal	1	20	0.3	0.5					Y_2O_3
INCONEL® MA 956†			Bal	20	4.5	0.5					Y_2O_3
Cast alloys**											
Alloy 713		Bal		12.5	6.1	0.8	4.2				
IN-100	15	Bal		10	5.5	4.7	3				
IN-738	8.5	Bal		16	3.4	3.4	1.7	2.6	0.9	Ta	
Mar M 247	10	Bal		8.3	5.5	1	0.7	10		Ta	
Mar-M 509	Bal	10		23.5				7		Ta	
X-40	Bal	10		25.5				7.5		0.7	Mn

*From Haynes International, Product Bulletin H-1064D, 1993.
†From Inco Alloys International, Product Handbook, 1988.
**From Sims, et al. [26] by permission of John Wiley & Sons, Inc.

Table 10
Common Application of Iron-, Nickel-, and Cobalt-based Superalloys

Wrought Alloys

HASTELLOY® C-4*	High temperature stability to 1,900°F. Excellent corrosion resistance.
HASTELLOY® C-22™*	Universal filler metal for corrosion-resistant welds. Resistance to localized corrosion, stress corrosion cracking, and oxidizing and reducing chemicals.
HASTELLOY® C-276*	Excellent resistance to oxidizing and reducing corrosives, mixed acids, and chlorine bearing hydrocarbons.
HASTELLOY® D-205™*	Superior performance in sulfuric acid of various concentrations.
HASTELLOY® S*	Low stress gas turbine parts. Excellent dissimilar filler metal.
HASTELLOY® W*	Aircraft engine repair and maintenance.
HASTELLOY® X*	Aircraft, marine, and industrial gas turbine engine combustors and fabricated parts.
HAYNES® 188*	Sulfidation resistant. Military and civilian aircraft engine combustors.
HAYNES® 214™*	Honeycomb seals demanding industrial heating applications.
HAYNES® 230™*	Gas turbine combustors and other stationary members, industrial heating, and chemical processing.
IN-625*	Aerospace, industrial heating, and chemical processing.
IN-718*	Extensive use in gas turbines.
Waspaloy*	Gas turbine components.
INCONEL® MA 754†	Mechanically alloyed for improved alloy stability. Gas turbine vanes.
INCONEL® MA 956†	Mechanically alloyed for improved alloy stability. Gas turbine combustors.
Cast Alloys	
Alloy 713	Turbine blades.
IN-100	Turbine blades.
IN-738	Turbine blades.
Mar-M 247	Turbine blades and vanes.
Mar-M 509	Turbine vanes.
X-40	Turbine vanes.

*Haynes International, Product Bulletin H-1064D, 1993.
†Inco Alloys International, Product Handbook, 1988.
HASTELLOY and HAYNES are registered trademarks of Haynes International, Inc.
C-22, D-205, 214, and 230 are trademarks of Haynes International, Inc. INCONEL and INCOLOY are registered trademarks of the Inco group of companies.

Aluminum Alloys

Aluminum alloys do not possess the high strength and temperature capability of iron-, nickel- or cobalt-based alloys. They are very useful where low density and moderate strength capability are required. Because of their relatively low melting point (less than 660°C), they can be readily worked by a number of different processes that metals with higher melting points cannot. Aluminum alloys are designated by their major alloying consituent. The common classes of alloying additions are listed in Table 11. Since alloy additions affect the melting range and strengthening mechanisms, a number of classes of alloys are generated that can have varying responses to heat treatment. Some alloys are solution heat treated and naturally aged (at room temperature), while some are solution treated and artificially aged (at elevated temperature). Table 12 lists several possible treatments for wrought aluminum alloys, and Table 13 lists typical applications.

Table 12
Common Al Alloy Temper Designations

O	Annealed.
F	As fabricated.
T1	Cooled from an elevated temperature shaping process and naturally aged to a substantially stable condition.
T2	Cooled from an elevated temperature shaping process, cold worked, and naturally aged to a substantially stable condition.
T3	Solution heat treated, cold worked, and naturally aged to a substantially stable condition.
T4	Solution heat treated and naturally aged to a substantially stable condition.
T5	Cooled from an elevated temperature shaping process and artifically aged.
T6	Solution treated and artificially aged.
T7	Solution treated and stabilized.
T8	Solution treated, cold worked, and artificially aged.
T9	Solution treated, cold worked, and artificially aged.
T10	Cooled from an elevated temperature shaping process, cold worked, and artificially aged.

From ASM Metals Handbook, Vol. 2, 9th Ed. [22].

Table 11
Major Alloying Elements for Aluminum Alloys and Compositions for Some Commonly Used Alloys

Series	Alloying element
1xxx	None 99.00% or greater Al
2xxx	Copper
3xxx	Manganese
4xxx	Silicon
5xxx	Magnesium
6xxx	Magnesium and silicon
7xxx	Zinc
8xxx	Other element
9xxx	Unused series

AA	Al	Si	Cu	Mn	Mg	Cr	Zn	Other
1050	99.50	—	—	—	—	—	—	—
1100	99.00	—	0.12	—	—	—	—	—
2014	93.5	0.8	4.4	0.8	0.5	—	—	—
2024	93.5	—	4.4	0.6	1.5	—	—	—
4032	85.0	12.2	0.9	—	1.0	—	—	0.9Ni
4043	94.8	5.2	—	—	—	—	—	—
5052	97.2	—	—	—	2.5	0.25	—	—
6063	98.9	0.4	—	—	0.7	—	—	—
7075	90.0	—	1.6	—	2.5	0.23	5.6	—

Adapted from ASM Metals Handbook, Vol. 2, 9th Ed. [22].

Table 13
Typical Applications and Mechanical Properties of Aluminum Alloys

1050	Chemical equipment, railroad tank cars
1100	Sheet metal work, spun hollow ware, fin stock
2014	Heavy duty forgings, plates and extrusions for aircraft fittings, wheels, truck frames
2024	Truck wheels, screw machine products, aircraft structures
4032	Pistons
4043	Welding electrode
5052	Sheet metal work, hydraulic tube, appliances
6063	Pipe railing, furniture, architectural extrusions
7075	Aircraft and other structures

Alloy	Temper	Tensile Strength (ksi)	Yield Strength (ksi)	Elongation in 50 mm (%)	Hardness HB (500 kg/10 mm ball)
1050	O	11	4	—	—
1100	O	13	5	35 (a)	23
2014	O	27	14	18 (b)	45
	T6	70	60	13 (b)	135
2024	O	27	11	20 (a)	47
	T3	70	50	18 (b)	120
4032	T6	55	46	9 (a)	120
4043	O	21	10	22 (b)	—
5052	O	28	13	25 (b)	36
6063	O	13	7	—	25
	T1	22	13	20 (b)	42
	T6	35	31	12 (b)	73
7075	O	38	15	17 (b)	60
	T6	83	73	11 (b)	150

Adapted from ASM Metals Handbook, Vol. 2, 9th Ed. [22].

Joining

Joining materials can be accomplished either mechanically, e.g., riveting, bolting, or metallurgically, e.g., brazing, soldering, welding. This section includes a brief discussion of metallurgical bonding.

Soldering

Solders are the lowest temperature metallurgical bonds that can be made. Typical materials that are soldered are wires and pipes. In a solder joint, the component pieces are not melted, only the solder filler metal. Soldering occurs at a temperature below 450°C (840°F). The metallurgical, physical, and chemical interaction of the elements, as well as the underlying thermodynamic and fluid dynamics of the solder, determine the properties of the solder joint.

A clean surface is required; the surface should be precleaned to remove any oil, pencil markings, wax, tarnish, and atmospheric dirt which can interfere with the soldering process. The surface may be cleaned with a flux which removes any adherent oxides and may further clean the surface. Fluxes may also serve to activate the surface. The type of flux used depends on the substrate and solder alloy. Most fluxes are proprietary, so experimentation is necessary to determine the effectiveness for the application. Removal of the oxide promotes wetting of the substrate with the solder alloy.

The joint strengths obtained by soldering depend on a number of factors, including the substrate material, solder composition, and joint geometry. Some typical joint geometries are depicted in Figure 5. Typically lead-tin solders are used. Table 14 lists a variety of Pb-Sn solders and their applications. Many of the solders have wide freezing ranges. This feature makes them useful for filling and wiping. An 80/20 Pb-Sn solder has a melting range of 170°F. This wide melting range allows one to work with it for an extended period of time. It can be used to fill dents in auto bodies.

The heat source for soldering is typically an iron, although torches, furnaces, induction coils, resistance, ultrasound, or hot dipping can be used to heat the joint.

Brazing

Brazing is related to soldering in that the substrate materials are not melted. The braze joint geometries are similar to soldering also. A metallurgical bond is formed between the two substrates via liquid enhanced diffusion. Intermetallic compounds may form between the braze and substrates.

Brazing may occur in several atmospheres including air, vacuum, and inert gas. The atmosphere used depends on the heat source and alloy. Heat sources can be torches, induction coils, furnaces, resistance heaters, etc.

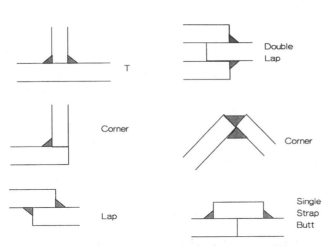

Figure 5. Typical solder joint geometries [36]. (*With permission, ASM International.*)

Table 14
Composition and Applications of Lead Tin Solders

| Composition | | Temperature (F) | | | |
Tin	Lead	Solidus	Liquidus	Melting Range	Uses
2	98	518	594	76	Side seams for can manufacturing.
5	95	518	594	76	Coating and joining metals.
10	90	514	570	56	
15	85	440	550	110	
20	80	361	531	170	Coating and joining metals, or filling dents and seams in automobile bodies.
25	75	361	511	150	Machine and torch soldering.
30	70	361	491	130	
35	65	361	477	116	General purpose and wiping solder.
40	60	361	460	99	Wiping solder for joining lead pipes and cable sheaths. For automobile radiator cores and heating units.
45	55	361	441	80	Automotive radiator cores and roofing seams.
50	50	361	421	60	General purpose.
60	40	361	374	13	Primarily used in electronic soldering applications where low soldering temperatures are required.
63	37	361	361	0	Lowest melting (eutectic) solder for electronic applications.

From MEI Metallurgy for the Non-Metallurgist, Lesson 9, *ASM International, 1987.*

Small steel assemblies can be furnace brazed. For furnace brazing of assemblies to be successful, the design of the parts must be such that the braze can be preplaced on the joint and remain in position during the braze cycle. A copper-based braze alloy is used because of the high strength of joint developed. The high brazing temperature (1,093° to 1,149°C or 2,000 to 2,100°F) necessary to melt the copper braze proves beneficial when the assembly needs to be heat treated after brazing. The operations entailed in furnace brazing are cleaning, brazing, and cooling.

Small steel assemblies, less than 5 pounds, are most efficiently brazed. Larger assemblies can be fabricated, but these may require specially designed and built furnaces.

Cleaning is typically accomplished by solvent cleaning, alkaline cleaning, or vapor degreasing. All alkaline compounds must be removed prior to placing assemblies in the brazing furnace. Adherent particles may be removed through mechanical means, such as wire brushing or light grinding.

Brazing of the components requires that they first be assembled and the braze applied. For multiple small articles, the components should be fitted, either through swaging or press fitting, such that no fixturing is required.

The articles are placed in the brazing furnace under an appropriate atmosphere to prevent oxidation and decarburization. When the assemblies reach a temperature higher than the melting range of the braze, the braze melts and flows into the joint via capillary action. Some diffusion occurs between the molten braze and the substrates, and the joint is formed. The heating cycle time is approximately 10–15 minutes, although longer times can be used to promote some diffusion and homogenization of the bond joint. Inadequate furnace heating can result in the braze melting but not flowing into the joint. This occurs because the assembly has not reached or exceeded the melting point of the braze. Increased superheat (temperature above the melting point) improves braze flow but may cause erosion.

Cooling of the assembly must be done under a protective atmosphere to prevent surface oxidation. The parts should be cooled to a temperature below about 150°C (300°F). The cooling typically occurs in a section of the furnace chamber that is not heated.

The furnaces used may be either batch or continuous. A batch type furnace requires an operator to place a tray of assemblies in the hot zone and move them to the cooling zone after the requisite braze cycle. In a continuous braze furnace, assemblies are placed on a chain link and the furnace pulls the assemblies through the hot zone and into the cooling zone of the furnace.

The atmosphere used can be either inert, protective, reducing, or carburizing. The selection of the gas atmosphere depends on the requirements of the parts and joint. If the atmosphere is incorrect, it can alter the surface chemistry of the parts and lead to rejected hardware, poor strength, or premature failure in service.

Steel assemblies can be torch brazed or induction brazed. In torch brazing, surface cleanliness is required, but because the protective atmosphere surrounding the flame is not always adequate, a flux may be necessary. Torch brazing can be fully manual, partially automated, or fully automated. The gases used are acetylene, natural gas, propane, and proprietary gas mixtures. Oxygen is principally used as a combustion agent because of its high heating rate. Lower grades of compressed oxygen, compressed air, or a blower can also be used to reduce costs.

Filler metals used in torch brazing are silver- or copper zinc-based. Silver alloys are used for steel-to-steel joints and most other metals except aluminum and magnesium. Copper zinc alloys can be used to join steels, and even nickel and cobalt alloys where corrosion resistance is not necessary. High temperature alloys like cobalt- and nickel-based superalloys can be brazed with Ni- or Co-based alloys also. The braze alloy selected is usually based on the base metal being brazed. The service temperature of the brazed assembly will generally be lower than the braze temperature. Diffusion heat treatments can be used to reduce the concentration of low melting point elements near the braze joint, which increases the braze remelt temperature, and possibly the service temperature.

The strength of a lap joint can be calculated using Equation 4:

$$x = \frac{ySw}{L} \tag{4}$$

where x is the length of the lap, y is the factor of safety, S is the tensile strength of the weaker member, w is the thickness of the weaker member, and L is the shear strength of the braze filler metal.

Induction heating with or without atmosphere can be used to make braze joints. The heat flux generated by an induction coil depends on the number of coils, distance between the coil and work piece, and geometry of the work piece.

Welding

Welding produces metallurgical bonds between the work pieces by melting them. The joints can be heterogenous if a filler metal is introduced or autogenous if none is introduced.

The need for filler metal is determined by the process that is used. There are several methods to introduce heat into the work pieces. Each process has its individual total heat input and concentration of heat input. Further, each process uses various methods to protect the molten metal and surrounding area from oxidation.

Joint geometry plays an important role in the ease of welding fabrication, generation of residual stress, and application. The typical joint geometries and weld types are shown in Figure 6. Joint preparation should include cleaning to remove any oils and cutting residue. Entrapped moisture can lead to hydrogen embrittlement also. The geometry of the joint should be designed so that there is easy access to the joint. The effect of residual stresses should be minimized. A poorly designed joint is shown in Figure 7.

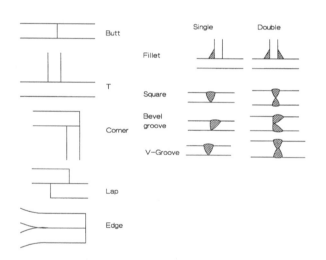

Figure 6. Typical joint geometries and weld types [36]. *(With permission, ASM International.)*

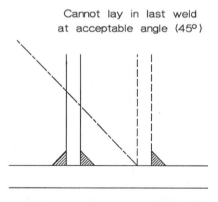

Figure 7. A poorly designed weld joint.

This joint design is poor because it does not allow welding of the second plate in an unobstructed manner or an appropriate angle.

The relationship of groove angle and root opening is shown in Figure 8. It is important to note that the root opening decreases with increasing bevel angle. The change in width is required for electrode access into the base of the joint. The selection for joint design depends on the base plate thickness and the amount of filler required to manufacture the joint.

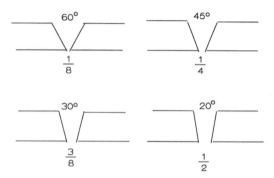

Figure 8. The relationship between groove angle and root opening for butt welds [36]. *(With permission, ASM International.)*

A number of processes are available for welding. The method selected depends on the joint requirements, material, and costs. Table 15 lists acronyms of the American Welding Society and uses for common engineering alloy classes.

Table 15
Common Welding Names and Applications

	Carbon Steel	Low-alloy Steel	Stainless Steel	Cast Iron	Nickel Alloys	Aluminum Alloys	Titanium Alloys	Copper Alloys
SMAW	All	All	All	All	All	*	*	*
GTAW	<1/4″	<1/4″	<1/4″	*	<1/4″	<3/4″	<3/4″	<1/4″
GMAW	>1/8″	>1/8″	<1/8″	1/8–3/4	All	<3/4″	<3/4″	<3/4″
EBW	All	All	All	*	All	All	All	All
LBW	<3/4″	<3/4″	<3/4″	*	<3/4″	<3/4″	<3/4″	*

*Process not applicable.
Adapted from ASM Metals Handbook, Vol. 6, 9th Ed. [36].

A brief description of the type of weldments made with the more common methods follows. Shielded metal arc welding (SMAW) is a portable and flexible welding method. It works well in all positions and can be done outside or inside. It is typically a manual process and is not continuous, as it relies on consumable electrodes that are from 12 to 18 inches long. The electrodes have a surface layer of flux on them which melts as the electrode is consumed and forms a slag over the weld. The slag protects the joint from oxidation and contamination while it solidifies and cools.

Gas metal arc welding (GMAW), also referred to as metal inert gas welding (MIG), is a continuous process that relies on filler wire fed through the torch. It can be used on aluminum, magnesium, steel, and stainless steel. A shielding gas, either argon, helium, or even carbon dioxide mixed with an inert gas, is used to protect the joint and heat-affected zone (HAZ) during the welding process.

Flux cored arc welding (FCAW) uses a continuous wire which has flux inside the wire. The electrode melts, fills the gap, and a slag is generated on the surface of the weld to protect it from oxidation. It is usually used only on steels. For additional protection around the weld, an auxilary shielding gas can be used.

Gas tungsten arc welding (GTAW), also referred to as tungsten inert gas (TIG), is a process that can be automated. It can be either autogenous and heterogenous, depending on whether filler wire is introduced; the tungsten electrode does not melt to fill the joint. An inert gas, typically argon, helium, or, more recently, carbon dioxide, is used to protect the joint from oxidation during welding. The filler wire selected for the joint should match the base metals of the joint materials. In some cases, the joint metal may be a different composition. In stainless steels, type 308 filler wire is used for 304 and 316 joints. A large number of metal alloys can be welded with GTAW, including carbon and alloy steels, stainless steels, heat-resistant alloys, refractory metals, titanium alloys, copper alloys, and nickel alloys. The nominal thickness that is easily welded is between 0.005 and 0.25 inch.

GMAW, SMAW, FCAW, and GTAW are all moderate rate of heat input techniques. This means that there is about a one-to-one ratio of weld penetration to weld width.

Figure 9 shows a typical weld depth to width of one, in addition to multiple passes which can be made on thick plates for weld metal build-up.

LASER (LBW) welding uses a concentrated coherent light source as a power supply. It provides unique characteristics of weld joints and can be used to weld foil (0.001 inch) as well as thick sections (1 inch). It is a high rate of input with deep penetration and narrow welds, shown schematically in Figure 9.

Another high rate of heat input is electron beam welding (EBW). It has a broad range of applicability and can weld thin foil, 0.001 inch thick, as well as plates, up to 9 inches thick. It has drawbacks in that it requires a high vacuum for the electron beam heat source, but these can be overcome for continuous welding uses. The beam will melt and vaporize the work piece. Metal is deposited aft of the beam, and a full penetration weld is made.

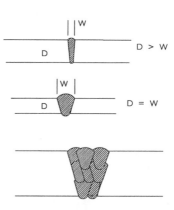

Figure 9. Weld depth to width greater than 1, typical of LASER or EBW (top). Weld depth to width approximately equal to 1, typical of SMAW, GTAW (middle). Multiple passes made on thick plate, typical of multipass GTAW (bottom).

Coatings

Coatings can be used for decoration or to impart preferential surface characteristics to the substrates. Coatings can be an effective and efficient method to treat the surface of a component to provide surface protection, while the substrate provides the mechanical and physical properties.

High temperature coatings can be applied by a number of methods. The approach used depends on the type of coating and the application. There are basically two types of coatings: overlay and diffusion. Overlay coatings are generally applied to the surface of the part and increase the over-all dimension of the part by the coating thickness. Methods to apply overlay coatings include chemical vapor deposition (CVD), physical vapor deposition (PVD), thermal spray deposition (TS), plasma spray deposition (PS), and high velocity oxygen fuel deposition (HVOF). Of the above listed methods, only CVD can coat in a non-line of sight. The others require that the coating area be visible. This limitation can pose problems for parts of complex geometry. Diffusion coatings may or may not be line of sight limited. There are several methods to apply diffusion coatings, the most common being the pack method although the use of CVD is growing. Table 16 compares diffusion and overlay coatings.

Table 16
Comparison Between Diffusion and Overlay Coatings

Diffusion Coatings	Overlay Coatings
Metallurgically reacts with base metal.	Thin metallurgical interaction. All thickness is added.
May detrimentally affect properties, especially fatigue.	Little effect on mechanical properties, although they increase cross-section without increasing load capabilities.
Approximately 50% of thickness is added.	
Internal coatings possible by some methods.	Line of sight limitations make complex geometries difficult to coat.
Limits on compositions.	Internal coatings not possible. Compositions by PVD, TS, P_s, HVOF nearly unlimited.

Adapted from Aurrecoechea [35]. Used with permission, Solar Turbines, Inc.

The compositions that can be applied by thermal spray processes and PVD are very wide. The chemistry depends on the application. CVD coating chemistries are limited by the type of precursor and the required chemical reaction to form the coating composition.

MCrAlY coatings are one family of coatings that can be used for hot corrosion and high temperature oxidation protection. The "M" stands for Fe, Ni, Co, or a combination of Ni and Co. Each element in the coating is present for a specific purpose. Typical coatings contain 6%–12% Al, 16%–25% Cr, and .3%–1% Y, balance Ni, Co, Fe, or Ni and Co. Table 17 lists the specific elements and the influence on the coating.

Aluminum provides high temperature oxidation resistance. It needs to be present in a sufficiently high concentration to be able to diffuse to the surface and react with the inward migrating oxygen. The activity and diffusivity of the Al is proportional to the concentration of Al. The overall oxidation rate and coating life is affected by the Al concentration. Excessive Al content can cause coating embrittleness which can lead to cracking and spalling of the coating.

Chromium is added to impart corrosion resistance; it also increases the activity of Al. This allows continuous aluminum oxide scales to form at lower Al concentrations than normally expected. The protection of the coating thus relies on the synergistic effect of Al and Cr additions.

Table 17
Overlay Coating and the Effect of Individual Elements on Coating Behavior

Coating Element	Amount	Effect
M (Fe, Ni, Co, Ni + Co)	Balance	Fe—Best oxidation and hot corrosion resistance, low temperature limitations Ni—Excellent high temperature oxidation resistance Co—Best hot corrosion resistance, not as good in high temperature oxidation Ni + Co—Best balance between oxidation and hot corrosion, mixed environments
Cr	16–25	Mainly hot corrosion resistance, synergistic effect with Al for oxidation resistance
Al	6–12	Oxidation resistance, although excess additions cause embrittlement
Y, Hf	0.3–1.0	Improved oxide adhesion by tying up S in alloy
Si, Pt, Pd		Hot corrosion
Ta, Pt, Pd		Oxidation

Adapted from Amdry Product Bulletin 961, 970, 995 [14].

Yttrium is added to improve the oxide adherence. Generally, oxides spall on coatings without reactive element additions. The method of improved adhesion is not fully understood, but experimentation has shown that the majority of the benefit is derived by tying up the sulfur inherently present in the alloys. Sulfur acts to poison the bond between the oxide scale and the coating. Yttrium or other reactive elements, Zr or Hf, also may promote the formation of oxide pegs which help mechanically key the oxide layer to the coating.

More advanced coatings may contain Hf and Si which act like Y to improve adherence. Hafnium, which acts chemically similar to yttrium, may be used in place of Y. Additions of Si can be used to improve the hot corrosion resistance of the coatings. Tantalum is sometimes added to improve both oxidation and corrosion resistance. Noble metals like platinum and palladium can be used similarly to chromium to improve both oxidation and corrosion resistance.

The major alloy element(s) affect the coatings in different ways. Iron (Fe) based MCrAlY coatings have superior oxidation resistance to the other types of coatings. They also tend to interact with the base metal and diffuse inward. Thus, they are limited in temperature to approximately 1,200–1,400°F. FeCrAlY coatings are suitable for high sulfur applications.

Cobalt-based alloys have superior hot corrosion resistance to NiCrAlY coatings due to the presence of cobalt which helps modify the thermochemistry of molten Na_2SO_4. This modification alters the oxygen and sulfur activities and lowers the rate of attack. CoCrAlY's do not have as good of oxidation resistance as NiCrAlY's, and have a temperature limit of 1,600°–1,700°F.

Nickel-based alloys are used for high temperature oxidation applications, such as aircraft coatings. They have outstanding oxidation and diffusional stability and can be used to temperatures of 1,800°F.

A mixture of both Ni and Co can be used for applications requiring a balance between hot corrosion and oxidation properties. These alloys are called NiCoCrAlY or CoNiCrAlY, the designation depending on which element is more prevalent.

Overlay coatings can be very thick, up to 0.1 inch, and can be used to refurbish mismachined parts. Further, ceramic coatings can be applied to provide reduced heat transfer and ultimately lower metal temperatures. Electron beam physical vapor deposition can be used to form vertically cracked thermal barrier coatings. The vertical cracks provide increased strain tolerance, as shown in Figure 10. This type of structure is possible to create with other methods, such as plasma spray, by careful control of the processing parameters.

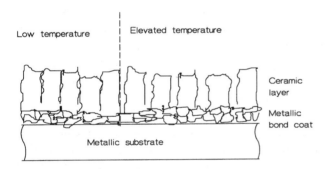

Figure 10. Schematic of a vertically cracked ceramic showing improved strain tolerance.

Diffusion coatings are less complex in terms of initial chemistry. Aluminum or chromium is diffused into a nickel- or cobalt-based alloy. This treatment results in an aluminum- or chromium-rich surface layer that adds some thickness to the part and also diffuses inward. The aluminum or chromium is applied to the surface either through some gas phase process or solid state diffusion process. A typical aluminide coating on a nickel substrate is shown schematically in Figure 11. The coating contains all of the elements present in the substrate. There is typically a "fin-

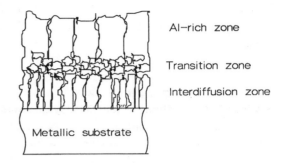

Figure 11. Schematic of a typical aluminide coating on a nickel substrate

ger" diffusion zone between the aluminum-rich surface and the substrate. The process requires a fairly high temperature which is generally determined by the heat treatment required for the base metal. Diffusion coatings are significantly less ductile than overlay coatings but provide excellent oxidation and moderate hot corrosion resistance depending on the coating composition. As is the case for overlay coatings, the amount of protection afforded to the substrate is determined by the aluminum and/or chromium content. The thickness of diffusion coatings is on the order of 0.001–0.003 inch.

Low temperature coatings consist of a number of possible metallic, inorganic, or organic compounds. Some possible metallic coatings are zinc plate, either galvanic or hot dipped, aluminum cladding, either by roll bonding or thermal spray, cadmium plating, or nickel plate. Zinc, aluminum, and cadmium coatings are useful for aqueous corrosion protection where moisture is in contact with the parts. Nickel plate can be used as a moderate temperature oxidation-resistant coating.

The coating selected needs to fulfill the application requirements. These include many of the same considerations as the substrate, e.g., temperature, active species, effect of failure, etc. An important consideration these days is the effect of coating-related processing on the environment. Plating processes which rely heavily on hazardous solutions are being phased out. Alternative methods of applying the decorative plate to materials are being sought. Thermal spray processing is one possible coating method. This has its own environmental hazards such as dust collection and cleaning solutions.

Organic coatings include substances like epoxies and paints. These can be used to protect materials from corrosion problems and as decoration for low temperature applications.

Corrosion

Corrosion is a material degradation problem that can cause immediate and long-term failures of otherwise properly designed components. Corrosion occurs because of oxidation and reduction of species in contact with one another. In an aqueous solution, the common reactants are hydrogen, oxygen, hydroxyl ions, a metal, and the metal's ions.

There are at least eight forms of corrosion attack. These are erosion-corrosion, stress corrosion, uniform, galvanic, crevice, pitting, intergranular attack, and selective leaching. Many of these types are based on the appearance of failed components.

Two electrochemical series can be used to determine the likelihood of corrosion. One is the electromotive force series (EMF) and the other is the galvanic series. The EMF series is based on a reversible equilibrium half-cell reaction from which the Gibb's Free Energy for reactions can be determined, while the galvanic series establishes the tendency for nonequilibrium reactions to occur in various solutions. Different galvanic series may be determined for seawater, fresh water, and industrial atmospheres (See Table 18).

The galvanic series indicates the tendency for corrosion to occur on two metals that are joined together either metallurgically (soldered, brazed, or welded) or mechanically (bolted, riveted, or adjacent) and is material combination and solution specific. In either case—EMF or galvanic—the tendency for corrosion is greater the wider the separation between the two materials. For instance, if one were to connect titanium and Inconel alloy 625, there would be little driving force for the reaction to occur. On the other hand, if one were to connect titanium and an aluminum alloy, there would be a large driving force for the reaction to occur. The EMF and galvanic series only address the likelihood of reaction and do not indicate the rate at which corrosion may proceed.

Erosion-Corrosion occurs when a corrosive medium is present that is in relative motion. It can cause rapid failure of a materials combination that had excellent life in laboratory testing but was tested under static conditions. Many metals rely on the formation of a protective film to decrease the corrosion rate. However, when the metals are placed in a solution that has relative motion—such as a pump impeller or pipe—the protective film may be removed by the fluid and accelerated corrosion can occur. Both the environment and the fluid velocity affect the corrosion rate, and

Table 18
Galvanic Series in Seawater at 77°F

Protected end (cathodic, or noble)
Platinum
Gold
Graphite
Titanium
Silver
HASTELLOY® alloy C
INCONEL® alloy 625
INCONEL® alloy 825
Type 316 stainless steel (passive)
Type 304 stainless steel (passive)
Type 410 stainless steel (passive)
Monel alloy 400
INCONEL® alloy 600 (passive)
Nickel 200 (passive)
Copper alloy C71500 (Cu 30% Ni)
Copper alloy C23000 (red brass 85% Cu)
Copper alloy C27000 (yellow brass 65% Cu)
HASTELLOY® alloy B
INCONEL® alloy 600 (active)
Nickel 200 (active)
Copper alloys C46400, C46500, C46600, C46700 (naval brass)
Tin
Lead
Type 316 stainless steel (active)
Type 304 stainless steel (active)
50-50 lead tin solder
Type 410 stainless steel (active)
Cast iron
Wrought iron
Low carbon steel
Aluminum alloys 2117, 2017, and 2024 in order
Cadmium
Aluminum alloys 5052, 3004, 3003, 1100, 6053 in order
Galvanized steel
Zinc
Magnesium alloys
Magnesium

Adapted from ASM Metals Handbook, Vol. 13, 9th Ed. [41].

there may be more than one minimum corrosion rate for a material combination. The following are general guidelines:

1. Any solution or additive that removes the protective scale increases erosion corrosion.
2. Increasing velocity generally increases corrosion rate.
3. High impingement angles increase erosion-corrosion (especially important at pipe elbows).
4. Soft metals are more susceptible to erosion corrosion since they are less resistant to mechanical wear.

Stress-Corrosion is a complicated interaction of both a specific corrosive environment and a tensile stress. The stresses may be external (due to applied loads) or residual (for example, differential cooling after welding). Unstressed materials may be relatively inert (non-reactive) in the specific environment but may crack catastrophically when a critical load is applied. The material combination is very important; for instance, brasses tend to crack in ammonia-containing solutions but not in chloride solutions. On the other hand, stainless steels tend to crack in chloride solutions but not in ammonia or many acids (nitric, sulfuric, acetic, or water). There is no one, particular crack morphology that occurs, although the crack is usually perpendicular to the stress axis. Cracks can be intergranular (between the grains), transgranular (across the grains), or single- or multi-branched.

The stress corrosion cracking phenomenon depends on the solution chemistry, temperature, metal composition and microstructure, and stress. In a given environment, the stress required to cause cracking may be as low as 10% of the yield strength or as high as 70%. Above the critical load for cracking, increasing the stress decreases the time required for cracking. Increasing the temperature decreases the time to cracking for 316 and 347 type stainless steels in sodium chloride (salt) solutions.

High strength, low alloy steels are more likely to exhibit stress corrosion cracking than low or medium strength steels. The presence of cold work increases the likelihood of stress corrosion cracking. The following stainless steels are ranked in increasing order of resistance to stress corrosion cracking: Type 304, types 316 and 347, and types 310 and 314 [10].

Tables describing the materials resistance to stress corrosion cracking can be found in Fontana [10].

Uniform corrosion occurs all over a surface or structure. A common example is unpainted steel, which turns rusty over the entire exposed surfaces. There are several methods to prevent uniform corrosion, including proper materials selection for the structure and, if necessary, coatings, addition of inhibitors (ionic species which preferentially oxidize; chromates in antifreeze) to the solution to reduce the corrosion rate, and cathodic protection.

Galvanic corrosion occurs when two or more dissimilar metals are coupled in a corrosive media. A potential difference is generated, the magnitude of which depends on the separation between the two metals on the galvanic series, shown in Table 18, or electromotive series. The larger the potential difference between the two metals, the higher the driving force for the reaction to occur; but the

overall rate is determined by local kinetics. Based on the separation argument, it would be acceptable to join Monel 400 and Nickel 200 (passive) in a salt water environment. On the other hand, it likely would be detrimental to join Monel 400 and low carbon steel, since they are widely separated.

The relative position of the two metals on the series dictates which metal will be oxidized (corroded) and which will be reduced. The lower the metal in the series, the more active and the more likely it will be oxidized (anode). The higher the position, the more noble and the more likely it will be reduced (cathode).

The galvanic cell can become polarized, having a lower concentration of ions being reduced, at the cathode. Polarization will decrease the corrosion rate of the anode since the overall solution must remain charge balanced.

It is difficult to predict whether or not galvanic corrosion will be a problem, so it is best to test the metal combinations in situations that simulate service. The following are general guidelines for preventing galvanic corrosion

Design systems that are susceptible to galvanic corrosion have a large anode and small cathode area ratio. Figure 12 shows favorable and unfavorable designs for dissimilar metals that are galvanically coupled. This design reduces the current density and limits the penetration. Fully isolate the materials using insulation appropriate for the solution and temperature. Apply coatings, but use with caution. If painting or coating is necessary, paint the cathode. In the event of a coating rupture, the favorable cathode-to-anode ratio is maintained. If a coating rupture occurs on a painted anode (component likely to corrode), then there is a large cathode and very small anode. This situation will lead to rapid attack of the anode, which can lead to penetration and subsequent failure. Avoid threaded fasteners for metals

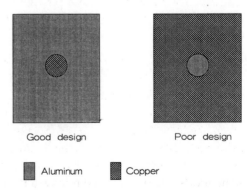

Good design   Poor design

▨ Aluminum   ▨ Copper

Figure 12. Favorable and unfavorable designs for dissimilar metals that are galvanically coupled.

which are far apart on the galvanic series, braze with a more noble alloy, or weld with the same metal. Design to replace anodic material. Introduce a third alloy that is more anodic than either of the component alloys as a replaceable anode for protection (sacrificial anode). Impress a current on the system so that no metal dissolution occurs.

Crevice corrosion is a caused by stagnant solutions in surface defects such as scratches, holes, gasket materials, lap joints, surface deposits, and beneath rivet and bolt heads. A crack or gap wide enough to alloy liquid penetration and stagnation (a few thousandths) but not provide for complete solution change creates an ideal environment for crevice corrosion. Fibrous gaskets promote wick action and stagnant solutions and, therefore, crevice corrosion. Stainless steels are very susceptible to crevice corrosion.

Crevice corrosion can be prevented by avoiding bolted construction in new designs. It is better to weld or braze so that no cracks are formed in the system. Welded construction should have full penetration welds and no porosity. The system should also be designed so that there are no stagnation points and the system can be fully drained. For existing systems, close the gaps using welds, brazes, or caulks. Inspect systems often to ensure that there are no deposits, and if there are, remove them. Remove any wet packing material. Use solid gaskets (such as teflon) rather than porous ones.

Pitting corrosion operates in a mechanism that is similar to crevice corrosion, but this insidious form of attack promotes the formation of its own crevice. It may also be influenced by gravity. It is difficult to inspect because corrosion products may form over the pits and make them difficult to detect. Further, the pits may not go straight through the wall; they may have a "worm hole" appearance.

Pitting corrosion can be prevented by appropriate selection of materials. In increasing order of pitting resistance, there are: 304 SS; 316 SS; HASTELLOY® F alloy, Duramet 20; HASTELLOY® C, Chloromet 3; and titanium. Proper heat treating can be used as a preventative measure since sensitized stainless steel is more susceptible to pitting than nonsensitized. Quenching, rapidly cooling, austenitic steels from above 1,800°F will help prevent sensitization. Polishing the surfaces of the parts makes them less likely to initiate pits, although once initiated, the pits tend to grow faster.

Intergranular corrosion is a preferential attack at the grain boundaries. It occurs for a number of reasons; for instance, grain boundaries are more active, grain boundaries can be enriched with either impurities or alloying additions,

or grain boundaries can be depleted in corrosion-resistant alloying additions.

Intergranular corrosion can be a problem in austenitic stainless steels with high carbon content (0.06%–0.08% C). Chromium is required at a minimum of 12% to ensure corrosion resistance; however, chromium carbides may form if the cooling rate through the sensitization range is not sufficiently rapid. A schematic of a sensitized austenitic stainless steel is shown in Figure 4. The Cr content around the grain boundaries is below the required minimum, and this variation leads to attack along the grain boundaries.

A related phenomenon is *weld decay,* which is accelerated corrosion at a distance away from the weld. It occurs because carbides precipitate in the heat-affected zone (since this area is heated to the susceptibility range). If welding is necessary, a moderate to high rate of heat input process should be selected.

Susceptibility of intergranular attack can be controlled by solutioning and water quenching parts through the susceptibility range. Stabilizers such as niobium or titanium can be used in the alloys (types 347 and 321 SS). These promote the formation of more stable carbides than chromium carbide. Ultra low carbon alloys can be used (<0.03% C). Castings should be poured into low carbon molds.

Welded stabilized alloys may exhibit a failure mode similar to weld decay. The rate of cooling may be too rapid at the weld/work piece interface for the high melting point

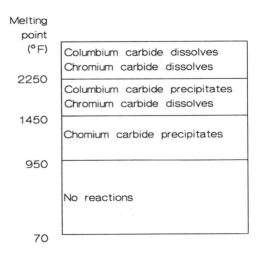

Figure 13. Schematic of the thermal cycle that occurs for a welded stabilized stainless steel [10]. (*Reprinted by permission of McGraw Hill, Inc.*)

carbides to reform on cooling, and only chromium carbides can precipitate. Figure 13 shows a schematic of what occurs as a stabilized alloy is welded. As the Cr and Cr-carbides are solutioned, it is possible for localized areas near the weld joint to fall below the critical Cr content required for corrosion resistance.

High strength aluminum alloys can suffer from a form of intergranular attack. Typically they are strengthened with copper. A potential difference (voltage) is established between the Cu_2Al precipitates and the surrounding alloy. A localized galvanic reaction occurs which results in what appears to be intergranular attack.

Selective leaching is a generic term that describes the removal of one constituent of the alloy by a corrosive process. A process termed *dezincification* is the most commonly observed. Dezincification occurs when brass, an alloy of zinc and copper, is dissolved and copper is plated back on the remaining structure. Dezincification results in a change in color of the parts and is visible to the naked eye.

Dezincification can be prevented by reducing the severity of the environment, reducing the zinc content of the alloy, or changing from brass to a cupronickel (copper-nickel) alloy.

Gray cast iron is also susceptible to selective leaching. In this case, the iron surrounding the graphite flakes is dissolved, leaving a spongy mass of rusted iron. The part appears to be surface rusted, but all of the metal may actually be dissolved, leaving a structure without integrity. This process occurs over a long period of time and can be prevented by using nodular (ductile) cast iron rather than gray cast iron.

Powder Metallurgy

Powder metallurgy is a material shaping process that starts with finely divided metal (ceramic powders can also be processed by typical powder metallurgical techniques), uses a compaction step to shape the powder, and applies heat to make a strong finished product. Nearly all types of metal alloys can be formed by powder metallurgy. In some instances, tungsten for example, it is the only method to produce a useful shape. This is because it is difficult if not impossible to melt and contain tungsten. If powder metallurgy is to be used, features of the final part must be incorporated early in the design stages to produce the parts at the lowest possible cost.

Undercuts and negative drafts cannot be pressed into a part using uniaxial compression. Feather edges should be avoided by incorporating a small step at the edge of the bevel. Central cut-outs should be rounded. Some of the salient features of designing powder metallurgy parts are illustrated in Figure 14.

There are basically three steps to producing powder metallurgy parts. Elemental or prealloyed powders are mixed with die lubricants or other additives. Blending is also performed when powders of the same composition from different lots or vendors are combined.

The mixed powders are then compacted into a green (unsintered) form. The compaction pressure can be applied uniaxially, the most common for simple two-dimensional shapes, or isostatically, a common method to produce complex shaped parts. The pressures used are as low as 10 tons per square inch or as high as 60 tons per square inch.

The pressing cycle includes a die filling, compaction, and ejection. The cycle is performed repeatedly until the number of parts required are processed.

Sintering is the final step. This process takes the green compacts and fires them at a high temperature to promote diffusion (atomic migration). The atoms in the powders move from one particle to another and form a metallurgical bond. The parts are generally not heated above the melting point of the powders, although some alloys are liquid phase sintered, like tungsten carbide cobalt for tool inserts. The liquid phase promotes densification and full density.

Powder metallurgy parts can be used after sintering, but sometimes additional steps such as coining, repressing, impregnating, machining, tumbling, plating, or heat treating are used to produce a completed part.

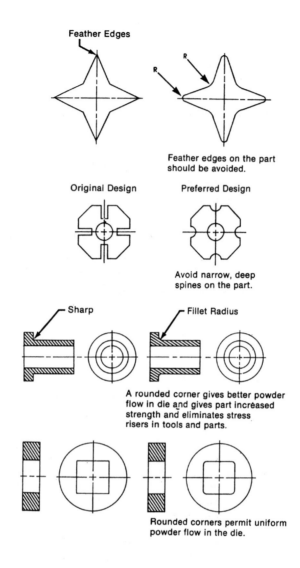

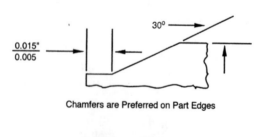

Figure 14. Powder metallurgy design considerations for successful implementation [13]. (*Reprinted by permission of Metal Powder Industries Federation.*)

POLYMERS

Use of polymers (plastics) has grown tremendously in recent years. To properly use these materials to replace metallic structures or to use them in new applications requires that certain information be considered prior to designing parts for the application.

To properly design and implement polymer parts use, the required properties that are of engineering importance must be defined. Often a list, such as that in Table 19, is generated and the material selection proceeds by process of elimination. Many polymeric materials can be eliminated from the list simply because they do not meet key/critical property requirements.

The material selection process must include specific recommendations, and one cannot simply request polypropylene or polyvinyl chloride as these terms are as nonspecific as requesting steel or an aluminum alloy. Various additives are made to the melt or powdered mixture to alter the key property requirements, such as flammability, plasticity, impact resistance, and toughness.

Fundamentally, there are two types of polymers: thermoplastics and thermosets. Thermoplastic polymers soften when they are heated and can be formed in this semiliquid state. Further, the process is reversible and it can be repeated a number of times. Thermoset polymers cross-link and change structure when heated. This process is a chemical reaction that is not reversible and results in a polymer that is permanently set once cooled.

Polymers can be either crystalline or amorphous. The mechanical properties resulting from either structure are different. Polymers need not be fully crystalline or amorphous. They can be fractionally crystalline. The polymer composition and process determines the degree of crystallinity. The degree of crystallinity can be determined using standard thermal techniques such as differential scanning calorimetry and x-ray diffraction.

Polymers are made of long organic molecules that can be combined to produce other polymers with mixed properties. Figure 15 shows how two polymers, A and B, can be combined to give properties that are mixed between homopolymer A and homopolymer B.

The mechanical response of polymers is viscoelastic; they can behave elastically or viscously or have a mix of both.

When a polymer behaves elastically it responds like a spring; it extends, instantaneously, as far as the load requires and then stops. The strain is recovered when the load is removed. The behavior is as indicated by Equation 5.

$$\gamma_e = \frac{\tau}{G} \tag{5}$$

where γ_e is the elastic shear strain, τ is the applied shear stress, and G is the elastic shear modulus.

If it is responding viscously, in a time dependent manner, then it elongates to failure. It can be modeled after a dashpot and responds as indicated in Equation 6. The strains introduced are not recoverable when the load is removed.

$$\gamma_v = \frac{\tau t}{\eta} \tag{6}$$

where γ_v is the viscous shear strain, t is time, and η is viscosity.

Table 19
Factors That Influence the Selection of Polymers for Engineering Applications

Mechanical	Service stress—type and magnitude
	Loading pattern—frequency and duration
	Fatigue resistance
	Strain limitations
	Impact resistance
Environmental	Temperature—range, maximum and minimum limits
	Corrosiveness—solvent, vapor, acid, & base attack
	Oxidation
	Exposure to sunlight
Electrical	Resistivity
	Antistatic properties
	Dielectric loss
Hazards	Toxicity—additives or degradation products
	Flammability
Appearance	Color
	Surface finish
	Long-term appearance
General weight	Space limitation
	Tolerances
	Life expectancy
	Permeability
Manufacturing	Process of choice
	Method of assembly
	Finishing requirements
	Quality control and inspection
Economics	Material cost
	Process cost—maintenance and fuel
	Capital costs—molds and equipment

Adapted from McCrum, et al. [11].

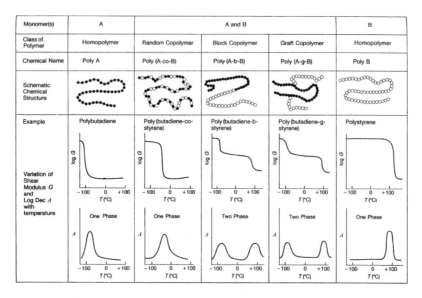

Figure 15. Effect of mixing (alloying) two homopolymers on thermal and mechanical properties [11]. (*Reprinted by permission of Oxford University Press.*)

If it has a combination of both properties, then it may be like a spring and dashpot in series, described by Equation 7. Systems like these will immediately recover the elastic portion of the deformation but will not recover the viscous portion.

$$\gamma = \tau \left(\frac{1}{G} + \frac{t}{\eta} \right) \qquad (7)$$

A more complex situation occurs if the viscous and elastic elements act in parallel—termed *anelastic* behavior. When this situation arises, the elongation behavior is time-dependent but relaxation is reversible. A model of this behavior is shown in Figure 16 and mathematically in Equation 8.

$$\gamma = \frac{\tau}{G} (1 - e^{-t/\gamma_v}) \qquad (8)$$

If it is viscoelastic, then all three elements can be incorporated. This type of behavior is illustrated in Figure 17 and Equation 9, which is the summation of the shears indicated in the equations above.

$$\gamma = \frac{\tau}{G_1} + \left(\frac{\tau}{G_2} \right)(1 - e^{-tG_2/\eta_2}) + \frac{t\tau}{\eta_3} \qquad (9)$$

Typical polymer names and applications are listed in Table 20.

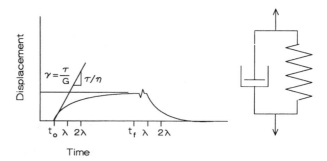

Figure 16. Behavior of a polymer with time-dependent elongation. This model element is also known as the Voight model.

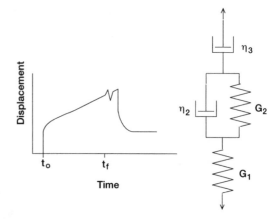

Figure 17. Viscoelastic material response that contains instantaneous, nonrecoverable time-dependent, and recoverable time-dependent elongation behaviors.

Table 20
Names and Applications of Polymers

Type	Name	Typical applications
Thermoset		
Phenolics "Bakelite"[a]	Electrical equipment	
Polyurethane	Sheet, tubing, foam, elastomers, fibers	
Amino resins (urea formaldehyde)	Dishes	
Polyesters	Fiberglass composites, coatings	
Epoxies	Adhesives, fiberglass composites, coatings	
Elastomers		
Butadiene/styrene	Tires, moldings	
Isoprene (natural rubber)	Tires, bearings, gaskets	
Chloroprene (Neoprene)	Structural bearings, fire resistant foam, power transmission belts	
Isobutene/Isoprene	Tires	
Silicones	Gaskets, adhesives	
Vinylidene fluoride/ hexafluoropropylene (Viton[b])	Seals, O-rings, gloves	
Thermoplastics		
Polyethylene	Clear sheet, bottles	
Polyvinyl chloride	Floors, fabrics, films	
Polypropylene	Sheet, pipe, coverings	
Polystyrene	Containers, foam	
ABS[c]	Luggage, telephones	
Acrylics[d] (Lucite[b])	Windows	
Cellulosics	Fibers, films, coatings, explosives	
Polyester, thermo-plastic type (Dacron[b], Mylar[b])	Magnetic tape, fibers, films	
Nylons	Fabrics, rope, gears, machine parts	
Polycarbonates (Lexan[e])	Machine parts, propellers	
Acetals	Hardware, gears	
Fluoroplastics (Teflon[b])	Chemical ware, seals, bearings, gaskets	

[a]*Trade name of Union Carbide.*
[b]*Trade name of Du Pont.*
[c]*Acrylonitrile-Butadiene-Styrene.*
[d]*Example: polymethyl methacrylate.*
[e]*Trade name of General Electric.*

Adapted from Harper [37], by permission of McGraw-Hill, Inc.

CERAMICS

Ceramics are generally brittle materials that have excellent compressive strength. The tensile strength is dictated by the presence of surface flaws (griffith cracks). The surface flaws are generated by the fabrication methods and subsequent machining operations.

There are two major types of ceramics: traditional and advanced. Traditional ceramics include refractory bricks, tableware, earthenware, whiteware, and common glass products. The advanced ceramics consist of both oxides and nonoxides. These ceramics may be used as structural ceramics and other technically advanced products. Some advanced ceramics and applications are listed in Table 21.

Ceramics can be formed by slip casting and powder metallurgical processes. Complex parts, like automotive turbochargers, can be formed by injection molding. Other methods include hot pressing and simple press and sinter operations. Often, the densification of the part is enhanced by adding lower melting compounds to the mixture prior to forming it. Yttrium oxide is a common addition to silicon nitride and silicon carbide to enhance the diffusivity.

Ceramics and glasses can be joined to metals, each other, and glasses. The joining technologies are just emerging for the most advanced materials. There are many physical

Table 21
Common Advanced Ceramics and Applications

Silicon nitride and silicon carbide	Structural applications, high temperature strength, oxidation resistance, automotive turbochargers
Boron carbide	Cutting and grinding wheels when composited with a metal
Complex oxides	Depends on the electronic properties required, also advanced high temperature superconducting oxides.
Glass ceramics	Forming seals between glasses and metals, cookware
Alumina	Refractory plates, labware
Mullite	Furnace tubes

From Engineered Materials Handbook, Volume 4 *[15]*.

properties that must be accounted for in selecting a joint combination. The two most significant are chemical compatibility and thermal expansion compatibility.

Ceramics can also be joined in the green state. This can greatly simplify the materials required to form the structure.

MECHANICAL TESTING

Tensile Testing

The tensile test is one of the simplest tests to perform. The gage length (uniformly reduced section) of the sample depends on the design requirements. Standard sizes are 0.5, 1, and 2 inches. The gage section length has the highest stress (load/area) because it is the smallest diameter. Tests can be performed at almost any temperature in almost any environment provided a suitable chamber is designed and built for testing. It is common to test at both room temperature and elevated temperature. Air is the most common environment to test in, although vacuum, argon, other inert

gases, hydrogen, or gas mixtures are all possible. In addition, one can test cathodically charged samples to ascertain the effect of hydrogen charging on the ductility.

The cross-section can have nearly any shape, although circular and rectangular are the most common. The cross-sectional area has some influence on the strength properties, with thinner samples having higher indicated properties. The shape is often dictated by the form of the material; flat plate will generally be tested as flats or sheets while bar stock may be tested in the round. Further, plate stock may be tested parallel to the rolling direction, long transverse, or short transverse. The properties will vary greatly with the orientation of the stress axis with respect to the rolling direction. Forgings are also grain orientation dependent, and more than one direction is tested to characterize the properties.

Strain is a normalized parameter which defines the unit length extension per length (in/in). There are two types of strain: elastic and plastic. Elastic strain is recoverable when the load (force, stress) is removed. Plastic strain is nonrecoverable. A typical stress-strain curve is shown in Figure 2. This figure also shows other important material characteristics.

Fatigue Testing

Fatigue testing is done to evaluate a materials response to cyclic loading. When a load less than the yield strength of the material is cycled, a sample is undergoing high-cycle fatigue or load-controlled fatigue. If a sample is plastically yielding, then the testing is termed low-cycle fatigue. Figure 18 shows areas that can be delineated for fatigue testing.

The limit between high-cycle and low-cycle fatigue can also be drawn based on the number of cycles sustained prior to failure—typically fewer than 10^4 cycles. The number of cycles to failure can be plotted against the applied maximum stress, mean stress, or stress range. A brief description of several test methods will enlighten the reader as to the significance of these terms. Fatigue tests are performed by applying a maximum and minimum stress, σ_{min} and σ_{max}. The σ_{min} is defined as the algebraically lowest stress and σ_{max} as the largest stress. Either of these stresses can be positive or negative. They can also be different magnitudes. The periodic loads can be fully reversed loading $\sigma_{min} = -\sigma_{max}$ for a mean stress ($[\sigma_{min} + \sigma_{max}]/2$) of zero (0), σ_{min} and σ_{max} both positive, and $\sigma_{min} < 0$ and $\sigma_{max} > 0$ but not equal. The loading does not have to be sinusoidal, it can be random, saw tooth, or almost any wave shape that can be created on waveform generator. The simplest waveforms to test are sinusoidal and sawtooth.

Fatigue failures are characterized by a flat, nearly featureless surface with "beachmarks" where the crack periodically stopped. A transition zone between the fatigue region and fast fracture region is apparent. The fast fracture region is more tortuous and may show signs of either ductile (cup and cone features) or brittle (cleavage or intergranular) fracture. Figure 18 shows a schematic of a fatigue failure. A more detailed discussion of fatigue is presented in a separate chapter.

The fatigue life of ferrous and nonferrous alloys is different. Ferrous alloys exhibit a fatigue limit (stress below which failure does not occur). Nonferrous alloys will fail at nearly any applied load, although the number of cycles

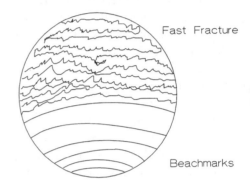

Figure 18. The beachmarks show major places where the crack was stopped. Finer details can be observed using high power optical or electron optical microscopy.

to failure can be very large. A limit of 10^8 cycles is used as a stopping point for most testing. This number of cycles is called the *endurance limit* of a material. Figure 19 shows a schematic of a stress-number of cycles to failure (S-N curve) plot of data for ferrous and nonferrous alloys. The

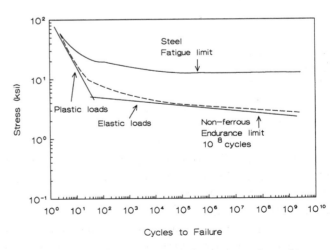

Figure 19. Fatigue terms for ferrous and nonferrous metals.

fatigue ratio (fatigue limit or fatigue strength for 10^8 cycles divided by the tensile strength) for most steels is 0.5. The fatigue ratio for nonferrous metals, such as nickel, copper, and magnesium, is about 0.35. These ratios are for smooth bars tested under zero mean stress. For notched samples, the ferrous fatigue ratio drops to 0.3–0.4 [4].

The fatigue life is directly affected by the surface preparation of both test samples and machine elements. Smooth surfaces increase the fatigue life. Surface treatments, such as carburizing and nitriding steels, increase life. Introduction of compressive surface stresses by shot peening increases life, especially if followed by light polishing to reduce surface roughness. Uneven peening can reduce life by introducing stress risers. Tensile stresses, which can be introduced by grinding and quenching, reduce fatigue life. Electroplating reduces fatigue life since the electroplate may have porosity, may introduce residual stresses, or may change the surface hardness. Residual stresses have the greatest influence on tests and materials that are loaded near the fatigue limit. Their influence is minimal at high loads.

There are many metallurgical factors that influence the fatigue resistance. The effect of solid solution alloying elements on fatigue has a parallel effect on the tensile strength of iron and aluminum alloys. Elements that promote wavy slip result in higher lives than elements that promote planar slip. Grain size can also affect the fatigue life, depending on whether initiation or propagation are the life-

limiting features. In alpha brasses, which have low stacking fault energies, grain size influences the fatigue life; in other Al and Cu alloys, which have high stacking fault energies, there is no such influence. In steels, a quenched and tempered martensitic structure has the optimum fatigue resistance, although ausformed steels (bainitic microstructure) with hardnesses in excess of HRC 40 are better than the same alloy in the quenched and tempered condition.

Suggested Reading

References for designing against fatigue failure which have extensive design examples include:

Ruiz, C. and Koenigsberger, F., *Design for Strength and Production.* New York: Gordon and Breach Science Publishers, Inc., 1970.

Juvinall, R. C., *Engineering Considerations of Stress, Strain, and Strength.* New York: McGraw-Hill Book Co., 1967.

Graham, J. A. (ed.), *Fatigue Design Handbook.* New York: Society of Automotive Engineers, 1968.

Osgood, C. C., *Fatigue Design,* 2nd Ed. New York: Permagon Press, 1982.

Boyer, H. E. (ed.), *Atlas of Fatigue Curves.* Metals Park, OH: American Society for Metals, 1985.

Hardness Testing

Hardness testing is a fairly simple method of conforming to quality control standards. It is also used to develop materials. A small sample is loaded by an indentor. The geometry of the indentor and the magnitude of the load is determined by the type of test that is performed and the material that is being tested.

A hardness test provides insight into several material properties. These are: strain hardening, since the test causes plastic deformation; plastic flow curve; and ultimate tensile strength. For heat-treated carbon and medium alloy steels, there is a direct correlation between the Brinell hardness test and the tensile strength. It is given in Equation 10:

$$UTS\ (\psi) = 500BHN \qquad (10)$$

where UTS (ψ) is the ultimate tensile strength and BHN is the Brinell hardness number.

The test can be conducted on either a macro or a micro scale. There are three indentor geometries: a ball, a brale, and a diamond pyramid. Table 22 lists a number of common hardness tests and the materials most likely to be tested. There are limits to the size of the samples that can be tested. Some general rules are that the hardness indentations be spaced at least 4 diameters apart, and the thickness of the material be 6 times the depth of the penetrator. ASTM E10 provides complete guidelines regarding the hardness test for Brinell testing [31], E18 for Rockwell testing [30], and E92 for Vickers hardness testing [32].

Table 22
Common Types of Hardness Tests, Indentors, and Applications

Test	Indentor	Load	Material
Brinell	10 mm ball	3,000 Kg	Cast iron and steel
Brinell	10 mm ball	500 Kg	Nonferrous alloys
Rockwell A	Brale	60 Kg	Very hard materials
Rockwell B	⅟₁₆″ ball	100 Kg	Brasses and low strength steels
Rockwell C	Brale	150 Kg	High strength steels
Rockwell N	Brale	15–45 Kg	Hard coatings, superficial hardness
Rockwell Y	½″ ball	15–45 Kg	Soft coatings, superficial hardness
Vickers	Diamond pyramid	50g–1 Kg	All materials
Knoop	Diamond pyramid	500 g	All materials

Creep and Stress Rupture Testing

Elevated temperature properties are measured using either creep or stress rupture tests. The tests measure different material properties. A creep test measures a material's resistance to elongation, while a stress rupture test measures the time for a sample to break into two separate pieces. Creep tests are run for long times and have small but finite elongations, typically 0.1, 0.2, 0.5, 1.0, 2.0, and 5.0. Further, they are run for long periods of time. Stress rupture tests are run for shorter times and high total strains, up to 50%. Both elongation and rupture time can be determined by a creep rupture test.

Creep tests are run at either constant stress or, more commonly, constant load. The main stages of creep and the differences between these two types of loading are shown in Figure 20. An initial strain is present due to elastic response of the material, (e.g., $\varepsilon = \sigma/E$). Primary creep is characterized by viscous flow of the material. Further, the creep rate decreases with increased time. Second-stage creep involves steady-state creep with the minimum creep rate. During this time period, the processes of hardening and recovery are occurring simultaneously at nearly the same rate. Tertiary creep is actually a geometric effect. It results because of localized necking of the sample which effectively increases the stress.

Creep rate increases with increasing stress at constant temperature and with increasing temperature at constant

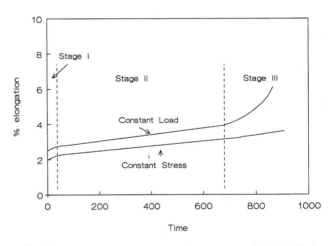

Figure 20. Stylized creep curve for constant load and constant stress. Also shown are the three stages of creep.

stress. Figure 21 shows the effect of increasing stress at constant temperature. Some typical creep limits for material applications are 1% in 10,000 hours (or a creep rate of 0.0001% per hour) for aircraft turbine parts and 1% in 100,000 hours (or a creep rate of 0.00001% per hour) for steam turbines and other similar equipment.

Stress rupture tests are typically conducted under constant load conditions and are terminated after samples break.

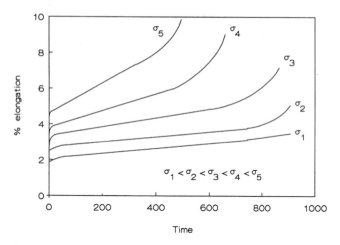

Figure 21. Effect of increasing stress on creep response at constant temperature.

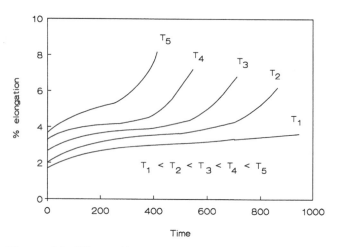

Figure 22. Effect of increasing temperature at constant load on stress rupture life.

They generally last less than 1,000 hours. They are inexpensive and require less complex instrumentation than creep tests. Stress rupture tests make it easy to characterize an alloy that is under development. The effect of temperature on the stress rupture life is shown schematically in Figure 22.

It is not always practical to test for the long times at lower temperature for design applications. Many methods have been devised that allow one to trade up in temperature for design data. One of the more common empirical equations is the Larson Miller parameter. It is provided in Equation 11.

$$P = (T + 460) * (\log [t] + A) \tag{11}$$

The time t can be either the time to reach a creep strain or the time to rupture, depending on whether one is modeling strain or stress rupture. The A in the equation can be either experimentally determined or is assumed to be 20. T is the temperature in °F.

FORMING

Metals can be shaped by a number of processes which include machining and plastic deformation. This section will briefly address metal shaping by plastic deformation. Plastic deformation is the application of force to change the shape of a material. Some of the more common types of plastic deformation are shown in Figure 23. Most of these methods can be used either hot or cold; extrusion is done hot. When used cold, these forming processes can increase strength with a resultant reduction in ductility, as shown schematically in Figure 24. When used hot, these processes can close off porosity from the casting stage.

In some cases the forming methods used can be primary, to rough a piece to near net shape, or secondary, to produce a finished part. Forging can produce simple pancakes

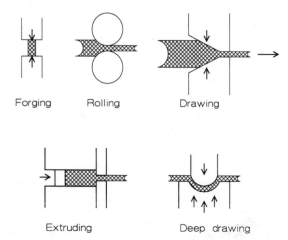

Figure 23. Five common metal forming methods [4]. (*Reprinted by permission of McGraw-Hill, Inc.*)

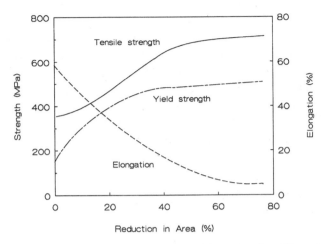

Figure 24. Influence of cold work on the ductility and strength of a metal.

or complex parts. It may require several intermediate dies and heat treatments to produce very complex parts. Rolling is used to produce flat plate, strip, and foil. Wire drawing, as the name implies, is used for wires. Tubes can also be drawn by including a plug in the die set-up. Extrusions are made of low melting point alloys, aluminum, copper, and lead because of the high forces needed to push the metal through the die and the high temperatures required to decrease the material strength. Deep drawing is used to form thin sections like doorknobs and beverage cans.

CASTING

The Forming section pertained to methods of getting metal into a usable form using solid-state techniques. Another option to obtain components is by solidifying metal in the desired shape. Casting technologies are varied for the many melting point materials that are used. The simplest use of casting is to form ingots that are subsequently shaped by forming processes. These castings can be very large, on the order of tons. Continuous casting is used in steel mills to form large, thick, flat slabs which are subsequently rolled into plate, sheet, and other simple shapes. These casting methodologies do not produce a final shape for use, rather, they produce raw material for other processing.

Some methods of producing final shapes through casting are sand molding, die casting, and investment casting. In sand molding, a carefully prepared molding sand containing a binder, which may be clay, is mixed with water. This mixture generates a material that has some strength to be molded. The sand is then either sculpted or molded into the desired shape. If it is molded, then a pattern must be fabricated. The pattern may be made of plastic, wood, or metal, depending on the number of pieces that will be cast. The pattern must have some draft or taper to allow it to be removed from the sand. Sprues, downspouts for the incoming metal, runners, channels to contain the molten metal, and gates to lead it into the die cavity must be cut into the sand or can be included in the mold. Internal passages can be formed by the use of cores.

Molten metal is poured into the cavity and allowed to solidify. The final product of a sand mold will have a coarse surface finish that depends on the size and texture of the sand used for molding. Often, castings will have pores and other defects that may make them unusable as is. They can be either repaired (welded) or scrapped.

Die casting (or pressure die casting) uses a permanent mold of tool steel or other high melting point alloy and injects molten metal into the cavity. The die may have water cooling to prevent its melting. The surface finish of a die cast is better than a sand cast part. Since the cooling rate is faster, a finer structure is produced.

Die casting is useful for making zinc, aluminum, and copper parts. The difficulty in producing quality parts increases with increased melting point, and metals with higher melting points than copper cannot be die cast. Many of the parts that are die cast can be produced more cheaply out of plastic and are being replaced by plastics.

Investment casting uses a sacrificial wax or plastic pattern that has all of the features of the finished part. The gates and sprue may be directly molded or attached by wax "welding." The mold, for a shell type casting, is made by assembling all of the components and then dipping the wax or plastic pattern in slurry of fine silica. Two or three slurry coats are applied, then several layers of a coarse slurry are used to give the shell strength. The wax is then melted out, the shell fired to eliminate all of the traces of the pattern and to strengthen the mold, and the molten metal poured into the mold. The metal solidifies and the shell is broken off.

Internal passages and other product features can be incorporated into the casting using cores. Excellent surface finish and dimensional control can be obtained. Complex turbine blades can be manufactured with this method. It is more expensive than other casting technologies.

A specialized form of investment casting is used to make single crystal and directionally solidified pieces. With these technologies, which are very important for materials that require long stress rupture and creep properties, the heat is preferentially extracted in a single direction. This promotes the growth of a single grain or a single grain orientation. The grain orientation selected depends on the crystal anisotropy and the property most important for the application.

Information about the castability of the various alloys can be found in *Principles of Metal Casting* [27] and the *ASM Metals Handbook,* Vol. 15, 9th Ed.

CASE STUDIES

Failure Analysis

Failure analysis entails the systematic investigation of why or how a component fails. Despite the best design, an improper material selection or a processing sequence can lead to a premature failure of said component. A detailed history is generally established. Temperature, expected environment, stresses, and strains are all important variables for the failure analyst to know. As one investigates various failures, documentation of the salient features is required. The methods used include photography, notetaking, videography, and the like. The examination of the fracture surfaces optically and electron optically are useful in determining the type of failure, e.g., brittle or ductile fracture, high or low cycle fatigue, environmentally assisted fracture, or wear.

Two operational failures and fixes will be discussed. Wear is one of the most important causes of failure, although many factors are usually involved. Piston rings, gears, and bearings are a few of the many parts where resistance to wear is required. Wear is probably the most easily recognized failure mode, as shown in Figure 25. Although wear may not be prevented, steps can be taken to reduce the rate and yield a long service life by the proper application of materials, lubrication, and design.

Often, improper application of steels, load distribution, heat treatment, and inadequate or faulty lubrication result in excessive wear and poor service life. High loads and speeds are capable of producing very high temperatures under which metal surfaces may actually melt. Friction is

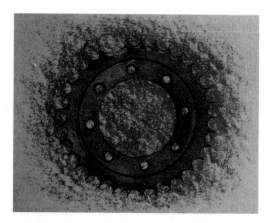

Figure 25. Excessive wear of gear teeth. (*Reprinted by permission of Republic Steel.*)

an important factor in producing temperatures that may cause the breakdown of hardened surfaces, such as those produced by carburizing. Therefore, special lubricants for specific applications involving very high unit pressures may be required.

The gear wear shown in Figure 25 was corrected by selecting a new material that was significantly harder than the 1020 rimmed steel with a Brinell hardness of 116. The worn teeth were driven by rollers in a chain link with a Brinell hardness of 401. An alloy steel with higher hardness was substituted, and the new sprocket was still in service after seven years [33].

Corrosion

The diagram in Figure 26 is a schematic of the lower end of a tube-and-shell heat exhanger made from mild steel. The unit was designed to heat oil in a chemical process plant. The oil was passed through the small tubes and the heat was supplied from steam which was injected into the shell. The unit had been in operation for only 2.5 years when one of the tubes perforated. When the tubes were extracted from the shell, it was found that they all had corroded on the outside over a distance of about 160 mm from the lower tube plate. On the worst-affected areas, attack had occurred to a depth of about 1.5 mm over regions measuring typically 10 mm by 20 mm. The corroded areas were light brown in color.

The heat exchanger was operated on a cyclic basis as follows. First, saturated steam was admitted to the shell at 180°C to heat a new batch of oil. The steam condensed on the surfaces of the tubes and the condensed water trickled down to the bottom of the shell, where it was drawn off via the condensate drain. When the oil was up to temperature, the steam supply was cut off and the pressure in the shell was dropped to atmospheric. The cycle was repeated when it was time to heat up a new batch of oil.

Based on the above observations and operating cycle, it is apparent that the corrosion product is red rust, i.e., hydrated Fe_2O_3. Of the three forms of iron oxide (FeO, Fe_3O_4), and Fe_2O_3), the latter has the highest ratio of oxygen to iron. It is the favored oxide in an oxygen-rich environment. When the oxygen concentration is low, the corrosion product consists of hydrated Fe_3O_4 (magnetite), which is black. But there was no evidence that this was present as a corrosion product. There is evidence, however, of oxygen in the condensate which presumably came from air dissolved in the make-up feed water to the boiler. This would have provided the oxygen needed for the cathodic reaction.

The design of the unit allows condensate to build up to the level of the drain. It is interesting that corrosion has only occurred in, or just above, the pool of condensate; it has not taken place farther up the tubes even though they would have been dripping with condensed steam. A likely scenario is that when the shell was let down to atmosphere, the water at the bottom of the shell was boiled off by the residual heat in the tube plate. This would have left either a concentrated solution or a solid residue containing most of the impurities that were originally dissolved in the condensate pool. With each cycle of operation, the concentration of impurities in the pool would have increased. A prime suspect is the carbonic acid, derived from carbon dioxide dissolved in the feed water. This would have made the liquid in the pool very acidic and given it a high ionic conductivity, both of which would have resulted in rapid attack. It can be seen from the electrochemical equilibrium diagram for iron [39], iron does not form a surface film in acid waters. Finally, the temperature is elevated so the rates of thermally activated corrosion processes should be high as well.

A simple design modification of moving the condensate drain from the side to the lowest point of the shell would prevent water from accumulating in the bottom of the shell [34].

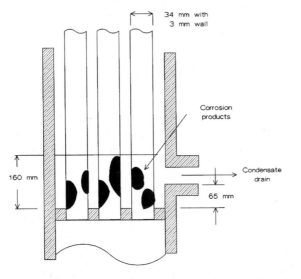

Figure 26. Schematic view of the lower end of a tube-and-shell heat exchanger made from mild steel.

REFERENCES

1. Bolz, R. E. and Tuve, G. L. (Eds.), *Handbook of Tables for Applied Engineering Science,* 2nd Ed. Boca Raton: CRC Press, 1984.

2. *ASM Metals Handbook: Properties and Selection—Irons and Steels,* Vol. 1, 9th Ed., ASM International, Materials Park, OH, 1978.

3. Callister, W. D., Jr., *Materials Science and Engineering, An Introduction.* New York: John Wiley & Sons, Inc., 1985.

4. Dieter, G. E., *Mechanical Metallurgy.* New York: McGraw-Hill, 1986.

5. Hertzberg, R. W., *Deformation and Fracture Mechanics of Engineering Materials,* 2nd Ed. New York: John Wiley & Sons, 1983.

6. Schackelford, J. F., *Introduction to Materials Science for Engineers,* 2nd Ed. New York: Macmillan Publishing, 1988.

7. Askeland, D. R., *The Science and Engineering of Materials.* Belmont, CA: Wadsworth, 1984.

8. Van Vlack, L. H., *Materials Science for Engineers.* Redding, MA: Addison Wesley, 1970.

9. Uhlig, H. H. and Revie, R. W., *Corrosion and Corrosion Control and Introduction to Corrosion Science and Engineering,* 3rd Ed. New York: John Wiley & Sons, Inc., 1985.

10. Fontana, M. G., *Corrosion Engineering.* New York: McGraw-Hill, 1986.

11. McCrum, N. G., Buckley, C. P., and Bucknall, C. B., *Principles of Polymer Engineering.* New York: Oxford University Press, 1988.

12. *Powder Metallurgy Design Solutions.* Metal Powder Industries Federation, Princeton, NJ, 1993.

13. German, R. M., *Powder Metallurgy Science.* Metal Powder Industries Federation, Princeton, NJ, 1984.

14. "Amdry MCrAlY Thermal Spray Powders Specially Formulated and Customized Alloys Provide Oxidation and Corrosion Resistance at Elevated Temperatures," Amdry Product Bulletin 961, 970, 995, Alloy Metals, Inc., 1984.

15. *Engineered Materials Handbook, Vol. 4: Ceramics and Glasses.* S. J. Schneider, Jr., Volume Chairman, ASM International, Materials Park, OH, 1991.

16. Davis, J. R. (Ed.), *ASM Materials Engineering Dictionary.* ASM International, Metals Park, OH, 1992.

17. Craig, B. D. (Ed.), *Handbook of Corrosion Data.* ASM International, Materials Park, OH, 1989.

18. McEvily, A. J. (Ed.), *Atlas of Stress-Corrosion and Corrosion Fatigue Curves.* ASM International, Materials Park, OH, 1990

19. Coburn, S. K. (Ed.), *Corrosion Source Book.* ASM International, Materials Park, OH, 1984.

20. Sedriks, A. J. (Ed.), *Corrosion of Stainless Steels.* New York: John Wiley & Sons, Inc., 1979.

21. Uhlig, H. H., *Corrosion Handbook.* New York: John Wiley & Sons, Inc., 1948.

22. *ASM Metals Handbook: Properties and Selection—Nonferrous Alloys and Pure Metals,* Vol. 2, 9th Ed., ASM International, Materials Park, OH, 1979.

23. Massalski, T. B., Okamoto, H., Subramanian, P. R., and Kacprzak, L. (Eds.), *Binary Alloy Phase Diagrams,* 2nd Ed., ASM International, Materials Park, OH, 1990.

24. Haynes International, Product Bulletin H-1064D, 1993.

25. Inco Alloys International, Product Handbook, 1988.

26. Sims, C. T., Stoloff, N. S., and Hagel, W. C. (Eds.), *Superalloys II High Temperature Materials for Aerospace and Industrial Power.* New York: John Wiley & Sons, Inc., 1987.

27. Heine, R. W., Loper, C. R., and Rosenthal, P. C., *Principles of Metal Casting,* 2nd Ed. St. Louis: McGraw-Hill, 1967.

28. Birks, N. and Meier, G. H., *Introduction to High Temperature Oxidation of Metals.* Great Britain: Edward Arnold, 1983.

29. *ASTM E112, Standard Method for Average Grain Size of Metallic Materials,* Volume 03.01, Metals-Mechanical Testing; Elevated and Low Temperature Test; Metallography, ASTM, 1992.

30. *ASTM E18, Standard Test Methods for Rockwell Hardness and Rockwell Superficial Hardness of Metallic Materials,* Volume 03.01, Metals-Mechanical Testing; Elevated and Low Temperature Test; Metallography, ASTM, 1992.

31. *ASTM E10, Standard Test Method for Brinell Hardness of Metallic Materials,* Volume 03.01, Metals-Mechanical Testing; Elevated and Low Temperature Test; Metallography, ASTM, 1992.

32. *ASTM E92, Standard Test Method for Vickers Hardness of Metallic Materials,* Volume 03.01, Metals-Mechanical Testing; Elevated and Low Temperature Test; Metallography, ASTM, 1992.

33. "Analysis of Service Failures," Republic Alloy Steels Handbook Adv. 1099R, Republic Steel Corporation, 1974.

34. Jones, D. R. H., *Engineering Materials 3, Materials Failure Analysis, Case Studies and Design Implications.* New York: Pergamon Press, 1993.

35. Aurrecoechea, J. M., "Gas Turbine Hot Section Coating Technology," Solar Turbines Incorporated, 1995.

36. *ASM Metals Handbook: Welding, Brazing, and Soldering,* Vol. 6., 9th Ed. ASM International, Materials Park, OH.

37. Harper, C. A. (Ed.), *Handbook of Plastics and Elastomers.* New York: McGraw-Hill, Inc., 1975.

38. *ASM Metals Handbook,* Vol. 15, 9th Ed., ASM International, Materials Park, OH, 1988.

39. Pourbaix, M., *Atlas of Electrochemical Equilibria in Aqueous Solutions,* National Association of Corrosion Engineers (NACE), Houston, TX, 1974.

40. *ASM Metals Handbook: Properties and Selection—Stainless Steels, Tool Materials, and Special Purpose Metals,* Vol. 3, 9th Ed., ASM International, Materials Park, OH, 1980.

41. *ASM Metals Handbook:* Corrosion, Vol. 13, 9th Ed., ASM International, Materials Park, OH, 1987.

13
Stress and Strain

Marlin W. Reimer, Development Engineer, Structural Mechanics Dept., Allison Engine Company

FUNDAMENTALS OF STRESS AND STRAIN

Introduction

Stress is a defined quantity that cannot be directly observed or measured, but it is the cause of most failures in manufactured products. Stress is defined as the force per unit area (σ) with English units of pounds per square inch (psi) or metric units of megapascals (mpa). The type of load, i.e., duration of load application, coupled with thermal conditions affects the ability of a structural component to resist failure at a particular magnitude of stress. Gas turbine airfoils under sustained rotating loads at high temperature may fail in creep rupture. Components subjected to cyclic loading may fail in fatigue. High speed rotating disks allowed to overspeed will burst when the average stress exceeds the rupture strength which is a function of the ductility and the ultimate strength of the material.

Conversely, *strain* is a measurable quantity. When the size or shape of a component is altered, the deformation in any dimension can be characterized by the deformation per unit length or strain (ε). Strain is proportional to stress at or below the proportional limit of the material. Hooke's law in one dimension relates stress to strain by the modulus of elasticity (E). Typical values for E at 70°F are listed in Table 1. At elevated temperatures, the modulus will decrease for the materials listed. Note that the ratio of modulus to density for the selected materials is relatively constant, i.e., $E/(\rho/g) \approx 10^8$:

$$\sigma = E\varepsilon$$

where σ = stress
  E = modulus of elasticity
  ε = strain

Table 1
Range of Elastic Modulus for Common Alloys

Material	Modulus (E) psi	Density (ρ/g) lb/in.^{2}
Aluminum alloys	$10.0 - 11.2 \times 10^6$	0.10
Cobalt alloys	$32.6 - 35.0 \times 10^6$	0.33
Magnesium alloys	$6.4 - 6.5 \times 10^6$	0.065
Nickel alloys	$28.0 - 31.5 \times 10^6$	0.30
Steel—carbon and low alloy	$30.7 - 31.0 \times 10^6$	0.28
Steel—stainless	$28.5 - 31.8 \times 10^6$	0.28
Titanium alloys	$15.5 - 17.9 \times 10^6$	0.16

Sources: Mil-Hdbk-5D *[1]*, Aerospace Structural Metals Handbook *[2]*.

Definitions—Stress and Strain

The following basic stress quantities are useful in the evaluation of many simple structures. They are depicted in Figures 1 through 4. Today, complex components with rapid changes in cross-section, multiple load paths, and stress concentrations are analyzed using finite element models. However, the basic equations supplemented by handbook solutions should be employed for preliminary calculations and to check finite element model results.

Basic Stress Quantities

Direct Stress: $\sigma = P/A$

where P = load
  A = area.

Bending Stress: $\sigma = Mc/I$

where M = moment
  I = area moment of inertia
  c = distance from neutral surface

Transverse Shear Stress: $\tau = VQ/It$

where V = shear force
  Q = first moment of the area
  I = area moment of inertia
  t = thickness of cross-section

Torsional Shear Stress: $\tau = Tr'/J$

where T = torque
  r' = distance from axis of shaft,
  J = polar moment of inertia

Figure 1. Direct stress.

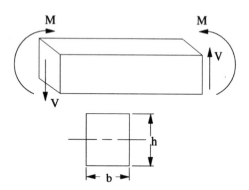

Figure 2. Bending stress.

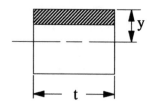

Figure 3. Transverse shear stress.

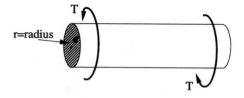

Figure 4. Torsional shear stress

Hooke's Law Equations

The proportionality of load to deflection in one dimension is written as:

$\sigma = E\varepsilon$ (for normal stress σ and strain ε)
$\tau = G\gamma$ (for shearing stress τ and strain γ)

Poisson's ratio (ν) is the constant for stresses below the proportional limit that relates strain in one dimension to an-

other. For an incompressible material, $\nu = 0.5$. Since actual materials are compressible, Poisson's ratio must be less than 0.5—typically $0.25 \le \nu \le 0.30$ for most metals.

Hooke's Law in three dimensions for normal stresses [3]:

$$\varepsilon_x = \frac{1}{E}\left[\sigma_x - \nu\left(\sigma_y + \sigma_z\right)\right]$$

$$\varepsilon_y = \frac{1}{E}\left[\sigma_y - \nu\left(\sigma_x + \sigma_z\right)\right]$$

$$\varepsilon_z = \frac{1}{E}\left[\sigma_z - \nu\left(\sigma_x + \sigma_y\right)\right]$$

Hooke's Law for shear stresses:

$$\tau_{xy} = G\gamma_{xy}$$

$$\tau_{yz} = G\gamma_{yz}$$

$$\tau_{xz} = G\gamma_{xz}$$

where G is the modulus of elasticity in shear.

The relationship between the shearing and tensile moduli of elasticity for an elastic material:

$$G = \frac{E}{2\left(1 + \nu\right)}$$

Von Mises Equivalent Stress

Most material strength data is based on uniaxial testing. However, structures are usually subjected to more than a uniaxial stress field. The Von Mises equivalent stress is generally used to evaluate yielding in a multiaxial stress field, allowing the comparison of a multiaxial stress state with the

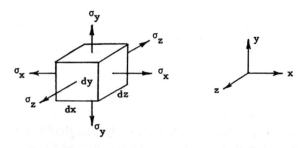

Figure 5. Three-dimensional normal stresses.

uniaxial material data. If the nominal equivalent stress is less than the yield strength, no gross yielding will occur.

$$\sigma_{equivalent} =$$

$$\left[\frac{[(\sigma_x - \sigma_y)^2 + (\sigma_y - \sigma_z)^2 + (\sigma_z - \sigma_x)^2 + 6(\tau_{xy}^2 + \tau_{yz}^2 + \tau_{zx}^2)]}{2} \right]^{-0.5}$$

Note that equivalent stresses are always positive. If the sum of the principal stresses σ_x, σ_y, and σ_z is positive, the equivalent stress is considered tensile in nature. A negative sum denotes a compressive stress.

Equilibrium

$$\Sigma F = 0$$

$$\Sigma M = 0$$

To successfully analyze a structural component, it is necessary to define the force balance on the part. A free body diagram of the part will assist in determining the path which various loads take through a structure. For example, in a gas turbine engine it is necessary to determine the separating force at axial splitline flanges between the engine cases to ensure the proper number of bolts and size the flange thicknesses. A free body diagram helps to isolate the various loads on the static structure connected to the case. The compressor case drawing in Figure 6 shows the axial vane and flange loads on the case. The pressure differential across the case wall would also contribute to the axial force balance if the case was conical in shape.

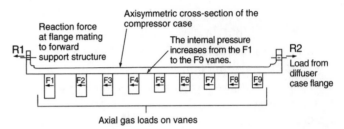

Figure 6. Free body diagram of a compressor case from a gas turbine engine.

Compatibility

Compatibility refers to the concept that strains must be compatible within a continuum, i.e., the adjacent deformed elements must fit together (see Figure 7). Boundary equations, strain-displacement, and stress equilibrium equations must be defined for the complete solution of a general stress problem.

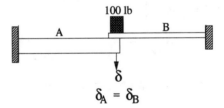

Figure 7. Compatibility.

Saint-Venant's Principle

Saint-Venant's principle states that if the forces acting on a local section of an elastic body are replaced by a statically equivalent system of forces on the same section of the body, the effect upon the stresses in the body is negligible except in the immediate area affected by the applied forces. The stress field remains unchanged in areas of the body which are relatively distant from the surfaces upon which the forces are changed. "Statically equivalent systems" implies that the two distributions of forces have the same resultant force and moment. Saint-Venant's principle allows simplification of boundary condition application to many problems as long as the system of applied forces is statically equivalent.

Superposition

The principle of superposition states that the stresses at a point in a body that are caused by different loads may be calculated independently and then added together, as long as the sum of the stresses remains below the proportional limit and remains stable. Application of this principle allows the engineer to break a more complex problem down into a number of fundamental load conditions, the solutions of which can be found in many engineering handbooks.

Plane Stress/Plane Strain

Often, for many problems of practical interest, it is possible to make simplifying assumptions with respect to the stress or strain distributions. For example, a spinning disk which is relatively thin (see Figure 8) is in a state of plane stress. The centrifugal body force is large with respect to gravity. No normal or tangential loads act on either the top or bottom of the disk. σ_z, τ_{rz}, and $\tau_{\theta z}$ are zero on these surfaces. Since the disk is thin, these stresses do not build up to significant values in the interior of the disk. Plane stress assumptions are valid for thin plates and disks that are loaded parallel to their long dimension. Thin plates containing holes, notches, and other stress concentrations, as well as deep beams subject to bending, can be analyzed as plane stress problems.

Another simplification can be made for long cylinders or pipes of any uniform cross-section which are loaded laterally by forces that do not vary appreciably in the longitudinal direction. If a long cylinder (see Figure 9) is subjected to a uniformly applied lateral load along its length and is constrained axially at both ends, the axial deflection (δ_z) at both ends is zero. By symmetry, the axial deflection at the center of the cylinder is also zero and the approximate assumption that δ_z is zero along the entire length of the cylinder can be made. The deformation of a large portion of the body away from the ends is independent of the axial coordinate z. The lateral and vertical displacements are a function of the x and y coordinates only. The strain components ε_z, γ_{xz}, and γ_{yz} are equal to zero and the cylinder is in a state of plane strain. A pipe carrying fluid under pressure is an example of plane strain.

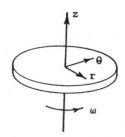

Figure 8. Thin spinning disk—an example of plane stress.

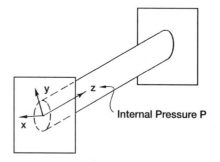

Figure 9. Pipe line—an example of plane strain.

Thermal Stresses

Thermal stresses are induced in a body when it is subjected to heating or cooling and is restrained such that it cannot expand or contract. The body may be restrained by external forces, or different parts of the body may expand or contract in an incompatible fashion due to temperature gradients within the body. A straight bar of uniform cross-section, restrained at each end and subjected to a temperature change ΔT, will experience an axial compressive stress per unit length of $E\alpha\Delta T$. α is the coefficient of thermal expansion. A flat plate of uniform section that is restrained at the edges and subjected to a uniform temperature increase ΔT will develop a compressive stress equal to $E\alpha\Delta T/(1-v)$. Additional miscellaneous cases for thermally induced stresses in plates, disks, and cylinders, are listed in Young [4] and Hsu [5]. Typical values for the coefficient of thermal expansion (α) for several common materials are listed in Table 2.

Design Hints

- If thermally induced stresses in a member exceed the capability of the material, increasing the cross-sectional area of the member will generally not solve the problem. Additional cross-section will increase the stiffness, and the thermally induced loads will increase almost as rapidly as the section properties. Often, the flexibility of the structure must be increased such that the thermal deflections can be accommodated without building up large stresses.
- Thermal stress problems can be minimized by matching the thermal growths of mating components through appropriate material selection.

- In situations where transient thermal gradients cause peak stresses, changes in the mass of the component, changes in the conduction path, addition of cooling flow, and shielding from the heat source may reduce the transient thermal gradients.

Table 2
Range of Coefficient of Thermal Expansion for Common Alloys

Material	Max. Recommended Temp. (°F)	α @ 600°F (in./in./°F)	α @ 1200°F (in./in./°F)
Aluminum alloys	300–600	$13.0–14.2\times10^{-6}$	—
Cobalt alloys	1,900–2,000	$7.0–7.7\times10^{-6}$	$7.8–8.7\times10^{-6}$
Magnesium alloys	300–600	$15.5–15.7\times10^{-6}$	—
Nickel alloys	1,400–2,000	$6.6–8.0\times10^{-6}$	$7.3–8.8\times10^{-6}$
Steel—carbon and low alloy	450–1,000	$7.1–7.3\times10^{-6}$	$7.7–8.3\times10^{-6}$
Steel—stainless	600–1,500	$6.0–9.7\times10^{-6}$	$6.7–10.3\times10^{-6}$
Titanium alloys	400–1,000	$5.0–5.4\times10^{-6}$	$5.5–5.6\times10^{-6}$

Sources: Mil-Hdbk-5D *[1]*, Aerospace Structural Metals Handbook *[2]*.

STRESS CONCENTRATIONS

The basic stress quantities used in design assume a constant or gradual change in cross-section. The presence of holes, shoulder fillets, notches, grooves, keyways, splines, threads, and other discontinuities cause locally high stresses in structural members. Stress concentration factors associated with the aforementioned changes in geometry have been evaluated mathematically and experimentally with tools such as finite element models and photoelastic studies, respectively.

The ratio of true maximum stress to the stress calculated by the basic formulas of mechanics, using the net section but ignoring the changed distribution of stress, is the factor of stress concentration (K_t). A concentrated stress is not significant for cases involving static loading (steady stress) of a ductile material, as the material will yield inelastically in the local region of high stress and redistribute. However, a concentrated stress is important in cases where the load is repeated, as it may lead to the fatigue failure of the component. Often components are subject to a combination of a steady stress (σ_s) due to a constant load and an alternating stress (σ_a) due to a fluctuating load such

that the stresses cycle up and down without passing through zero (see Figure 10). Note that the steady stress and the mean stress (σ_m) may not have the same value. The steady stress can have any value between the maximum and minimum stress values. The damaging effect of a stress concentration is only associated with the alternating portion of the stress cycle. Hence, it is common practice to apply only any existing stress concentration to the alternating stress [6]. A good example of this situation is a shaft transmitting a steady state torque that is also subject to a vibratory torsional

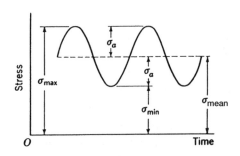

Figure 10. Fluctuating stress.

load which may be ±5% of the steady state torque. For stress concentration features such as shoulder fillets, the K_t would be applied to the vibratory or alternating stress.

Design Hint

- Eliminate unnecessary stress concentrations. Avoid abrupt changes in section where stress concentrations cannot be relieved by a tolerable degree of local plastic deformation. All fillet radii should be made as generous as is practicable.
- When possible, keep hole locations away from areas of high nominal stress. For example, in high speed rotating disks such as turbine wheels (see Figure 11), holes near the bore will be in a region of high hoop stress. Peak stresses at holes in the web of a rotating disk may also be affected by the radial stresses due to thermal gradients, rotational speed, and bending in the web due to eccentric loads on the rim of the disk. If web holes are unavoidable, try to locate the holes in the most biaxial stress field in the web, i.e., where the tensile hoop stress and the tensile radial stress are nearly equal. It may also be necessary to increase the thickness of the section around the holes to compensate for the stress concentration.
- The use of corrosion-resistant materials helps prevent stress concentrations caused by the pitting that may accompany typical corrosive attack.

- In certain situations, the clever removal of material reduces the effect of stress concentrations such as flange bolt holes. Scalloping flanges as shown in Figure 12 cut the hoop stress path, thus decreasing the effect of the holes on the peak stress.

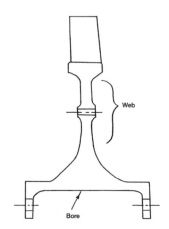

Figure 11. Axisymmetric cross-section of a turbine wheel.

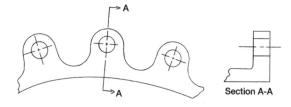

Figure 12. Scalloped flange.

Determination of Stress Concentration Factors

A first approximation for the stress concentration due to a single small hole in a plate (Figure 13) subjected to a uniaxial stress field is $K_t = 3$. In a biaxial stress field with equal stresses (σ_0) of the same sign, the same hole (Figure 14) would cause a maximum stress equal to twice the nominal stress ($K_t = 2$). Conversely, for a biaxial stress field with equal stresses of opposite sign, $K_t = 4$. This latter situation would occur at a small hole in a thin cylinder subjected to pure torsion where σ_{max} equals four times the nominal torsional shear stress (τ). *Stress Concentration Factors* by R. E. Peterson [7] is the best source of numerical values of K_t for grooves, notches, shoulder fillets, holes, and certain other miscellaneous design elements.

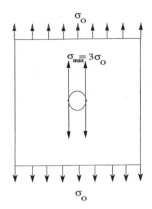

Figure 13. Small hole in a plate subject to a uniaxial stress field.

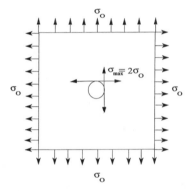

Figure 14. Small hole in a plate subject to a biaxial stress field.

Example: For the hollow shaft in Figure 15, determine the maximum equivalent stress at the shoulder fillet. The shaft is subjected to an axial tensile load and torque.

T = 1,600 in. lb
r = 0.07 in.
D = 1.50 in.
d_1 = 0.45 in.
d = 0.70 in.
P = −5,000 lb.

- Determine the torsional K_t from Figures 16 and 17:

r/d = 0.10
D/d = 2.14
d_1/d = 0.643

thus from Figure 16:

K_{solid} = 1.41

From Figure 17:

$$\frac{K_{solid} - 1}{K_{hollow} - 1} = 0.77$$

thus K_{hollow} = 1.53

- Determine the axial K_t from Figure 18:
r/d = 0.07
D/d = 2.14

thus K_t = 2.22

- Determine the nonconcentrated axial and torsional shear stresses:

$$\frac{P}{A} = \frac{-5000}{\Pi/2\,(0.70^2 - 0.45^2)} = -11,072\ \text{psi}$$

$$\tau = \frac{1600\,\frac{.70}{2}}{\frac{\Pi}{32}(.70^4 - .45^4)} = 28,650\ \text{psi}$$

- Assuming full notch sensitivity, calculate the peak equivalent stress:

$$\sigma_{equivalent} =$$
$$\left[\frac{[2\,(2.22\,(-11,072))]^2 + 6\,(1.53\,(28,650))^2}{2}\right]^{-0.5}$$
$$= 79,803\ \text{psi}$$

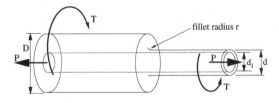

Figure 15. Hollow shaft subject to axial load and torque.

(*text continued on page 305*)

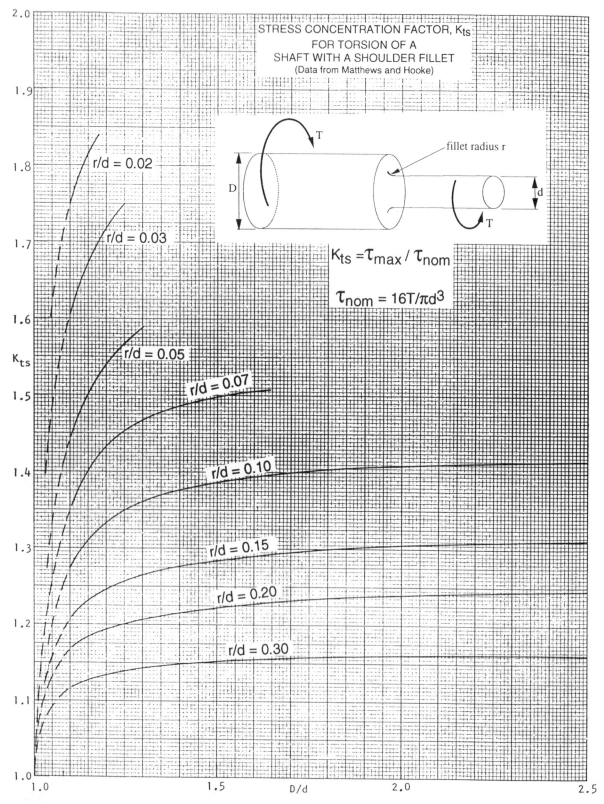

Figure 16. Stress concentration factor for torsion of a shaft with a shoulder fillet. (From *Stress Concentration Factors* by R. E. Peterson [7]. Reprinted by permission of John Wiley & Sons, Inc.)

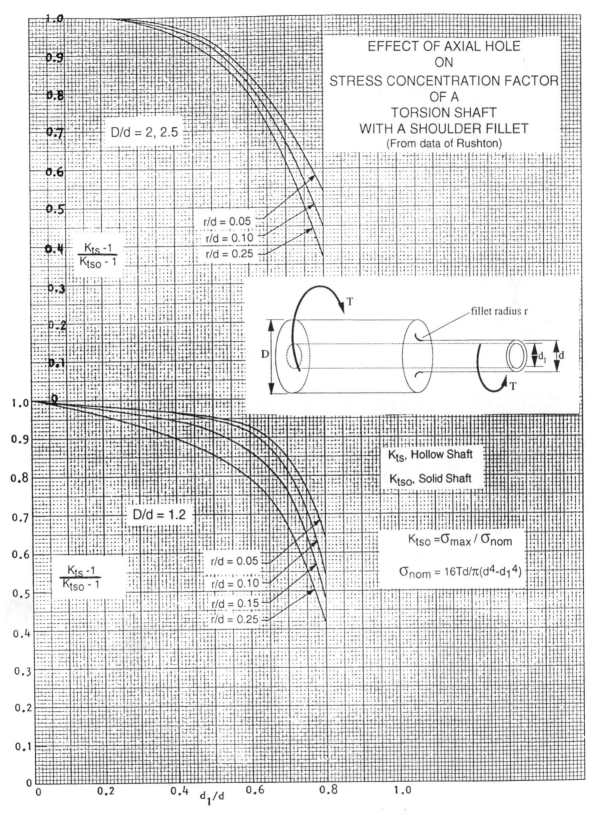

Figure 17. Effect of axial hole on stress concentration factor of a torsion shaft with a shoulder fillet. (From *Stress Concentration Factors* by R. E. Peterson [7]. Reprinted by permission of John Wiley & Sons, Inc.)

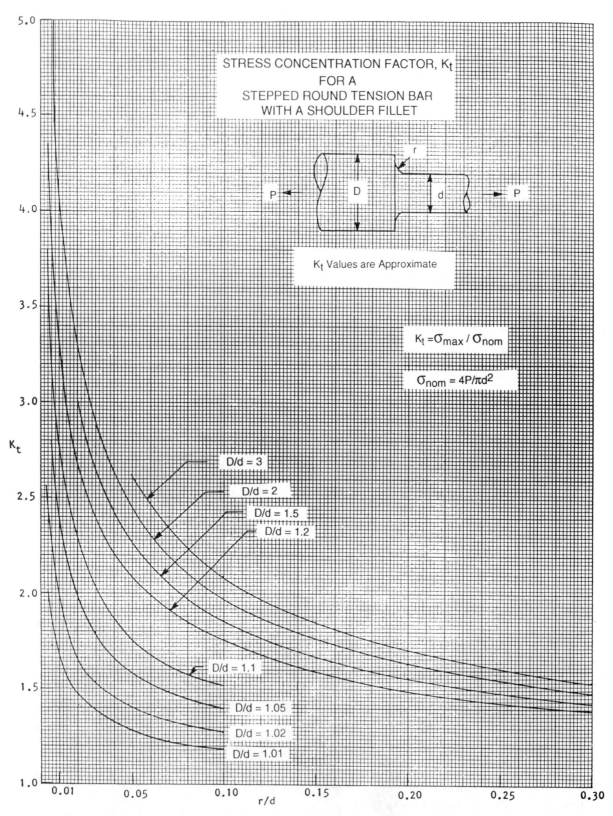

Figure 18. Stress concentration factor for a stepped round tension bar with a shoulder fillet. (From *Stress Concentration Factors* by R. E. Peterson [7]. Reprinted by permission of John Wiley & Sons, Inc.)

(*text continued from page 301*)

DESIGN CRITERIA FOR STRUCTURAL ANALYSIS

Comprehensive design criteria should always be developed early in the design process to ensure that the component or system will meet the functional requirements. The lack of a published design criteria leads to a lack of communication between the design functions and will contribute to a situation in which the design goals become a moving target. This often leads to slipped schedules due to the redesign efforts needed to correct the functional deficiencies.

General Guidelines for Effective Criteria

An effective structural design criteria must address the functional requirements as they relate to *cyclic, time dependent,* and *time independent* load conditions.

In applications where cyclic loading is expected, the low cycle fatigue life (crack initiation life) requirement must be specified. The requirement should not only define the total number of cycles but also the nature of the cycle or mission, i.e., the operating points within the mission. If significant vibratory loads are present, the high cycle fatigue life must be considered as well. The endurance strength, i.e., the alternating stress below which the cyclic life exceeds 10^7 cycles, is usually used to define a Goodman diagram. Depending upon the application, fatigue crack growth requirements may need to be included. These requirements would specify the minimum flaw or crack size that can be detected by the inspection technique employed and the inspection interval in terms of number of cycles and/or hours.

Time dependent criteria generally include creep and stress rupture, although fatigue crack growth can have time dependence. Creep requirements should specify an allowed growth over the life of the component based on the dimensional tolerances between mating parts. Stress rupture is usually only a problem for high-temperature applications where temperatures exceed 50% of the melting temperature of the material. In such cases a combination of temperature and stress over a period of time will lead to failure.

Time independent load criteria include limit and ultimate loads. Limit loads are defined as the maximum expected operational loads. The average cross-sectional stresses imposed by limit loads are compared to the yield strength of the material. Under normal operating conditions, this protects the design from gross yielding. Local yielding in an area of stress concentration is permissible. Ultimate loads are defined as any load conditions in which the structure should tolerate a single application without catastrophic failure, i.e., the load path remains effective after one application, but the part will be inspected and probably replaced after such an event.

Strength Design Factors

The average material strength properties are seldom used to define allowable stresses. Usually, the properties are degraded on a statistical basis to account for scatter. Aircraft applications usually degrade the creep, yield, and ultimate strengths by three standard deviations (-3 σ). Assuming a normal distribution of properties, "three standard deviations" is equivalent to saying that only 1 out of 741 parts could possess lower strength properties. Frequently, a more stringent requirement is placed on the low cycle fatigue properties, i.e., -3.72σ or 1 out of 10,000.

In addition to scatter in material properties, it is sometimes prudent to factor in the scatter associated with dimensional tolerances if the fit between mating components has a large effect on the stress levels.

A *factor of safety* (F.S.) is defined as the allowable strength divided by the calculated or measured stress, whereas the *margin of safety* (M.S.) equals the factor of safety minus one.

$$F.S. = F_{allowable}/F_{calculated}$$
$$M.S. = (F_{allowable} / F_{calculated}) - 1$$

Required factors of safety may vary widely between industries and applications. For example, if weight is not a consideration and material cost is low, large factors of safety can be employed to reduce the risk of failure and avoid costly test programs. Factors of safety based on yield criteria range between 1.5 and 4.0 depending upon the uncertainty associated with the materials and load conditions. Test rig hardware should be designed with larger fac-

tors of safety to reduce the risk of rig failure prior to failure of the tested components. However, overdesign of products is not a luxury permitted in industries where weight is minimized to provide improved energy and material costs, as in the automotive and aircraft industries.

BEAM ANALYSIS

A beam is commonly regarded as a structural member which is much longer than its cross-sectional dimensions and is subjected to loads applied transverse to its longitudinal axis. Beams are classified with respect to the type of support applied to the beam, as shown in Figure 19. Some examples include:

- Simple beam with pinned ends—rotation of the ends is allowed but translation is restrained such that vertical and horizontal reactions are developed.
- Simple beam with rollers—rotation of ends is allowed and only vertical reactions are developed.
- Simple beam with overhang—beam overhangs supports either at one or both ends.
- Cantilever beam—one end is built into a wall preventing rotation and transverse motion such that a moment and reactions are developed.
- Continuous beam—a beam that has more than two simple supports.

Beams are said to be statically determinant when the reactions at the supports can be determined by use of the equations of static equilibrium. If the number of reactions exceeds the number of equations of static equilibrium, the beam is statically indeterminate and additional equations based upon the deformations of the beam must be used to solve for the reactions.

In general, the maximum fiber stress occurs at the point farthest from the neutral axis at the beam section possessing the largest bending moment. The maximum shear stress is usually at the neutral axis of the section subject to the greatest shear load. This may not be true if the width of the section at the neutral axis is wider than other points in the section. For a rectangular section the maximum shear stress is 1.5 times the average (V/A), while for a solid circular section it is 1.33 times the average.

Extensive tables for calculating shear forces, moments, reactions, slopes, and deflections for straight beams are found in Young [4] and Hsu [8]. A few of the most common beam end conditions are found in the section on Beams at the end of this chapter.

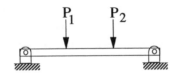

(a) simple beam: pinned ends

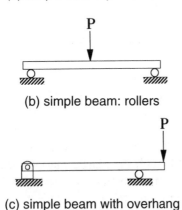

(b) simple beam: rollers

(c) simple beam with overhang

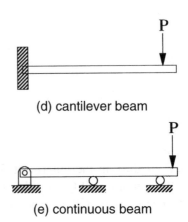

(d) cantilever beam

(e) continuous beam

Figure 19. Examples of beam classifications.

Limitations of General Beam Bending Equations

Often, practicing engineers neglect consideration of the assumptions associated with the flexural formula $\sigma = Mc/I$ and the transverse shearing stress equation $\tau = VQ/It$ and will use these equations indiscriminately. A review of the assumptions behind these equations is helpful.

- The beam is made of an isotropic and homogeneous material with the same modulus of elasticity in tension and compression.
- The beam is straight or has at least a radius of curvature that is more than 10 times the depth. Depending upon the beam cross-section, a radius of curvature equal to 10 introduces error in bending stress (Mc/I) calculations of roughly 5% to 12% [6]. The inside fiber will error on the low side while the outside fiber will error on the high side. Tables providing bending stress correction factors for curved beams are found in Young [4], and Seely and Smith [6].
- The beam has a uniform cross-section and an axis of symmetry in the plane of bending. If the beam cross-section changes gradually and is statically determinant, the flexural formula is adequate for approximate stress calculations. Abrupt changes in cross-section create stress concentrations with high local stresses. Young [4] provides a table for reaction and deflection coefficients for tapered beams.
- All loads and reactions are perpendicular to the axis of the beam and lie in the same plane.
- The beam is long as compared to its depth:
  length/depth > 8 for compact sections of metal beams [4]
  length/depth > 15 for beams with thin webs [4]
- The beam is not disproportionately wide. Since the formula VQ/It averages the shear stress across the thickness t, it is accurate only if the thickness is not too great as compared to its depth. For a rectangular beam where the depth is two times the thickness, the error is only 3%. For a square beam, the error climbs to 12%. For a beam with a thickness-to-depth ratio of 4, the error is 100% [9]!
- The maximum stress does not exceed the proportional limit, i.e., Hooke's Law applies.
- The loads on the beam must be static loads, i.e., impact loads accelerate the beam, thus the stresses and strains would not be predicted by the flexural formula.
- The beam must be free of residual stresses due to thermal gradients, heat treatment, cold working, etc.

Short Beams

The deflection due to shearing stresses is neglected for a beam that is long compared to its depth. However, for a short, heavily loaded beam, the deflection due to shear loads can be significant. The shear stress contribution to deflection is a function of the modulus of rigidity G. In short beams where the material has a small value of G as compared to E, such as wood, the shear contribution to deflection is much more significant. For metal beams, the deflection due to shear stress can still be neglected unless the length/depth (l/d) ratio is very small.

The stress distribution for short, deep beams is no longer linear. However, for a l/d ratio above 3, the flexural formula provides a reasonable approximation of the maximum bending stress [4].

Plastic Bending

For a beam made of a ductile material such as structural steel or wrought aluminum, the beam's resistance to failure under static ultimate loads is determined by calculating the fully plastic moment (M_p). The moment necessary to produce yielding in the outermost fibers only is known as the elastic moment (M_e). As additional moment is applied to the beam, more of the cross-section becomes inelastically strained. At the point where the entire cross-section is inelastically strained, the section is said to be fully plastic. The plastic moment is determined by multiplying the yield strength by the plastic section modulus. Young [4], provides a table with the plastic moduli for a number of cross-sections. For the rectangular section of Figure 20, the plastic moment is equal to $\sigma_{yield} \times (bh^2/4)$, or $1.5 \times M_e$.

Table 3
Plastic Moments for Common Beam Cross-sections

Cross-section	Plastic Moment
solid rectangle	$1.5 \times M_e$
solid circle	$1.698 \times M_e$
I-section	$1.15 \times M_e$
diamond	$2.0 \times M_e$
hourglass	$1.333 \times M_e$

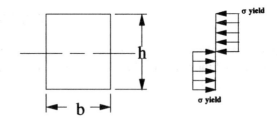

Figure 20. Plastic moment for a rectangular cross-section.

Torsion

The torsional stress and angular twist calculations for a beam with a circular cross-section and length (L) are straightforward applications of Tr'/J and TL/JG, respectively. For noncircular cross-sections, the torsional shear stress and twist cannot be obtained from a simple equation. The maximum torsional shear stress for rectangular sections occurs at the center of each side. A rectangular cross-section where the length exceeds the thickness, has a maximum shear stress of $3T/bt^2$. As the dimensions b and t approach the same magnitude, the following expressions apply if end effects are small (see also Figure 21).

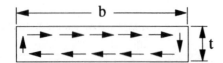

Figure 21. Rectangular cross-section under torsional load.

$$\tau = T/\alpha bt^2$$

$$\phi = TL/\beta bt^3 G$$

where α and β are obtained from Table 4.

Table 4
Constants for Torsional Shear Stress and Angular
Deflection Equations for Noncircular Cross-sections

b/t	1.00	1.50	1.75	2.00	2.50	3.0	4.0	6.0	8.0	10.0	∞
α	0.208	0.231	0.239	0.246	0.258	0.267	0.282	0.299	0.307	0.313	0.333
β	0.141	0.196	0.214	0.229	0.249	0.263	0.281	0.299	0.307	0.313	0.333

Source: [10] Reprinted from Aircraft Structures by D. J. Perry and J. J. Azar with permission of McGraw-Hill Book Company.

If a cross-section is composed of several rectangular elements such as shown in Figure 22, the shear constant $K = \beta bt^3$ for each element can be calculated. The torque applied to the overall cross-section can then be distributed over each rectangular element based on the ratio of K_1 to K_{total}.

$$K_1 = \beta_1 b_1 t_1^3$$

$$K_{total} = \beta_1 b_1 t_1^3 + \beta_3 b_2 t_2^3 + \beta_3 b_3 t_3^3$$

The maximum shear stress for section 1 becomes

$$\tau_1 = T K_1/\alpha b_1 t_1^2 K_{total}$$

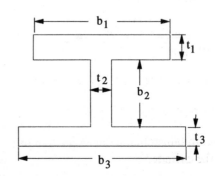

Figure 22. Cross-section consisting of multiple rectangular areas.

PRESSURE VESSELS

Cylindrical pressure vessels can be classified as either "thin-walled" or "thick-walled." If the wall thickness-to-ra- dius ratio of a cylinder is 1/10 or less, it is considered to be thin-walled.

Thin-walled Cylinders

For thin-walled cylinders (Figure 23), the tangential stress due to an internal pressure can be assumed to be uni- formly distributed across the wall thickness. The hoop stress equals $pD/2t$ where p = internal pressure, D = di- ameter, and t = wall thickness. For a closed cylinder, an axial stress $\sigma_z = pD/4t$ will also be induced because of the pres- sure acting on the ends of the cylinder.

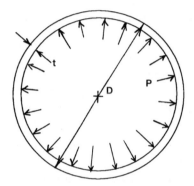

Figure 23. Thin-walled cylinder.

Thick-walled Cylinders

For a thick-walled cylinder (Figure 24), the radial and tangential stresses are a function of the radius. The radial and hoop stresses can be obtained from the equations list- ed in Table 5 for a disk of uniform thickness under inter- nal and/or external pressures. The sum of σ_t and σ_r at all points through the wall is constant.

$$\sigma_r + \sigma_t = 2(p_i r_i^2 + p_o r_o^2)/(r_o^2 - r_i^2)$$

For a closed cylinder the axial stress equals:

$$\sigma_z = (p_i r_i^2 + p_o r_o^2)/(r_o^2 - r_i^2) \ or \ \sigma_z = (\sigma_r + \sigma_t)/2$$

The maximum tangential stress will occur at the inner di- ameter of the cylinder.

$$\sigma_{tmax} = (p_i r_i^2 - p_o r_o^2 + r_o^2(p_i - p_o))/(r_o^2 - r_i^2)$$

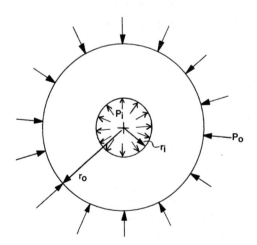

Figure 24. Thick-walled cylinder.

PRESS FITS BETWEEN CYLINDERS

If one cylinder is shrink-fit onto another, such as a bearing race on a shaft (Figure 25), the pressure force p developed between the two cylinders equals:

$$p = \frac{\Delta r}{\dfrac{r_2}{Es}(1 - v_s) + \dfrac{r_2}{E_{br}}\left(\dfrac{r_3^2 + r_2^2}{r_3^2 - r_2^2} + v_{br}\right)}$$

which simplifies to the following if the bearing race and shaft are made of the same material and the shaft is solid.

$$p = \frac{E\,(r_3^2 - r_2^2)\,\Delta r}{2r_2 r_3^2}$$

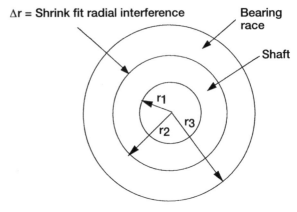

Figure 25. Press fit between cylinders.

ROTATING EQUIPMENT

The design of rotating disks and shafts can present a design challenge, particularly when high rotational speeds are coupled with significant transmitted torque, large number of start-stop cycles, desire for lightweight design, and elevated temperature operation. The mechanical stresses in the hoop and radial directions are a function of the square of the angular velocity (ω). A 10% increase in the rotational speed will increase the stress level by 21%. Thermal gradients in turbine disks have a significant impact on the total stress range experienced in the disk and will have a big effect on the low cycle fatigue life of the disk. Where interference fits known as radial pilots are required between shafting components, the internal or external pressure loads should be included in the calculation of peak stresses.

Rotating Disks

In the detail design of high speed rotating disks, such as those found in the compressor and turbine sections of gas turbine engines for aircraft, finite element models are employed to analyze the sometimes rather complex shape of the disks. However, hand calculation of mechanical stresses is important for preliminary sizing and for checking the validity of a finite element model. The average hoop stress in the disk is evaluated against yield, creep, and ultimate (burst) strength requirements. The average radial stress through a section of the disk web (Figure 26) is evaluated against creep rupture (if higher temperatures are involved) and ultimate strength. The average equivalent stress in the web should meet yield strength requirements to avoid excessive deformation which could lead to contact between rotating and static components. Finally, the fatigue life associated with the peak equivalent stress must be compared to the fatigue life requirements for the design.

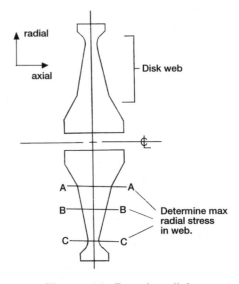

Figure 26. Rotating disk.

Mechanical Stresses

Table 5 presents the equations for calculating the radial and tangential stresses and radial deflections for disks of uniform thickness under internal/external pressure loads and rotation. If a disk is under a combination of these load conditions, the equations can be superimposed on one another. The following points are noted for the rotating rings and disks in Figure 27:

1. A relatively thin ring (thickness t is small compared to the mean radius r) with a mass density of ρ and rotating at ω radians/second will have an average hoop stress of $\sigma = \rho r^2 \omega^2$.
2. For a rotating solid circular disk of uniform thickness, the maximum hoop and radial stresses are equal and occur at the center of the disk:

$$\sigma_{rmax} = \sigma_{tmax} = \rho\omega^2 r_o^2 (3 + v)/8$$

3. For a rotating circular disk with a center hole of radius r_i, the maximum radial stress occurs at $r = (r_i r_o)^{0.5}$ and the maximum tangential stress occurs at the inner radius r_i.

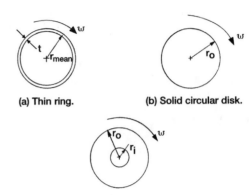

(a) Thin ring. (b) Solid circular disk.

(c) Circular disk with center hole.

Figure 27. Rotating rings and disks.

$$\sigma_{rmax} = (3 + v)\,\rho\omega^2(r_o - r_i)^2/8$$

$$\sigma_{tmax} = \rho\omega^2[(3 + v)r_o^2 + (1 - v)\,r_i^2]/4$$

Thermal Stresses in Disks

Tensile and compressive hoop stresses are induced in disks with radial thermal gradients. If the outer diameter of the disk is hotter than the inner diameter, then the outer

Table 5
Stress Equations for Disks of Uniform Thickness

LOAD	RADIAL STRESS	TANGENTIAL STRESS	AVERAGE TANGENTIAL STRESS	RADIAL DEFLECTION
SOLID DISK				
EXTERNAL PRESSURE (P_o) + ROTATION	$-P_o + \dfrac{3+v}{8}\,\rho\omega^2\left(r_o^2 - r^2\right)$	$-P_o + \dfrac{3+v}{8}\,\rho\omega^2\left(r_o^2 - \dfrac{1+3v}{3+v}r^2\right)$	$-P_o + \dfrac{\rho r_o\omega^2}{3}$	$-P_o r\dfrac{1-v}{E} + \dfrac{1-v}{8E}\,\rho\omega^2 r\left[(3+v)r_o^2 - (1+v)r^2\right]$
DISK WITH CENTER HOLE				
INTERNAL PRESSURE (P_i)	$P_i\dfrac{r_i^2}{r_o^2 - r_i^2}\left(1 - \dfrac{r_o^2}{r^2}\right)$	$P_i\dfrac{r_i^2}{r_o^2 - r_i^2}\left(1 + \dfrac{r_o^2}{r^2}\right)$	$\dfrac{P_i r_i}{r_o - r_i}$	$\dfrac{P_i r}{E}\dfrac{r_i^2}{r_o^2 - r_i^2}\left[(1-v) + (1+v)\dfrac{r_o^2}{r^2}\right]$
EXTERNAL PRESSURE (P_o)	$-P_o\dfrac{r_o^2}{r_o^2 - r_i^2}\left(1 - \dfrac{r_i^2}{r^2}\right)$	$-P_o\dfrac{r_o^2}{r_o^2 - r_i^2}\left(1 + \dfrac{r_i^2}{r^2}\right)$	$\dfrac{-P_o r_o}{r_o - r_i}$	$\dfrac{-P_o r}{E}\dfrac{r_o^2}{r_o^2 - r_i^2}\left[(1-v) + (1+v)\dfrac{r_i^2}{r^2}\right]$
ROTATION ONLY	$\dfrac{\rho\omega^2(3+v)}{8}\left(r_o^2 + r_i^2 - \dfrac{r_o^2 r_i^2}{r^2} - r^2\right)$	$\dfrac{\rho\omega^2(3+v)}{8}\left(r_o^2 + r_i^2 + \dfrac{r_o^2 r_i^2}{r^2} - \dfrac{1+3v}{3+v}r^2\right)$	$\dfrac{\rho\omega^2}{3}\left(r_o^2 + r_o r_i + r_i^2\right)$	$\dfrac{\rho\omega^2 r}{E}\dfrac{(3+v)(1-v)}{8}\left(r_o^2 + r_i^2 + \dfrac{1+v}{1-v}\dfrac{r_o^2 r_i^2}{r^2} - \dfrac{1+v}{3+v}r^2\right)$

r_o = outer radius, r_i = inner radius, r = radius for calculation, ρ = mass density, ω = rotational velocity, v = Poisson's ratio.

fibers will be in compression while the inner fibers are in tension. A reverse gradient will have the opposite effect on the sign of the hoop stresses in the disk. The radial thermal gradient also increases the radial stresses in the web of the disk. In high temperature applications, the transient thermal gradients may be severe and affect the fatigue life of the component.

Yield Stress Criteria

A yield stress criteria for a rotating disk ensures that no gross deformation will occur under the limits of normal operation. Under the combined effects of rotation and thermal gradients, the disk average tangential stress and the maximum web average equivalent stress, i.e., averaged through the thickness of the disk web, must remain under the yield strength of the material. In aircraft engine structures, it is typical practice to use the -3 sigma yield strength of the material.

Ultimate/Burst Stress Criteria

An ultimate stress criteria for a rotating disk prevents the disk from bursting below a predetermined rotational velocity. A disk allowed to burst will cause tremendous damage to surrounding structure and present a safety hazard for any nearby personnel. Usually, a burst disk will rupture into several large chunks and numerous small fragments. If tangential stresses cause the burst, the disk will rupture through the bore or center of the disk (Figure 28a). A radial burst would leave the center of the disk intact and the disk would rupture circumferentially through the web of the disk (Figure 28b).

The rupture stress at which a disk bursts is a function of the ultimate tensile strength and the ductility of the material. Ductility is measured by the percentage elongation present in the material specimen at fracture. A material with less than 5% elongation is usually designated as brittle, while a material with more than 5% elongation is ductile. Cast materials, as a rule, have much lower ductility than forged materials. As noted in the following procedure, ductile castings with greater than 5% elongation actually exhibit a higher percentage of utilizable ultimate strength than very ductile forged materials as long as the peak tangential stress is not more than twice the average tangential stress of the disk. However, a forged material has higher ultimate strength, thus providing higher rupture strength. The ultimate strength used in the burst calculation should be degraded by the scatter (three standard deviations are often used for aircraft components) associated with the material data.

Tangential Burst Calculation:

1. Calculate average tangential stress for the disk.
2. Calculate maximum tangential stress for the disk.
3. Determine average rupture strength, i.e., average stress at which disk burst occurs.
   - For a typical cast material:

Elongation	Avg. Tangential Stress/ Max Tangential Stress	Rupture Strength
5 to 8%	≥0.7	0.95 × Ult.
≈1%	0.7	0.80 × Ult.
≈1%	0.2	0.45 × Ult.
≈8%	0.2	0.80 × Ult.

   - For high ductility materials, such as forgings, the rupture strength is estimated to be 85% to 90% of the ultimate strength.
4. Tangential Burst Speed = (Rupture Strength/Avg. Tangential Stress)$^{0.5}$ (100) (Rotational Speed).

The average tangential stress in a turbine disk does not change appreciably when thermal gradients are applied to the disk. Thus, even with thermal effects the average tangential stress is approximately proportional to the square of the rotational speed.

Radial Burst Calculation:

1. Calculate the maximum average radial stress through the thickness of the disk web for rotation only ($\sigma_{rotation}$).

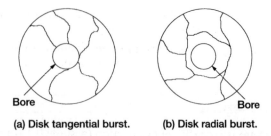

Bore   Bore

(a) Disk tangential burst.   (b) Disk radial burst.

Figure 28. Rotating disk burst failure modes.

2. If thermal gradients are present in the disk, calculate the maximum average radial stress through the thickness of the disk web due to temperature ($\sigma_{thermal}$).

3. Typically, a factor of .95 is applied to the ultimate strength regardless of ductility for a radial burst calculation. Since thermal stresses do not ratio with speed, the allowable rupture strength is degraded by the magnitude of the radial stress due to temperature.

4. Radial Burst Speed = $([0.95$ Ultimate Strength $- \sigma_{thermal}]/\sigma_{rotation})^{0.5}$ (100) (Rotational Speed).

Rotating Shafts

Stresses Due to Torsion, Thrust, and Bending Loads

Rotating shafts may be subjected to a combination of torsion, bending, and axial thrust loads. Since these loads are often a combination of static and alternating loads, both fatigue and static strength criteria are considered in shafting design.

Select sections as illustrated in Figure 29 that include holes, shoulder fillets, splines, minimum section properties, and other sources of stress concentration for stress and life analysis. Calculate the maximum operating torque, bending moment, and axial load at each section.

$$T = 63{,}025(HP)/N$$

where   T = torque (in. lb)
         HP = horsepower
         N = rpm

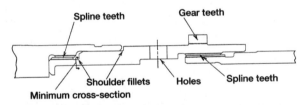

Figure 29. Shafting features requiring stress analysis.

Moments may be imparted to the shaft by gear loads, rectilinear accelerations normal to the shaft or gyroscopic loads. Gyroscopic loads are always considered for aircraft applications of rotating shafting.

$$M_{gyro} = I\omega\Omega$$

where I = polar moment of inertia of the shaft components (in. lb sec^2)
     ω = angular velocity of shaft (radians/sec)
     Ω = precession speed (radians/sec) or angular velocity of the shaft spin axis about an axis normal to the spin axis

A gyroscopic moment will cause an alternating bending stress in the shaft. The torsional stress will remain steady. Experimental results [11] indicate that the bending fatigue strength is not sensitive to the torsional mean stress until it exceeds the torsional yield strength by as much as 50%.

Bearings should be located near the largest masses connected to the shafting to minimize the bending moments associated with rectilinear accelerations and gyroscopic effects. Nonconcentrated loads should be limited to 85% of the yield strength.

Mechanical Couplings

Flanged. This type of coupling is used to transmit torque through shear in the attaching bolts or by friction between the adjacent flanges.

- Figure 30 depicts two mating shaft flanges connected with "body bound" bolts such that the torque is transmitted solely through the bolts. The bolts themselves pilot the radial position of the flange. With this type of assembly, it is necessary to ream the bolt holes on assembly or use the same fixture for both parts. The bolt shear stress is:

$$\tau = T/NAR$$

where T = torque (in. lb)
     N = number of bolts
     A = bolt shear area (in.^{2})
     R = bolt circle radius (in.)

- Figure 31 shows a rotating flanged joint where a circumferential flange pilot is used to control alignment and the torque is transmitted by friction between the bolted flange surfaces. The bolt tensile stress required to drive the torque by friction is:

$$\sigma_a = T/NAR\mu$$

where μ = coefficient of friction (assume 0.1 for metal on metal contact)

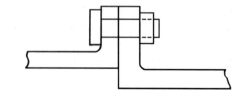

Figure 30. Bolts transmit torque.

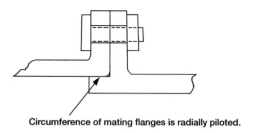

Circumference of mating flanges is radially piloted.

Figure 31. Friction between mating surfaces transmits torque.

Splined. The splined coupling in Figure 32 is a torque-carrying connection with axial or helical teeth on the outer diameter and inner diameter of the male and female members, respectively. The involute spline profile is common and similar to gear teeth. Splines differ from gears in that there is no rolling action and all teeth may contact at the same time. When failure occurs, splines usually fail by fretting or fatigue. The two basic types of splines with respect to fit are fixed (nonworking) and flexible or floating (working) splines. Fixed splines do not allow relative motion between the internal and external teeth. A radial press fit between the internal and external teeth or on a radial pilot surface of the shaft adjacent to the spline prevents relative motion. Flexible splines permit some degree of rocking motion and can accommodate axial movement between the mating shafts. Case-hardened surfaces should be used with this spline type. Positive lubrication is required at all times for flexible splines to minimize wear and fretting. The rim or supporting ring of material beneath the spline teeth should have a minimum thickness of 1.25 times the total tooth height.

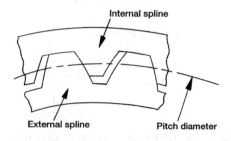

Figure 32. Splined coupling.

Highly loaded aircraft splines will have spline lengths ranging from 50% to 100% of the pitch diameter. For flexible splines, a wide face width has limited value in extending the load-carrying capacity. Generally flexible splines are misaligned (since accommodation of misalignment is one of its main functions), thus the contact load between the teeth tends to load the ends of the spline. Increased length does not relieve the intensity of the loading but will increase the time required for the wear in the center of the spline to catch up with the wear at the ends. The number of spline teeth is governed by cost and manufacturing considerations because doubling the number of teeth does not have a large effect on the tooth stresses, i.e., the teeth become half as big when doubled, but there are twice as many teeth to carry the load—the two effects tend to cancel each other.

A complete evaluation of a splined coupling should include the tooth shear stress, torsional shear stress, tangential or bursting stress due to the radial load between the mating spline teeth and rpm, and the bearing or contact stresses between the teeth. Allowable spline stresses depend upon the material, severity of loading, number of torque cycles, and whether the spline is fixed or flexible. Fixed splines have a greater ability to carry the contact stress because there is no relative motion between the mating teeth.

Curvic®. A fixed curvic® (a registered name by Gleason Works in Rochester, New York) is a face spline with teeth which are spaced circumferentially about the face and possess a characteristic curved shape when viewed in a plane perpendicular to the coupling axis (Figure 33). A curvic®

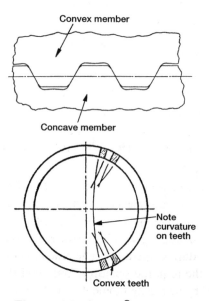

Figure 33. Curvic® coupling.

coupling between shafting members must be bolted to-gether with sufficient load to handle the axial separation loads due to shaft operating thrust loads, axial reaction from rotor torque, axial inertia loads, and bending loads due to lateral inertia loads, angular accelerations, and gyro-scopic loads. The tooth shear and contact stresses govern the required size of the curvic® coupling. The design fac-tors governing curvic design can be obtained from the *Gleason Curvic Coupling Design Manual* [12].

FLANGE ANALYSIS

Mating cylindrical cases such as those found in aircraft engine applications may have axial as well as circumfer-ential splitline flanges. These cases are subject to internal gas pressures. Under relatively low pressures, the flush type flange shown in Figure 34 is adequate. However, the undercut type flange joint of Figure 35 is recommended for those nongasketed high pressure applications where gas leakage must be minimized. A rigorous stress and deflec-tion analysis of flanges can be somewhat complex de-pending upon the geometry, thermal gradients, and load con-ditions. The following guidelines offer assistance in the preliminary sizing of typical splitline flanges.

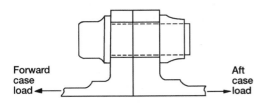

Figure 34. Flush flange.

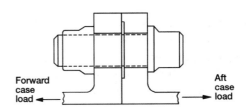

Figure 35. Undercut flange.

Flush Flanges

- Design splitline joint such that the flanges do not sep-arate (may leak) under proof test loading which is often specified as two times the maximum operating pressure load.

- Never exceed a five-bolt diameter separation between bolts to ensure that leakage is minimized.

- Account for the difference in thermal expansion coef-ficients for bolt and flange materials under operating conditions when calculating the required cold assem-bly bolt load.

- If the case moment at the junction of the flange and case is not significant, then the minimum bolt load required to react the case load at operating conditions can be cal-culated using Figure 36. Assume the flanges open at point D during operation. The reaction per bolt at point A and the required operating bolt load F_B can be cal-culated by summing moments about point A. F_{case} is the case load per bolt.

$$\Sigma M_A = 0$$

$$F_{case} (a + b) = F_B (a)$$

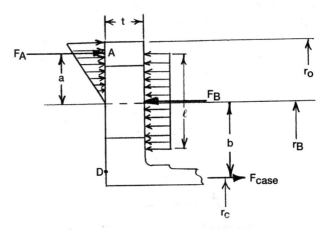

Figure 36. Calculation of operating bolt load for flush flange.

where $a = 2/3(r_O - r_B)$
$b = r_B - r_C$
$\therefore F_B = F_{case}(a + b)/a$

- The bending moment between bolts is calculated as follows:

$$M_B = F_A(a) - F_B \frac{\ell}{8}$$

Remember to use the maximum permissible bolt load to calculate the bending moment. The flange bending stress σ_b per bolt equals:

$$\sigma_b = Mc/I = 6M/ht^2$$

where h is the distance between bolt holes.

Undercut Flanges

- Undercut flanges minimize the clamping load requirements. The bolt clamping force is applied through two narrow contact lands between the mating flanges. The local stiffness of the flange is reduced by the undercut. During assembly, the reduced stiffness allows slightly more deflection under bolt preload, hence it provides some additional margin for thermal mismatch between the flange and bolt materials.
- The flange thickness and bolt load must be designed such that the undercut area does not bottom out against the mating flange surface. This can be accomplished by limiting the bending stress in the flange to the yield strength of the flange material.
- The undercut width should be approximately 50% wider than the bolt diameter, and the outer contact land width equal to 20% of the bolt diameter.
- If the case moment at the junction of the flange and the case is not significant, then the minimum bolt load at operating conditions can be calculated using Figure 37. The mating flanges should remain in contact at points A and D. As long as the preload F_D on the inner contact land D exceeds the case load (F_{case}), the bending moment is dependent upon F_D. First calculate the required operating preload F_D to prevent separation at D.

$$F_D = F_{case}(c/b)$$

where $b = r_b - r_f + (r_f - ri)/3$
$c = r_b - r_c$

Then sum moments about the bolt circle point B to solve for F_A.

$$\Sigma M_B = 0$$

$$F_A(a) = F_D(b)$$

where $a = r_e - r_b + (r_o - r_e)/3$

The minimum operating bolt load equals:

$$F_B = F_A + F_D$$

The bending moment and stress at the bolt circle are calculated in a manner identical to the flush flange.

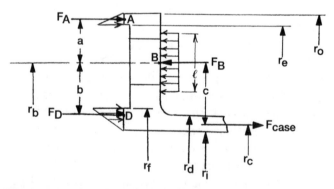

Figure 37. Calculation of operating bolt load for undercut flange.

MECHANICAL FASTENERS

The selection of appropriate mechanical fasteners is not an insignificant consideration in the design of certain products. Two to three million fasteners are used in the construction of a single jumbo jet. Choice of the correct fasteners is a function of the parts being joined, space limitations, severe operating loads which include static, cyclic, and thermally induced loads, and the assembly and maintenance requirements. Fasteners are designated as to whether the application is predominately shear or tension.

Threaded Fasteners

Threaded fasteners include screws, bolts, and studs. Several fundamental quantities which apply to screw threads include the following:

pitch: the distance between adjacent thread forms.

pitch diameter: the diameter to an imaginary line drawn through the thread profile such that the width of the thread tooth and groove are equal.

major diameter: the largest diameter of the screw thread.

minor diameter: the smallest diameter of the screw thread.

thread tensile area: the tensile area of a screw thread is based upon the experimental evidence that an unthreaded rod with a diameter equal to the mean of the pitch and minor diameters will have the same strength as the threaded rod.

proof load: the maximum tensile load that a bolt can tolerate without incurring a permanent set.

Bolted Joints

Bolted joints which often resist a combination of external tensile and shear loads are the focus of this section. Bolt material should be strong and tough, whereas the nut material should be relatively soft, i.e., more ductile. A soft nut allows some plastic yielding which results in a more even load distribution between the engaged threads. Three full threads are required to develop the full bolt strength, and good design practice dictates that the bolt extend two full threads beyond the outer end of the nut.

The bolts and the clamped joint members must possess similar thermal coefficients of expansion to minimize load fluctuation in different thermal operating environments. A larger length-to-diameter (L/D) ratio will allow bolt flexibility to offset any difference in thermal expansion. In higher temperature environments, the bolt material must be selected for resistance to creep to prevent loss of preload.

A bolt should be relatively flexible as compared to the joint members being clamped together. Bolts with the largest possible L/D ratio decrease the potential for vibration loosening of the bolt. A L/D > 8 would effectively prevent this occurrence [13]. Bolt stiffness is a function of the effective length of the bolt. For a large L/D ratio, the thread engagement has a small effect upon the effective length. The effect is more significant for short bolts. Typically one-half the nut or hole threads are assumed to carry some load and contribute to the effective length. The spring rate for a bolt equals AE/L, where A = area, E = modulus of elasticity, and L = effective bolt length. The clamped components act as compressive springs in series such that the total spring rate of the members is $1/K_t = 1/K_1 + 1/K_2 + \ldots + 1/K_n$. The actual effective stiffness of each member is difficult to obtain without experimentation or finite element modeling, as the bearing force between the clamped components spreads out in a nonuniform manner. For the sake of approximation, use a cylinder with an outer diameter of three times the bolt diameter and an inner diameter equal to the bolt diameter to represent the components clamped together by the bolt. If the bolt and clamped components have the same modulus E, this assumption infers that the clamped components are eight times as stiff as the bolt.

Bolt Preload

In a bolted joint under tension, the bolt preload has two functions: keeping the clamped parts together and increasing the fatigue resistance of the joint The preload is proportional to the torque applied to the bolt head. This relationship between torque and preload is dependent on the actual coefficient of friction between the bolt and the mating components. The coefficients of friction for the bolt threads and bearing surfaces of the bolt head and nut range from 0.12 to 0.20 depending upon the material and lubrication. Approximately 50% of the assembly torque is used to overcome friction between the bearing face of the nut and mating clamped component. Another 40% is used to overcome thread friction, and the balance produces bolt tension.

Depending upon the application, maximum bolt preload recommendations range from 75% to 100% of the proof load [14]. Using 100% of the proof strength reduces the number of bolts and generally reduces the alternating load on the bolt, i.e., increases the fatigue life. However, for joints that experience substantial cyclic loading, a high preload may actually lower fatigue life because of the high mean stress. Applications which demand repeated assembly and disassembly are not good candidates for the 100% proof load specification, as the bolts will experience some yield in service and should not be reused. The 100% goal also requires more sophisticated assembly equipment to guarantee that the bolts are not overloaded.

The lower end of the range (75% to 80%) is much more widely used, as it provides an adequate margin of safety for traditional methods of assembly where a torque wrench is used to meet a specified torque. The combination of tensile and torsional stresses at the outer surface of a bolt often reach the yield strength at 80% of the proof load.

As pointed out in the previous section, the clamped joint components are generally several times more stiff than the bolts. An external static tensile load applied to the joint will extend the bolt and relieve the compression in the joint members. If the joint opens, the bolt will feel 100% of the external load. Assuming that the bolt and joint members are in the elastic range and the joint does not separate, the degree of external load experienced by the bolt is proportional to the ratio of bolt to joint stiffness. For example, if the joint members are eight times as stiff as the bolt, the bolt will feel approximately 11% of the external load. A cyclic external load is split between the bolt and joint members in the same fashion. Thus, the bolt usually only experiences a fraction of the cyclic load applied to the joint. A higher bolt preload will lower the effect of the cyclic load on the total bolt load. As a rule, the cyclic load in the bolt should be less than 25% of its yield strength.

In general, joint fatigue can occur when the alternating stress amplitude in the bolt exceeds the fatigue strength of the bolt or if the joint opens and the bolt experiences the full external load.

Thread and fillet rolling after heat treatment will increase the fatigue strength of a bolt by creating a residual compressive stress.

Common Methods for Controlling Bolt Preload

Torque. A specified range of torque is applied to the nut or bolt by some form of a calibrated torque wrench. In terms of application, this method is the simplest and is used where threads assemble into blind holes. Since the preload is a function of the coefficients of friction, the preload may vary by 25% [15].

Bolt elongation. This method can be used when the overall length of the bolt can be measured with a micrometer after assembly to achieve an accuracy of 3% to 5% for the preload [15]. The required bolt elongation (δ) can be calculated using the bolt stiffness:

$$\delta = PL/AE$$

where P is the required bolt preload.

Nut Rotation. The nut rotation method requires a calculation of the fractional number of turns of the nut required to develop the desired preload. The nut rotation is measured from the snug or finger-tight condition. This method controls the preload to within 15% of desired levels [15].

Strain gages. Special fasteners with strain gages located inside the fastener or on its surface can be used to control the preload to within one percent [15]. Due to expense, this degree of control is usually used for design development.

Pins

A pin is a simple and inexpensive fastener for situations where the joint is primarily loaded in shear. The two broad classes of pins are semipermanent and quick release. The semipermanent class includes the standard machine pins which are grouped in four categories: dowel, taper, clevis, and cotter. In general, semipermanent pins should not be aligned such that the direction of vibration loads parallels the pin axis. Also, the shear plane of the pin should not lie more than one diameter from the end of the pin. Specific design data for each type of pin is available in vendor catalogs.

Rivets

Rivets are permanent shear fasteners in which the rivet material is deformed to provide some clamping or retaining ability. Rivets should not be used as tension fasteners because the formed head is not capable of sustaining tensile loads of any magnitude. There are two families of rivets: tubular and blind. As the name suggests, the blind rivets require access to only one side of the components being assembled.

In terms of design and analysis, riveted joints are treated exactly like bolted joints that are loaded in shear. These joints can fail by shear of the rivets, tensile failure of the joined members, crushing (bearing stress) failure of the rivet or joined members, or shear tear-out. For rivet shear stress calculation, the nominal diameter (D) of the rivet is used for the area calculation. The tensile stress in the joined members is based on the net area (area with holes removed), with the stress concentration effects included for cyclic loads. The bearing stress between the rivets and joined members is calculated using the projected rivet area $A = tD$, where t is the thickness of the thinnest joined member. Shear tear-out is avoided by maintaining an edge distance greater than one and one-half diameters.

Additional design tips include:

1. Use washers to reinforce riveted joints in brittle material and thin sections.
2. When joining thick and thin members, position the rivet head against the thin section.

WELDED AND BRAZED JOINTS

Welding is defined as a group of metal joining processes which allow parts to coalesce along their contacting surface by application of heat, pressure, or both. A filler metal with a melting point either approximately the same or below that of the base metals may be used. Welded joints should be designed such that the primary load transfer produces shear load rather than a tension load in the weld. Sharp section changes, crevices, and other surface irregularities should be avoided at welded joints.

Fillet and butt welds are common weld forms found in machine components and pressure vessels. Fillet welds should be between 1.0 and 1.5 times the thickness of the thinnest material in the joint. For fillet and butt welds, the average shear stress is calculated using the weld throat area (see Figure 38). If the joint is subject to fatigue loads, the appropriate stress concentration factor is applied to the nominal cyclic stress. The reinforcement shown for the butt weld will cause a stress concentration, thus it is often necessary to grind this extra material off if the joint is subject to cyclic loads.

Depending upon the geometry and type of welding process, it may be difficult to guarantee full weld penetration. Often either larger factors of safety are used to compensate for this potential or the effective weld area is reduced. Welding codes generally have conservatism built into the allowable stresses. Both the strength of the weld metal and the joined parent materials in the welded condition must be determined.

Brazing is defined as a group of metal joining processes where the filler material is a nonferrous metal or an alloy whose melting point is lower than that of the metals to be joined. The brazing process spreads the filler material between closely fitted surfaces by capillary attraction. The strength of the brazed joint depends upon the surface area of the joint and the clearance between the parts being joined. A lap of four times the thickness of the thinnest part being joined is typically specified for brazing.

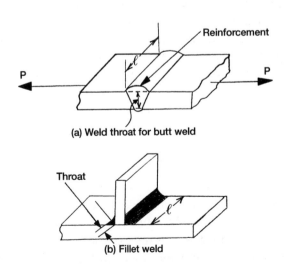

Figure 38. Weld throat area.

CREEP RUPTURE

Creep is plastic deformation which increases over time under sustained loading at generally elevated temperatures. Stress rupture is the continuation of creep to the point where failure takes place. Metallic and nonmetallic materials vary in their susceptibility to creep, but most common structural materials exhibit creep at stress levels below the proportional limit at elevated temperatures which exceed one-third to one-half of the melting temperature. A few metals, such as lead and tin, will creep at ordinary temperatures.

The typical strain-time diagram in Figure 39 for a material subject to creep illustrates the three stages of creep behavior. After the initial elastic deformation, the material exhibits a relatively short period of primary creep (stage 1), where the plastic strain rises rapidly at first. Then the strain versus time curve flattens out. The flatter portion of the curve is referred to as the secondary or steady state creep (stage 2). This is the stage of most importance to the en-

gineer in the design process. The final or tertiary creep stage (stage 3) is characterized by an acceleration of the creep rate, which leads to rupture in a relatively short period of time.

High stresses and high temperatures have comparable effects. Quantitatively, as a function of temperature, a logarithmic relationship exists between stress and the creep rate. A number of empirical procedures are available to correlate stress, temperature, and time for creep in commercial alloys. The Larson-Miller parameter (K_{LM}) is an example of one of these procedures. The general form of the Larson-Miller equation is:

$$K_{LM} = (0.001)(T + 460)(\log ct + 20)$$

where

T = temperature in °F.
t = time (hours)
c = empirical parameter relating test specimens to design
    component

Creep strength is specified as the stress corresponding to a given amount of creep deformation over a defined period of time at a specified temperature, i.e., 0.5% creep in 10,000 hours at 1,200°F. The degree of creep that can be tolerated is a function of the application. In gas turbine engines, the creep deformation of turbine rotating components must be limited such that contact with the static structure does not occur. In such high temperature applications, stress rupture can occur if the combination of temperature and stress is too high and leads to fracture. As little as a 20° to 30°F increase in temperature or a 10% increase in stress can halve the creep rupture life. *Mark's Handbook* [16] provides some creep rate information for steels.

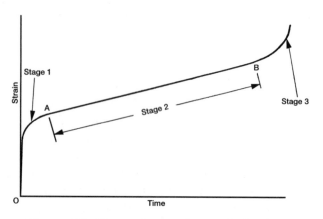

Figure 39. Three stages of creep behavior.

FINITE ELEMENT ANALYSIS

Over the last 25 years, the finite element method (FEM) has become a standard tool for structural analysis. Advances in computer technology and improvements in finite element analysis (FEA) software have made FEA both affordable and relatively easy to implement. Engineers have

access to FEA codes on computers ranging from mainframes to personal computers. However, while FEA aids engineering judgment by providing a wealth of information, it is not a substitute.

Overview

FEM has its origins in civil engineering, but the method first matured and reached a higher state of development in the aerospace industry. The basis of FEM is the representation of a structure by an assemblage of subdivisions, each of which has a standardized shape with a finite number of degrees of freedom. These subdivisions are finite elements. Thus the continuum of the structure with an infinite number of degrees of freedom is approximated by a number of finite elements. The elements are connected at nodes, which are where the solutions to the problem are calculated. FEM proceeds to a solution through the use of stress and strain equations to calculate the deflections in each element produced by the system of forces coming from adjacent elements through the nodal points. From the deflections of the nodal points, the strains and stresses are calculated. This procedure is complicated by the fact that the force at each node is dependent on the forces at every other node. The elements, like a system of springs, deflect until all the forces balance. The solution to the problem requires that a large number of simultaneous equations be solved, hence the need for matrix solutions and the computer.

Each FEA program has its own library of one-, two-, and three-dimensional elements. The elements selected for an analysis should be capable of simulating the deformations to which the actual structure will be subjected, such as bending, shear, or torsion.

The Elements

One-dimensional Elements

The term one-dimensional does not refer to the spatial location of the element, but rather indicates that the element will only respond in one dimension with respect to its local coordinate system. A truss element is an example of a one-dimensional element which can only support axial loads. See Figure 40.

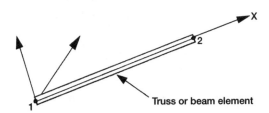

Figure 40. One-dimensional element.

Two-dimensional Elements

A general two-dimensional element can also span three-dimensional space, but displacements and forces are limited to two of the three dimensions in its local coordinate system. Two dimensional elements are categorized as plane stress, plane strain, or axisymmetric.

Plane stress problems assume a small dimension in the longitudinal direction such as a thin circular plate loaded in the radial direction. As a result the shear and normal stresses in the longitudinal direction are zero. Plane strain problems pertain to situations where the longitudinal dimension is long and displacements and loads are not a function of this dimension. The shear and normal strains in the longitudinal direction are equal to zero. Axisymmetric elements are used to model components which are symmetric about their central axes, i.e., a volume of revolution. Cylinders with uniform internal or external pressures and turbine disks are examples of axisymmetric problems. Symmetry permits the assumption that there is no variation in stress or strain in the circumferential direction.

Two-dimensional elements may be triangular or quadrilateral in shape. Lower order linear elements have only corner nodes while higher order isoparametric elements may have one or two midsides per edge. The additional edge nodes allow the element sides to conform to curved boundaries in addition to providing a more accurate higher order displacement function. See Figure 41.

Three-dimensional Elements

Three-dimensional solid elements are used to model structures where forces and deflections act in all three directions or when a component has a complex geometry that does not permit two-dimensional analysis. Three-dimensional elements may be shell, hexahedra (bricks), or tetrahedra; and depending upon the order may have one or two midside nodes per edge. See Figure 42.

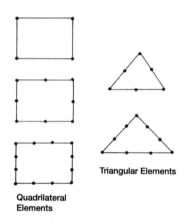

Triangular Elements

Quadrilateral
Elements

Figure 41. Two-dimensional elements.

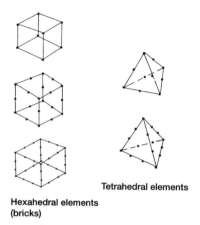

Tetrahedral elements

Hexahedral elements
(bricks)

Figure 42. Three-dimensional elements.

Modeling Techniques

The choice of elements, element mesh density, boundary conditions, and constraints are critical to the ability of a model to provide an accurate representation of the physical part under operating conditions.

- Element mesh density is a compromise between making the mesh coarse enough to minimize the computation time and fine enough to provide for convergence of the numerical solution. Until a "feel" is developed for the number of elements necessary to adequately predict stresses, it is often necessary to modify the mesh density and make additional runs until solution convergence is achieved. Reduction of solution convergence error achieved by reducing element size without changing element order is known as *h-convergence.*
- Models intended for stress prediction require more elements than those used for thermal or dynamic analyses. Mesh density should be increased near areas of stress concentration, such as fillets and holes (Figure 43). Abrupt changes in element size should be avoided, as the mesh density transitions away from the stress concentration feature.
- Compared to linear corner noded elements, fewer higher order isoparametric elements are required to model a structure. In general, lower order 2D triangular elements and 3D tetrahedral solid elements are not adequate for structural analysis. Some finite element codes use an automated convergence analysis technique

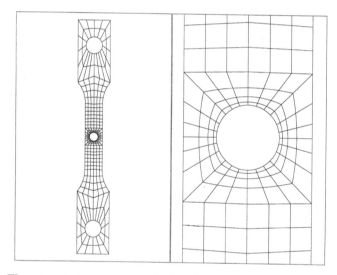

Figure 43. Increase mesh density near stress concentrations.

known as the p-convergence method. This method maintains the same number of elements while increasing the order of the elements until solution convergence is achieved or the maximum available element order is reached.

- Convergence of the maximum principal stress is a much better indicator than the maximum Von Mises equivalent stress. The equivalent stress is a local measure and does not converge as smoothly as the maximum principal stress.

- Elements with large aspect ratios should be avoided. For two-dimensional elements, the aspect ratio is the ratio of the larger dimension to the smaller dimension. While an aspect ratio of one would be ideal, the maximum allowable element aspect ratio is really a function of the stress field in the component. Larger aspect ratios with a value of 10 may be acceptable for models of components such as cylinders subjected only to an axial load. Generally, the largest aspect ratio should be on the order of 5.
- Highly distorted elements should be avoided. Two-dimensional quadrilateral and three-dimensional brick elements should have corners which are approximately right angled and resemble rectangles and cubes respectively as much as possible, particularly in regions of high stress gradient. The angle between adjacent edges of an element should not exceed 150° or be less than 30°. Many current finite element modeling codes have built-in options which permit identification of elements with sufficient distortion to affect the model's accuracy.
- Symmetry in a component's geometry and loading should be considered when constructing a model. Often, only the repeated portion of the component need be modeled. A section of a shaft contains three equally spaced holes. A solid model of the shaft containing one hole or even one-half of a hole (Figure 44), if the holes are loaded in a symmetric manner, must be modeled to perform the analysis. Appropriate constraints which define the hoop continuity of the shaft

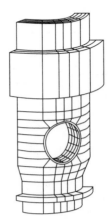

Figure 44. Sector model of a shaft cross-section containing three holes.

must be applied to the nodes on the circumferential boundaries of the model.
- A number of FEA modeling codes have automated meshing features which, once the solid geometry is defined, will create a mesh at the punch of a button. This greatly speeds the production of a model, but it cannot be assumed that the model that is created will be free of distorted elements. Auto mesh programs are prone to creating an excessive number of elements in areas where the stress field is fairly uniform and such mesh density is unwarranted. The analyst must use available mesh controls and diagnostic tools to minimize these potential problems.

Advantages and Limitations of FEM

Generally, the finer the element mesh, the more accurate the analysis. However, this also assumes that the model is loaded appropriately to mimic the load conditions to which the part is exposed. It is always advisable to ground the analysis with actual test results. Once an initial correlation between the model and test is established, then subsequent modifications can be implemented in the model with relative confidence. In many instances, the FEA results predict relative changes in deflection and stress between design iterations much better than they predict absolute deflections and stresses.

CENTROIDS AND MOMENTS OF INERTIA FOR COMMON SHAPES

Key to table notation: A = area (in.^{2}); I$_1$ = moment of inertia about axis 1-1 (in.^{4}); J$_0$ = polar moment of inertia (in.^{4}); c—denotes centroid location; α and β are measured in radians.

Rectangle

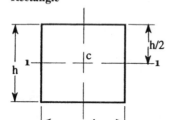

$$A = bh$$
$$I_1 = \frac{bh^3}{12}$$

Trapezoid

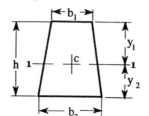

$$A = h(b_1 + b_2)/2$$
$$I_1 = h^3(b_1^2 + 4b_1b_2 + b_2^2)/(36(b_1 + b_2))$$
$$y_1 = h[(2b_2 + b_1)/3(b_2 + b_1)]$$
$$y_2 = h[(b_2 + 2b_1)/3(b_2 + b_1)]$$

Circle

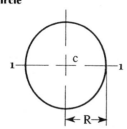

$$A = \pi R^2$$
$$I_1 = \frac{\pi R^4}{4}$$
$$J_o = \frac{\pi R^4}{2}$$

Circular Sector

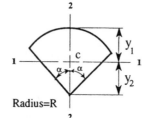

$$A = \alpha R^2$$
$$I_1 = \frac{R^4}{4}[\alpha + \sin\alpha \cos\alpha - 16\sin^2\alpha/9\alpha]$$
$$I_2 = \frac{R^4}{4}[\alpha - \sin\alpha \cos\alpha]$$
$$y_1 = R[1 - 2\sin\alpha/3\alpha]$$
$$y_2 = 2R\sin\alpha/3\alpha$$

Semicircle

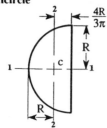

$$A = \frac{\pi R^2}{2}$$
$$I_1 = \frac{\pi R^4}{8}$$
$$I_2 = 0.1098 R^4$$

Solid Ellipse

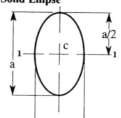

$$A = \frac{\pi ba}{4}$$
$$I_1 = \frac{\pi ba^3}{64}$$

Hollow Circle

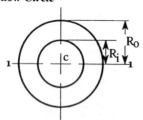

$$A = \pi(R_o^2 - R_i^2)$$
$$I_1 = \frac{\pi}{4}(R_o^4 - R_i^4)$$
$$J_o = \frac{\pi}{2}(R_o^4 - R_i^4)$$

Hollow Ellipse

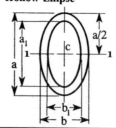

$$A = \frac{\pi}{4}(ba - b_1a_1)$$
$$I_1 = \frac{\pi}{64}(ba^3 - b_1a_1^3)$$

Triangle

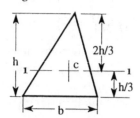

$$A = \frac{bh}{3}$$
$$I_1 = \frac{bh^3}{36}$$

Thin Annulus

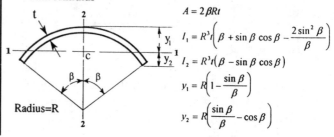

$$A = 2\beta Rt$$
$$I_1 = R^3t\left(\beta + \sin\beta \cos\beta - \frac{2\sin^2\beta}{\beta}\right)$$
$$I_2 = R^3t(\beta - \sin\beta \cos\beta)$$
$$y_1 = R\left(1 - \frac{\sin\beta}{\beta}\right)$$
$$y_2 = R\left(\frac{\sin\beta}{\beta} - \cos\beta\right)$$

BEAMS: SHEAR, MOMENT, AND DEFLECTION FORMULAS FOR COMMON END CONDITIONS

Key to table notation: P = concentrated load (lb.); W = uniform load (lb./in.); M = moment (in. lb.); V = shear (lb.); R = reaction (lb.); y = deflection (in.); θ = end slope (radians); E = modulus (psi); I = moment of inertia (in.⁴). Loads are positive upward. Moments which produce compression in the upper surface of beam are positive.

1. Cantilever - End Load

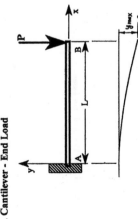

$R_A = P$
$V = -P$

$M = P(x - L)$
$M_{max} = -PL$, (at A)

$y_{max} = \dfrac{-PL^3}{3EI}$, (at x = L)

$\Theta = \dfrac{-PL^2}{2EI}$

2. Cantilever - Intermediate Load

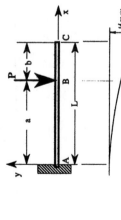

$R_A = P$
$V = -P$, (A to B)
$V = 0$, (B to C)

$M = -P(a - x)$, (A to B)
$M = 0$, (B to C)
$M_{max} = -Pa$, (at A)

$y_{max} = -\left(\dfrac{P}{6EI}\right)\left(3a^2L - a^3\right)$, (at B)

$\Theta = \dfrac{-Pa^2}{2EI}$, (B to C)

3. Cantilever - Uniform Load

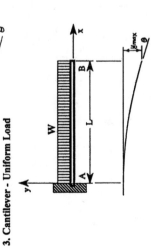

$R_A = WL$
$V = -W(L - x)$

$M = WLx - \dfrac{W(L^2 + x^2)}{2}$

$M_{max} = \dfrac{-WL^2}{2}$, (at A)

$y_{max} = \dfrac{-WL^4}{8EI}$, (at B)

$\Theta = \dfrac{-WL^3}{6EI}$

4. Cantilever - End Moment

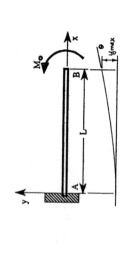

$R_A = 0$

$V = 0$

$M = M_0$

$M_{max} = M_0$

$y_{max} = \dfrac{M_0 L^2}{2EI}$

$\Theta = \dfrac{M_0 L}{EI}$, (at B)

5. End Supports - Intermediate Load

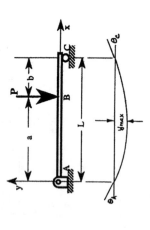

$R_A = \dfrac{Pb}{L}$, $R_C = \dfrac{Pa}{L}$

$V = \dfrac{Pb}{L}$, (A to B)

$V = \dfrac{-Pa}{L}$, (B to C)

$M = \dfrac{Pb}{L}x$, (A to B)

$M = \dfrac{Pa}{L}(L-x)$, (B to C)

$M_{max} = \dfrac{Pab}{L}$, (at B)

$y_{max} = \dfrac{-Pb(L^2 - b^2)^{3/2}}{9\sqrt{3}LEI}$, at $x = \sqrt{(L^2 - b^2)/3}$

$\Theta_A = \dfrac{-Pb(L^2 - b^2)}{6LEI}$

$\Theta_C = \dfrac{Pa(L^2 - a^2)}{6LEI}$

6. End Supports - Uniform Load

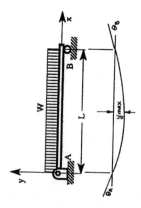

$R_A = \dfrac{WL}{2}$; $R_B = \dfrac{WL}{2}$

$V = \dfrac{WL}{2}\left(1 - \dfrac{2x}{L}\right)$

$M = \dfrac{WL}{2}\left(x - \dfrac{x^2}{L}\right)$

$M_{max} = \dfrac{WL^2}{8}$, (at L/2)

$y_{max} = \dfrac{-5WL^4}{384EI}$, (at L/2)

$\Theta_A = \dfrac{-WL^3}{24EI}$; $\Theta_B = \dfrac{WL^3}{24EI}$

7. One End Supported and One End Fixed - Intermediate Load

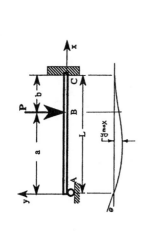

$$R_A = \frac{P}{2}\left(\frac{3b^2L - b^3}{L^3}\right)$$

$$R_C = P - R_A$$

$$V = R_A, \text{ (A to B)}$$

$$V = R_A - P, \text{ (B to C)}$$

$$M_C = \frac{P}{2}\left(\frac{b^3 + 2bL^2 - 3b^2 L}{L^2}\right)$$

$$M = R_A x \text{ (A to B)}$$

$$M = R_A x + P(a-x), \text{ (B to C)}$$

$$(+)M_{max} = R_A(a), \text{ (at B when a = .366L)}$$

$$(-)M_{max} = -M_C, \text{ (when b = .4227L)}$$

$$y_{max} = -.0098\frac{PL^3}{EI}, \text{ (at B when b = .586L)}.$$

$$\Theta_A = \frac{P}{4EI}\left(\frac{b^3}{L} - b^2\right)$$

8. One End Supported and One End Fixed - Uniform Load

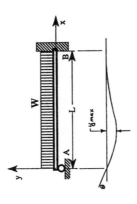

$$R_A = \frac{3WL}{8}; \quad R_B = \frac{5WL}{8}$$

$$V = WL\left(\frac{3}{8} - \frac{x}{L}\right)$$

$$M_B = \frac{WL^2}{8}$$

$$M = WL\left(\frac{3x}{8} - \frac{x^2}{2L}\right)$$

$$(+)M_{max} = \frac{9WL^2}{128}, \text{ (at x = 3L/8)}$$

$$(-)M_{max} = \frac{-WL^2}{8}, \text{ (at B)}$$

$$y_{max} = -.0054\frac{WL^4}{EI}, \text{ (for x = .4215L)}.$$

$$\Theta_B = \frac{-WL^3}{48EI}, \text{ (at A)}$$

9. Both Ends Fixed - Intermediate Load

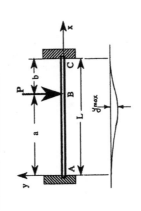

$$R_A = \frac{Pb^2}{L^3}(3a+b)$$

$$R_C = \frac{Pa^2}{L^3}(3b+a)$$

$$V = R_A, \text{ (A to B)}$$

$$V = R_A - P, \text{ (B to C)}$$

$$M_A = \frac{Pab^2}{L^2}$$

$$M_C = \frac{Pa^2b}{L^2}$$

$$M = \frac{-Pab^2}{L^2} + R_A x, \text{ (A to B)}$$

$$M = \frac{-Pab^2}{L^2} + R_A x - P(x-a), \text{ (B to C)}$$

$$(+)M_{max} = \frac{-Pab^2}{L^2} + R_A a, \text{ (at B)}$$

$$= \frac{PL}{8}, \text{ (for a = L/2)}$$

$$(-)M_{max} = M_A = M_A = -.1481PL, \text{ (for a = L/3)}$$

$$(-)M_{max} = M_C = M_C = -.1481PL, \text{ (for a = 2L/3)}$$

If a > b

$$y_{max} = \frac{-2P}{3EI}\left(\frac{a^3b^2}{(3a+b)^2}\right),$$

$$\left(\text{at x} = \frac{2aL}{3a+b}\right).$$

If a < b

$$y_{max} = \frac{-2P}{3EI}\left(\frac{a^2b^3}{(3b+a)^2}\right),$$

$$\left(\text{at x} = L - \frac{2bL}{3b+a}\right).$$

10. Both Ends Fixed - Uniform Load

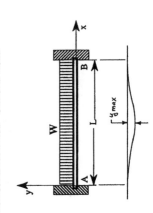

$$R_A = R_B = \frac{WL}{2}$$

$$V = \frac{WL}{2}\left(1 - \frac{2x}{L}\right)$$

$$M_A = M_B = \frac{WL^2}{12}$$

$$M = \frac{WL}{2}\left(x - \frac{x^2}{L} - \frac{L}{6}\right)$$

$$(+)M_{max} = \frac{WL^2}{24}, \text{ (for x = L/2)}$$

$$(-)M_{max} = \frac{-WL^2}{12}, \text{ (at A and B)}$$

$$y_{max} = \frac{-WL^4}{384EI}, \text{ (for x = L/2)}$$

REFERENCES

1. Dept. of Defense and Federal Aviation Administration, *Mil-Hdbk-5D, Metallic Materials and Elements for Aerospace Vehicle Structures,* Vol. 1–2, Philadelphia: Naval Publications and Forms Center, 1983.

2. *Aerospace Structural Metals Handbook,* Vol. 1–5, 1994 ed. W. Brown Jr., H. Mindlin, and C. Y. Ho (Eds.). West Lafayette, IN: CINDAS / USAF CRDA Handbooks Operations Purdue University.

3. Wang, C., *Applied Elasticity.* New York: McGraw-Hill Book Co., 1953, pp. 30–31.

4. Young, W. C., *Roark's Formulas for Stress and Strain,* 6th Ed. New York: McGraw-Hill Book Co., 1989.

5. Hsu, T. H., *Stress and Strain Data Handbook.* Houston: Gulf Publishing Co., 1986, pp. 364–366.

6. Seely, F. B. and Smith, J. O., *Advanced Mechanics of Materials,* 2nd Ed. New York: John Wiley & Sons, Inc., 1952, p. 415.

7. Peterson, R. E., *Stress Concentration Factors.* New York: John Wiley & Sons, Inc., 1974.

8. Hsu, T. H., *Stuctural Engineering & Applied Mechanics Data Handbook, Volume 1: Beams.* Houston: Gulf Publishing Co., 1988.

9. Higdon, A., Ohlsen, E. H., Stiles, W. B., and Weese, J. A., *Mechanics of Materials,* 2nd Ed. New York: John Wiley & Sons, Inc., 1967, p. 236.

10. Perry, D. J. and Azar, J. J., *Aircraft Structures,* 2nd Ed. New York: McGraw-Hill Book Co., 1982, p. 313.

11. Shigley, J. E., *Mechanical Engineering Design,* 3rd Ed. New York: McGraw-Hill Book Co., 1977, p. 208.

12. Gleason Works, *Gleason Curvic® Coupling Design Manual.* Rochester, NY: Gleason Works, 1973.

13. *Machine Design 1993 Basics of Design Engineering Reference Volume,* Vol. 65, No. 13, June 1993, p. 271.

14. Dann, R. T. "How Much Preload for Fasteners?" *Machine Design,* Aug. 21, 1975, pp. 66–69.

15. Franco, P. R. "Are Your Fasteners Really Reliable?" *Machine Design,* Dec. 10, 1992, pp.66–70.

16. MacGregor, C. W. and Symonds, J. "Mechanical Properties of Materials" in *Marks' Standard Handbook for Mechanical Engineers,* 8th ed. T. Baumeister, E. A. Avallone, and T. Baumeister III (Eds.). New York: McGraw-Hill Book Co., 1978, pp. 5–11.

14
Fatigue

J. Edward Pope, Ph.D., Senior Project Engineer, Allison Advanced Development Company

INTRODUCTION

Fatigue is the failure of a component due to repeated applications of load, which are referred to as *cycles*. An example of fatigue failure can be generated using a paper clip. Bending it back and forth will cause failure in only a few cycles. It has been estimated that up to 90% of all design-related failures are due to fatigue. This is because most design problems are worked out in the development stage of a product, but fatigue problems may not appear until many cycles have been applied. By this time, the product may already be in service.

In the 1840s, the railroad industry pushed the limits of engineering design, much as the aerospace industry does today. It was noted by those in the field that axles on railroad cars failed after repeated loadings. At this time, the concept of ultimate stress was well understood, but this type of failure was clearly something new and puzzling. The phenomenon was termed *fatigue* because it appeared that the material simply became tired and failed. August Wohler, a German railroad engineer, performed the first thorough investigation of fatigue in the 1850s and 1860s. He showed that fatigue life was related to the applied load.

The basic principles discovered by Wohler are still valid today, although much additional knowledge has been gained. Many of these lessons have been learned the hard way. Some of the more notable fatigue problems include:

- World War II liberty ships, which sometimes broke in half.
- Two Comet aircraft, the world's first passenger jet, lost due to fatigue failure which originated at the corner of a window.
- Several USAF F-111 aircraft, lost in the 1960s due to the unforgiving nature of titanium.

STAGES OF FATIGUE

Fatigue failure generally consists of three stages (see Figure 1):

1. Crack initiation (may be multiple initiation sites)
2. Stable crack growth
3. Unstable crack growth (fast fracture)

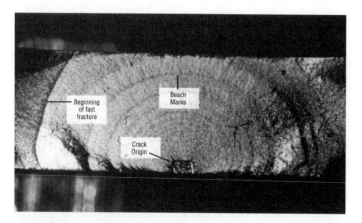

Figure 1. Typical fatigue fracture surface. (*Courtesy of A. F. Grandt, Jr.*)

Although cracks may be created during manufacturing, they generally do not initiate until after a considerable period of usage. Cracks commonly form at metallurgical defects such as voids or inclusions, or at design features such as fillets, screw threads, or bolt holes. A crack can initiate at any highly stressed location.

After a crack has initiated, it will grow for a while in a stable manner. During this stage, the crack will grow a very short distance during each load cycle. This creates patterns known as "beach marks" because they resemble the patterns left in sand by wave action along a beach. As the crack becomes larger, it usually grows at an increasingly rapid rate.

Final failure occurs very quickly. For small components, it happens when the cross-sectional area has been reduced by the crack so much that the applied stress exceeds the ultimate strength of the material. In larger components, fast fracture occurs when the fracture toughness of the material has been exceeded, even though the remaining cross-sectional area is still large enough to keep the applied stress well below the ultimate strength. This will be explained in the section on crack propagation.

DESIGN APPROACHES TO FATIGUE

Fatigue is dealt with in different ways, depending on the application:

1. Infinite life design
2. Safe life design
3. Fail safe design
4. Damage tolerant design

In Wohler's original work on railroad axles, he noted that there is a stress below which failure will not occur. This stress level is referred to as the *fatigue strength* or *endurance limit*. The simplest and most conservative design approach is to keep the stress below this level and is called *infinite life design*. For some applications, the cycles accumulate so rapidly that this is virtually the only approach. A gear tooth undergoes one cycle each time it meshes with another gear. If the gear rotates at 4,000 rpm, each tooth will experience nearly a quarter million cycles during every hour of operation. Vibratory stresses must also be kept below the endurance limit, since these cycles mount up even faster. This type of loading is referred to as *high cycle fatigue,* or HCF. The term *low cycle fatigue,* or LCF, is used to describe applications in which the load is applied more slowly, such as in steam turbines. One cycle is applied when the engine is started and stopped, and the engine may run continuously for months at a time.

In the aerospace business, the excessive weight required to design for infinite life is prohibitive. With the *safe life design* approach, a life is calculated which will cause a small percentage of the parts (typically 1 out of 10,000) to initiate a crack. All parts are removed from service when they reach the design life, even though the vast majority show no evidence of cracking. This approach has been used in the aircraft and turbine engine industries. When the design life is calculated, the analysis must account for significant scatter in the applied loads and fatigue properties of the materials.

In some instances, design precautions can be taken such that the failure of a particular component will not be catastrophic. This is known as *fail safe design*. After failure, the component can be replaced. This often involves redundant systems and multiple load paths. An obvious example of this approach is a multi-engine plane. If one engine fails, the others can still provide power to keep the plane flying. In the design of the aircraft and engine, it is necessary to ensure that debris from the failure of one engine will not take out vital systems. In one airline accident, fragments from a turbine wheel burst in one engine knocked out all three hydraulic systems which were placed closely together at one point along the fuselage.

Damage tolerant design assumes that newly manufactured parts may have cracks already in them. The design life is based on the crack growth life of the largest crack that may escape detection during inspection. This approach has been championed by the U.S. Air Force for many years. It puts a greater emphasis on the crack growth properties of the material, while the safe life approach emphasizes the crack initiation properties. It also requires good inspection capability.

CRACK INITIATION ANALYSIS

The first step in calculating crack intitiation life is to determine the stresses in a component. Life is related to the range of stress, as shown in Figure 2 (Life can also be related to strain range). These are known as *S-N curves,* and are plotted on log-log paper. While the alternating stress is the major factor in determining life, the ratio of the minimum stress to the maximum stress, also known as the *R ratio,* is a secondary factor. For a given alternating stress, increasing the R ratio will decrease the crack initiation life. For example, a component with stresses varying from 50 to 100 ksi will have a lower life than a component with stresses varying from 0 to 50 ksi.

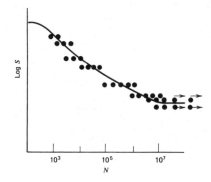

Figure 2. Typical S-N (log stress versus log life) plot [14]. *(Reprinted with permission of John Wiley & Sons, Inc.)*

Residual Stresses

The designer can sometimes use the R ratio effect to his advantage. Surface treatments such as shot-peening and carburizing create residual stresses. Residual stresses are sometimes referred to as *self-stresses*. Figure 3 shows how the residual stress varies below the surface. One seldom gets something for nothing, and residual stresses are no exception. Although the stress at the surface is significantly compressive, it is counterbalanced by tensile stress below the surface. Fortunately, this condition is generally a very good trade because most cracks initiate at the surface.

Residual stresses can also reduce crack initiation life. Improper machining, such as grinding burns, can cause large tensile residual stresses. While in service, a component may be exposed to a compressive stress that is large enough to cause yielding, leaving a tensile residual stress. Tensile residual stresses increase the R ratio, and therefore lower crack initiation life.

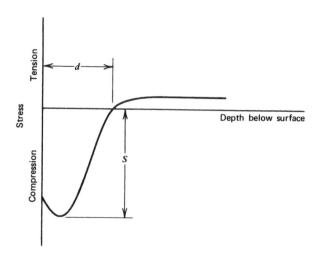

Figure 3. Typical distribution of residual stress under a shot-peened surface [14]. (*Reprinted with permission of John Wiley & Sons, Inc.*)

Notches

In most cases, cracks initiate at some kind of notch or stress concentrator. Typical examples include:

- Fillets
- Bolt holes
- Splines
- Fitted assemblies
- Keyways

Figure 4 shows how the negative effects of notches can be lessened. These strategies can be summed up as follows:

- Allow the stress to flow smoothly through the component (think of stresses as flowing water).
- Provide generous fillets and avoid sharp corners.
- Increase the cross-section where the notch occurs. The stress concentration factor will be just as high, but the nominals tress it is applied to will be lowered.

The first example in 4 may seem strange at first. How can the stress concentration effect of a hole be lowered by drilling more holes? The two smaller holes provide for a smoother flow of stress around the larger hole. Figure 5 shows that this could significantly lower the stress concen-

tration factor. Because, as a rule of thumb, a 10% decrease in stress doubles the crack initiation life, this could lead to a dramatic improvement in the durability of the component. A similar effect could be achieved by creating an elliptical hole instead of a round one. It should be pointed out that neither strategy should be applied unless the single round hole results in insufficient crack initiation life. Extra holes mean extra expense, and no one wants to drill elliptically shaped holes. The best design is the one that meets the criteria at the lowest cost. Keep the design simple whenever possible.

Fatigue Notch Factor

The stress concentration factor is normally represented by K_t and relates the peak stress to the nominal stress:

$$\sigma_{peak} = K_t \sigma_{nominal}$$

For crack initiation life calculations, the fatigue notch factor (K_f) is applied to the nominal stress rather than the stress concentration factor. These two factors are related by the equation:

$$K_f = 1 + q (K_t - 1)$$

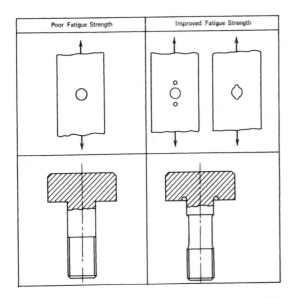

Figure 4. Good and bad design practices [15].

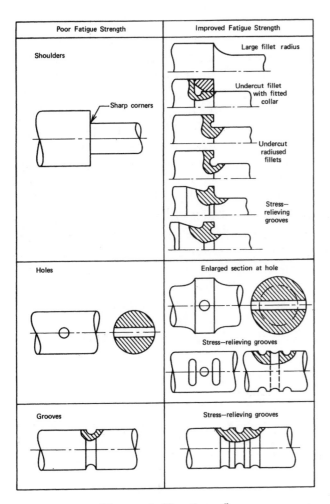

Figure 4. (Continued)

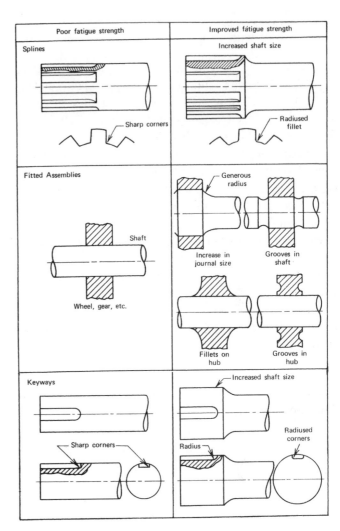

Figure 4. (Continued)

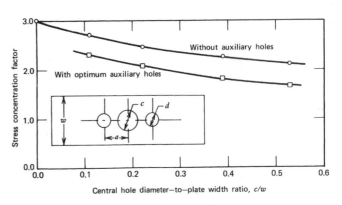

Figure 5. Stress concentration factors with and without auxiliary holes [16]. (*Reprinted with permission of Society for Experimental Mechanics.*)

The notch sensitivity factor "q" is a material property which varies with temperature. Its value ranges from 1 (fully notch sensitive) to 0 (notch insensitive). The value of q should be assumed to be 1 if it is not known. This will give a conservative estimate of crack initiation life.

Localized Yielding

Cracks generally initiate at a notch where there is some localized plastic yielding. Typically, the designer has only elastic stresses from a finite element model. Fortunately, elastic stresses are sufficient to make a good approximation of the true stresses as long as the yielding is localized. There are two methods for making this approximation:

• Neuber method
• Glinka method

Figure 6 graphically compares the two methods. Each diagram shows a plot of stress versus strain. For both methods, the area under the elastic stress-strain curve (A1) is calculated. With the Neuber method [1] a point on the true stress-strain curve is found such that the triangular area A2 equals A1. With the Glinka method [2], a point on the true stress-strain curve is found such that the area under that curve (A3) equals A1. The Neuber method is more commonly used in industry, and has an advantage in that it can

be solved directly. The Glinka method is slightly more accurate, but the calculation is a little more difficult because the area under the stress strain curve must be integrated. Both methods require an iterative solution. The user should always remember that these methods are limited to cases where there is only localized yielding. For situations involving large-scale yielding, plastic finite element analysis is required. The Ramberg-Osgood equation can be used to define the relationship between true stress and true strain:

$$\varepsilon_{true} = \left\{ \frac{\sigma_{true}}{E} \right\} + \left\{ \frac{\sigma_{true}}{K} \right\}^{(1/n)}$$

where: ε_{true} = true strain
σ_{true} = true stress
E = modulus of elasticity
K = monotonic strength coefficient
n = strain hardening exponent

The parameters K and n have different values for the initial monotonic loading and the stabilized cyclic loading. The monotonic and cyclic behaviors may be very different, as Figure 7 illustrates. In general, if the ratio of the ultimate strength to the .2% yield strength is high (>1.4), the material will cyclically harden (Waspaloy in Figure 7). If the ratio of the ultimate strength to the .2% yield strength is low (<1.2), the material will cyclically soften (SAE 4340 in Figure 7) [3]. For most crack initiation analysis, adequate results can be obtained by using the monotonic K and n on the initial loading, and the cyclic values for all subsequent loadings. The transitional behavior from monotonic to cyclic is seldom significant.

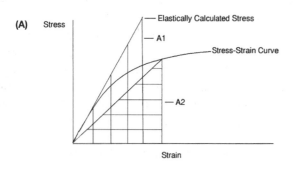

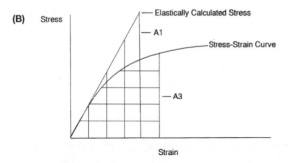

Figure 6. Neuber (A) and Glinka (B) methods of computing true stresses from elastically calculated variables.

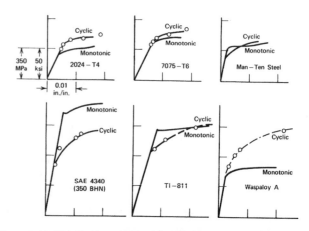

Figure 7. Typical monotonic and cyclic stress-strain curves. (*Copyright American Society for Testing and Materials. Reprinted with permission.*)

Real World Loadings

For tensile bars, the cycles which are applied are quite obvious. In real world applications, the loading can be quite complex. For fatigue analysis, the loading is assumed to be a combination of cycles. The largest one is referred to as the major cycle, and all others are minor cycles.

Cycle Counting

Consider Figure 8(A), which represents the stresses that might occur at a particular location on a compressor wheel of a gas turbine engine during a typical mission. It is obvious that the major cycle will have a range from 0 to 50 ksi. However, the stress is not monotonically increased to 50 ksi and then monotonically decreased back to 0 ksi. This erratic path between start-up and shut-down contains several minor cycles. A method is required to determine the cyclic content of this mission. The best known method is rainflow counting, which was named because it resem-

bles the flow of rainwater off pagoda roofs in Japan. The simplest method to learn is the range-pair method, which gives the same result. To use the range-pair method, start by counting the small cycles, such as the 25–30 ksi (1) and 30–35 ksi (2) in A of Figure 8. After these are eliminated, the mission looks like B. After the 10–40 ksi (3) cycle is counted and removed, all that remains is the 0–50 ksi major cycle in C. Therefore, the final result is:

(1) 0–50 ksi major cycle:
(1) 10–40 ksi minor cycle:
(1) 30–35 ksi minor cycle:
(1) 25–30 ksi minor cycle.

No matter what method is used to count cycles, it is important to get the major cycle correct. Fatigue is nonlinear; one 0–50 ksi cycle is far more damaging than a 0–25 ksi plus a 25–50 ksi cycle.

It should be pointed out that the cycle counting should always be done on stress. Sometimes in the gas turbine industry, cycle counting is based on power setting or rotational speed. This can lead to errors if stress is not a linear function of these variables.

The analyst should not be worried if different parts that are subjected to the same mission wind up having very different cyclic content. This usually occurs because the maximum and minimum stress values occur at different points in the mission. For one component, the extreme stress values may be due to thermal gradients which occur during transient operation. For another component, they may be due to rotational speed and occur when rpm is at a maximum or minimum. Different locations of the same component can also have different cyclic content.

Life Calculations with Minor Cycles

Once the cyclic content has been determined, the crack initiation life of this mixture must be determined. This is done with Miner's Rule [4], which is based on the idea that a certain amount of damage is done on each cycle. The damage is the reciprocal of the crack initiation life. When the sum of the damage for all applied cycles equals one, crack initiation is assumed to occur. For example:

During each mission, a component is subjected to:

1 major cycle of 0–100 ksi
15 minor cycles of 0–85 ksi

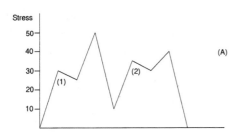

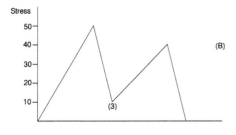

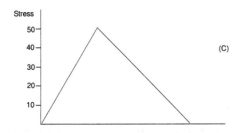

Figure 8. Evaluating cyclic content of a mission by the range-pair method.

If the crack initiation life is:

10,000 cycles at 0–100 ksi

100,000 cycles at 0–85 ksi

then the damage done during each mission is:

1/10,000 + 15/100,000 = . 00025

Therefore, the component will last 1/. 00025 or 4,000 missions. Conservatism can be added to this calculation by assuming that failure occurs at values lower than one.

There are numerous articles in technical journals pointing out cases where Miner's Rule is inadequate. Most of these cases involve sequence loadings which occur infrequently in real world applications. These are cases where a large cycle is applied for a while, then a smaller cycle is applied until a crack initiates (A of Figure 9). In this case Miner's Rule over estimates the crack initiation life. For cases where the smaller cycles are applied first, Miner's Rule underestimates the crack initiation life (B of Figure 9). In most field applications, the applied loading consists of a mission which is repeated until the component is retired from service. Therefore, the smaller cycles are mixed among the larger ones (C of Figure 9). In this case, Miner's Rule is perfectly satisfactory. Keep in mind that the scatter in crack initiation life is quite high. It is considered good correlation when the calculated life is with a factor of two of the actual life. It is a waste of time to attempt overly precise calculations. Fatigue lives should never be considered more than ballpark estimates.

Multiaxial Stresses

For applications involving uniaxial loading, the value to use for stress is obvious. In many applications, a multiaxial state of stress occurs. To determine the maximum stress point in a mission, calculate the equivalent stresses at the critical points. Since equivalent stress is always a positive value, it must be determined whether each is a tensile or compressive stress. If the first invariant (sum of principal stresses) is positive, then the equivalent stress is considered to be tensile. Otherwise, it is assumed to be compressive.

When determining the stress range between two conditions, it should be calculated from the ranges of the individual stress components [5]:

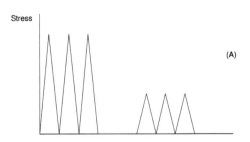

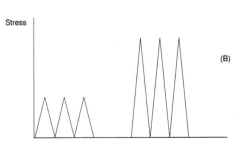

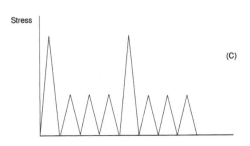

Figure 9. Miner's Rule will overestimate crack initiation life in case A, underestimate life in case B, and provide a reasonable estimate in case C.

$$\Delta\sigma_{eq} = \sqrt{.5 \times \left\{ \{\Delta\sigma_x - \Delta\sigma_y\}^2 + \{\Delta\sigma_y - \Delta\sigma_z\}^2 + \{\Delta\sigma_z - \Delta\sigma_x\}^2 + 6\{\Delta T_{xy}^2 + \Delta T_{yz}^2 + \Delta T_{zx}^2\} \right\}}$$

where: $\Delta\sigma_x = \sigma_x$ (at condition 1) $- \sigma_x$ (at condition 2)
         $\Delta\sigma_y = \sigma_x$ (at condition 1) $- \sigma_y$ (at condition 2)
         $\Delta\sigma_z = \sigma_x$ (at condition 1) $- \sigma_z$ (at condition 2)
         $\Delta T_{xy} = T_{xy}$ (at condition 1) $- T_{xy}$ (at condition 2)
         $\Delta T_{yz} = T_{yz}$ (at condition 1) $- (T_{yz}$ (at condition 2)
         $\Delta T_{zx} = T_{zx}$ (at condition 1) $- T_{zx}$ (at condition 2)

S_{min} is calculated by:

$$S_{min} = S_{max} - \Delta\sigma_{eq}$$

If this results in a cycle with a compressive mean stress, then:

$$S_{max} = .5 \times \Delta\sigma_{eq}$$

$$S_{min} = -.5 \times \Delta\sigma_{eq}$$

These values can then be used to calculate life based on uni-axial test data.

Vibratory Stresses

In many applications, components are subjected to vibratory loading. This type of loading is referred to as high cycle fatigue, or HCF. Since the number of applied cycles mounts very rapidly (a component vibrating at 1,000 Hertz will accumulate 3. 6 million cycles in one hour of operation), the vibratory stress must be less than the endurance limit, or failure will occur. Testing to determine the endurance limit is expensive, and is usually only done at an R ratio of zero. If the endurance limit at R = 0 is not available, it can be estimated to be 45% of the ultimate strength of the material. In field usage, the vibratory stress generally occurs on top of a steady-state stress.

To estimate the endurance limit at R ratios above zero, the Goodman Diagram is used. This requires only two parameters:

• Endurance limit at R = 0
• Ultimate tensile strength

Figure 10 illustrates how the Goodman Diagram is constructed. The ultimate strength is plotted along the X axis (A), while the endurance limit is plotted on the Y axis (B). These points are then connected by a straight line (C). A component should not fail as long as the combination of vibratory and constant stress is below this line. Points above this line will have less than infinite crack initiation life. There are alternatives to the Goodman Diagram, but most of these are refinements of the line connecting points A and B in figure 10. The Goodman Diagram is most commonly used in industry, and is a little more conservative than the others.

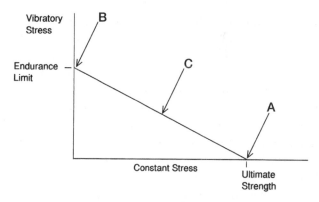

Figure 10. Goodman diagram.

Temperature Interpolation

Fatigue testing is done at discrete temperatures, and the operating temperature of the component will fall within the range of test values. To interpolate between S-N curves:

• Calculate fatigue life at both temperatures.
• Interpolate between temperatures based on log of life.

$$\log(N_t) = \log(N_{t1}) + \{\log(N_{t2}) - \log(N_{t1})\} \{(t - t_1)/(t_2 - t_1)\}$$

where: t = component temperature at critical location
t_1 = lower Walker LCF model temperature
t_2 = higher Walker LCF model temperature
N_t = component life:

N_{t1} = life at t_1
N_{t2} = life at t_2

If fatigue data is available at only one temperature, it can be scaled to another temperature by using the ratio of the ultimate strengths at the two temperatures.

$$\sigma_{allowable}(T_2) = \sigma_{allowable}(T_1) \times \frac{\sigma_{ult}(T_2)}{\sigma_{ult}(T_1)}$$

If it is necessary to use this method to scale from a lower to a higher temperature, the result should be used with extreme caution.

Material Scatter

Since designers want to ensure that only a very small fraction of their components fail, material scatter must be taken into account. This is generally done by specifying that -3σ material properties should be used. Only one out of about 800 specimens should have fatigue properties below this level. Three rules of thumb are commonly used to account for material scatter and estimate minimum material properties:

1. The simplest is to divide the calculated life by a factor of 3.
2. For cast materials, use 70% of the calculated allowable stress.
3. For forged materials, use 85% of the calculated allowable stress.

Estimating Fatigue Properties

Often, an engineer has to make preliminary design decisions when no fatigue data is available. Fortunately, the *Modified Universal Slopes Equation* [6] allows rough life estimates to be made based only on values from tensile bar data:

$$\Delta_\varepsilon = .0266 D^{.155} \left(\frac{\sigma_{ult}}{E}\right)^{-.53} N_f^{-.56} + 1.17 \left(\frac{\sigma_{ult}}{E}\right)^{.832} N_f^{-.09}$$

where: Δ_ε = total strain range
$\quad\quad\quad N_f$ = fatigue life
$\quad\quad\quad D$ = ductility
$\quad\quad\quad \sigma_{ult}$ = ultimate tensile strength

The ductility and ultimate strength values used in this equation should be at the temperature for which the fatigue life will be calculated. This equation should be used only in the sub-creep temperature range (up to about ½ the absolute melting temperature of the material). The original Universal Slopes Equation was developed by Manson and Hirschberg in 1965. The modified version shown here is slightly better. Both models were compared with test data from 47 different engineering materials, which included steel, aluminum, and titanium alloys. The modified version was shown to be slightly better, although the difference is quite small [6]. The life estimates from this equation were within a factor of 10 of the lives for the test specimens. (This correlation is better than it might first appear, since the ratio of average life to minimum life is approximately 3.)

A useful feature of this equation is that it can show the effects of processes or operational factors that alter the ductility or ultimate strengths. Exposure to nuclear radiation reduces ductility, and this model shows the accompanying reduction in fatigue life. The effect of ductility and ultimate strength on fatigue life depends on the life region of interest. Below 1,000 cycles, ductility has the dominant influence upon crack initiation life. For greater lives, ultimate strength is more important.

CRACK PROPAGATION ANALYSIS

Most of the crack growth analysis that is commonly performed in industry is based on *linear elastic fracture mechanics* (LEFM). As a general rule, LEFM is considered to be valid as long as the plastic zone in front of the crack is less than 1⁄10 the crack length. This is adequate for most analyses since components with large amounts of cyclic plastic yielding do not have enough life for practical applications. Elastic-plastic fracture mechanics is quite complex, and is well beyond the scope of this book.

There are three modes of crack propagation (Figure 11):

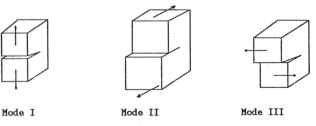

Mode I   Mode II   Mode III

Figure 11. Crack propagation modes.

1. Mode I is relative displacement of the crack surfaces perpendicular to the crack plane, which opens the crack.
2. Mode II is relative displacement of the crack surfaces in the crack plane and perpendicular to the crack front, which shears the crack.
3. Mode III is relative displacement of the crack surfaces in the crack plane and parallel to the crack front, which tears the crack.

Mode I is the most common and the only one considered in this book.

Crack propagation analysis is a relatively new discipline. The U. S. Air Force has mandated its usage to ensure damage tolerance. In a nutshell, USAF requires that crack propagation analysis:

• Assumes that new components have cracks which are at the limit of your company's inspection techniques. If a .050-inch crack is the smallest crack you can reliably detect, an initial flaw size of .050 inches must be assumed. Design engineers and inspectors always like to say things like, "We can detect cracks as small as .010 inches long." The crack propagation analyst should always reply with, "What I really need to know is the size of the largest crack you might miss."
• Shows that these cracks will not grow to failure until the component has been inspected twice. If the component will be inspected every 1,000 cycles, the analysis must show a crack propagation life of 2,000 cycles.
• Is only required on items which are "fracture critical." "Fracture-critical components are those that would cause loss of the aircraft if they should fail.

K—The Stress Intensity Factor

The stress intensity factor (K) is crucial to fracture mechanics. It is calculated by the formula [7]:

$$K = \sigma \sqrt{\pi a}\, B$$

where: σ = stress
        a = crack length
        B = a geometry factor

The parameter K defines the stress field directly ahead of the crack tip, perpendicular to the crack plane (Figure 12).

$$\sigma_y = \frac{K}{\sqrt{2\pi r}}$$

This equation is only valid a short distance ahead of the crack. Two cracks with very different geometries and very different loadings will have similar stress fields near their crack tips if they have the same stress intensity factors. The stress intensity factor also determines the behavior of the crack:

• If K_{max} exceeds the fracture toughness, fast fracture occurs.
• If ΔK is less than the ΔK_{th} (threshold stress intensity factor), no crack growth occurs.
• If ΔK is greater than ΔK_{th}, but K_{max} is less than the fracture toughness, stable crack growth under cyclic loading occurs.

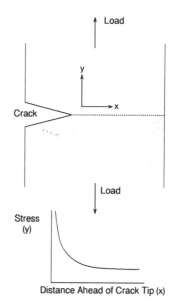

Figure 12. Stress distribution ahead of crack tip.

The stress intensity factor K should not be confused with the unrelated stress concentration factor K_t. The units for K are:

(stress) $\times$ *(length)*$^{.5}$

For the English system, this is typically *(ksi) (inches)*$^{.5}$. For the metric system, it is typically *(mpa) (meters)*$^{.5}$. The conversion factor from the metric to the English system is:

1 *(mpa) (meters)*$^{.5}$ = .91 *(ksi) (inches)*$^{.5}$

Fast Fracture

As stated earlier, fast fracture will occur when the stress intensity factor exceeds the fracture toughness. Fracture toughness is a material property that is dependent on *temperature* and *component thickness*.

Figure 13 shows that decreasing temperature decreases the fracture toughness. This is due to the lower ductility at lower temperatures. Figure 14 shows that increasing thickness decreases the fracture toughness. This is because thin specimens are subject to plane stress, which allows much more yielding at the crack tip. As specimen thickness is in-

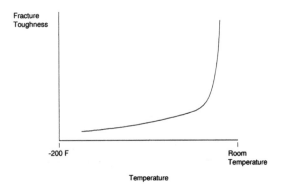

Figure 13. Effect of decreasing temperature on fracture toughness.

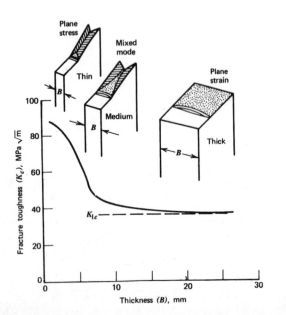

Figure 14. Effect of specimen thickness on fracture toughness [14]. (*Reprinted with permission of John Wiley & Sons, Inc.*)

creased, it asymptotically approaches a minimum value. This is the value that is quoted as the fracture toughness. The actual fracture toughness for thin specimens may be somewhat greater than K_{IC}.

Figure 15 shows how fracture toughness varies with yield strength for aluminum, titanium, and steels. Note that for all three, alloys which had very high yield strengths had relatively low values of fracture toughness. Also, because the stress intensity factor is a function of crack length to the .5 power, reducing the fracture toughness by a factor of 2 will reduce the the critical crack size by a factor of 4. Therefore, a designer who considers specifying an alloy with a high yield strength should realize he may be significantly reducing the critical crack size. In field service, this could result in sudden failures instead of components being replaced after cracks were discovered.

Threshold Stress Intensity Factor

If the stress intensity factor range does not exceed the threshold value, then a crack will not propagate. ASTM defines the threshold value to be where the crack growth per cycle(da/dn) drops below 3×10^{-9} inches per cycle. Unlike fracture toughness, the threshold stress intensity factor, ΔK_{th}, is dependent upon the R ratio.

Threshold behavior is very important for components which must endure millions of cycles. In analysis, it can be used to make very conservative estimates. If the applied stress is known, a flaw size can be assumed, and the stress intensity factor can be calculated. As long as the threshold value exceeds ΔK, no crack growth can occur. If the stress

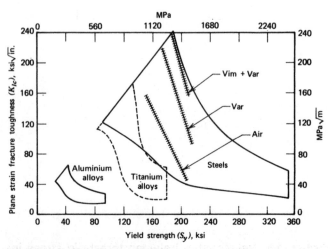

Figure 15. Fracture toughness versus yield strength for various classes of materials [18].

intensity factor range exceeds ΔK_{th}, then more detailed analysis must be made to determine the number of cycles the component can be expected to last in service.

Sometimes, after many parts have been released for field service, a problem will be discovered in the manufacturing process. The question is then raised, "Do we need to replace these parts immediately, or can we safely wait and replace them during the next overhaul?" Obviously, the first approach is safer, but may be extremely costly. However, if the parts are left in service and fail, the consequences can be devastating. Therefore, any analysis which is done must err on the conservative side. In these instances, the stress intensity factor is often calculated and compared to the ΔK_{th}.

Estimates of threshold stress intensity factor. Testing to obtain ΔK_{th} is quite time-consuming and, consequently, quite expensive. Cracks have to be propagated in a specimen, and then the load is slowly decreased until the crack stops propagating. Fortunately, there are some ranges available for different classes of materials at room temperature. These are shown in Figures 16 through 20. It may be necessary to adjust these values to use them at elevated temperatures, which may be done by scaling the threshold value based on the ratio of Young's modulus. For example, a threshold value for steel is needed at:

• An R ratio of .4
• A temperature of 500 degrees

From Figure 16, the minimum threshold stress intensity factor is 4.0 at room temperature. If Young's modulus is 25×10^6 at 500 degrees and 30×10^6 at room temperature, the minimum stress intensity factor at 500 degrees should be $4.0 \times (25/30) = 3.33$.

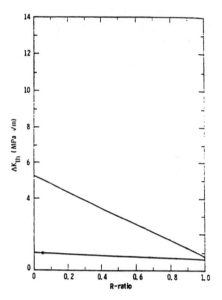

Figure 17. Relationship between threshold stress intensity range and R ratio in aluminum alloys [17]. (*With permission of Elsevier Science Ltd.*)

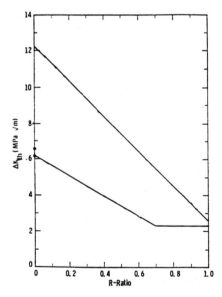

Figure 16. Relationship between threshold stress intensity range and R ratio in iron alloys [17]. (*With permission of Elsevier Science Ltd.*)

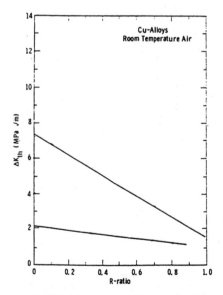

Figure 18. Relationship between threshold stress intensity range and R ratio in copper alloys [17]. (*With permission of Elsevier Science Ltd.*)

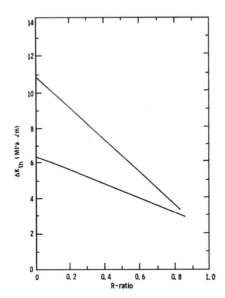

Figure 19. Relationship between threshold stress intensity range and R ratio in nickel alloys [17]. (*With permission of Elsevier Science Ltd.*)

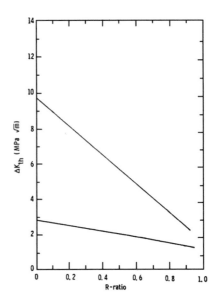

Figure 20. Relationship between threshold stress intensity range and R ratio in titanium alloys [17]. (*With permission of Elsevier Science Ltd.*)

Crack Propagation Calculations

If the stress intensity factor is below the fracture toughness, and the range of the stress intensity factor is greater than the threshold value, the crack will grow in a stable manner. This is often referred to as "Subcritical crack growth." Three approaches are commonly used to relate the crack growth rate to the stress intensity factor (C_n indicates the material constant):

• Paris law [8]:

$$\frac{da}{dn} = C_1(\Delta K)^{C_2}$$

This law assumes that the data can be fit as a straight line on a log-log plot. This usually gives a fair definition of the curve. It usually does not model the crack growth rate well at low and high values of ΔK.
• Modified Paris law: This model seeks to overcome the limitations of the Paris law by using three sets of coefficients which are used over three ranges of stress intensity factors.
• Hyperbolic sine model: This model strives to use one relationship that is applicable over the entire range of stress intensity factors, and accurately models the crack growth rate at low and high values of ΔK.

$$\log\left(\frac{da}{dn}\right) = C_1 \sinh\left(C_2[\log(\Delta K) + C_3]\right) + C_4$$

The analyst should realize that because cracks tend to grow at a continually increasing rate, most of the life occurs when the cracks are quite small. Therefore, it is important to accurately model the crack growth rate at small values of ΔK, but usually not important at near-fracture values. The exponent of the Paris law can be quite useful for determining the effect of a change in stress on crack growth life. Since ΔK is proportional to the applied stress range, The change in crack growth life may be estimated by:

$$\text{Life}_2 = \text{Life}_1\left(\frac{\sigma_1}{\sigma_2}\right)^{C_2}$$

If the stress is increased 10%, and the Paris law exponent is 4, the crack growth rate will be increased by $(1.10)^4$, which means it will grow 1.46 times faster. Therefore, the crack propagation life will be reduced by a factor of (1/1.46) to 68% of its previous value.

Crack growth under cyclic loading for a given material is dependent upon three variables:

1. ΔK
2. R ratio (K_{min}/K_{max})
3. temperature

The crack growth data is typically shown on a log-log plot, such as shown in Figure 21. The crack propagation rate increases with increased stress intensity factor range and higher R ratios.

In the absence of more specific data, Barsom [9] recommends using these rather conservative equations:

• Ferritic-pearlitic steels:

$$\frac{da}{dn} \, (\text{in.}/\text{cycle}) = 3.6 \times 10^{-10} \, (\Delta K)^{3.00}$$

• Martensitic steels:

$$\frac{da}{dn} \, (\text{in.}/\text{cycle}) = 6.6 \times 10^{-9} \, (\Delta K)^{2.25}$$

• Austentitic stainless steels:

$$\frac{da}{dn} \, (\text{in.}/\text{cycle}) = 3.0 \times 10^{-10} \, (\Delta K)^{3.25}$$

When the minimum stress intensity factor is negative, a value of 0 should be used to calculate ΔK. This is because crack propagation does not occur unless the crack is open at the tip. If the amount of compression is small (R > −.5), then crack growth data at R = 0 (or .05) may be used with no significant loss in accuracy. If the amount of compression is large, (R < −1), it may be wise to obtain data at the appropriate R ratio. Crack propagation at compressive R ratios is seldom done because the most commonly used test specimen, known as the compact tension (CT) specimen, can only be tested in tension.

Estimation of K

For a fairly uniform stress field, the analyst can estimate K quite easily. Common crack types are shown in Figure 22. Approximate values of β are:

β	Crack Type
0.71	corner or surface cracks
1.00	center cracked panels
1.12	through cracks

These values will change when the crack approaches a free surface. Typically, this is not a significant effect until the crack is about 40% of the way through the section for corner and surface cracks. It is much more significant for center and through cracks because of the loss of cross-sectional area.

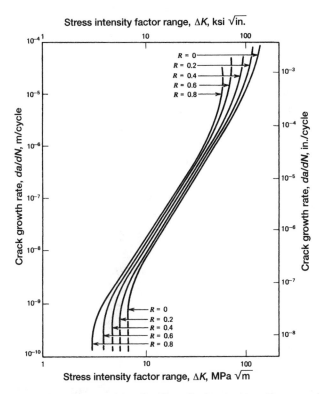

Figure 21. Increasing the R ratio increases the crack growth rate [14]. (*Reprinted with permission of John Wiley & Sons, Inc.*)

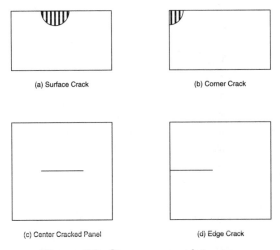

(a) Surface Crack   (b) Corner Crack

(c) Center Cracked Panel   (d) Edge Crack

Figure 22. Common crack types.

Computer codes which calculate crack growth use two approaches:

1. The cycle-by-cycle approach is the simplest, but it can be excessively time-consuming for slow-growing cracks. With this method, the stress intensity factors and crack growth for one cycle is calculated. The crack length is then increased by this amount and the process is repeated until the desired crack length is reached or the fracture toughness is exceeded.
2. With the step method, the number of cycles to grow the crack a certain distance (or step) is calculated after the crack growth rate is determined. This method generally requires significantly fewer iterations than the cycle-by-cycle method. Care must be taken in selecting the step size. If the step is too large, accuracy can be lost. If the step is too small, too much computer time will be required. For initial crack sizes around .015 inches, .001 makes a good step size.

Any engineer can write a simple computer code to perform these iterations. All that is required is a stress intensity factor solution and a relationship between the stress intensity factor range and crack growth rate.

For simple cases, the crack growth life can be calculated by simple integration. For example:

- If $da/dn = 2.0 \times 10^{-12}(\Delta K)^5$
- and $K = 50\sqrt{\pi a}\ .71$
- Solving for dn and integrating from initial crack size A_i and final crack size A_f gives:

$$N = \frac{1}{(-1.5)(2 \times 10^{-12})(50\sqrt{\pi}\ .71)^5}(A_f^{-1.5} - A_i^{-1.5})$$

- For an initial crack size of .015 inches and a final crack size of .050 inches, the crack growth life would be 153,733 cycles.

Plastic Zone Size

Because a crack is assumed to be infinitely sharp, the elastic stresses are always infinite at the tip, but drop off very quickly. Yielding always occurs in the region ahead of the crack tip, which is referred to as the *plastic zone*. The size of the plastic zone under plane stress conditions can be estimated by:

$$r_p = \frac{1}{2\pi}\left(\frac{\Delta K}{\sigma_y}\right)^2$$

Under plane strain, the plastic zone is approximately one-fourth as large. The plastic zone is important for a number of reasons:

- For LEFM calculations to be valid, the crack length should be at least 10 times the length of the plastic zone.
- Anyone testing to determine crack growth properties should realize that large and sudden changes in the loads can affect the plastic zone ahead of the crack and significantly alter the crack growth properties.
- It is possible with a single overload to significantly reduce the crack growth rate, or even arrest the crack. This can occur because the overload causes additional yielding in front of the crack, which inhibits its future growth through the region. Keep in mind that during the single overload, the crack grew at a greater rate, so it is not certain what the net effect of the overload will be.

Creep Crack Growth

Creep crack growth (da/dt) occurs when a tensile stress is applied for an extended time at a high temperature. This process should not be confused with conventional creep, which is an inelastic straining of material over time. Creep crack growth can be detected by metallurgical investigation of the crack surfaces:

- When cyclic crack growth dominates, the crack grows across the grains (transgranular).
- When creep crack growth dominates, the crack grows along the grain boundaries (intergranular).

Several points should be made comparing creep and cyclic crack growth:

- The rate of creep crack growth is related to the steady-state K, while cyclic crack growth rate is based on ΔK.
- Creep crack growth is time dependent, while cyclic crack growth is not.
- The threshold value for creep crack growth is much higher than it is for cyclic crack growth.
- Temperature has a much greater influence on creep crack growth than on cyclic crack growth.

INSPECTION TECHNIQUES

Several inspection methods are available, each with its own positive and negative aspects.

Fluorescent Penetrant Inspection (FPI)

With fluorescent penetrant inspection (FPI), a fluorescent dye penetrant is smeared on a surface to be checked for cracks. The surface is then cleaned off and placed under a black light. Lines will be observed where the dye seeped into cracks. The advantages of this method are that it is simple and requires no elaborate test apparatus. The disadvantages are that it is applicable only to surface cracks, may only be used on a relatively smooth surface (rough surfaces will give many false indications of cracks), and is very op-erator-dependant. The human observer is the weak link in this system. Studies show that crack detection capability varies widely from person to person. These studies also show that a given person's capability will vary from one day to another. If the surface crack is in a residual compressive stress field, it will be very difficult to detect because the crack faces will be pressed tightly together. This will make it difficult for the dye penetrant to seep into the crack.

Magnetic Particle Inspection (MPI)

Magnetic particle inspection (MPI) is similar to FPI, but it can be used only on ferrous metals. With this method, a liquid containing magnetic particles is applied to the surface being tested. A magnetic field is then produced in the component by induction or passing an electric component through it. Surface or near-surface cracks will disrupt the magnetic flux lines and cause the magnetic particles to collect around them. MPI is more reliable than FPI, especially for detecting cracks in residual compression and cracks which are filled with foreign matter. These may block the dye penetrant from entering the crack, but the magnetic field is still disrupted. MPI also has limited capability to detect cracks just below the surface. In general, MPI is superior to FPI and should be used when possible.

Radiography

Radiography utilizes penetrating radiation (typically x-rays) to detect cracks. The basic concept is that where less material exists, less radiation will be absorbed. The unabsorbed radiation is measured after passing through the test article (Figure 23). Radiography is used to detect internal cracks or voids, but it is not good at detecting cracks oriented perpendicular to the radiation beam. This is because there is very little difference in the amount of absorbed radiation between parallel paths. It is also widely used to inspect welds and determine whether two pieces have been joined together solidly or are merely attached at the surface (Figure 24).

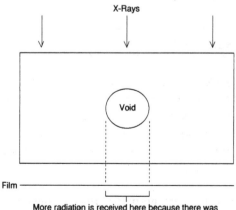

Figure 23. Radiographic inspection for cracks.

Figure 24. Radiography is often used to inspect welds.

Ultrasonic Inspection

Ultrasonic inspection utilizes high-frequency sound waves which are reflected by discontinuities. The return signal is measured and analyzed to detect cracks. A great deal of energy is reflected at both surfaces of the component. This creates two dead zones where acoustic reflections caused by cracks cannot be distinguished from those caused by the component surfaces. The return signal for an uncracked component will look like Figure 25: Peak A represents the echo from the front wall, and peak B represents the back wall echo. Crack A in Figure 26 would reflect a portion of the signal and cause another peak between A and B. Crack B would reflect some of the wave at an angle and, therefore, would not cause another peak. Its existence could be deduced, however, because the back wall echo would be significantly curtailed. Any increase in the return signal between these two peaks indicates a discontinuity in

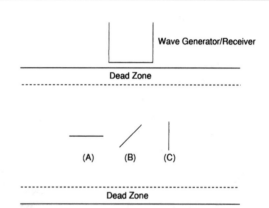

Figure 26. Effect of crack orientation on detectability: (A) crack reflects signal back to receiver; (B) crack deflects signal, reducing back wall echo; (C) crack cannot be detected.

the component. Crack C would be extremely difficult to detect, since it would have little effect on the return signal because of its orientation.

Ultrasonic inspection is very accurate and can be used to inspect thick sections. It is limited to detecting internal cracks away from specimen surfaces. It requires a flat surface through which to apply the ultrasonic energy. Reference standards are required, making this method unusable for one-of-a-kind inspections. For some materials, grain boundaries, precipitates, or other internal inhomogeneities reflect so much acoustic energy, that crack detection is extremely difficult.

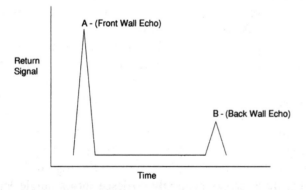

Figure 25. Ultrasonic return signal for an uncracked component.

Eddy-Current Inspection

Eddy-current inspection can be performed on materials that conduct electricity. This method is based on the principle that cracks distort the eddy-currents which occur in a sample when current is passed through a nearby coil. Both surface and near-surface cracks may be detected reliably. Perhaps the most important feature of eddy-current inspection is that it can be automated. This improves accuracy significantly, and reduces cost if the inspections will be done in volume. The negative aspects are that special machinery and reference standards (showing the signals for cracked and uncracked parts) are necessary, making it impractical for one-of-a-kind inspections.

Evaluation of Failed Parts

If a failure occurs, all relevant information about the event should be recorded as quickly as possible. Seemingly minor details may help pinpoint the cause of the failure. If a turbine wheel fails when power is being increased or decreased, it may indicate that thermal gradients are responsible. Any failed parts that can be retrieved should be handled with extreme care. The parts should not be cleaned until they have been examined thoroughly, because surface debris may yield important clues. Paint on a cracked surface might indicate that the crack occurred during the manufacturing process, and was present when the component was placed into service. Oxidation on a cracked surface may indicate how long that surface was exposed to the environment. Important regions should be photographed for future reference. From the primary crack (the one responsible for failure), attempts should be made to determine:

- Crack origin (there may be more than one)
- Critical crack size
- Crack growth rate (if striations are present)

The critical crack size and crack growth rates may be used to make rough estimates of the loads present. Secondary cracks (ones not responsible for failure) may also be used for this purpose. The surfaces of the primary crack are often too damaged for meaningful evaluation after failure. Secondary cracks, whose surfaces are more protected, may be more useful for post-failure examination. Examining the microstructure of the component, including regions not near the fracture surface, can indicate what conditions it was subjected to during operation. The microstructure of some materials changes when they are subjected to high temperatures. Armed with this knowledge, it might be possible to determine that the operating temperature exceeded a certain level. Small specimens may be

(a) Tension   (b) Bending

Figure 27. Failure surfaces of round bars subjected to (A) tensile loading and (B) bending.

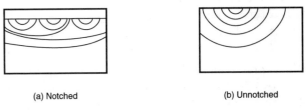

(a) Notched   (b) Unnotched

Figure 28. Failure surfaces of notched and unnotched rectangular bars.

cut from a failed component and tested to determine if the material properties are within specifications. Keep in mind that different regions of the same component can have significantly different properties.

The fracture surface may also indicate the type of load. Figure 27 shows failure surfaces for round bars subjected to tensile loading and to bending. In bending, there is a tendency to initiate cracks at the top and bottom, since the stresses will be highest at these two locations. Figure 28 compares the cross-section of notched and unnotched rectangular bars. The notch creates a local area of high stress, and multiple cracks tend to initiate in this region. Eventually, these cracks generally coalesce into a single large crack. For cases of uniform stress, the tendency is towards a single crack.

NONMETALLIC MATERIALS

This chapter has focused on metals, which are widely used for structural components. A few comments will be made about plastics, composites, and ceramics as well. The properties of plastics vary greatly, but the same methodology that is used for metals can often be used for plastics. Dr. Grandt has done numerous experiments on polymethylmethacrylate and polycarbonate [10]. His research indicates that crack growth in these plastic materials can be related to the stress intensity factor, just as it is in metals.

Composites are being used more frequently for structural components. Fatigue of composite materials is a very complicated subject, which will not be dealt with in this book. Complications in fatigue analysis arise because:

- The material is not homogeneous.
- Residual stresses are present due to the difference in thermal expansion coefficients between the fibers and matrix materials.
- Once cracks initiate, the fibers often act as barriers to crack growth.

Compared to metals, fatigue strengths are generally higher (relative to their ultimate strengths) for composites under uniaxial tensile loading (R > 0). The opposite is true when compressive or fully reversed (R < -1)loading is applied [11]. Inspection of composites is difficult because cracking is more likely to occur internally than it is for metals.

Ceramics offer strength, light weight, and outstanding temperature resistance, but have been shunned in the past due to their brittleness. A great deal of effort is currently being expended to develop ceramics with improved toughness. If we assume that ceramic materials have no ductility, then some simple analysis can be done. When ceramic components are manufactured, they always have some flaws in them. Since their ductility is assumed to be negligible, no subcritical crack growth can occur. If the load is increased to the point that the stress intensity factor at one of these inherent defects exceeds the fracture toughness, fast fracture will occur. At a given stress level, a certain percentage of ceramic components will fail immediately, while the rest will not fail no matter how many times the load is applied. This makes the standard S-N curves that are used

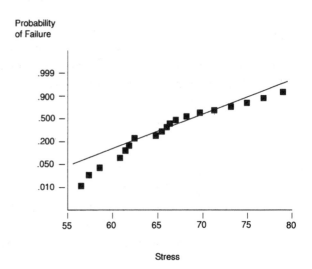

Figure 29. Typical probability of failure versus stress plot.

to calculate lives for metals useless. Instead, a probabilistic method is applied. The *probability of survival* (POS) is plotted against strength (Figure 29). Since larger components have more material, and therefore a greater probability of containing a large flaw, failure must be normalized to account for size. In test specimens, the failures are generally classified as either surface (failures originating at the surface) or volume (failures originating internally). Separate Weibull curves are calculated for each type of failure. When this information is used to calculate the life of a component, its total probability of survival is the product of its POS for surface and volume flaws:

$$POS_{total} = POS_{surface} \times POS_{volume}$$

A positive aspect of the lack of ductility in ceramics is that they can be "proof-tested." Because an applied load to a component with no ductility will do no damage unless it causes fast fracture, parts can be tested at a load equal to the highest load (multiplied by an appropriate safety factor) that they will experience in service. Those that survive should not fail in service unless they are subjected to an even greater load. This means that if the specimens in Figure 29 are subjected to a stress of 60 ksi, the three weakest specimens would fail, but those remaining would be undamaged. This approach should not be used with metals, because they undergo subcritical crack growth.

FATIGUE TESTING

Fatigue testing can be difficult and can yield misleading data if not done correctly. Numerous small companies specialize in generating fatigue data, and their services might be useful to those without proper facilities or experience. Companies that decide to generate their own fatigue data should carefully review the pertinent ASTM guidelines:

1. Low Cycle Fatigue (ASTM Standard E606-80).
2. High Cycle Fatigue (ASTM Standard E466-82).
3. Statistical Analysis of Linear or Linearized Stress-Life and Strain-Life Fatigue Data (ASTM Standard E739-91)
4. Plane-Strain Fracture Toughness Test Method (ASTM Standard E399-90).
5. Fatigue Crack Growth and Threshold Crack Growth Test Method(ASTM Standard E647-91).
6. Creep Crack Growth Test Method (ASTM Standard E1457-92).
7. Surface Fatigue Crack Growth Test Method (ASTM Standard E740-88).

It is particularly important to exercise care when testing to determine ΔK_{th}. High loads are required to initiate a crack for testing purposes. If these loads are not reduced gradually after initiation but before taking measurements to determine ΔK_{th}, the tests may show the threshold value to be much higher than it actually is. Obviously, this could have very serious consequences.

Every effort should be made to keep the test specimens as similar to the actual hardware as possible. Seemingly unimportant details, such as how a surface is machined, may induce residual stresses or create small cracks which drastically alter the fatigue life. Figure 30 shows how the surface factor (C_f) is related to tensile strength and machining operations [12]. If it is necessary to use test data based on specimens with a different surface finish than actual components, the calculated life should be corrected:

$$N_{actual} = N_{calculated} \times \frac{C_f \text{ (actual hardware)}}{C_f \text{ (test specimens)}}$$

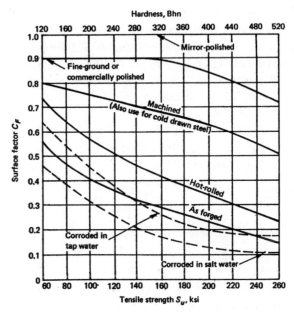

Figure 30. Surface factors for various machining operations [13]. (*Reprinted with permission of Prentice-Hall, Inc.*)

where: N_{actual} = actual life of component:
$N_{calculated}$ = calculated life based on test specimen data

Environmental Effects

Fatigue can be accelerated significantly by aggressive environments. This is especially true when the loads are applied and maintained for long periods of time. These effects are difficult to quantify. The best rule is to attempt to simulate environmental conditions during fatigue testing as closely as possible. Aggressive environments may include everything from salt air for carrier-based aircraft to nuclear radiation for electrical generating plants.

LIABILITY ISSUES

Because fatigue analysis involves calculating component lives, the analyst is likely to be involved in litigation at some time during his career. When writing reports, several items should be remembered:

- Be accurate. If it is necessary to make assumptions (and usually it is), state them clearly.
- The plaintiffs will have access to all of your reports, memos, photographs, and computer files. Nothing is sacred.
- Do not "wave the bloody arm." This refers to unnecessarily describing the results of component failure. Say "this component does not meet the design criteria," instead of "this component will fail, causing a crash, which could kill hundreds of people" (you may verbally express this opinion to gain someone's attention).

- Limit the report to your areas of expertise. If you decide to discuss issues outside your area, document your sources.
- Do not make recommendations unless you are sure they will be done. If you receive a report or memo that makes recommendations which are unnecessary or inappropriate, explain in writing why they should not be followed and what the proper course of action should be. If they are appropriate, make sure they are carried out. This is known as "closing the loop."
- Avoid or use with extreme care these words: *defect, flaw, failure.*
- If errors are detected in your report after it is published, correct it in writing immediately.

Engineers should not avoid writing reports for fear they may be used against them in a law suit. If a report is accurate and clearly written, it should help the defense.

REFERENCES

1. Neuber, H., "Theory of Stress Concentration for Shear-Strained Prismatical Bodies with Arbitrary Nonlinear Stress Strain Laws, "Trans. ASME, *J. Appl. Mech.,* Volume 28, Dec. 1961, p. 544.

2. Glinka, G., "Calculation of Inelastic Notch-Tip Strain-Stress Histories Under Cyclic Loading," *Engineering Fracture Mechanics,* Volume 22, No. 5, 1985, pp. 839–854.

3. Smith, R. W., Hirschberg, M. H. , and Manson, S. S., "Fatigue Behavior of Materials Under Strain Cycling in Low and Intermediate Life Range," NASA TN D-1574, April 1963.

4. Miner, M. A., "Cumulative Damage in Fatigue," Trans. ASME, *J. Appl. Mech.,* Volume 67, Sept. 1945, p. A159.

5. Marin, J., "Interpretation of Fatigue Strengths for Combined Stresses," presented at The American Society of Mechanical Engineers, New York, Nov. 28–30, 1956.

6. Muralidharan, U. and Manson, S. S., "A Modified Universal Slopes Equation for the Estimation of Fatigue Characteristics of Metals," *Journal of Engineering Materials and Technology,* Volume 110, Jan. 1988, pp. 55–58.

7. Irwin, G. R., "Analysis of Stresses and Strains Near the End of a Crack Transversing a Plate," Trans ASME, *J. Appl. Mech.,* Volume 24, 1957, p. 361.

8. Paris, P. C. and Erdogan, F., "A Critical Analysis of Crack Propagation Law," Trans. ASME, *J. Basic Engr.,* Volume 85, No. 4, 1963, p. 528.

9. Barsom, J. M., "Fatigue-Crack Propagation in Steels of Various Yield Strengths," Trans. ASME, *J. Eng. Ind.,* Ser. B, No. 4, Nov. 1971, p. 1190.

10. Troha, W. A., Nicholas, T., Grandt, A. F., "Observations of Three-Dimensional Surface Flaw Geometries During Fatigue Crack Growth in PMMA," *Surface-Crack Growth: Models, Experiments, and Structures,* ASTM STP 1060, 1990, pp. 260–286.

11. McComb, T. H., Pope, J. E., and Grandt, A. F., "Growth and Coalesence of Multiple Fatigue Cracks in Polycarbonate Test Specimens," *Engineering Fracture Mechanics,* Volume 24, No. 4, 1986, pp. 601–608.

12. Stinchcomb, W. W., and Ashbaugh, N. E., *Composite Materials: Fatigue and Fracture,* Fourth Volume, ASTM STP 1156, 1993.

13. Deutschman, A. D., Michels, W. J., and Wilson, C. E., *Machine Design Theory and Practice.* New Jersey: Prentice Hall, 1975, p. 893.

14. Fuchs, H. O. and Stephens, R. I., *Metal Fatigue in Engineering.* New York: John Wiley & Sons, Inc., 1980.

15. Mann, J. Y., *Fatigue of Materials.* Victoria, Australia: Melbourne University Press, 1967.

16. Erickson, P. E. and Riley, W. F., *Experimental Mechanics,* Vol. 18, No. 3, Society of Experimental Mechanics, Inc., 1987, p. 100.

17. Liaw, et al., "Near-Threshold Fatigue Crack Growth," *Actametallurgica,* Vol. 31, No. 10, 1983, Elsevier Science Publishing, Ltd., Oxford, England, pp. 1582–1583.

18. Pellini, W. S., "Criteria for Fracture Control Plans," NRL Report 7406, May 1972.

Recommended Reading

Metal Fatigue in Engineering by H. O. Fuchs and R. I. Stephens is an excellent text on the subject of fatigue. Most of the chapters contain "dos and don'ts" in design that provide excellent advice for the working engineer. *Metals Handbook, Volume 11: Failure Analysis and Prevention* by the American Society for Metals deals with metallurgical aspects, failure analysis, and crack inspection methods. *Analysis and Representation of Fatigue Data* by Joseph B. Conway and Lars H. Sjodahl explains how to regress test data so that it can be used for calculations. *Composite Material Fatigue and Fracture* by Stinchcomb and Ashbaugh, ASTM STP 1156, deals with the many complications that arise in fatigue calculations of composites. *Stress Intensity Factors Handbook,* Committee on Fracture Mechanics, The Society of Materials Science, Japan, by Y. Murakami is the most complete handbook of stress intensity factors, but is quite expensive. The *Stress Analysis of Cracks Handbook,* by H. Tada, P. Paris, and G. Irwin, is not as complete nor as expensive.

15

Instrumentation

Andrew J. Brewington, Manager, Instrumentation and Sensor Development, Allison Engine Company

INTRODUCTION

The design and use of sensors can be a very challenging field of endeavor. To obtain an accurate measurement, not only does the sensor have to possess inherent accuracy in its ability to transfer the phenomenon in question into a readable signal, but it also must:

- be stable
- be rugged
- be immmune to environmental effects
- possess a sufficient time constant
- be minimally intrusive

Stability implies that the sensor must consistently provide the same output for the same input, and should not be confused with overall accuracy (a repeatable sensor with an unknown calibration will consistently provide an output that is always incorrect by an unknown amount). *Ruggedness* suggests that the environment and handling will not alter the sensor's calibration or its ability to provide the correct output. *Immunity to environmental effects* refers to the sensor's ability to respond to only the measurand (item to be measured) and not to extraneous effects. As an example, a pressure sensor that changes its output with temperature is not a good sensor to choose where temperature changes are expected to occur; the temperature-induced output will be mixed inextricably with the pressure data, resulting in poor data. *Sufficient time constant* suggests that the sensor will be able to track changes in the measurand and is most critical where dynamic data is to be taken.

Of the listed sensor requirements, the most overlooked and probably the most critical is the concept of *minimal intrusion*. This requirement is important in that the sensor must not alter the environment to the extent that the measurand itself is changed. That is, the sensor must have sufficiently small mass so that it can respond to changes with the required time constant, and must be sufficiently low in profile that it does not perturb the environment but responds to that environment without affecting it. To properly design accurate sensors, one must have an understanding of material science, structural mechanics, electrical and electronic engineering, heat transfer, and fluid dynamics, and some significant real-world sensor experience. Due to these challenges, a high-accuracy sensor can be rather expensive to design, fabricate, and install.

Most engineers are not sufficiently trained in all the disciplines mentioned above and do not have the real-world sensor experience to make sensor designs that meet all the application requirements. Conversely, if the design does meet the requirements, it often greatly exceeds the requirements in some areas and therefore becomes unnecessarily costly. Luckily, many of the premier sensor manufacturers have design literature available based on research and testing that can greatly aid the engineer designing a sensor system. Sensor manufacturers can be found through listings in the *Thomas Registry* and *Sensor* magazine's "Yearly Buying Guide" and through related technical societies such as the Society for Experimental Mechanics and the Instrument Society of America. A good rule of thumb is to trust the literature provided by manufacturers, using it as a design tool; however, the engineer is cautioned to use common sense, good engineering judgment, and liberal use of questions to probe that literature for errors and inconsistencies as it pertains to the specific objectives at hand. See "Resources" at the end of this chapter for a listing of some vendors offering good design support and additional background literature useful in sensor design and use.

It is important to understand the specific accuracy requirements before proceeding with the sensor design. In many instances, the customer will request the highest accuracy possible; but if the truth be known, a much more reasonable accuracy will suffice. At this point, it becomes an economic question as to how much the improved accuracy would be worth. As an example, let us say that the customer requests a strain measurement on a part that is operating at an elevated temperature so that he can calculate how close his part is to its yield stress limit in service. That customer will undoubtedly be using the equation:

$$\sigma = E\varepsilon$$

where ε is strain, E is the material's modulus of elasticity, and σ is the stress. Depending on the material in question, the customer may have a very unclear understanding of E at temperature (that is, his values for E may have high data scatter, and the variation of modulus with temperature may not be known within 5–10%). In addition, he will, by necessity, be using a safety factor to ensure that the part will survive even with differing material lots and some customer abuse. In a case such as this, an extremely accurate, high-cost strain measurement (which can cost an order of magnitude higher than a less elaborate, less accurate measurement) is probably not justified. Whether the strain data is

0.1% accurate or 3% accurate probably will not change the decision to approve the part for service.

Although there are a wide variety of parameters that can be measured and an even wider variety of sensor technologies to perform those measurements (all with varying degrees of vendor literature available), there are a few basic measurands that bear some in-depth discussion. The remainder of this chapter deals with:

- fluid (gas and liquid) temperature measurement
- surface temperature measurement

- fluid (gas and liquid) total and static pressure measurement
- strain measurement
- liquid level and fluid (gas and liquid) flow measurement

These specific measurands were chosen due to their fundamental nature in measurement systems and their wide use, with consideration given to the obvious scope limitations of this handbook.

TEMPERATURE MEASUREMENT

Temperature measurement can be divided into two areas: fluid (gas and/or liquid) measurement and surface measurement. Fluid measurement is the most difficult of the two because (1) it is relatively easy to perturb the flow (and therefore, the parameter needing to be measured) and (2) the heat transfer into the sensor can change with environmental conditions such as fluid velocity or fluid pressure.

After these two measurement areas are investigated, a short section of this chapter will be devoted to an introduction to some common temperature sensors. Because the sensing device is located directly at the measurand location, it is important to understand some of the sensor limitations that will influence sensor attachment design.

Fluid Temperature Measurement

Fluid temperature measurement can be relatively easy if only moderate accuracy is required, and yet can become extremely difficult if high accuracy is needed. High accuracy in this case can be interpreted as $\pm0.2°F$ to $\pm10°F$ or higher depending on the error sources present, as will be seen later. In measuring fluid temperature, one is usually interested in obtaining the total temperature of the fluid. Total temperature is the combination of the fluid's static temperature and the extra heat gained by bringing the fluid in question to a stop in an isentropic manner. This implies stopping the fluid in a reversible manner with no heat transfer out of the system, thereby recovering the fluid's kinetic energy. Static temperature is that temperature that would be encountered if one could travel along with the fluid at its exact velocity. For isentropic flow (adiabatic and reversible), the total temperature (T_t) and the static temperature (T_s) are related by the equation:

$$T_s/T_t = 1/[1 + \tfrac{1}{2}(\gamma - 1)M^2]$$

where γ is the ratio of specific heats (c_p/c_v) and equals 1.4 for air at 15°C. M is the mach number. The isentropic flow

tables are shown in Table 1 for $\gamma = 1.4$ and provide useful ratios for estimating total temperature measurement errors.

In measuring T_t, there are three "configuration," or physical, error sources independent of any sensor-specific errors that must be addressed. These are radiation, conduction, and flow velocity-induced errors. Each of these errors is driven by heat transfer coefficients that are usually not well defined. As a result, it is not good practice to attempt to apply after-the-fact corrections for the above errors to previously obtained data. One could easily over-correct the data, with the result being further from the truth than the original, unaltered data. Instead, it is better to assume worst-case heat transfer conditions and design the instrumentation to provide acceptable accuracy under those conditions.

Radiation error is governed by:

$$q = \varepsilon A \sigma \left(T_{surf}^4 - T_\infty^4\right)$$

where q is the net rate of heat exchange between a surface of area A, emissivity ε, and temperature T_{surf} and its surroundings at temperature T_∞ (σ being the Stefan-Boltzmann constant and equal to 5.67×10^{-8} W/m$^2 \times$ K^4). It is appar-

Table 1
Isentropic Flow Tables ($\gamma = 1.4$)

M	p_s/p_t	T_s/T_t	A/A^*	M	p_s/p_t	T_s/T_t	A/A^*
0	1.0000	1.0000	∞	.60	.7840	.9328	1.1882
.01	.9999	1.0000	57.8738	.61	.7778	.9307	1.1767
.02	.9997	.9999	28.9421	.62	.7716	.9286	1.1657
.03	.9994	.9998	19.3005	.63	.7654	.9265	1.1552
.04	.9989	.9997	14.4815	.64	.7591	.9243	1.1452
.05	.9983	.9995	11.5914	.65	.7528	.9221	1.1356
.06	.9975	.9993	9.6659	.66	.7465	.9199	1.1265
.07	.9966	.9990	8.2915	.67	.7401	.9176	1.1179
.08	.9955	.9987	7.2616	.68	.7338	.9153	1.1097
.09	.9944	.9984	6.4613	.69	.7274	.9131	1.1018
.10	.9930	.9980	5.8218	.70	.7209	.9107	1.0944
.11	.9916	.9976	5.2992	.71	.7145	.9084	1.0873
.12	.9900	.9971	4.8643	.72	.7080	.9061	1.0806
.13	.9883	.9966	4.4969	.73	.7016	.9037	1.0742
.14	.9864	.9961	4.1824	.74	.6951	.9013	1.0681
.15	.9844	.9955	3.9103	.75	.6886	.8989	1.0624
.16	.9823	.9949	3.6727	.76	.6821	.8964	1.0570
.17	.9800	.9943	3.4635	.77	.6756	.8940	1.0519
.18	.9776	.9936	3.2779	.78	.6691	.8915	1.0471
.19	.9751	.9928	3.1123	.79	.6625	.8890	1.0425
.20	.9725	.9921	2.9635	.80	.6560	.8865	1.0382
.21	.9697	.9913	2.8293	.81	.6495	.8840	1.0342
.22	.9668	.9904	2.7076	.82	.6430	.8815	1.0305
.23	.9638	.9895	2.5968	.83	.6365	.8789	1.0270
.24	.9607	.9886	2.4956	.84	.6300	.8763	1.0237
.25	.9575	.9877	2.4027	.85	.6235	.8737	1.0207
.26	.9541	.9867	2.3173	.86	.6170	.8711	1.0179
.27	.9506	.9856	2.2385	.87	.6106	.8685	1.0153
.28	.9470	.9846	2.1656	.88	.6041	.8659	1.0129
.29	.9433	.9835	2.0979	.89	.5977	.8632	1.0108
.30	.9395	.9823	2.0351	.90	.5913	.8606	1.0089
.31	.9355	.9811	1.9765	.91	.5849	.8579	1.0071
.32	.9315	.9799	1.9219	.92	.5785	.8552	1.0056
.33	.9274	.9787	1.8707	.93	.5721	.8525	1.0043
.34	.9231	.9774	1.8229	.94	.5658	.8498	1.0031
.35	.9188	.9761	1.7780	.95	.5595	.8471	1.0022
.36	.9143	.9747	1.7358	.96	.5532	.8444	1.0014
.37	.9098	.9733	1.6961	.97	.5469	.8416	1.0008
.38	.9052	.9719	1.6587	.98	.5407	.8389	1.0003
.39	.9004	.9705	1.6234	.99	.5345	.8361	1.0001
.40	.8956	.9690	1.5901	1.00	.5283	.8333	1.000
.41	.8907	.9675	1.5587	1.01	.5221	.8306	1.000
.42	.8857	.9659	1.5289	1.02	.5160	.8278	1.000
.43	.8807	.9643	1.5007	1.03	.5099	.8250	1.001
.44	.8755	.9627	1.4740	1.04	.5039	.8222	1.001
.45	.8703	.9611	1.4487	1.05	.4979	.8193	1.002
.46	.8650	.9594	1.4246	1.06	.4919	.8165	1.003
.47	.8596	.9577	1.4018	1.07	.4860	.8137	1.004
.48	.8541	.9560	1.3801	1.08	.4800	.8108	1.005
.49	.8486	.9542	1.3595	1.09	.4742	.8080	1.006
.50	.8430	.9524	1.3398	1.10	.4684	.8052	1.008
.51	.8374	.9506	1.3212	1.11	.4626	.8023	1.010
.52	.8317	.9487	1.3034	1.12	.4568	.7994	1.011
.53	.8259	.9468	1.2865	1.13	.4511	.7966	1.013
.54	.8201	.9449	1.2703	1.14	.4455	.7937	1.015
.55	.8142	.9430	1.2550	1.15	.4398	.7908	1.017
.56	.8082	.9410	1.2403	1.16	.4343	.7879	1.020
.57	.8022	.9390	1.2263	1.17	.4287	.7851	1.022
.58	.7962	.9370	1.2130	1.18	.4232	.7822	1.025
.59	.7901	.9349	1.2003	1.19	.4178	.7793	1.026

Source: John [27], adapted from NACA Report 1135, "Equations, Tables, and Charts for Compressible Flow," AMES Research Staff.

ent from this equation that if either the absolute value of T_s − T_{sur} is large or both T_s and T_{sur} are large, then the radiation error can be significant. This is a rule that holds for all conditions but can be further exacerbated by those situations where extremely slow flow exists. In this situation, it becomes difficult to maintain sufficient heat transfer from the fluid to the sensor to overcome even small radiative flux. Obviously, radiative heat transfer can never raise the sensor temperature above that of the highest temperature body in the environment. If all of the environment exists within a temperature band that is a subset of the accuracy requirements of the measurement, radiation errors can be summarily dismissed.

Conduction errors are present where the mounting mechanism for the sensor connects the sensor to a surface that is not at the fluid's temperature. Since heat transfer by conduction can be quite large, these errors can be considerable. As with radiation errors, conditions of extremely slow flow can greatly compound conduction error because heat transfer from the fluid is not sufficiently large to help counter the conduction effect.

Velocity-induced errors are different from radiation and conduction in that some fluid velocity over the sensor is good while even the smallest radiation and conduction effects serve to degrade measurement accuracy. Fluid flow over the sensor helps overcome any radiation and conduction heat transfer and ensures that the sensor can respond to changes in fluid temperature. However, as mentioned earlier, total temperature has a component related to the fluid velocity. A bare, cylindrical sensor in cross flow will "recover" approximately 70% of the difference between T_t and T_s. At low mach numbers, the difference between T_t and T_s is small, and an error due to velocity (i.e., the amount not recovered) of 0.3 $(T_t - T_s)$ may be perfectly acceptable. For higher-velocity flows, it may be necessary to slow the fluid. This will cause an exchange of velocity (kinetic energy) for heat energy, raising T_s and hence the sensor's indicated temperature, T_i (T_t remains constant).

A shrouded sensor, as shown in Figure 1, can serve the purpose of both slowing fluid velocity and acting as a radiation shield. With slower velocity, T_s is higher, so $T_i = 0.7$ $(T_t - T_s)$ is higher and closer to T_t. The shield, with fluid scrubbing over it, will also attempt to come up to the fluid temperature. Most of the environmental radiation flux that would have been in the field of view of the sensor now can only "see" the shield and, therefore, will affect only the shield temperature. In addition, the sensor's field of view is now limited to a small forward-facing cone of the original environment, with the rest of its field of view being the shield and/or sensor support structure. Since the shield and support structure are at nearly the same temperature as the sensor, there is little driving force behind any shield-sensor radiation exchange, and the sensor is protected from this error. Conduction effects are minimized by the slender nature of the sensor (note that the sensor has a "length divided by diameter" [L/D] ratio of 10.5).

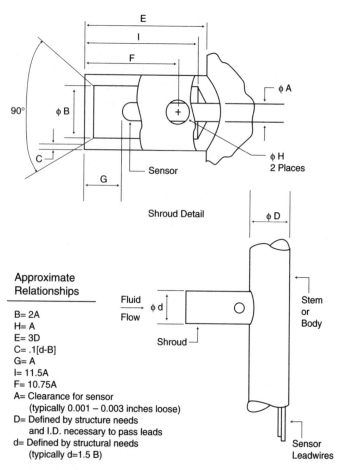

Approximate Relationships

B= 2A
H= A
E= 3D
C= .1[d-B]
G= A
I= 11.5A
F= 10.75A
A= Clearance for sensor
   (typically 0.001 – 0.003 inches loose)
D= Defined by structure needs
   and I.D. necessary to pass leads
d= Defined by structural needs
   (typically d=1.5 B)

Figure 1. Parametric design: single-shrouded total temperature probe.

Figure 1 shows a general sensor configuration suited for mach 0.3 to 0.8 with medium radiation effects. This design is somewhat complicated to machine and would be considerably more expensive than the sensor configuration shown in Figure 2. Differing fluid velocities and environment temperatures would require changing Figure 1 by altering bleed hole diameters (H), adding other concentric radiation shields, and/or lengthening sensor L/D ratios. In Figure 2, the sensor hangs in a pocket cut from a length of support tube. This arrangement offers some radiation shielding (but decidedly inferior to that in Figure 1) and some velocity recovery. The placement of the sensor within the cutout will greatly influence the flow velocity over the sensor and hence its recovery. In fact, depending on flow environmental conditions (vibration, flow velocity, particles within the flow, etc.) the sensor may shift within the pocket during use causing a change in reading that does not correspond to a change in fluid conditions. The probe in Figure 2, therefore, is better suited for mach 0.1 to 0.4 in areas with low radiation effects.

Figure 3 shows a compromise probe configuration in terms of cost and performance. It is designed for mach 0.1 to 0.4 with medium radiation effects. The perforations will slow the flow somewhat less than the probe in Figure 1 and will reduce radiation effects better than the probe in Figure 2. This configuration does, however, have a significant advantage where flow direction can change. While the probe in Figure 3 has stable recovery somewhat independant of flow yaw angle, the probe in Figure 2 is very susceptible to pitch angle variation and moderately susceptible to yaw variations. By comparison, the probe in Figure 1 is rather insensitive to yaw and pitch variations up to ±30.

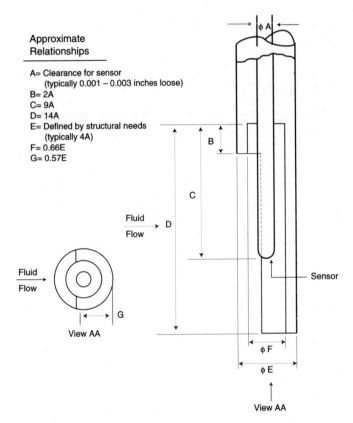

Approximate
Relationships

A= Clearance for sensor
   (typically 0.001 – 0.003 inches loose)
B= 2A
C= 9A
D= 14A
E= Defined by structural needs
   (typically 4A)
F= 0.66E
G= 0.57E

Figure 2. Parametric design: half-shielded total temperature probe.

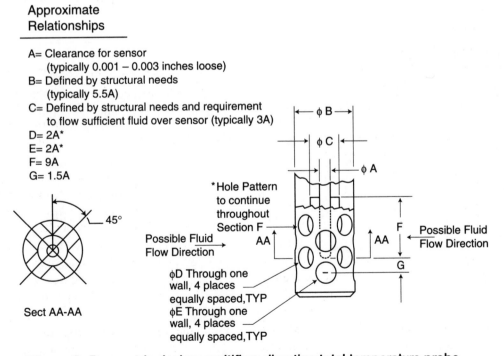

Approximate
Relationships

A= Clearance for sensor
   (typically 0.001 – 0.003 inches loose)
B= Defined by structural needs
   (typically 5.5A)
C= Defined by structural needs and requirement
   to flow sufficient fluid over sensor (typically 3A)
D= 2A*
E= 2A*
F= 9A
G= 1.5A

Figure 3. Parametric design: multiflow direction total temperature probe.

Surface Temperature Measurement

Surface temperature measurement can be somewhat easier than fluid temperature measurement due to fewer configuration error sources. Radiation effects can, to a large extent, be ignored, as a sensor placed on a surface will see the same radiative flux as the surface beneath it would if the sensor were not present. The only exception to this would occur in high radiative flux environments where the sensor has a significantly different emissivity than that of the surface to be measured. Error sources, then, for surface temperature measurement are constrained to conduction and velocity-induced effects.

Conduction errors occur when the sensor body contacts an area of different temperature than that being measured. The sensor then acts as an external heat transfer bridge between those areas, ultimately altering the temperature to be measured. As with fluid temperature measurement, a sufficient sensor L/D ratio (between 8 and 15) will help ensure that conduction errors are minimized.

Velocity errors are present when the sensor body rests above the surface to be measured and, acting like a fin, transfers heat between the surface and the surrounding fluid. This can occur in relatively low-flow velocities but is obviously worse with increasing fluid speed. Even at low speeds the sensor can serve to trip the flow, disrupting the normal boundary layer and increasing local heat transfer between

fluid and surface. Sensors that are of minimal cross-section or are embedded into the surface of interest minimize velocity errors. Embedding is preferred over surface mounting because of the superior heat transfer to the sensor along the increased surface area of the groove (see Figure 4).

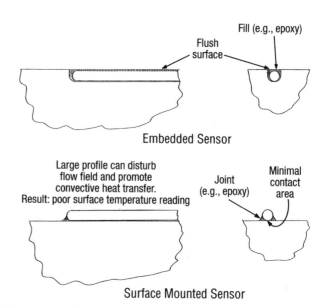

Figure 4. Embedded versus surface mounting technique for surface temperature measurement.

Common Temperature Sensors

The most common temperature sensor is the *thermocouple* (T/C). In a T/C, two dissimilar metals are joined to form a junction, and the remaining ends of the metal "leads" are held at a reference (known) temperature where the voltaic potential between those ends is measured. When the junction and reference temperatures are not equal, an electromotive force (emf) will be generated proportional to the temperature difference. The single most important fact to remember about thermocouples is that emf will be generated only in areas of the T/C where a temperature gradient exists. If both the T/C junction and reference ends are kept at the same temperature T_1, and the middle of the sensor passes through a region of temperature T_2, the emf generated by the junction end of the T/C as it passes from T_1 to T_2 will be directly canceled by the voltage generated by the lead end of the T/C as it passes from T_2 to T_1. Both

voltages will be equal in magnitude but opposite in sign, with the net result being no output (see Example 1). Further explanation of thermocouple theory, including practical usage suggestions, can be found in Dr. Robert Moffat's *The Gradient Approach to Thermocouple Circuitry* [2].

Thermocouples are inexpensive and relatively accurate. As an example, chromel-alumel wire with special limits of error has a 0.4% initial accuracy specification. T/Cs can be obtained in differing configurations from as small as sub-0.001-inch diameter to larger than 0.093-inch diameter and can be used from cryogenic to 4,200°F. However, If very high accuracy is required, T/Cs can have drawbacks in that output voltage drift can occur with temperature cycles and sufficient time at high temperature, resulting in calibration shifts.

Two other commonly used temperature sensors are resistance temperature devices (RTDs) and thermistors, both

Example 1
The Gradient Approach to Thermocouple Circuitry

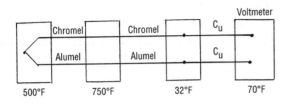

Example of a Type K (Chromel-Alumel) thermocouple with its junction at 500°F and reference temperature of 32°F where a splice to the copper leadwires is made. In this example, the thermocouple passes through a region of higher temperature (750°F) on its way to the 32°F reference.

The voltage (E) read at the voltmeter can be represented as a summation of the individual emfs (ε) generated along each discrete length of wire. The emf generated by each section is a function of the thermal emf coefficient of each material and the temperature gradient through which it passes. Therefore:

$$E_{IND} = \int_{70°F}^{32°F} \varepsilon_{CU} + \int_{32°F}^{750°F} \varepsilon_{CHR} + \int_{750°F}^{500°F} \varepsilon_{CHR} + \int_{500°F}^{750°F} \varepsilon_{AL}$$
$$+ \int_{750°F}^{32°F} \varepsilon_{AL} + \int_{32°F}^{70°F} \varepsilon_{CU}$$

Rearranging and expanding, we see:

$$E_{IND} = \left[\int_{70°F}^{32°F} \varepsilon_{CU} - \int_{70°F}^{32°F} \varepsilon_{CU} \right] + \left[\int_{32°F}^{500°F} \varepsilon_{CHR} + \int_{500°F}^{750°F} \varepsilon_{CHR} \right.$$

$$\left. - \int_{500°F}^{750°F} \varepsilon_{CHR} \right] + \left[\int_{500°F}^{750°F} \varepsilon_{AL} - \int_{500°F}^{750°F} \varepsilon_{AL} + \int_{500°F}^{32°F} \varepsilon_{AL} \right]$$

$$E_{IND} = \int_{32°F}^{500°F} \varepsilon_{CHR} + \int_{500°F}^{32°F} \varepsilon_{AL}$$

If the far left temperature zone was at 32°F instead of 500°F, all equations would remain the same but the final form could be further reduced to the following:

$$E_{IND} = \int_{32°F}^{32°F} \varepsilon_{CHR} + \int_{32°F}^{32°F} \varepsilon_{AL} = 0$$

Theory based on Moffat [2].

of which have sensing elements whose resistance changes in a repeatable way with temperature. RTDs are usually constructed of platinum wire, while thermistors are of integrated circuit chip design. RTDs can be used from −436°F to +2,552°F, while thermistors are usually relegated to the −103°F to +572°F range. Each of these sensors can be very accurate over its specified temperature range, but both are sensitive to thermal and mechanical shock. Thermistors do have an advantage in very high resistance changes with temperature, however, those changes remain linear over a relatively small temperature range.

One other surface temperature measurement technique that bears mention is *pyrometery*, which can be used to measure surface temperatures from +1,200°F to +2,000°F. When materials get hot they emit radiation in various amounts at various wavelengths depending on temperature. Pyrometers use this phenomenon by nonintrusively measuring the emitted radiation at specific wavelengths in the infrared region of the spectra given off by the surface of interest and, provided the surface's emissivity is known, inferring it's temperature. The equation used is

$$P = \varepsilon \sigma T^4$$

where P is the power per unit area in W/m², ε is the emissivity of the part, σ is the Stefan-Boltzmann constant (5.67 ∗ 10⁻⁸ W/(m²K⁴)), and T is the temperature in K. Pyrometers use band-pass filters to allow only specific wavelength photons to reach silicon or InGaAs photodiodes, which then convert the incoming photons to electrons yielding a current that is proportional to the temperature of the part in question. These sensors are not influenced by the above-mentioned physical error sources (because they are nonintrusive) but can be greatly affected by incorrect emissivity assessments, changes in emissivity over time, and reflected radiation from other sources such as hot neighboring parts or flames.

PRESSURE MEASUREMENT

Pressure measurement can be divided into two areas: total pressure and static (or cavity) pressure. In most cases it won't be practical to place a pressure transducer directly into the fluid in question or even mount it directly to the flow-containing wall because of the vibration, space, and temperature limitations of the transducer. Instead, it is common practice to mount the open end of a tube at the sensing location and route the other end of the tube to a separately mounted transducer. Due to this consideration, the remainder of this section concentrates on tube mounting design considerations. Pressure transducers can be chosen as stock vendor supplies that simply meet the requirements in terms of accuracy, frequency response, pressure range, over-range, sensitivity, temperature shift, nonlinearity and hysterisis, resonant frequency, and zero offset, and will not be further discussed here.

Total Pressure Measurement

As with temperature, fluid pressure readings can be static or total. Static pressure (P_s) is the pressure that would be encountered if one could travel along with the fluid at its exact velocity, and total pressure (P_t) is that pressure found when flow is stopped, trading its kinetic energy for pressure rise above P_s. P_s and P_t are related by the equation:

$$P_t/P_s = [1 + \tfrac{1}{2} (\gamma - 1)M^2]^{\gamma/(\gamma - 1)}$$

where γ is the ratio of specific heats (c_p/c_v) and equals 1.4 for air at 15°C. M is the mach number. See Table 1 for tabular form of this equation.

The most common method of measuring P_t is to place a small tube (pressure probe) within the fluid at the point of interest and use the tube to guide pressure pulses back to an externally mounted pressure transducer. Error sources for this arrangement include inherent errors within the pressure transducer, response time errors for nonconstant flow conditions, and errors based on incorrect tube alignment into the flow and/or configuration.

In order to minimize response time lags, the pressure transducer should be mounted as close as practical to the point of measurement. Also, the tube's inner cross-sectional area should not be significantly smaller than the pressure transducer's reference volume, located immediately in front of its measuring diaphragm. Additionally, increases in the tube's cross-sectional area between the sensing point and the transducer will slow response time. Finally, the tubing should be seamless when possible and have minimal bends. All necessary bends should be constructed with a minimum inner radius of 1.5 times the tube's outer diameter (for annealed metallic tubing).

For the tube to correctly recover the full P_t, it is critical that the sensor (tube) face directly into the flow. Often it may not be possible to know flow direction accurately, or the flow angle is known to change during operation. In these cases, modifications to the tube sensing end must be used to correct for flow angle discrepancies. In Figure 5, four tube end arrangements are shown: (a) shows a sharp-edged impact tube; (b) adds a shield; (c) is a tube with an inner to outer diameter ratio of 0.2 and a 15° chamfer; and (d) is a cylinder in cross flow with a capped end and a small hole in its wall. As shown in Figure 0.2, each of these arrangements has a differing ability to accept angled flow and still transfer P_t accurately to its transducer.

While the head modifications compensate for improper flow angle, another error can occur if pressure gradients exist within the flow. In the subsonic flow regime discussed here, the flow can sense and respond to the presence of the pressure probe within it. As a result, the flow will turn and shift toward the lower pressure area when presented with the blockage of the pressure probe. By ensuring that the length of tube along the flow direction is at least three times the width of the body to which the tube is mounted (with the body perpendicular to the flow direction), this effect can be minimized (see Figure 6).

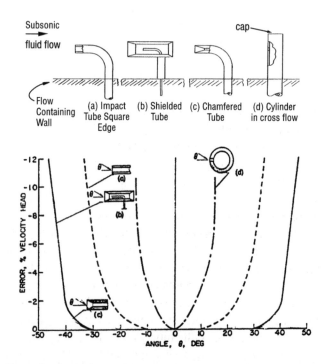

Figure 5. Total pressure probe, tube sensing end variations, and error with respect to flow angle [1]. (*Courtesy of Instrument Society of America. Reprinted by permission.*)

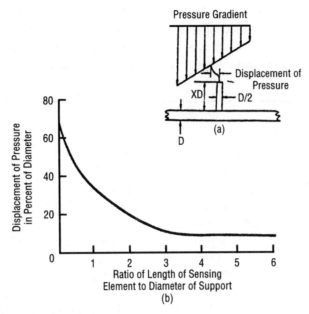

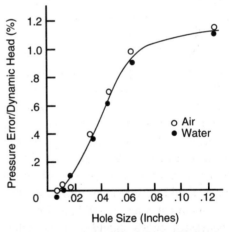

Figure 6. Total pressure probe errors of pressure gradient displacement due to sensing tube length [1]. (*Courtesy of Instrument Society of America. Reprinted by permission.*)

Static/Cavity Pressure Measurement

While it is difficult to measure static pressure (P_s) accurately within the flow (as any intrusive sensor will recover a significant portion of $P_t - P_s$, and the P_s probes that have been designed are sensitive to flow angle), it is relatively easy to measure P_s using a hole in the wall that contains the flow. Either the pressure transducer can be directly mounted to the wall or, more commonly, a tube will be placed flush with the inner wall at the sensing end with a pressure transducer connected to the opposite end.

Error sources in obtaining accurate static pressure measurements fall into the same categories as those of total pressure, with inherent errors caused by the pressure transducer, response-time errors for non-constant flow conditions, and errors based on incorrect tube alignment at the wall. Response-time errors are very similar to those found in total pressure measurement systems. To reduce response-time errors, keep all tubing lengths as short as possible and minimize bending. For all necessary bends, keep a minimum inner radius of 1.5 times the tube outer diameter (for annealed, seamless, metallic tubing). Finally, minimize all increases in the tube cross-sectional area between the sensing point and the sensor.

Static pressure errors related to configuration are somewhat more complex. As shown in Figure 7, the size of the static pressure port diameter (tube inner diameter in most

Figure 7. Errors in static pressure reading as a function of hole size for air and water [4]. (*Reprinted by permission of ASME.*)

cases) can produce errors and must be balanced against practical machining considerations and flow realities. While a 0.010-inch inner tube diameter may provide a very accurate reading, it may not be practical to obtain tubing of that size or to machine the required holes. In addition, if the flow field consists of highly viscous oil or air with high particulate count (soot, rust, etc.) then a 0.010-inch diameter orifice would impede pressure pulse propagation and/or would plug completely.

Not only is static pressure port diameter a consideration, but changes in that port diameter along its length close to the opening to the flow field can also be a source of error. It is a good rule of thumb not to allow changes in the static pressure port diameter to occur within a length of 2.5 times the static pressure port diameter itself. For example, if a 0.020-inch diameter hole is added to a pipe for the purpose of measuring static pressure in the pipe, then the 0.020 hole should remain that size, with no interruptions or steps for at least 0.050 inches away from the opening to the pipe. A length of 3.0 to 5.0 times the hole diameter is preferred where practical. See *ASME Power Test Codes, Supplement on Instruments and Apparatus:* Part 5, Measurement of Quantity of Materials, Chapter 4: Flow Measurement, copyright 1959.

A final effect to be considered concerns that of orifice edge and hole angle with respect to the flow path (see Figure 8). It is best to keep the hole perpendicular to the flow and retain sharp edges. Failure to remove burrs created during hole machining can give negative errors of 15–20% of dynamic head, while failure to completely remove the burrs (e.g., burr area cannot be detected by touch but is visibly brighter than surrounding area) can give negative errors up to 2% of the dynamic head. For these reasons, the note to "remove burrs but leave sharp edges" should always be used when calling out the machining of static pressure

ports holes on a drawing. See "Influence of Orifice Geometry on Static Pressure Measurement," R. E. Rayle, ASME Paper Number 59-A-234.

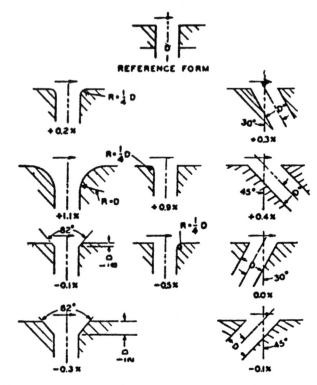

Figure 8. Effect of orifice edge form on static pressure measurements (variation in percent of dynamic head) [4]. (*Reprinted by permission of ASME.*)

STRAIN MEASUREMENT

It is often necessary to discern the stress within a component of interest. As there are no practical ways to obtain stress information directly, it is customary to measure strain (ε) and, using the material's known modulus of elasticity (E), calculate the stress (σ) via the equation:

$$\sigma = E\varepsilon$$

Strain is simply the change in length (ΔL) of a material divided by the length over which that change is measured (gauge length, L). As an example, if the original length between two known points on a surface of interest is 1.0000 inches, and the length measured under loading is found to

be 1.0001 inches, then the change in length is 0.0001 and the gauge length is 1.0000. The strain is therefore:

$$\Delta L/L = 0.0001/1.0000 = 0.0001 \text{ strain}$$

As strain numbers are usually very small, it is customary to use the units of microstrain ($\mu\varepsilon$), which are 10^6 times normal strain values. The above example would be read as $100\mu\varepsilon$.

The following sections highlight the electrical resistance strain gauge and its common data acquisition system. Additionally, some effort is made to discuss compensation techniques to provide a customer-oriented output useful in a variety of conditions.

The Electrical Resistance Strain Gauge

The most common strain measurement transducer is the electrical resistance strain gauge. In this sensor, an electrical conductor is bonded to the surface of interest. As the surface is strained, the conductor will become somewhat longer (assuming the strain field is aligned longitudinally with the conductor) and the cross-sectional area of the conductor will decrease. Additionally, the specific resistivity of the material may change. The summation of these three effects will result in a net change in resistance of the conductor, which can be measured and used to infer the strain in the surface. The relationship that ties this change in resistance to strain level is:

$$GF = [\Delta R/R]/\varepsilon$$

where GF is the gauge factor of the specific gauge, ΔR is the change in gauge resistance, R is the initial gauge resistance, and ε is the strain in inches/inch (not $\mu\varepsilon$).

Electrical resistance strain gauges can be purchased in a variety of sizes as fine wire grids (e.g., 0.001-inch diameter) but are more commonly available as thin film foil patterns. These foil gauges offer high repeatability, a wide variety of grid sizes and orientations, and a multitude of solder tab arrangements. Multiple gauge alloys are available, each with characteristics suited for a trade-off between fatigue life, stability, temperature range, etc. The gauges can also be purchased with self temperature compensation (STC) which serves to match the general coefficient of thermal expansion of the part to which the gauge will be bonded, thereby reducing the "apparent strain" (see the Full Wheatstone Compensation Techniques section). In addition to multiple alloys, there are multiple gauge backing materials from which to choose. The gauge backing serves to both electrically isolate the gauge from ground and to transfer the strain to the alloy grid. Finally, gauges can be purchased with the grid exposed or fully encapsulated for grid protection. Once these choices are made, it is still necessary to pick the proper cement, leadwire, and solder.

As was stressed in the introduction, relying on the technical expertise of a competent vendor is critical in obtaining usable results with an unfamiliar sensor system. This is particularly true with strain gauge application. There are so many variables, choices, and error sources that, without solid technical counseling, the chances for obtaining poor data are relatively high. It is beyond the scope of this chapter to go through the finer points of gauge application, especially with the excellent vendor literature available. However, some common failure points in gauge application include the areas of improper cleanliness of the part (and the hands of the gauge application technician), improper part surface finish, and poor solder joints or incomplete flux removal.

Keeping the gauge area on the part clean and free of oxides is critical to obtaining a good gauge bond. Once the area is clean, install the gauge in a timely manner so as not to allow the area to pick up dirt. Perform the cleaning and gauge application in a draft-free, air-conditioned area when possible. This will provide the air with some humidity control and filtering. Not only does the part need to be cleaned, but it is also good practice to have the technician wash his hands prior to beginning each gauge application. This will reduce contamination of the gauge area with dirt, oil, and salts from the skin.

Additionally, if the part surface has a rough surface finish in the gauge area, an inconsistent adhesive line thickness can exist across the gauge. This can yield poor strain transfer to the gauge, especially if the part is subject to temperature excursions where thermal expansion mismatch between the part and cement can cause unwanted grid deflection. Problems can also exist, however, if the part has a surface finish that is smooth like glass. In this case, insufficient tooth may exist to obtain maximum cement adhesion, resulting in gauge slippage at high strains or under high cycle fatigue. Again, follow the manufacturer's recommendations (usually, 60 μin, rms is the recommended surface finish).

Finally, it is common practice to use some flux to aid in soldering the lead wires to the strain gauge tabs to assure proper solder wetting. However, flux residue that is not completely removed can serve to corrode the metals and eventually cause shorts to ground. What is insidious about this failure mode is how slowly it works. Flux residue can go unnoticed as the gauge is checked, covered with protective coatings, and delivered to test. Then, during the test phase when critical data are being taken, the gauge can develop intermittent signal spikes and drop-outs, eventually resulting in low resistance to ground.

Electrical Resistance Strain Gauge Data Acquisition

One common method for measuring the gauge resistance changes caused by strains is the Wheatstone Bridge completion circuit. This circuit can have one, two, or four active legs corresponding to single-gauge configuration (i.e., ¼ bridge), two-gauge configuration (i.e., ½ bridge), and four-gauge configuration (full bridge). Figure 9 shows ¼, ½, and full bridge arrangements together with their representative output equations. Let us examine the full bridge configuration.

Bridge/Strain Arrangement	Description	Output V_0/V (mV/V) at Gage Factor, F, when ϵ expressed as microstrain
	Single active gage in uniaxial tension or compression.	$\dfrac{V_0}{V} = \dfrac{F\epsilon \times 10^{-3}}{4 + 2F\epsilon \times 10^{-6}}$
	Two active gages with equal and opposite strains — typical of bending-beam arrangement.	$\dfrac{V_0}{V} = \dfrac{F\epsilon}{2} \times 10^{-3}$
	Four active gages in uniaxial stress field—two aligned with maximum principal strain, two "Poisson" gages (column).	$\dfrac{V_0}{V} = \dfrac{F\epsilon\,(1 + \nu) \times 10^{-3}}{2 + F\epsilon\,(1 - \nu) \times 10^{-6}}$
	Four active gages with pairs subjected to equal and opposite strains (beam in bending or shaft in torsion).	$\dfrac{V_0}{V} = F\epsilon \times 10^{-3}$

Figure 9. Full, one-half, and one-quarter active bridge arrangements with output voltage equations. (*Courtesy of Measurements Group, Inc., Raleigh, NC.*)

In a simple, uniform cantilever beam with a single load on the free end deflecting the beam downward, the top surface of the beam is in tension and the bottom of the beam is in compression (Figure 10). As shown, the neutral axis of the beam is located along the beam's centerline which implies that the strain on the beam's top surface is equal in magnitude to the strain on the beam's bottom surface. To determine the strain in the beam, gauges #1 and #2 should be placed on the top surface of the beam and gauges #3 and #4 should be placed on the bottom. All four of the gauges should be oriented longitudinally along the beam at the same distance from the fixed end. The gauges should then be wired as shown in Figure 11.

From Figure 10 we can see that the calculated strain along either surface's outer fibers at the strain gauge loca-

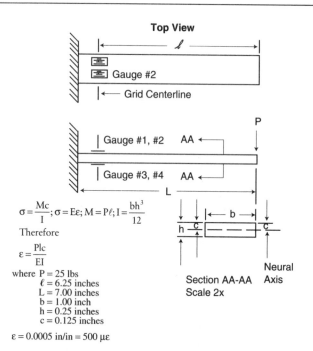

$$\sigma = \frac{Mc}{I}; \ \sigma = E\epsilon; \ M = P\ell; \ I = \frac{bh^3}{12}$$

Therefore

$$\epsilon = \frac{P\ell c}{EI}$$

where P = 25 lbs
ℓ = 6.25 inches
L = 7.00 inches
b = 1.00 inch
h = 0.25 inches
c = 0.125 inches

ϵ = 0.0005 in/in = 500 $\mu\epsilon$

Figure 10. Simple uniform cantilever beam with full bridge to measure bending.

tion equals 500$\mu\epsilon$. Each single 350 Ω gauge (with a gauge factor of 2.1), placed longitudinally in this location, would see a resistance change of:

$$\Delta R = R(\epsilon) \ GF$$

$$\Delta R = (350)(0.0005)(2.1) = 0.3675 \ ohms$$

Therefore, the two gauges in compression each read 349.6325 under load while the two gauges in tension each read 350.3675 under load. With an input voltage of 5.0 volts, the output voltage equals 5.25 mV (see Figure 11).

To have the Wheatstone Bridge perform properly, the bridge must be balanced. That is, each leg must have the same resistance, otherwise, a voltage output will be present under zero strain conditions. Unbalanced situations occur due to inherent resistance differences between gauges in the bridge coupled with resistance differences due to different length internal bridge wires. Balancing can be performed external to the bridge with the use of many readout devices; however, it can also be handled within the bridge circuitry, simplifying future data acquisition concerns. Special resistors can be bonded within the circuit and then trimmed to leave the bridge output at just a few microstrain under zero load.

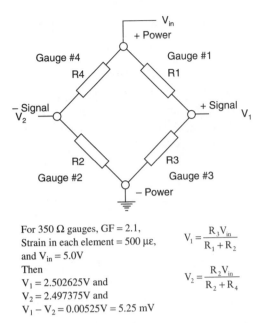

For 350 Ω gauges, GF = 2.1,
Strain in each element = 500 µε,
and V_{in} = 5.0V

Then
V_1 = 2.502625V and
V_2 = 2.497375V and
$V_1 - V_2$ = 0.00525V = 5.25 mV

$$V_1 = \frac{R_3 V_{in}}{R_1 + R_2}$$

$$V_2 = \frac{R_2 V_{in}}{R_2 + R_4}$$

Figure 11. Wheatstone wiring diagram for beam in bending; calculations for beam in Figure 12.

Full Wheatstone Bridge Compensation Techniques

If the beam in Figure 10 is to be used at elevated (or reduced) temperatures, further compensation for "apparent strain" (ε_{app}) and "modulus" may be used to optimize the data for the customer. Apparent strain compensation helps account for strain output resulting from no-load temperature excursions. These extraneous readings are caused by (1) temperature coefficient of resistance (TCR) changes in the gauges and wiring of the bridge and (2) differences in thermal expansion between the component being instrumented and the gauge itself. Modulus compensation attempts to account for additional extraneous strain output at temperature caused by both changes in the strain sensor's gauge factor (GF) with temperature and changes in component modulus of elasticity (E), allowing the data to represent only the strain due to load. This last compensation allows the design engineer to use the data without having to know specific time-temperature history to input varying E and GF values to get true stress.

As further clarification of this compensation issue, it may be viewed as follows: If the part can be taken through a temperature excursion in a no-load condition and the gauge output remains essentially zero, then the gauge requires no (further) ε_{app} compensation. If the part can be taken through a temperature excursion under load and the gauge output represents accurately the load conditions in-

dependent of temperature, then the sensor requires no (further) modulus compensation.

Apparent strain can be corrected by (1) using the correct STC (self temperature compensating) gauges, (2) compensating with the addition of special wire segments or trimmable bondable foil patterns within the bridge itself, or (3) both. The compensating wire or foil pattern used in (2) and (3) is chosen such that its TCR will serve to balance the undesirable TCR changes in the bridge wiring and gauges. This compensation step is required if the STC gauges don't accurately match the part's coefficient of thermal expansion, or if higher accuracy is required than offered by the generalized STC gauges.

Compensation by addition of wire or foil resistors can be accomplished by starting with a balanced bridge bonded to the component of interest. Attach the bridge to a quality strain output measurement device (it should read very close to zero microstrain if balanced well) and place temperature sensors on the component in various locations. Place the component in an oven, or other environmental chamber, that can duplicate the expected operating temperature. Subject the component to a temperature that will stabilize the gauge/epoxy system (preferably 25° to 50°F above the expected operating temperature). The part should remain at this temperature until the strain output varies by less than 2 microstrain per hour for two consecutive hours. Return the component to room temperature. When the part is cooled and isothermal, record the strain output. Finally, subject the component to its operating temperature and record the strain output change from the last reading. This is the uncompensated ε_{app}.

At this point resistors of appropriate TCR need to be added to the correct bridge legs to balance the uncompensated ε_{app}. It is known that

$$\Delta R = R(\varepsilon)\ GF$$

Rewriting and expanding this equation, we see that:

$$R_{comp} = R(\varepsilon_{app})\ GF/\Delta T\ (TCR)$$

where R_{comp} is the additional compensating resistance to be added, R is the bridge resistance, ΔT is the change in temperature from room temperature to operating temperature, and TCR is the temperature coefficient of resistance of R_{comp} (e.g., $TCRB_{alco}$ = 0.0025°F^{-1}). Returning to Figure 11, if the ε_{app} reading showed that V_2 had a higher potential than V_1, then we can infer that resistance should be added to the leg containing either R_4 or R_3. After the resistor has been correctly inserted, it is good practice to recheck the ε_{app} at temperature.

Unlike apparent strain compensation, which involves placing a special resistor within the bridge circuit itself, modulus compensation involves placing two resistors, one in each of the power legs leading to the bridge. These resistors should be placed as close to the bridge as possible so as to be within the same thermal environment, thereby responding to temperature changes with resistance changes that adjust the bridge input voltage. Although it is best to test the bridge output at temperature under known load to calculate modulus compensation, it is often not practical. Therefore, the following formula may be used to estimate the resistances required:

$$R_{mc} = \frac{R(GF_2 E_1 - GF_1 E_2)}{GF_1 E_2 + (GF_1 \times E_1 \times TCR\,(T_2 - T_1)) - GF_2 E_1}$$

where R_{mc} = the resistance to be split equally between the positive power and negative power legs (Ω), R = the bridge resistance (Ω), T_1 = room temperature (°F), T_2 = bridge operating temperature (°F), E_1 = the component material modulus of elasticity at T_1 (psi), E_2 = the component material modulus of elasticity at T_2, GF_1 = the gauge factor of the bridge gauges at T_1, GF_2 = the gauge factor of the bridge gauges at T_2, and TCR = the temperature coefficient of resistance of the compensating resistors (°F^{-1}).

LIQUID LEVEL AND FLUID FLOW MEASUREMENT

The measurement of liquid level and fluid flow rate is required in virtually all aspects of industrial process control and power generation/conversion. Due to the widespread need for these measurements and the similar nature of virtually all liquid storage and fluid delivery (piping) systems, common sensor solutions are readily available from a well-established vendor base. These sensor packages can meet a wide variety of price and accuracy requirements and can be tailored to almost any liquid, regardless of properties. In addition, much work has been done by the technical societies to standardize on measurement methodologies and practices.

As a result of these readily available systems and standards, this chapter will not stress fundamental theory or practices necessary to custom-design specialized sensor packages. Instead, it will simply outline some of the more common level and flow systems available, detailing cost, accuracy, and environmental considerations where applicable and suggesting general rules to use in making a selection. The sensor vendors will be familiar with the vast majority of level and flow problems and be able to recommend "off-the-shelf" solutions. Should highly specialized cases occur, additional information on the design and development of custom systems can be obtained from the reference literature.

Liquid Level Measurement

Level sensors relay information regarding the interface between a liquid and a gas (often air) or the interface between two liquids of differing specific gravities. The sensors can be point-specific or continuous-reading. Point-specific level sensors trip a switch when the liquid interface reaches a set point. These switches might activate alarms, open valves, turn on pumps, etc. Continuous-reading level sensors relay values corresponding to interface position within an operating range. The information from continuous-reading sensors may be used as diagnostic data or as input to a control system.

Level sensors rely on measuring buoyant forces, pressures, timing of ultrasonic pulses, etc., to infer the location of the interface in question. These sensors range from simple floats that move with the interface (as used in a sump

pump), to more complex sensors measuring capacitance fluctuations as the fluid changes height between conductive "plates." In this chapter we will quickly compare the following most common level sensors: float, displacer, delta-pressure systems, capacitance, and ultrasonic.

The simple *float sensor* relies on the buoyant force of the liquid in question to maintain the float at a specific relation to the liquid interface. In Figure 12 (a) a float switch is attached to a motion transducer providing a continuous reading system. In the sump pump reference above, the float is attached to a switch, creating a point-specific system. These switches can be designed so that a permanent magnet on the end of the float rod actuates a hermetically sealed reed switch, making a very rugged system. Float sensors are generally economical, rugged, repeatable, and us-

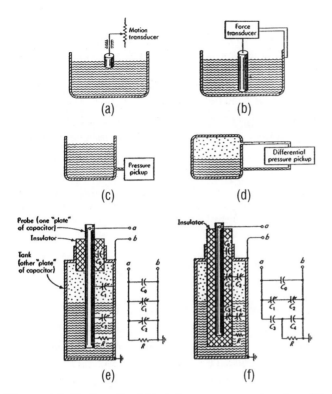

Figure 12. Various liquid level measurement meters [9]. (*Reprinted by permission of McGraw-Hill Book Co.*)

able with a wide variety of fluids. They are relatively insensitive to shock and vibration. Some particularily inexpensive float switches, however, may fail after a time in service due to their economical construction. If very low-cost float switches are chosen, it may be prudent to install two independent sensor systems to minimize risks.

The *displacer sensor,* Figure 12 (b), is related to the simple float in that it relies on the fluid's buoyant force and is usable with many fluids. Unlike the float, however, it does not follow the liquid interface. Instead, the changing interface alters the buoyant force on the quasi-stationary displacer, and that change in force is read on an attached force transducer. The displacer's construction allows for accurate resolution of small level changes, but its force transducer can make it more sensitive to shock and vibration.

Delta-pressure systems, Figures 12 (c) and (d), use the hydrostatic pressure of the liquid to infer the interface location. The pressure force of the column of fluid is a function of the liquid level times the fluid's specific weight. With the open container situation, the pressure gauge should be chosen to measure "gauge pressure." If an absolute pressure gauge were chosen, it would measure the weight of both the liquid and the column of air above it (barometric pressure); changes in the barometer would adversely affect the repeatability of the system.

The pressure vessel situation requires the use of a differential pressure sensor to compensate for the column of vapor over the liquid interface. Both of these techniques can be used in limited slurries and corrosives if proper precautions such as liquid seals, water purge, or air purge are used to isolate the pressure meter from the process.

Capacitance-based sensors use a probe and the wall of the vessel as two plates of a capacitor with the liquid medium between as a dielectric. As the liquid level changes, the effective capacitance between the "plates" is altered in a repeatable manner. For fluids that are nonconducting (dielectric constant < 4) the probe can be inserted directly into the liquid—see Figure 14 (e). For conductive fluids, the probe must be insulated to keep the liquid from short-circuiting the system—see Figure 12 (f). Should the vessel be nonconducting, a second probe can be inserted into the fluid to act as the second capacitor plate.

Capacitance level sensors can be used with many common liquids (including many liquid metals and corrosives) and granular solids. They can be operated through a wide temperature and pressure range. In the case where fluid turbulence is possible, the probe(s) must be designed so as not to move. Even small changes in the probe's position will alter the effective distance between capacitor plates and will appear to the sensor as a level change!

Ultrasonic level sensors transmit and then read high-frequency pressure pulses to infer interface levels. In the case of noncontacting sensors, Figure 13 (a), the ultrasonic beam strikes the fluid surface and the echo is detected by the sensor, creating a continuous reading sensor. As in sonar, the timing of the signals indicates the distance between the probe and the "surface" that created the echo. Some of these sensors utilize additional software that creates a running average of the distance signals being received. New signals are compared to this running average and to new signals; in this way, spurious signals and noise can be

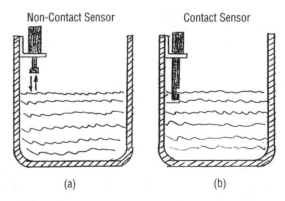

Figure 13. Ultrasonic liquid level measurement sensors.

identified and rejected. As noncontacting sensors, these ultrasonic sensors are of great value when highly corrosive liquids and/or adverse environmental conditions exist. Their accuracy and repeatability, however, drop as a function of the distance between probe and the liquid interface.

Contact ultrasonic sensors, Figure 13 (b), contain a gap between the point at which the high-frequency pulse is generated and a point farther along the probe where that pulse is detected. The sensor is positioned so that the gap is located in a known point in the fluid vessel. When the liquid rises to the point where it fills this gap, the signal eas-

ily propagates from the pulse generator to the detector. When the liquid level falls, the signal is attenuated, resulting in a relay switching. In order to compensate for some surface turbulence and agitation, delay logic is built into the probe circuitry so the relay will not switch immediately.

These contact sensors are usable with most fluids, however, highly aerated fluids can cause the signal to remain attenuated even when the fluid is filling the gap. Additionally, highly viscous liquids may cause errors should the fluid remain in the gap after the liquid interface level has dropped.

Fluid Flow Measurement

There are an exhaustive number of flow meters available on the market to fit virtually any flow measurement need. The most common flow meters can be divided into the four main areas of differential pressure, positive displacement, velocity, and mass. Differential flow meters include orifice, target, venturi, flow nozzle, pitot tube, elbow, and variable-area meters. Piston, oval gear, and rotary vane meters are types of positive displacement flow meters. Velocity meters include turbine, vortex, electromagnetic, and ultrasonic variations. Finally, Coriolis meters measure the mass rate of flow directly and are the most common of the mass meters.

The most common flow meter in use today is the orifice meter, primarily because of its wide pressure and temperature operating range, low cost, and reasonable accuracy. In Table 2, a comparison of all of the above meters is outlined, stressing summary characteristics. This table can be used in making an initial selection of the few meters that should be pursued in greater detail for the required application. Should an orifice meter, venturi, or flow meter be selected, ASME has some excellent literature available to aid in the selection and design (e.g., *Fluid Meters*, 6th ed., ASME, 1971, and *ASME Power Test Codes, Supplement on Instruments and Apparatus:* Part 5, Measurement of Quantity of Materials, Chapter 4: Flow Measurement, copyright 1959). Additional information on the selection and design of custom systems can be obtained from the reference literature.

Orifices are thin, flat plates of metal placed between flanges of the flow-carrying pipe so as to block flow. These plates have specific configuration openings machined into them that allow the flow to pass through. Static pressure taps on each side of the plate allow the measurement of the dif-

ferential pressure caused by the flow obstruction. By knowing the size and configuration of the opening in the orifice plate and the two pressure readings, the flow rate can be deduced. The orifice is especially subject to errors due to wear and abrasion.

Target meters measure the force imparted on a target placed in the flow. The force is the upstream pressure minus the downstream pressure integrated over the area of the target, and is therefore directly related to the flow rate.

Venturi meters are a section of piping that has a gradually inward-sloping entrance section (approximately 21° included angle), a straight section of reduced diameter, and a gradually sloping exit section (approximately 8° included angle). Static pressure measurements taken prior to the entrance section and at the reduced diameter "throat" allow calculation of the flow rate. The venturi offers reduced abrasion characteristics when compared to the orifice meter.

Flow nozzles are literally nozzles inserted concentrically into the flow path to gradually reduce the pipe area and then dump the flow back to the original pipe diameter. Flow nozzles possess the same general accuracy and resistance to abrasion advantages of the venturi while offering a shorter installed length and a cheaper price. These nozzles, however, are more costly than orifices, have pressure recovery performance between that of the orifice and venturi, and are somewhat difficult to install.

Pitot tubes use total and static pressure readings at a single flow stream, via an inserted probe, to determine flow velocity and flow rates. As stated in the Static Pressure Measurement section, it is difficult to measure P_s in a flow field due to flow angle sensitivity of the inserted probes. If the flow field is very well established, pitot tubes that measure total pressure at their tips and static pressure along the

Table 2

Comparison of Flow Meters

Flow Meter	GAS clean	GAS dirty	LIQUIDS clean	LIQUIDS dirty	LIQUIDS viscous	LIQUIDS corrosive	LIQUIDS slurry	Approx. max. temperature °F	Approx. max. pressure (psig)	Accuracy uncalibrated including transmitter	Reynolds number	Pipe size, in.	Required upstream pipe, diam.	Pressure loss	Range	Relative cost
Orifice, square-edged	•		•	x		x		to 1,000 (transmitter to 250)	to 6000	±1-2%URV*	$R_D > 1{,}000$	> 1.5	10-30	medium	4:1	low
Target	•	•	•	•	•	x	x	"	"	±1-5%URV	$R_D > 100$	> .5-4	10-30	medium	4:1	medium
Venturi	•	x	•	x	x	x	x	"	"	±1-2% URV	$R_D > 75{,}000$	> 2	5-20	low	4:1	medium
Flow nozzle	•	x	•	x	x	x	x	"	"	±1-2%URV	$R_D > 10{,}000$	> 2	10-30	medium	4:1	medium
Pitot tube	•		•			x		"	"	±3-5%URV	no limit	> 3	20-30	very low	3:1	low
Elbow	•	x	•	x		x	x	"	"	±5-10%URV	$R_D > 10{,}000$	> 2	30	very low	3:1	low
Electromagnetic			•	•	•	•	•	360	≤1500	±0.5% of rate	no limit	.1-72	5	none	40:1	high
Mass flow meter (coriolis)	x		•	•	•	•	x	< 570	≤2000	±0.4% of rate	no limit		none	low	10:1	high
Positive displacement	•		•			x		gas: 250 liquid: 600	≤1400	gas: ±1% URV liquid: ±0.5% of rate	≤8,000 cSt	<12	none	high	10:1	medium
Turbine	•		•			x		500	≤3000	gas: ±0.5% of rate liquid: ±1% or rate	≤2-15 cSt	.25-24	5-10	high	20:1	high
Ultrasonic (time of flight)	•		•		x	•		500	pipe rating	±1% URV to ±5% of rate	no limit	>.5	5-30	none	20:1	high
Ultrasonic (Doppler)				•	x	•	•	250	pipe rating	±5% of URV	no limit	>.5	5-30	none	10:1	high
Variable area	•		•		•	x		glass: ≤400 metal: ≤1000	glass: 350 metal: 720	±0.5% of rate to ±1% URV	to highly viscous fluid	≤3	none	medium	10:1	low
Vortex	•	x	•	x		x		≤750	≤1500	±0.5-1.5% of rate	> 10,000	.5-16	10-20	medium	10:1	high

• Designed for this application

x Normally applicable

(no symbol) Not designed for this application

*URV: Upper range value of the flow (formerly full-scale flow rate)

Sources: Adapted from Miller [5], by permission of McGraw-Hill Book Co., and Plant Engineering, Nov. 21, 1984, copyright © Cahners Publishing Co., by permission.

probe bodies can be used. When flow profiles change within the pipe, a single-point pitot tube can show errors due to both improper P_s values and the fact that the single-point measurement no longer represents the true average velocity. Averaging pitot tubes, which sample multiple locations in the flow, are often used as a more accurate measure of the flow field velocity.

Elbow meters use the principle that fluid flowing along the outer path of a turn will exert a centrifugal force that is proportional to fluid velocity. Static pressure taps are placed in the inner radius and outer radius of the elbow.

Variable-area meters have a float, suspended in a vertical section of slightly diverging "pipe" diameter. The diverging flow section allows flow to pass by the float on the sides. The float's position in that section is a function of the flow rate and always seeks the equilibrium point where the differential pressure on the float is balanced by the force of gravity on the float.

Positive displacement flow meters separate the flow into established sections of known volume as the flow passes through the meter. By counting the throughput frequency of those sections, the flow rate is deduced.

Turbine meters utilize a free-wheeling rotor placed within the flow stream. As flow rate increases, the rotor speed increases. By measuring the frequency of the rotor, the flow rate can be calculated. Potential error sources for turbine meters include calibration shifts due to blade wear, bearing wear (and its associated increase in friction), and overspeed when packets of vapor enter the meter.

Vortex meters utilize the approximately sinusoidal pressure (and velocity) changes that are caused by moving vortices shed by a bluff-body placed in the flow stream. The frequency of the vortex shedding is a function of the fluid velocity and, therefore, can be used to determine fluid flow rate.

Electromagnetic flow meters use the principle of induction whereby a conductor moving in a magnetic field generates a voltage proportional to its velocity through the field. With this sensor, the fluid acts as the conductor and the meter houses the electromagnet that creates the magnetic field. The walls of the pipe contain electrodes that sense the induced voltage, allowing the fluid velocity to be calculated.

Ultrasonic (time of flight) meters send high-frequency pressure waves across the pipe at an acute angle to the flow. The time it takes for the pulse to return is a function of the velocity of the flow as it speeds or slows the signal. The signal is proportional to the average velocity along the line of the pressure pulse. Often, multiple pressure pulse paths are utilized to provide a better average flow rate.

Ultrasonic (Doppler) meters rely on particulate matter within the flow to reflect the pressure pulse back to the receiver. The frequency of the collected signal is interrogated to determine its shift and, by the Doppler principle, the fluid flow rate is calculated.

Coriolis meters cause the fluid to both translate and rotate about a point which results in Coriolis acceleration. A common form of this meter uses a U-shaped flow tube that is fixed at the ends and vibrated at its natural frequency like a cantilever. When the tube is moving upward the fluid flowing into the tube resists the movement by pressing down on the tube. The fluid flowing out of the tube has been forced up and resists moving back down to exit by pressing up on the tube. As the fluid entering presses down and the fluid exiting presses up, the tube twists. During the downward motion of the U-shaped tube, the forces are reversed and the angle of twist reverses also.

REFERENCES

1. Gettelman, C. C. and Krause, L. N., "Considerations Entering into the Selection of Probes for Pressure Measurement in Jet Engines," ISA 1952 Proceedings—Paper No. 52-12-1.
2. Moffat, R. J., "The Gradient Approach to Thermocouple Circuitry," originally published by Reinhold Publishing Company in *Temperature—Its Measurement and Control in Science and Industry, Vol. 3,* Part 2.
3. Krause, L. N. and Gettelman, C. C., "Effect of Interaction Among Probes, Supports, Duct Walls, and Jet Boundaries on Pressure Measurements in Ducts and Jets," ISA 1952 Proceedings—Paper 52-12-2.
4. Rayle, R. E., "Influence of Orifice Geometry on Static Pressure Measurements," ASME Paper 59-A-234.
5. Miller, R. W., *Flow Measurement Engineering Handbook,* 2nd Ed. New York: McGraw-Hill Book Co., 1989.

6. Avallone, E. A. and Baumeister, T., III, *Mark's Standard Handbook for Mechanical Engineers,* 9th Ed. New York: McGraw-Hill Book Co., 1978.

7. Buckwith, T. G., Buck, N. L., and Marangoni, R. D., *Mechanical Measurements,* 3rd Ed., Redding, MA: Addison-Wesley Publishing Co., Inc., 1982.

8. Benedict, R. P., *Fundamentals of Temperature, Pressure, and Flow Measurements,* 3rd Ed. New York: John Wiley & Sons, Inc., 1984.

9. Doebelin, E. O., *Measurement Systems Application and Design,* 4th Ed. New York: McGraw-Hill Book Co., 1990.

10. *Fluid Meters: Their Theory and Application,* 6th Ed., ASME, 1971.

11. *ASME Power Test Codes, Supplement on Instruments and Apparatus:* Part 5, Measurement of Quantity of Materials, Chapter 4: Flow Measurement, copyright 1959.

12. Cheremisinoff, N. P. and Cheremisinoff, P. N., *Flow Measurement for Engineers and Scientists.* New York: Marcel Dekker, Inc., 1988.

13. *Flow and Level Handbook,* Volume 28, Omega Engineering, Inc., 1992.

14. *Temperature Handbook,* Volume 28, Omega Engineering, Inc., 1992.

15. ANSI/MC96.1-1982, "Temperature Measurement Thermocouples," ISA, 1982.

16. ANSI/ASME: MFC-2M-1983 (R-1988), "Measurement Uncertainty for Fluid Flow in Closed Conduits," Bk. No. K00112, 1983, p. 71.

17. ASME:MFC-3M-1989, "Measurement of Fluid Flow in Pipes Using Orifice, Nozzle, and Venturi," Bk. No. K000113, 1985, p. 63.

18. ANSI/ASME:MFC-4M-1986, "Measurement of Gas Flow by Turbine Meters," Bk. No. K0018, 1986, p. 18.

19. ANSI/ASME: MFC-5M-1985 (R-1989), "Measurement of Liquid Flow in Closed Conduit Using Transit Time Ultrasonic Flowmeters," Bk. No. K0015, 1985, p. 14.

20. ASME/ANSI: MFC-6M-1987, "Measurement of Liquid Flow in Pipes Using Vortex Flow Meters," Bk. No. K00117, 1987, p. 11.

21. ASME: PTC 19.3, "Instruments and Apparatus: Temperature Measurement," (R 1986), Bk. No. C00035, p. 118.

22. ISA-RP16.5, "Installation, Operation, Maintenance Instructions for Glass Tube Variable Area Meters (Rotameters)," 1961, p. 6.

23. ISA-RP31.1 (ANSI/ISA RP31.1-1977), "Specification, Installation, and Calibration of Turbine Flowmeters," 1977, p. 21.

24. ISO: R541-1967, "Measurement of Fluid Flow by Means of Orifice Plates and Nozzles."

25. ISO: 3966-1977, "Measurement of Fluid Flow in Closed Circuits—Velocity Area Method Using Pitot Static Tubes."

26. Bentley, J. P., *Principles of Measurement Systems,* 2nd Ed., White Plains, NY: Longman Scientific & Technical, 1983.

27. John, James, E. (Ed.), *Gas Dynamics,* 2nd Ed. Needham Heights, MA: Allyn & Bacon, Inc., 1984.

Resources

The following sensor vendors offer excellent technical support:

OMEGA Engineering, Inc. (temperature, pressure, level, flow, data systems, etc.)
P.O. Box 2349
Stamford, CT 06906, USA
800-872-9436

Watlow Gordon (temperature measurement products)
5710 Kenosha Street
Richmond, IL 60071, USA
815-678-2211

Measurements Group, Inc. (strain sensors, data systems, etc.)
P.O. Box 27777
Raleigh, NC 27611, USA
919-365-3800

16
Engineering Economics

Lawrence D. Norris, Senior Technical Marketing Engineer—Large Commercial Engines, Allison Engine Company, Rolls-Royce Aerospace Group

TIME VALUE OF MONEY: CONCEPTS AND FORMULAS

The value of money is not constant, but changes with time. A dollar received today is worth more than a dollar received a month from now, for two reasons. First, the dollar received today can be invested immediately and earn interest. Second, the purchasing power of each dollar will decrease during times of inflation, and consequently a dollar received today will likely purchase more than a dollar next month. Therefore, in the capital budgeting and decision-making process for any engineering or financial project, it is important to understand the concepts of *present value, future value,* and *interest.* Present value is the "current worth" of a dollar amount to be received or paid out in the future. Future value is the "future worth" of an amount invested today at some future time. Interest is the *cost* of borrowing money or the *return* from lending money. Interest rates vary with time and are dependent upon both risk and inflation. The rate of interest charged, and the rate of return expected, will be higher for any project with considerable risk than for a "safe" investment or project. Similarly, banks and investors will also demand a higher rate of return during periods of monetary inflation, and interest rates will be adjusted to reflect the effects of inflation.

Simple Interest vs. Compound Interest

Simple interest is interest that is only earned on the original principal borrowed or loaned. Simple interest is calculated as follows:

$$I = (P)\,(i)\,(n)$$

where: I = simple interest
  P = principal (money) borrowed or loaned
  i = interest rate per time period (usually years)
  n = number of time periods (usually years)

Compound interest is interest that is earned on both the principal and interest. When interest is compounded, interest is earned each time period on the original principal *and* on interest accumulated from preceding time periods. Exhibits 1 and 2 illustrate the difference between simple and compound interest, and show what a dramatic effect interest compounding can have on an investment after a few years.

Exhibit 1
Comparison Between $5,000 Invested for 3 Years at 12% Simple and 12% Compound Interest

Simple Interest

End of Year	Interest Earned	Cumulative Interest Earned	Investment Balance
0	$0.00	$0.00	$5,000.00
1	$5,000.00 × .12 = $600.00	$600.00	$5,600.00
2	$5,000.00 × .12 = $600.00	$1,200.00	$6,200.00
3	$5,000.00 × .12 = $600.00	$1,800.00	$6,800.00

Compound Interest (Annual Compounding)

End of Year	Interest Earned	Cumulative Interest Earned	Investment Balance
0	$0.00	$0.00	$5,000.00
1	$5,000.00 × .12 = $600.00	$600.00	$5,600.00
2	$5,600.00 × .12 = $672.00	$1,272.00	$6,272.00
3	$6,272.00 × .12 = $752.74	$2,024.64	$7,024.64

Exhibit 2
Comparison of Simple vs. Compound Interest

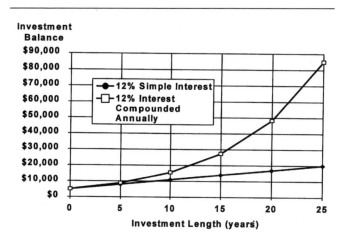

Nominal Interest Rate vs. Effective Annual Interest Rate

Almost all interest rates in the financial world now involve compound rates of interest, rather than simple interest. Compound interest rates can be quoted either as a *nominal rate* of interest, or as an *effective annual rate* of interest. The nominal rate is the stated annual interest rate compounded periodically (at the stated compounding time interval). The effective annual rate is the rate that produces the same final value as the nominal rate, when compounded only once per year. When the compounding period for a stated nominal rate is one year, then the nominal rate *is the same* as the effective annual rate of interest. The following formula can be used to convert nominal interest rates to effective annual rates:

$$i_{eff} = \left(1 - \frac{i_{nom}}{c}\right)^c - 1$$

where: i_{eff} = effective annual rate
i_{nom} = nominal rate
c = number of compounding periods per year

Note: interest rates are expressed as fractions (.12) rather than percentage (12%).

For *continuous* compounding, natural logarithms may be used to convert to an effective annual rate:

$$i_{eff} = e^{i_{cont}} - 1$$

where: i_{cont} = continuous compounding rate

Note: interest rates are expressed as fractions (.12) rather than percentage (12%).

Present Value of a Single Cash Flow To Be Received in the Future

How much is an amount of money to be received in the future worth today? Stated differently, how much money would you invest today to receive this cash flow in the future? To answer these questions, you need to know the *present value* of the amount to be received in the future. The present value (PV) can easily be calculated by *discounting* the future amount by an appropriate compound interest rate. The interest rate used to determine PV is often referred to as the *discount rate, hurdle rate,* or *opportunity cost of capital.* It represents the opportunity cost (rate of return) foregone by making this particular investment rather than other alternatives of comparable risk.

The formula for calculating present value is:

$$PV = \frac{FV}{(1+r)^n}$$

where: PV = present value
FV = future value

r = discount rate (per compounding period)
n = number of compounding periods

Example. What is the present value of an investment that guarantees to pay you $100,000 three years from now? The first important step is to understand what risks are involved in this investment, and then choose a discount rate equal to the rate of return on investments of comparable risk. Let's say you decide that 9% effective annual rate of interest (9% compounded annually) is an appropriate discount rate.

$$PV = \frac{\$100,000}{(1+.09)^3} = \$77,218$$

Thus, you would be willing to invest $77,218 today in this investment to receive $100,000 in three years.

Future Value of a Single Investment

Rearranging the relationship between PV and FV, we have a simple formula to allow us to calculate the future value of an amount of money invested today:

$$FV = PV \, (1 + r)^n$$

where: FV = future value
PV = present value
r = rate of return (per compounding period)
n = number of compounding periods

Example. What is the future value of a single $10,000 lump sum invested at a rate of return of 10% compounded monthly (10.47 effective annual rate) for 5 years? Note that monthly compounding is used in this example.

$$FV = \$10,000 \, (1 + .10/12)^{60} = \$16,453$$

The $10,000 investment grows to $16,453 in five years.

The Importance of Cash Flow Diagrams

Many financial problems are not nearly as simple as the two previous examples, which involved only one *cash outflow* and one *cash inflow*. To help analyze financial problems, it is extremely useful to diagram cash flows, since most problems are considerably more complex than these previous examples, and may have multiple, repetitive, or irregular cash flows. The *cash flow diagram* is an important tool which will help you understand an investment project and the timing of its cash flows, and help with calculating its PV or FV. The cash flow diagram shown represents the earlier example (in which PV was calculated for an investment that grew to $100,000 after annual compounding at 9% for 3 years).

The cash flow diagram is divided into equal time periods, which correspond to the compounding (or payment) periods of the cash flows. Note that this diagram is also presented from the investor's (lender's) point of view, since the initial investment is a cash outflow (negative) and the final payoff is a cash inflow (positive) to the investor. The dia-

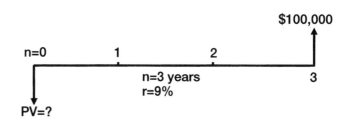

gram could also have been drawn from the borrower's point of view, in which case the PV would be a cash inflow and therefore positive. Before drawing a cash flow diagram, you must first choose which viewpoint (lender or borrower) the diagram will represent. It is then extremely important to give each cash flow the correct sign (positive or negative), based upon whichever point of view is chosen. Other examples of cash flow diagrams will be presented in additional sample problems in this chapter.

Analyzing and Valuing Investments/Projects with Multiple or Irregular Cash Flows

Many investments and financial projects have multiple, periodic, and irregular cash flows. A cash flow diagram will help in understanding the cash flows and with their analysis. The value of the investment or project (in current dollars) may then be found by discounting the cash flows to determine their present value. Each cash flow must be discounted individually back to time zero, and then summed to obtain the overall PV of the investment or project. This *discounted cash flow formula* may be written as:

$$PV = \sum_{i=1}^{n} \frac{C_i}{(1+r)^i}$$

where: PV = present value
C_i = cash flow at time i
n = number of periods
r = discount rate of return (opportunity cost of capital)

Example. How much would you be willing to invest today in a project that will have generated positive annual cash flows of $1,000, $1,500, $2,400, and $1,600, respectively, at the end of each of the first 4 years? (Assume a 10% annual rate of return can be earned on other projects of similar risk.)

Step 1: Cash flow diagram:

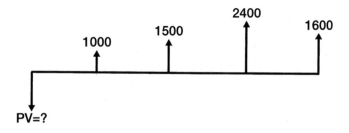

Step 2: Discount cash flows into current dollars (present value):

$$PV = \frac{1,000}{(1+.1)^1} + \frac{1,500}{(1+.1)^2} + \frac{2,400}{(1+.1)^3} + \frac{1,600}{(1+.1)^4}$$
$$= \$5,044.74$$

Thus, you should be willing to invest a maximum of $5,044.74 in this project.

Perpetuities

A *perpetuity* is an asset that provides a periodic fixed amount of cash flow in perpetuity (forever). Endowments and trust funds are often set up as perpetuities, so that they pay out a fixed monetary amount each year to the beneficiary, without depleting the original principal. For valuation purposes, some physical assets such as an oil well, natural gas field, or other long-lived asset can also be treated as a perpetuity, if the asset can provide a steady cash flow stream for many years to come. The cash flow diagram for a perpetuity is represented here:

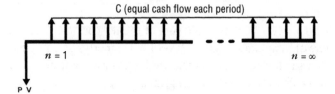

The value of a perpetuity can be found by this simple formula:

$$PV_{perpetuity} = \frac{C}{r}$$

where: C = cash flow per time period
r = discount rate of return (per time period)

A *growing perpetuity* is a cash flow stream that grows at a constant rate. The formula to value a growing perpetuity expands the basic perpetuity formula to include a growth rate for the cash flow stream:

$$PV = \frac{C_1}{r-g}$$

where: C_1 = initial cash flow
r = discount rate of return (per time period)
g = growth rate of cash flows (per time period)

Example. An existing oil well has been producing an average of 10,000 barrels per year for several years. Recent estimates by geologists show that this well's oil field is not large enough to justify additional drilling, but can maintain the existing well's production rate for hundreds of years to come. Assume that the recent price of oil has averaged $20 per barrel, oil prices are historically estimated to grow 3% a year forever, and your discount rate is 10%. Also assume that maintenance costs are insignificant, and the land has no value other than for the oil. What is the current monetary value of this oil field and well, if the owner were to sell it?

Solution:

$$PV = \frac{(10,000 \times 20)}{.10 - .03} = \$2,857,143$$

The value of the oil field and well should be $2,857,143 based upon these assumptions.

Future Value of a Periodic Series of Investments

Both individual investors and companies often set aside a fixed quantity of money every month, quarter, or year, for some specific purpose in the future. This quantity of money then grows with compound interest until it is withdrawn and used at some future date. A cash flow diagram for a periodic investment is shown.

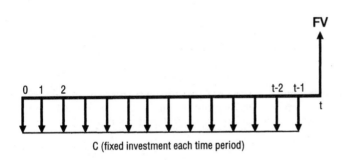

C (fixed investment each time period)

It can easily be calculated how much these periodic, fixed investments each time period will accumulate with compounding interest to be in the future. In other words, we want to determine the future value (FV), which can be calculated using the following formula:

$$FV = \frac{C}{r}[(1+r)^t - 1]$$

where: C = cash flow invested each time period
r = fractional compound interest rate (each time period)
t = number of time periods (length of investment)

Example. A self-employed contract engineer has no company-paid retirement plan, so he decides to invest $2,000 per year in his own individual retirement account (IRA), which will allow the money to accumulate tax-free until it is withdrawn in the future. If he invests this amount every year for 25 years, and the investment earns 14% a year, how much will the investment be worth in 25 years?

Solution:

$$FV = \frac{\$2,000}{.14}[(1+.14)^{25} - 1] = \$363,742$$

The periodic investment of $2,000 each year over 25 years will grow to $363,742 if a 14% rate of return is achieved. This example illustrates the "power of compounding." A total of only $50,000 was invested over the 25-year period, yet the future value grew to over 7 times this amount.

Annuities, Loans, and Leases

An *annuity* is an asset that pays a constant periodic sum of money for a fixed length of time. Bank loans, mortgages, and many leasing contracts are similar to an annuity in that they usually require the borrower to pay back the lender a constant, periodic sum of money (principal and interest) for a fixed length of time. Loans, mortgages, and leases can therefore be considered "reverse" annuities, as shown by the cash flow diagrams.

Cash Flow Diagrams: Annuity

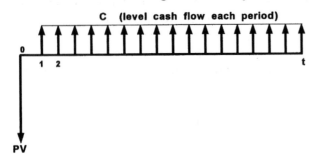

C (level cash flow each period)

PV

Cash Flow Diagrams: Reverse Annuity

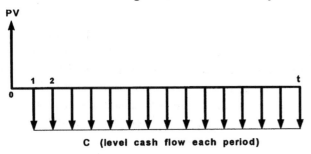

C (level cash flow each period)

The relationship between the value (PV) of the annuity/loan/lease, its periodic cash flows, and the interest or discount rate is:

$$PV = \frac{C}{r}\left[1 - \frac{1}{(1+r)^t}\right]$$

where: C = cash flow received/paid each time period
r = interest or discount rate (per time period)
t = number of time periods (time length of the annuity)

This formula may be used to determine the PV of the annuity, or conversely, to determine the periodic payouts or payments when the PV is known. It should be noted that the future value (FV) at time "t" will be zero. In other words, all the principal of the annuity will have been expended, or all the principal of the loan or lease will have been paid off, at the final time period. Financial calculators are also pre-programmed with this formula, and can simplify and expedite annuity, loan, and other cash flow problem calculations.

Example. Oil pipeline equipment is to be purchased and installed on a working pipeline in a remote location. This new equipment will have an operational life of 5 years and will require routine maintenance once per month at a fixed cost of $100. How much would the oil company have to deposit now with a bank at the remote location to set up an annuity to pay the $100 a month maintenance expense for 5 years to a local contractor? Assume annual interest rates are 9%.

Solution:

$$PV = \frac{100}{(.09/12)}\left[1 - \frac{1}{(1+.09/12)^{60}}\right] = \$4,817.34$$

Example. Apex Corp. borrows $200,000 to finance the purchase of tools for a new line of gears it's producing. If the terms of the loan are annual payments over 4 years at 14% annual interest, *how much are the payments?*

$$\$200,000 = \frac{C}{.14}\left[1 - \frac{1}{(1+.14)^4}\right]$$

Solution:

C = $68,640.96 per year

Gradients (Payouts/Payments with Constant Growth Rates)

A *gradient* is similar to an annuity, but its periodic payouts increase at a constant rate of growth, as illustrated by its cash flow diagram. A gradient is also similar to a growing perpetuity, except that it has a fixed time length. The cash flow diagram for a gradient loan or lease is simply the inverse of this cash flow diagram, since the loan/lease payments will be negative from the borrower's point of view. The PV of a gradient may be found by:

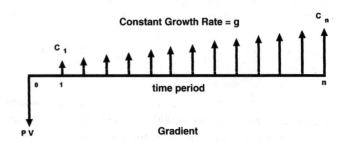

Constant Growth Rate = g

time period

PV

Gradient

$$PV = \frac{C_1(1 - x^n)}{(1 + r)(1 - x)} \qquad \text{when } g \neq r$$

or

$$PV = \frac{C_1 n}{(1 + r)} \qquad \text{when } g = r$$

where: C_1 = initial cash flow at the end of first time period
r = rate of return (discount rate)
n = number of time periods
$x = (1 + g)/(1 + r)$
g = fractional growth rate of cash flow per period

Analyzing Complex Investments and Cash Flow Problems

No matter how complex a financial project or investment is, the cash flows can always be diagrammed and split apart into separate cash flows if necessary. Using the discounted cash flow (DCF) formula, each cash flow can be discounted individually back to time zero to obtain its present value (PV), and then summed to determine the overall valuation of the project or investment. It may also be possible to simplify things if you can identify a series of cash flows in the complex investment. Cash flow series can be discounted using either annuity or perpetuity formulas, and then added together with the PV of any singular cash flow.

When analyzing any complex cash flow problem, it is important to remember the basic time value of money (TVM) concept (i.e., *the value of money changes with time*). Therefore, if you wish to add or compare any two cash flows, they both must be from the same time period. Cash flows in the future must be discounted to their present value (PV) before they can be added or compared with current cash flows. Alternatively, current cash flows can be compounded into the future (FV) and then added or compared with future cash flows.

Example. An automotive components company has been asked by a large automobile manufacturing company to increase its production rate of components for one particular car model. Sales of this car model have been much higher than expected. In order to increase its production rate, the components company will have to purchase additional specialized machinery, but this machinery will easily fit into its existing facility. The automobile manufacturer has offered to sign a contract to purchase all the additional output that can be produced for 4 years, which will increase annual net revenues by $300,000 without any additional administrative overhead or personnel expense. The added machinery will cost $450,000 and will require maintenance every 6 months at an estimated cost of $20,000.

The machinery will be sold at the end of the 4-year period, with an estimated salvage value of $80,000. *If the company requires a minimum rate of return of 20% on all investments, should it sign the contract, purchase the machinery, and increase its production rate?* (Depreciation and tax effects are not considered in this example.)

Step 1: From the point of view of the automotive components company, the cash flow diagram will be:

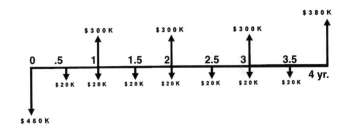

Step 2: Separate the overall cash flow diagram into individual components and series, and then determine the PV of each future cash flow or series of flows by discounting at the required annual rate of return of 20%:

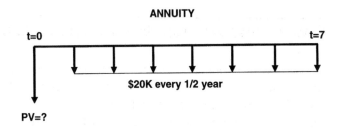

$$PV = \frac{\$20,000}{.20/2}\left[1 - \frac{1}{(1 + .20/2)^7}\right] = \$97,368 \text{ (negative)}$$

ANNUITY

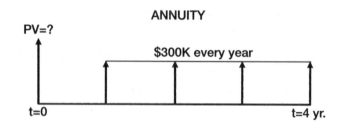

$$PV = \frac{\$300,000}{.2}\left[1 - \frac{1}{(1+.2)^4}\right] = \$776,620 \quad \text{(positive)}$$

SINGLE CASH FLOW FROM SALVAGE VALUE

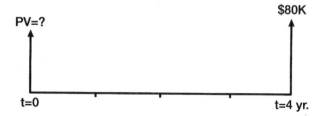

$$PV = \frac{\$80,000}{(1+.2)^4} = \$38,580 \text{ (positive)}$$

Step 3: Sum the PV of all cash flows to obtain the project's overall PV from all *future* cash flows:

$$PV = \$776,620 + \$38,580 - \$97,368$$
$$= \$717,832 \text{ (positive)}$$

Step 4: Compare the PV of the *future* cash flows with the project's *initial* cash flow (initial investment):

The $717,832 PV of the project's future cash inflows (positive) is clearly greater than the $450,000 (negative) initial investment. Thus, this project has a positive *net present value* (NPV) because the PV of the future cash flows are greater than the initial investment. A positive NPV means that the rate of return on the investment is greater than the rate (20%) which was used to discount the cash flows.

Conclusion: The company should buy the additional machinery and increase its production rate.

DECISION AND EVALUATION CRITERIA FOR INVESTMENTS AND FINANCIAL PROJECTS

How should you decide whether or not to make an investment or go forward with a financial project? If you also have a number of alternative investment options, how do you decide which is the best alternative? How do you evaluate an investment's performance some time period after the investment has been made and the project is up-and-running? There are a number of analysis techniques that can help with these questions. Four separate methods are presented here, although each method varies in its effectiveness and complexity. There are also additional methods and techniques that can be found in other engineering economics, finance, and accounting textbooks. However, most modern books and courses in these fields of study recommend the *net present value* (NPV) method as the most effective and accurate technique of evaluating potential investments and financial projects. Accounting measures such as *rate of return* (ROR) on an investment or *return on equity* (ROE) are useful to evaluate a project's financial results over a specific time period, or to compare results against competitors.

Payback Method

The payback method is a simple technique that determines the number of years before the cash flow from an investment or project "pays back" the initial investment. Obviously, the shorter the length of time it takes a project to pay off the initial investment, the more lucrative the project.

Example. A company wants to add a new product line and is evaluating two types of manufacturing equipment to produce this new line. The first equipment type costs $700,000 and can produce enough product to result in an estimated $225,000 in additional after-tax cash flow per year. The second type of equipment is more expensive at $900,000, but has a higher output that can bring an estimated $420,000 per year in after tax cash flows to the company. Which type of equipment should the company buy?

Solution:

Equipment Type 1
Payback Period = $700,000/$225,000 per year
= 3.11 years

Equipment Type 2
Payback Period = $900,000/$420,000 per year
= 2.14 years

Based upon this payback analysis, equipment type 2 should be purchased.

The advantage of the payback method is its simplicity, as shown in the example. However, there are several disadvantages that need to be mentioned. First, the payback method does not evaluate any cash flows that occur after the payback date, and therefore ignores long-term results and salvage values. Secondly, the payback method does not consider the time value of money (TVM), but assumes that each dollar of cash flow received in the second year and beyond is equal to that received the first year. Thirdly, the payback method has no way to take into account and evaluate any risk factors involved with the project.

Accounting Rate of Return (ROR) Method

The *accounting rate of return* for an existing company or project can be easily calculated directly from the company's or project's quarterly or annual financial accounting statements. If a potential (rather than existing) project is to be evaluated, financial results will have to be projected. Since ROR is an accounting measure, it is calculated for the same time period as the company or project's income statement. The rate of return on an investment is simply the net income generated during this time period, divided by the book value of the investment at the start of the time period.

$$R.O.R. = \frac{\text{Net Income}}{\text{Net Investment (book value)}}$$

Net income is an accounting term which is defined as "revenues minus expenses plus gains minus losses" (i.e., the "bottom line" profit or loss on the income statement). Net investment is the "book value" of the investment, which is the original investment amount (or purchase price) less any depreciation in value of the investment.

Example. Company XYZ invested $500,000 in plant and equipment four years ago. At the end of year 3, the company's financial statements show that the plant and equipment's value had depreciated $100,000, and had a listed book value of $400,000. If company XYZ has a net income of $60,000 during its 4th year of operation, what is its ROR for this time period?

Solution:

$$R.O.R. = \frac{60,000}{400,000} = 0.15 \quad (15\%)$$

Like the payback method, the advantage of the accounting ROR is that it is a simple calculation. It is especially simple for an existing company or project when the accounting statements are already completed and don't have to be projected. One disadvantage of ROR is that it doesn't consider TVM when it is used for more than one time period. It also uses accounting incomes rather than cash flows, which are not the same (see Accounting Fundamentals in this chapter). Another disadvantage is that ROR is a very subjective method, because the ROR is likely to vary considerably each time period due to revenue and expense fluctuations, accumulating depreciation, and declining book values. (Some textbooks recommend averaging ROR over a number of time periods to compensate for fluctuations and declining book values, but an "averaged" ROR still does not consider TVM.)

In general, accounting measurements such as ROR are best used for evaluating the performance of an *existing* company over a specific annual or quarterly time period, or for comparing one company's performance against another. Net present value (NPV) and internal rate of return (IRR) are superior methods for evaluating *potential* investments and financial projects, and in making the final decision whether to go ahead with them.

Internal Rate of Return (IRR) Method

Internal rate of return (IRR) is an investment profitability measure which is closely related to net present value (NPV). The IRR of an investment is that rate of return which, when used to discount an investment's future cash flows, makes the NPV of an investment equal zero. In other words, when the future cash flows of an investment or project are discounted using the IRR, their PV will exactly equal the initial investment amount. Therefore the IRR is an extremely useful quantity to know when you are evaluating a potential investment project. The IRR tells you the exact rate of return that will be earned on the original investment (or overall project with any additional investments) if the projected future cash flows occur. Similarly, when analyzing past investments, the IRR tells you the exact rate of return that was earned on the overall investment. The IRR decision rule for whether or not to go ahead with any potential investment or project being considered is simple: *If the IRR exceeds your opportunity cost of capital (rate of return that can be earned elsewhere), you should accept the project.*

The following example will make it clear how the IRR of an investment is calculated. It is highly recommended that a cash flow diagram be drawn first before trying to calculate the IRR. Particular attention should be paid to make sure cash inflows are drawn as positive (from the investor's point of view) and cash outflows are drawn as negative. It is also recommended that a financial calculator (or computer) be used to save time, otherwise an iteration procedure will have to be used to solve for the IRR.

Example. Acme Tool and Die Company is considering a project that will require the purchase of special machinery to produce precision molds for a major customer, Gigantic Motors. The project will last only four years, after which Gigantic Motors plans to manufacture its own precision molds. The initial investment for the special machinery is $450,000, and the machinery will also require a partial overhaul in three years at an estimated cost of $70,000. Projected after-tax cash flows resulting from the project at the end of each of the four years are $140,000, $165,000, $185,000, and $150,000, respectively. These projections are based upon a purchase agreement that Gigantic is prepared to sign, which outlines the quantities of molds it will buy. Expected salvage value from the machinery at the end of the four-year period is $80,000.

Should Acme proceed with this project, if its opportunity cost of capital is 23%?

Step 1: Cash flow diagram:

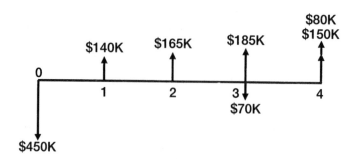

Step 2: Write an equation which sets the original investment equal to the PV of future cash flows, using the discounted cash flow formula. Solve for r, which is the IRR. (This is a simple process with a financial calculator or computer, but otherwise requires iteration with any other calculator).

$$450,000 = \frac{140,000}{(1+r)} + \frac{165,000}{(1+r)^2} + \frac{(185,000 - 70,000)}{(1+r)^3}$$
$$+ \frac{(150,000 + 80,000)}{(1+r)^4}$$

Solving for r (IRR):

IRR = 0.1537 (15.37%)

Step 3: Apply decision rule to results: Since the IRR for this project is 15.37%, and Acme's required rate of return (opportunity cost of capital) is 23%, Acme should not proceed with this project.

IRR has a number of distinct advantages (as does NPV) over both the payback and ROR methods as a decision-making tool for evaluating potential investments. The first advantage is that the IRR calculation takes into account TVM. Secondly, IRR considers all cash flows throughout the life of the project (including salvage values), rather than looking at only one time period's results like accounting ROR does, or the years until the initial investment is paid off like the payback method. Therefore, IRR is an objective criterion, rather than a subjective criterion, for making decisions about investment projects. Thirdly, IRR uses actual cash flows rather than accounting incomes like the ROR method. Finally, since IRR is calculated for the entire life of the project, it provides a basis for evaluating whether the overall risk of the project is justified by the IRR, since the IRR may be compared with the IRR of projects of similar risk and the opportunity cost of capital.

The IRR method has several disadvantages compared to the NPV method, though only one disadvantage is mentioned here for purposes of brevity. Further information about potential problems with the IRR method (compared to NPV) may be obtained from most finance textbooks. One major problem with IRR is the possibility of obtaining multiple rates of return (multiple "roots") when solving for the IRR of an investment. This can occur in the unusual case where cash flows change erratically from positive to negative (in large quantities or for sustained time periods) *more than once* during the life of the investment. The graph in the following example illustrates how multiple IRR roots can occur for investments with these types of cash flows. The second graph is for comparison purposes, and typifies the IRR calculation and graph for most investments. (Normally, NPV declines with increasing discount rate, thus giving only one IRR "root".) When multiple "roots" are obtained using the IRR method, it is best to switch to the NPV method and calculate NPV by discounting the cash flows with the opportunity cost of capital.

Example. A project under consideration requires an initial investment/cash flow of $3,000 (negative) and then has cash flows of $26,000 (positive) and $30,000 (negative) at the end of the following two years. A graph of NPV versus discount rate shows that two different IRR "roots" exist. (IRR is the point on the curve that intersects the hor-

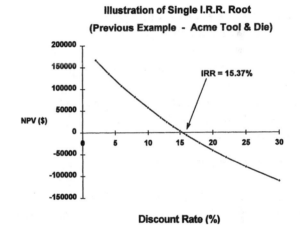

izontal axis at NPV = 0.) The second graph is from the previous example, showing that most "normal" investments have only one IRR root.

The IRR for this example is both 37.1% and 629.6%, as NPV equals zero at these two discount rates.

Net Present Value (NPV) Method

Net present value (NPV), as its name suggests, calculates the net amount that the discounted cash flows of an investment exceed the initial investment. Using the discounted cash flow (DCF) formula, the future cash flows are discounted by the rate of return offered by comparable investment alternatives (i.e., the opportunity cost of capital), and then summed and added to the initial investment amount (which is usually negative).

$$NPV = C_0 + \sum_{i=1}^{n} \frac{C_i}{(1+r)^i}$$

where: C_o = initial cash flow (negative for a cash "outflow")
C_i = cash flow in time period i
n = number of time periods
r = opportunity cost of capital/discount rate

Though similar to the IRR method, NPV does not calculate an investment's exact rate of return, but instead calculates the exact dollar amount that an investment exceeds, or fails to meet, the *expected* rate of return. In other words, if an investment provides a rate of return exactly equal to the opportunity cost of capital, then the NPV of this investment will be zero because the discounted future cash flows equal the initial investment. Thus, NPV provides an excellent decision criterion for investments. An investment with a positive NPV should be accepted, since it provides a rate of return above the opportunity cost of capital. By the same reasoning, an investment with a negative NPV should be rejected. NPV also does *not* suffer from any of the drawbacks of the payback, accounting ROR, or IRR methods. Because of this, NPV is the method most recommended by financial experts for making investment decisions. IRR is still useful to determine the exact rate of return for an investment, but NPV has none of the problems that IRR may have with unusual investments (such as the "multiple root" problem illustrated previously).

The following example shows how NPV is used to decide between investment alternatives.

Example. A company must choose between two alternative manufacturing projects, which require different amounts of capital investment and have different cash flow patterns. The company's opportunity cost of capital for projects of similar risk is 15%. Which project should it choose?

Project	C_0	C_1	C_2	C_3	C_4
A	−$125,000	$31,000	$43,000	$48,000	$51,000
B	−$210,000	$71,000	$74,000	$76,000	$79,000

Solving with the NPV formula:

$$NPV(A) = -\$4,809$$

$$NPV(B) = \$2,833$$

Conclusion: The company should choose project B. Despite a higher initial cost than A, Project B earns $2,833 more than its required rate of return of 15%. Incidentally, IRR also could have been used, since future cash flows are all positive, and therefore no multiple roots exist. Project A's IRR is 13.23%; Project B's IRR is 15.65%.

SENSITIVITY ANALYSIS

The previous sections of this chapter show how to analyze investments and financial projects. These analysis methods can help quantify whether or not an investment is worthwhile, and also help choose between alternative investments. However, no matter what method is used to analyze a financial project, that method, and the decision about the project, will ultimately rely upon the *projections* of the project's future cash flows. It is, therefore, extremely important that the future cash flows of any financial project or investment under consideration be forecast as accurately and realistically as possible. Unfortunately, no matter how precise and realistic you try to be with your projections of cash flows, the future is never certain, and actual results may vary from what is projected. Thus, it is often useful to perform a *sensitivity analysis* on the investment or financial project.

Sensitivity analysis is a procedure used to describe analytically the effects of uncertainty on one or more of the parameters involved in the analysis of a financial project. The objective of a sensitivity analysis is to provide the decision maker with quantitative information about what financial effects will be caused by variations from what was projected. A sensitivity analysis, once performed, may influence the decision about the financial project, or at least show the decision maker which parameter is the most critical to the financial success of the project. The NPV method lends itself nicely to sensitivity analysis, since the discount rate is already fixed at the opportunity cost of capital. One parameter at a time can be varied in steps (while the other parameters are held constant) to see how much effect this variance has on NPV. A financial calculator or computer spreadsheet program will greatly expedite the multiple, repetitive calculations involved in this analysis. Plotting the results graphically will help show the sensitivity of NPV to changes in each variable, as illustrated in the following example.

Example. A-1 Engineering Co. is contemplating a new project to manufacture sheet metal parts for the aircraft industry. Analysis of this project shows that a $100,000 investment will be required up front to purchase additional machinery. The project is expected to run for 4 years, resulting in estimated after-tax cash flows of $30,000 per year. Salvage value of the machinery at the end of the 4 years is estimated at $32,000. A-1's opportunity cost of capital is 15%. Perform a sensitivity analysis to determine how variations in these estimates will affect the NPV of the project.

Step 1: Cash flow diagram:

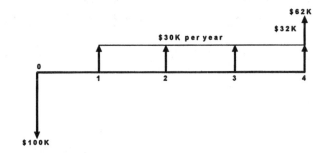

Step 2: Calculate baseline NPV:

$$NPV = -100,000 + \frac{30,000}{(1+.15)^1} + \frac{30,000}{(1+.15)^2} + \frac{30,000}{(1+.15)^3}$$

$$+ \frac{62,000}{(1+.15)^4} = \$3,945.45$$

Step 3: Change salvage value, annual cash flow revenues, and initial investment amount in steps. Calculate NPV for each step change.

Step 4: Plot values to illustrate the NPV sensitivity to changes in each variable.

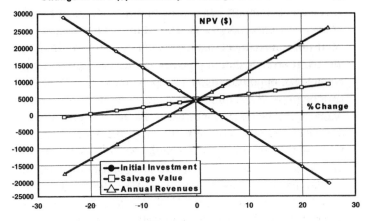

Conclusion: The graph shows that this project's NPV is equally sensitive to changes in the required initial investment amount (inversely related), and the annual revenues (after-tax cash flow) from the project. This is easily seen from the slope of the lines in the graph. The graph also shows that NPV is not affected nearly as much by changes in salvage value, primarily because the revenues from salvage value occur at the end of the fourth year, and are substantially discounted at the 15% cost of capital. In conclusion, this sensitivity analysis shows A-1 Machinery that it cannot accept anything more than a 5% increase in the initial cost of the machinery, or a 5% decrease in the annual cash flows. If these parameters change more than this amount, NPV will be less than zero and the project will not be profitable. Salvage value, however, is not nearly as critical, as it will take more than a 20% decrease from its estimated value before the project is unprofitable.

DECISION TREE ANALYSIS OF INVESTMENTS AND FINANCIAL PROJECTS

Financial managers often use *decision trees* to help with the analysis of large projects involving sequential decisions and variable outcomes over time. Decision trees are useful to managers because they graphically portray a large, complicated problem in terms of a series of smaller problems and decision branches. Decision trees reduce abstract thinking about a project by producing a logical diagram which shows the decision options available and the corresponding results of choosing each option. Once all the possible outcomes of a project are diagrammed in a decision tree, objective analysis and decisions about the project can be made.

Decision trees also allow *probability theory* to be used, in conjunction with NPV, to analyze financial projects and

investments. It is often possible to estimate the probability of success or failure for a new venture or project, based upon either historical data or business experience. If the new venture is a success, then the projected cash flows will be considerably higher than what they will be if the project is not successful. The projected cash flows for both of these possible outcomes (success or failure) can then be multiplied by their probability estimates and added to determine an *expected* outcome for the project. This expected outcome is the *average* payoff that can be expected (for multiple projects of the same type) based upon these probability estimates. Of course, the *actual* payoff for a single financial project will either be one or the other (success or failure), and not this average payoff. However, this average expected payoff is a useful number for decision making, since it incorporates the probability of both outcomes.

If there is a substantial time lapse between the time the decision is made and the time of the payoff, then the time value of money must also be considered. The NPV method can be used to discount the expected payoff to its present value and to see if the payoff exceeds the initial investment amount. The standard NPV decision rule applies: *Accept only projects that have positive NPVs.*

The following example illustrates the use of probability estimates (and NPV) in a simple decision tree:

Example. A ticket scalper is considering an investment of $10,000 to purchase tickets to the finals of a major outdoor tennis tournament. He must order the tickets one year in advance to get choice seats at list price. He plans to resell the tickets at the gate on the day of the finals, which is on a Sunday. Past experience tells him that he can sell the tickets and double his money, but only if the weather on the day of the event is good. If the weather is bad, he knows from past experience that he will only be able to sell the tickets at an average of 70% of his purchase price, even if the event is canceled and rescheduled for the following day (Monday). Historical data from the *Farmer's Almanac* shows that there is a 20% probability of rain at this time of year. His opportunity cost of capital is 15%. Should the ticket scalper accept this financial project and purchase the tickets?

Step 1: Construct the decision tree (very simple in this example).

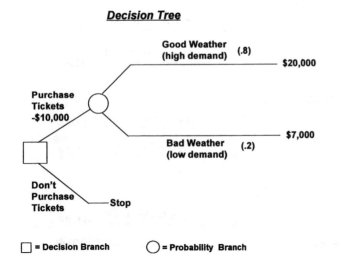

Decision Tree

$\square$ = Decision Branch $\bigcirc$ = Probability Branch

Step 2: Calculate the expected payoff from purchasing the tickets:

Expected payoff = (probability of high demand) × (payoff with high demand) + (probability of low demand) × (payoff with low demand)
= (.8 × 20,000) + (.2 × 7,000) = $17,400

Step 3: Calculate NPV of expected payoff:

$$NPV = -10,000 + \frac{\$17,400}{(1+.15)^1} = \$5,130$$

Remember that the cash flow from selling the tickets occurs one year after the purchase. NPV is used to discount the cash flow to its present value and see if it exceeds the initial investment.

Conclusion: The ticket scalper should buy the tickets, because this project has a positive NPV.

The preceding example was relatively simple. Most financial projects in engineering and manufacturing applications are considerably more complex than this example, therefore, it is useful to have a procedure for setting up a decision tree and analyzing a complex project.

Here is a procedure that can be used:

1. Identify the decisions to be made and alternatives available throughout the expected life of the project.
2. Draw the decision tree, with branches drawn first for each alternative at every decision point in the project. Secondly, draw "probability branches" for each possible outcome from each decision alternative.
3. Estimate the probability of each possible outcome. (Probabilities at each "probability branch" should add up to 100%.)
4. Estimate the cash flow (payoff) for each possible outcome.
5. Analyze the alternatives, starting with the most distant decision point and working backwards. Remember the time value of money and, if needed, use the NPV method to discount the expected cash flows to their present values. Determine if the PV of each alternative's expected payoff exceeds the initial investment (i.e., positive NPV). Choose the alternative with the highest NPV.

The following example illustrates the use of this procedure:

Example. Space-Age Products believes there is considerable demand in the marketplace for an electric version of its barbecue grill, which is the company's main product. Currently, all major brands of grills produced by Space-Age and its competitors are either propane gas grills or charcoal grills. Recent innovations by Space-Age have allowed the development of an electric grill that can cook and sear meat as well as any gas or charcoal grill, but without the problems of an open flame or the hassle of purchasing propane gas bottles or bags of charcoal.

Based upon test marketing, Space-Age believes the probability of the demand for its new grill being high in its first year of production will be 60%, with a 40% probability of low demand. If demand is high the first year, Space-Age estimates the probability of high demand in subsequent years to be 80%. If the demand the first year is low, it estimates the probability of high demand in subsequent years at only 30%.

Space-Age cannot decide whether to set up a large production line to produce 50,000 grills a month, or to take a more conservative approach with a small production line that can produce 25,000 grills a month. If demand is high, Space-Age estimates it should be able to sell all the grills the large production line can make. However, the large production line puts considerably more investment capital at risk, since it costs $350,000 versus only $200,000 for the small line. The smaller production line can be expanded with a second production line after the first year (if demand is high) to double the production rate, at a cost of an additional $200,000.

Should Space-Age initially invest in a large or small production line, or none at all? The opportunity cost of capital for Space-Age is 15%, and cash flows it projects for each outcome are shown in the decision tree. Cash flows shown at the end of year 2 are the PV of the cash flows of that and all subsequent years.

Steps 1–4 of procedure: (Construct decision tree and label with probabilities and cash flow projections).

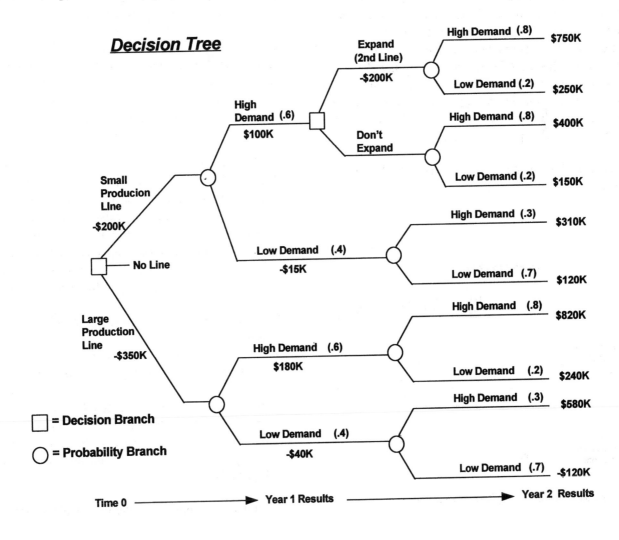

Decision Tree

- Expand (2nd Line) −$200K
  - High Demand (.8) $750K
  - Low Demand (.2) $250K
- Don't Expand
  - High Demand (.8) $400K
  - Low Demand (.2) $150K
- High Demand (.6) $100K
- Low Demand (.4) −$15K
  - High Demand (.3) $310K
  - Low Demand (.7) $120K
- Small Production Line −$200K
- No Line
- Large Production Line −$350K
  - High Demand (.6) $180K
    - High Demand (.8) $820K
    - Low Demand (.2) $240K
  - Low Demand (.4) −$40K
    - High Demand (.3) $580K
    - Low Demand (.7) −$120K

☐ = Decision Branch
○ = Probability Branch

Time 0 ——→ Year 1 Results ——→ Year 2 Results

Step 5: Analysis of alternatives:

(a) Start the analysis by working backwards from the R/H side of the decision tree. The only decision that Space-Age needs to make at the start of year 2 is deciding whether or not to expand and install a second production line if demand is high the first year (if the initial decision was for a small production line).

Expected payoff of expanding with a second small production line (at the end of year 2):

$$= (.8 \times \$750 \text{ K}) + (.2 \times \$250 \text{ K}) = \$650,000$$

Expected payoff without the expansion (at the end of year 2):

$$= (.8 \times \$400 \text{ K}) + (.2 \times \$150 \text{ K}) = \$350,000$$

To decide if expansion is worthwhile, the NPVs of each alternative (at end of year 1) must be compared:

$$NPV \text{ (with 2nd line)} = -\$200 \text{ K} + \frac{\$650 \text{ K}}{(1+.15)^1}$$
$$= \$365,217$$

$$NPV \text{ (without 2nd line)} = \$0 + \frac{\$350 \text{ K}}{(1+.15)^1}$$
$$= \$304,348$$

These NPV amounts show that Space-Age *should clearly expand with a second small production line* (if its initial decision at time 0 was to open a small production line and the demand was high the first year). Now that this decision is made, the $365,217 NPV of this expansion decision will be used to work backwards in time and calculate the overall NPV of choosing the small production line.

(b) The expected payoff (at the end of year 2) if the demand is low the first year with the small production line:

$$= (.3 \times \$310 \text{ K}) + (.7 \times \$120 \text{ K}) = \$177,000$$

This expected payoff must be discounted so that it can be added to the projected cash flow of year 1:

$$PV = \frac{\$177,000}{(1+.15)} = \$153,913$$

(c) The expected payoff (at the end of year 1) for the small production line can now be determined:

$$= [.6 \times (\$100,000 + \$365,217)] + [.4 \times (-\$15,000 + \$153,913)] = \$334,695$$

Notice that the cash flow for year 1 is the sum of the first year's projected cash flow plus the discounted expected payoff from year 2.

(d) Calculate the NPV (at time 0) of the small production line option:

$$NPV = \$200,000 + \frac{\$334,695}{(1+.15)} = \$91,039$$

(e) Calculate the NPV (at time 0) of the large production line option. (Since there are no decisions at the end of year 1, there is no need to determine intermediate NPVs, and the probabilities from each branch of each year can be multiplied through in the NPV calculation as follows:

$$NPV = -\$350 \text{ K} + \frac{(.6)(\$180 \text{ K}) + (.4)(-\$40 \text{ K})}{(1+.15)^1}$$
$$+ \frac{(.6)[(.8)(\$820 \text{ K}) + (.2)(\$240 \text{K})] + (.4)[(.3)(\$580 \text{ K}) + (.7)(-\$120 \text{ K})]}{(1+.15)^2}$$

Solving: NPV = $76,616

(f) *Conclusion:* Both the NPVs of the large and small production lines are positive, and thus both production lines would be profitable based upon the expected payoffs. However, the NPV of the small production line is higher than the NPV of the large production line ($91,039 vs. $76,616). Therefore, Space-Age's decision should be to set up a small production line, and then expand with a second line if the demand for electric grills is high in the first year of production.

ACCOUNTING FUNDAMENTALS

Regardless of whether a person is managing the finances of a small engineering project or running a large corporation, it is important to have a basic understanding of accounting. Likewise, it is also impossible to study engineering economics without overlapping into the field of accounting. This section, therefore, provides a brief overview of accounting fundamentals and terminology.

Accounting is the process of measuring, recording, and communicating information about the economic activities of accounting entities (financial projects, partnerships, companies, corporations, etc.). Public corporations, and most other types of companies, issue financial statements from their accounting data on a periodic basis (usually quarterly and yearly). These financial statements are necessary because they provide information to outside investors and creditors, and assist a company's own management with decision making. Four primary types of financial statements are normally issued:

- Balance sheet
- Income statement
- Statement of changes in retained earnings
- Statement of cash flows

The *balance sheet* provides information (at one particular instant in time) about financial resources of a company, and claims against those resources by creditors and owners. The balance sheet's format reflects the *fundamental accounting equation:*

Assets = Liabilities + Owner's Equity

Assets are items of monetary value that a company possesses, such as plant, land, equipment, cash, accounts receivable, and inventory. *Liabilities* are what the company owes its creditors, or conversely, the claims the creditors have against the company's assets if the company were to go bankrupt. Typical liabilities include outstanding loan balances, accounts payable, and income tax payable. *Owner's equity* is the monetary amount of the owner's claim against the company's assets. (Owner's equity is also referred to as stockholder's equity for corporations that issue and sell stock to raise capital.) Owner's equity comes from two sources: (1) the original cash investment the owner used to start the business, and (2) the amount of profits made by the company that are not distributed (via *dividends*) back to the owners. These profits left in the company are called *retained earnings.*

There are numerous asset, liability, and owner's equity *accounts* that are used by accountants to keep track of a company's economic resources. These accounts are shown in the following "balance sheet" illustration, which is a typical balance sheet for a small manufacturing corporation.

Notice in this illustration that the balance sheet's total assets "balance" with the sum of liabilities and owner's equity, as required by the fundamental accounting equation. There are also a number of additional *revenue* and *expense* accounts that are maintained, but not shown in the balance sheet. The balances of these revenue and expense accounts are closed out (zeroed) when the financial statements are prepared, and their amounts are carried over into the income statement. The *income statement* is the financial statement that reports the profit (or loss) of a company over a specific period of time (quarterly or annually). The income statement reflects another basic and fundamental accounting relationship:

Revenues – Expenses = Profit (or Loss)

A sample "income statement" for our typical small corporation is illustrated.

The income statement simply subtracts the company's expenses from its revenues to arrive at its *net income* for the period. Large companies and corporations usually separate their income statements into several sections, such as income from operations, income or loss from financial activities and discontinuing operations, and income or loss from extraordinary items and changes in accounting principles.

Companies frequently show an expense for depreciation on their income statements. *Depreciation* is an accounting method used to allocate the cost of asset "consumption" over the life of the asset. The value of all assets (building, equipment, patents, etc.) depreciate with time and use, and depreciation provides a method to recover investment capi-

XYZ Company
Balance Sheet
As of March 31, 1994

ASSETS

Current Assets:

Cash	$25,000.00	
Accounts Receivable	$268,000.00	
Inventory	$163,000.00	
Total Current Assets		$456,000.00

Property, Plant, & Equipment:

Building	$600,000.00		
less Accumulated Depreciation	($250,000.00)	$350,000.00	
Equipment	$300,000.00		
less Accumulated Depreciation	($125,000.00)	$175,000.00	
Land		$100,000.00	
Total Property, Plant, & Equipment			$625,000.00
Total Assets			**$1,081,000.00**

Liabilities

Current Liabilties:

Accounts Payable	$275,000.00	
Salaries Payable	$195,000.00	
Interest Payable	$13,000.00	
Income Tax Payable	$36,000.00	
Total Current Liabilities	$519,000.00	

Long-Term Liabilities:

Notes Payable	$260,000.00	
Total Liabilities		$779,000.00

Owner's (Stockholder's) Equity

Capital Stock	$200,000.00	
Retained Earnings	$102,000.00	
Total Owner's Equity		$302,000.00
Total Liabilities & Owner's Equity		**$1,081,000.00**

tal by expensing the portion of each asset "consumed" during each income statement's time period. Depreciation is a *noncash expense,* since no cash flow actually occurs to pay for this expense. The *book value* of an asset is shown in the balance sheet, and is simply the purchase price of the asset minus its accumulated depreciation. There are several methods used to calculate depreciation, but the easiest and most frequently used method is *straight-line depreciation,* which is calculated with the following formula:

$$\text{Depreciation Expense} = \frac{\text{Cost} - \text{Salvage Value}}{\text{Expected Life of Asset}}$$

Straight-line depreciation allocates the same amount of depreciation expense each year throughout the life of the asset. *Salvage value* (residual value) is the amount that a company can sell (or trade in) an asset for at the end of its useful life. "Accelerated" depreciation methods (such as sum-of-the-years'-digits and declining balance methods) are available and may sometimes be used to expense the cost of an asset faster than straight-line depreciation. Howev-

```
                          XYZ Company
                        Income Statement
                For the Quarter Ended March 31, 1994

Sales Revenues                                        $2,450,000.00
Less: Cost of Goods sold                             ($1,375,000.00)
Gross Margin                                          $1,075,000.00
Less: Operating Expenses
    Selling expenses              $295,000.00
    Salaries expense              $526,000.00
    Insurance expense              $32,000.00
    Property Taxes                 $27,000.00
    Depreciation, Equipment        $35,000.00
    Depreciation, Building         $70,000.00         ($985,000.00)
Income from Operations                                   $90,000.00
Other Income
    Sale of Assets                 $16,000.00
    Interest Revenue                $2,000.00           $18,000.00
                                                       $108,000.00
Less: Interest Expense                                 ($22,000.00)
Income before Taxes                                     $86,000.00
Less: Income Tax Expense                               ($34,400.00)
Net Income                                              $51,600.00
```

```
                          XYZ Company
            Statement of Changes in Retained Earnings
                For the Quarter Ended March 31, 1994

Retained Earnings, Dec. 31, 1993              $50,400.00
Net Income, Quarter 1, 1994                   $51,600.00
Retained Earnings, March 31, 1994            $102,000.00
```

er, the use of accelerated depreciation methods is governed by the tax laws applicable to various types of assets. A recent accounting textbook and tax laws can be consulted for further information about the use of accelerated depreciation methods.

The profit (or loss) for the time period of the income statement is carried over as retained earnings and added to (or subtracted from) owner's equity, in the "statement of changes in retained earnings" illustration.

If the company paid out any dividends to its stockholders during this time period, that amount would be shown as a deduction to retained earnings in this financial statement. This new retained earnings amount is also reflected as the retained earnings account balance in the current balance sheet.

The last financial statement is the *statement of cash flows,* which summarizes the cash inflows and outflows of the company (or other accounting entity) for the same time period as the income statement. This statement's purpose is to show where the company has acquired cash (inflows) and to what activities its cash has been utilized (outflows) during this time period. The statement divides and classifies cash flows into three categories: (1) cash provided by or used by operating activities; (2) cash provided by or used by investing activities; and (3) cash provided by or used by financing activities. Cash flow from operations is the cash generated (or lost) from the normal day-to-day operations of a company, such as producing and selling goods or services. Cash flow from investing activities includes selling or purchasing long-term assets, and sales or purchases of investments. Cash flow from financing activities covers the issuance of long-term debt or stock to acquire capital, payment of dividends to stockholders, and repayment of principal on long-term debt.

The statement of cash flows complements the income statement, because although income and cash flow are related, their amounts are seldom equal. Cash flow is also a better measure of a company's performance than net income, because net income relies upon arbitrary expense allocations (such as depreciation expense). Net income can also be manipulated by a company (for example, by reducing R&D spending, reducing inventory levels, or changing accounting methods to make its "bottom line" look better in the short-term. Therefore, analysis of a company's cash flows and cash flow trends is frequently the best method to evaluate a company's performance. Additionally, cash flow rather than net income should be used to determine the rate of return earned on the initial investment, or to determine the value of a company or financial project. (Value is determined by discounting cash flows at the required rate of return.)

The statement of cash flows can be prepared directly by reporting a company's gross cash inflows and outflows over a time period, or indirectly by adjusting a company's net income to net cash flow for the same time period. Adjustments to net income are required for (1) changes in assets and liability account balances, and (2) the effects of non-cash revenues and expenses in the income statement.

For example, in comparing XYZ Company's current balance sheet with the previous quarter, the following information is obtained about current assets and liabilities:

- Cash increased by $10,000
- Accounts receivable decreased by $5,000
- Inventory increased by $15,000
- Accounts payable increased by $25,000
- Salaries payable increased by $5,000

Additionally, the balance sheets show that the company has paid off $166,600 of long-term debt this quarter. This in-

XYZ Company
Statement of Cash Flows
For the Quarter Ended March 31, 1994

Net Income		$51,600.00
Adjustments to reconcile net income to cash flow from op. activities:		
Decrease in accounts receivable	$5,000.00	
Increase in inventory	($15,000.00)	
Increase in accounts payable	$25,000.00	
Decrease in salaries payable	$5,000.00	
Depreciation expense, equipment	$35,000.00	
Depreciation expense, building	$70,000.00	
Total Adjustments:		$125,000.00
Net cash flow from operating activities:		$176,600.00
Cash flows from financing activities:		
Retirement of long-term debt (notes payable)	($166,600.00)	
Net cash flow from financing activities:		($166,600.00)
Net Increase in cash:		$10,000.00
Cash account balance, Dec. 31, 1993		$15,000.00
Cash account balance, March 31, 1994		$25,000.00

formation from the balance sheets, along with the non-cash expenses (depreciation) from the current income statement, are used to prepare the "statement of cash flows." The example demonstrates the value of the statement of cash flows because it shows precisely the sources and uses of XYZ Company's cash. Additionally, it reconciles the cash account balance from the previous quarter to the amount on the current quarter's balance sheet.

REFERENCES AND RECOMMENDED READING

Engineering Economics

1. Newman, D. G., *Engineering Economic Analysis.* San Jose, CA: Engineering Press, 1976.
2. Canada, J. R. and White, J. A., *Capital Investment Decision Analysis for Management and Engineering.* Englewood Cliffs, NJ: Prentice-Hall, 1980.
3. Park, W. R., *Cost Engineering Analysis: A Guide to Economic Evaluation of Engineering Projects,* 2nd ed. New York: Wiley, 1984.
4. Taylor, G. A., *Managerial and Engineering Economy: Economic Decision-Making,* 3rd ed. New York: Van Nostrand, 1980.
5. Riggs, J. L., *Engineering Economics.* New York: McGraw-Hill, 1977.
6. Barish, N., *Economic Analysis for Engineering and Managerial Decision Making.* New York: McGraw-Hill, 1962.

Finance

1. Brealey, R. A. and Myers, S. C., *Principles of Corporate Finance,* 3rd ed. New York: McGraw-Hill, 1988.
2. Kroeger, H. E., *Using Discounted Cash Flow Effectively.* Homewood, IL: Dow Jones-Irwin, 1984.

Accounting

1. Chasteen, L.G., Flaherty, R.E., and O'Conner, M.C. *Intermediate Accounting,* 3rd ed. New York: McGraw-Hill, 1989.
2. Eskew, R.K. and Jensen, D.L., *Financial Accounting,* 3rd ed. New York: Random House, 1989.

Additional References

Owner's Manual from any Hewlett-Packard, Texas Instruments, or other make business/financial calculator.

Note: A business or financial calculator is an *absolute must* for those serious about analyzing investments and financial projects on anything more than an occasional basis. In addition to performing simple PV, FV, and mortgage or loan payment calculations, modern financial calculators can instantly calculate NPVs and IRRs (including multiple roots) for the most complex cash flow problems. Many calculators can also tabulate loan amortization schedules and depreciation schedules, and perform statistical analysis on data. The owner's manuals from these calculators are excellent resources and give numerous examples of how to solve and analyze various investment problems.

Appendix

Lawrence D. Norris, Senior Technical Marketing Engineer—Large Commercial Engines, Allison Engine Company, Rolls-Royce Aerospace Group

Conversion Factors

Category	Multiply	By	To Obtain
Acceleration	ft/sec^2	0.3048	m/sec^2
	in/sec^2	0.0254	m/sec^2
	m/sec^2	3.2808	ft/sec^2
	m/sec^2	39.3701	in/sec^2
Angle	degrees	6.0000	minutes
	degrees	1.745328×10^{-2}	radians
	degrees	2.777778×10^{-3}	revolutions
	degrees	3600.00	seconds
	minutes	0.1667	degrees
	radians	57.2958	degrees
	revolutions	360	degrees
	seconds	2.77778×10^{-4}	degrees
Area	ft^2	0.0929	m^2
	in^2	6.4516×10^{-4}	m^2
	yard^2	0.8361	m^2
	mile^2	2.590×10^6	m^2
	acres	4.047×10^3	m^2
	m^2	10.7643	ft^2
	m^2	1550.0	in^2
	m^2	1.1960	yard^2
	m^2	3.8610×10^{-7}	mile^2
	m^2	2.471×10^{-4}	acres
	acres	0.4047	hectares
	acres	4.356×10^4	ft^2
	acres	4.047×10^3	m^2
	acres	1.562×10^{-3}	mile^2
Density	gram/cm^3	1000	kg/m^3
	lb$_m$/ft^3	16.0185	kg/m^3
	lb$_m$/gallon (U.S. liquid)	119.826	kg/m^3
	slug/ft^3	515.379	kg/m^3
	kg/m^3	0.001	gram/cm^3
	kg/m^3	0.0624	lb$_m$/ft^3
	kg/m^3	8.3454×10^{-3}	lb$_m$/gallon (U.S. liquid)
	kg/m^3	1.9403×10^{-3}	slug/ft^3
Energy/Work	Btu	1055.06	joule
	erg	1.000×10^{-7}	joule
	ft-lb	1.3558	joule
	kilowatt-hour	3.600×10^6	joule
	calorie	4.1859	joule
	newton-meter	1.000	joule
	watt-second	1.000	joule
	joule	9.4781×10^{-4}	Btu
	joule	1.000×10^7	erg

(table continued on next page)

Category	Multiply	By	To Obtain
Energy/Work (cont'd)	joule	0.7376	ft-lb
	joule	2.7778×10^{-7}	kilowatt-hour
	joule	0.2389	calories
	joule	1.000	newton-meter
	joule	1.000×10^{7}	dyne centimeters
	joule	1.000	watt-second
	Btu	2.930×10^{-4}	kilowatt-hours
	kilowatt-hours	3412.97	Btu
	Btu	253.0	calories
	calorie	3.953×10^{-3}	Btu
	kilowatt-hour	2.655×10^{6}	ft-lb
	ft-lb	3.766×10^{-7}	kilowatt-hour
Force	dyne	1.000×10^{-5}	newton (N)
	pound (lb)	4.4482	newton
	newton (N)	1.000×10^{5}	dyne
	newton	0.2248	pound (lb)
Length	foot (ft)	0.3048	meter (m)
	inch (in)	2.540×10^{-2}	meter
	yard	0.9144	meter
	micron	1.000×10^{-6}	meter
	mile (mi)	1.6093×10^{3}	meter
	mile, nautical (nm)	1.8520×10^{3}	meter
	meter (m)	3.2808	foot (ft)
	meter	39.3701	inch (in)
	meter	1.0936	yards
	meter	1.000×10^{6}	micron
	meter	6.2139×10^{-4}	mile (mi)
	meter	5.3996×10^{-4}	mile, nautical (nm)
	mile, nautical (nm)	1.15076	mile
	mile	0.86896	mile, nautical
	mile	5280	foot
	mile	1760	yard
Mass	pound mass (lb$_m$)	0.4536	kilogram (kg)
	slug	14.5939	kilogram
	ounce	2.83495×10^{-2}	kilogram
	ton (metric)	1000.00	kilogram
	ton (2000 lb$_m$)	907.185	kilogram
	kilogram (kg)	2.2046	lb$_m$
	kilogram	0.0685	slug
	kilogram	35.2739	ounce
	kilogram	0.0010	ton (metric)
	kilogram	1.1023×10^{-3}	ton (2000 lb$_m$)
	pound mass	16	ounce
	ounce	0.06250	pound mass
Moment (Force)	foot-pound (ft-lb)	1.35582	newton-meter (N-m)
	dyne-centimeter	1.000×10^{-7}	N-m
	newton-meter (N-m)	0.73756	foot-pound (ft-lb)
	newton-meter	1.000×10^{7}	dyne-centimeter

Category	Multiply	By	To Obtain
Moment of Inertia (Area)	meter^4	115.8618	foot^4
	foot^4	8.63097×10^{-3}	meter^4
	meter^4	2.40251×10^6	in^4
	in^4	4.16232×10^{-7}	meter^4
Moment of Inertia (Volume)	meter^5	380.1239	foot^5
	foot^5	2.63072×10^{-3}	meter^5
	meter^5	9.45870×10^7	in^5
	in^5	1.05723×10^{-8}	meter^5
Moment of Inertia (Mass)	m^2-kg	23.73034	ft^2-lb
	ft^2-lb	4.214013×10^{-2}	m^2-kg
	m^2-kg	3417.171	in^2-lb
	in^2-lb	2.926397×10^{-4}	m^2-kg
Power	Btu/hour	0.29307	watt (W)
	erg/second	1.000×10^{-7}	watt
	ft-lb/second	1.35582	watt
	horsepower (HP)	745.699	watt
	calories/second	4.186	watt
	joule/second	1.000	watt
	watt (W)	3.41214	Btu/hour
	watt	1.000×10^7	erg/second
	watt	0.737562	ft-lb/second
	watt	1.34102×10^{-3}	horsepower
	watt	0.2389	calories/second
	watt	1.000	joule/second
	ft-lb/second	1.818×10^{-3}	horsepower
	horsepower	550.06	ft-lb/second
	ft-lb/second	0.32394	calories/second
	calories/second	3.087	ft-lb/second
	horsepower	42.426	Btu/minute
	Btu/minute	0.02357	horsepower
Pressure and Stress	atmosphere	1.01325×10^5	pascal (Pa)
	bar	1.000×10^5	pascal
	cm. of Hg (0°C)	1333.22	pascal
	in. of Hg (0°C)	3.386×10^3	pascal
	in. of H$_2$O (4°C)	249.10	pascal
	dyne/cm^2	0.10000	pascal
	lb/ft^2	47.88026	pascal
	lb/in^2 (psi)	6894.757	pascal
	kilogram/cm^2 (kg/cm^2)	9.8067×10^4	pascal
	newton/meter^2 (n/m^2)	1.000	pascal
	pascal (Pa)	1.000	n/m^2
	pascal	9.86923×10^{-6}	atmosphere
	pascal	1.000×10^{-5}	bar
	pascal	7.50064×10^{-4}	cm. of Hg (0°C)
	pascal	2.953×10^{-4}	in. of Hg (0°C)
	pascal	4.014×10^{-3}	in. of H$_2$O (4°C)
	pascal	10.000	dyne/cm^2

(table continued on next page)

Category	Multiply	By	To Obtain
Pressure and Stress (cont'd)	pascal	0.020885	lb/ft^2
	pascal	1.450377×10^{-4}	lb/in^2 (psi)
	pascal	1.0197×10^{-5}	kg/cm^2
	bar	0.9869	atmosphere
	atmosphere	1.0132	bar
	in. of Hg (0°C)	0.4912	lb/in^2 (psi)
	lb/in^2 (psi)	2.0358	in. of Hg (0°C)
	in. of H2O (4°C)	0.03613	lb/in^2 (psi)
	lb/in^2 (psi)	27.678	in. of H$_2$O (4°C)
Velocity	feet/second (ft/sec)	0.30480	meters/sec (m/sec)
	inch/second (in/sec)	2.5400×10^{-2}	m/sec
	kilometer/hour (km/hr)	0.27777	m/sec
	knot (nautical mi/hr)	0.514444	m/sec
	miles/hour (mi/hr)	0.447040	m/sec
	meters/sec (m/sec)	3.28084	ft/sec
	m/sec	39.38008	in/sec
	m/sec	3.60010	km/hr
	m/sec	1.94385	knot
	m/sec	2.23694	mi/hr
Volume	foot^3	2.831685×10^{-2}	meter^3
	gallon (U.S. liquid)	3.785412×10^{-3}	meter^3
	imperial gallon (U.K. liquid)	4.546087×10^{-3}	meter^3
	inch^3	1.638706×10^{-5}	meter^3
	cord	3.62456	meter^3
	board foot	2.359737×10^{-3}	meter^3
	liter	1.000×10^{-3}	meter^3
	quart (U.S. liquid)	9.463529×10^{-4}	meter^3
	barrel (U.S. liquid)	0.1589873	meter^3
	centimeter^3 (cm^3)	1.000×10^{-6}	meter^3
	fluid ounce (U.S. liquid)	2.957353×10^{-5}	meter^3
	bushel (U.S. dry)	3.523907×10^{-2}	meter^3
	peck (U.S. dry)	8.809768×10^{-3}	meter^3
	meter^3	35.314662	foot^3
	meter^3	264.1720	gallon (U.S. liquid)
	meter^3	219.9694	gallon (U.K. liquid)
	meter^3	6.102376×10^4	inch^3
	meter^3	0.275896	cord
	meter^3	423.77604	board foot
	meter^3	1000.00	liter
	meter^3	1056.688	quart (U.S. liquid)
	meter^3	6.289811	barrel (U.S. liquid)
	meter^3	1.000×10^6	centimeter^3
	meter^3	3.381402×10^4	fluid ounce (U.S.)
	meter^3	28.37759	bushel (U.S. dry)
	meter^3	113.5104	peck (U.S. dry)
	imperial gallon (U.K. liquid)	1.20095	gallon (U.S. liquid)
	gallon (U.S. liquid)	0.83267	gallon (U.K. liquid)
	gallon (U.S. liquid)	3.785	liter
	liter	0.2642	gallon (U.S. liquid)

Systems of Basic Units

Designation	English (FPS)	System Metric (MKS)	International (SI)
Length	foot (ft)	meter (m)	meter (m)
Mass	pound (lb$_m$)	kilogram (kg)	kilogram (kg)
Time	second (sec)	second (s)	second (s)
Electric Current	ampere (A)	ampere (A)	ampere (A)
Temperature	degree Fahrenheit (°F)	degree Celsius (°C)	degree Kelvin (°K)
Luminous intensity	candela (cd)	candela (cd)	candela (cd)

Decimal Multiples and Fractions of SI Units

Factor	Prefix	Symbol	Factor	Prefix	Symbol
10^1	deka	D (da)	10^{-1}	deci	d
10^2	hecto	h	10^{-2}	centi	c
10^3	kilo	k	10^{-3}	milli	m
10^6	mega	M	10^{-6}	micro	μ
10^9	giga	G	10^{-9}	nano	n
10^{12}	tera	T	10^{-12}	pico	p
10^{15}	peta	P	10^{-15}	femto	f
10^{18}	exa	E	10^{-18}	atto	a

Temperature Conversion Equations

(°F)	(°C)	(°K)	(°R)
(°F) = 9/5(°C) + 32	(°C) = 5/9(°F − 32)	(°K) = °C + 273.15	(°R) = 9/5(°C) + 491.67
= 9/5(°K − 255.37)	= °K − 273.15	= 5/9(°F + 459.67)	= °F + 459.67
= °R − 459.67	= 5/9(°R − 491.67)	= 5/9(°R)	= 9/5(°K)

Absolute zero temperature = −273.15°C = −459.67°F = 0.00°K = 0.00°R
Freezing point of water = 0.00°C = +32.00°F = +273.15°K = +491.67°R
Boiling point of water = +100.00°C = +212.00°F = +373.15°K = +671.67°R

Index